METHODS IN MOLECULAR BIOLOGY

For further volumes:
http://www.springer.com/series/7651

For over 35 years, biological scientists have come to rely on the research protocols and methodologies in the critically acclaimed *Methods in Molecular Biology* series. The series was the first to introduce the step-by-step protocols approach that has become the standard in all biomedical protocol publishing. Each protocol is provided in readily-reproducible step-by-step fashion, opening with an introductory overview, a list of the materials and reagents needed to complete the experiment, and followed by a detailed procedure that is supported with a helpful notes section offering tips and tricks of the trade as well as troubleshooting advice. These hallmark features were introduced by series editor Dr. John Walker and constitute the key ingredient in each and every volume of the *Methods in Molecular Biology* series. Tested and trusted, comprehensive and reliable, all protocols from the series are indexed in PubMed.

Histone Methyltransferases

Methods and Protocols

Edited by

Raphaël Margueron and Daniel Holoch

Institute Curie, INSERM U934/CNRS UMR 3215, Paris, France

Humana Press

Editors
Raphaël Margueron
Institute Curie, INSERM
U934/CNRS UMR 3215
Paris, France

Daniel Holoch
Institute Curie, INSERM U934/CNRS UMR 3215
Paris, France

ISSN 1064-3745 ISSN 1940-6029 (electronic)
Methods in Molecular Biology
ISBN 978-1-0716-2480-7 ISBN 978-1-0716-2481-4 (eBook)
https://doi.org/10.1007/978-1-0716-2481-4

This Humana imprint is published by the registered company Springer Science+Business Media, LLC, part of Springer Nature.
The registered company address is: 1 New York Plaza, New York, NY 10004, U.S.A.

Preface

The first histone methyltransferase was described by the laboratory of Michael Stallcup in 1999 [1], shortly followed by the first report of site-specific histone methylation by the laboratory of Thomas Jenuwein in 2000 [2]. It is fair to say in hindsight that these were watershed discoveries in the effort to understand how cells regulate the transcription of their genomes and how organisms coordinate key developmental processes. The two decades that followed saw the identification of an extensive array of histone methyltransferases and substrate residues across diverse eukaryotic lineages, as well as vigorous research into their functions as regulators of transcription, genome stability, and cell identity. The depth of conservation of histone methyltransferases underscores the scale of the biological questions to be unraveled, and one can safely predict that many exciting discoveries in this field still lie ahead.

Meanwhile, a host of methodological advances have facilitated investigations into the molecular mechanisms, cellular activities, and biological roles of these enzymes. This volume of the *Methods in Molecular Biology* series has aimed to assemble a wide-ranging assortment of protocols to help researchers approach histone methyltransferases from a variety of angles depending on their interests and expertise.

In the book's opening chapter, Erlendson and Freitag set the stage for the methods proper by reviewing the phylogenetic diversity and evolutionary history of histone methyltransferases, of which the well-studied SET-domain-containing lysine methyltransferases are but a subset. The authors concentrate on a few intriguing examples of crosstalk between different histone methylations and enzymes with multiple substrate specificities.

Next, the first techniques presented in the volume are essential protocols to assay histone methyltransferase and histone demethylase activities in vitro. In Chapter 2, Idigo and Voigt offer a universal procedure for histone methyltransferase assays, with a variety of readout choices depending on the expected catalytic activity of the enzyme and the reagents available in the lab. Likewise, Shengjiang Tu shares his practical expertise on the in vitro detection of histone demethylase activity in Chapter 3, again providing a range of readout options to suit different experimental goals and enzymatic properties.

The succeeding chapters go deeper into methodologies for studying the impact of specific biochemical and metabolic conditions on histone methyltransferase activity in vitro and in cells. In Chapter 4, Guohong Li and colleagues present a protocol for generating chromatin substrates containing histone variants and investigating the effect of these variants on the physical structure of chromatin and the activity of histone methyltransferases. Chapter 5, by Popoca and Lee, delves into the intriguing automethylation capability of certain histone methyltransferases and strategies for characterizing such automethylation events and their functional importance for the regulation of enzymatic function. The critical influence of metabolic cues on histone methyltransferases and other histone-modifying enzymes has come to be increasingly recognized in recent years, and Raphaël Rodriguez and colleagues in Chapter 6 share their state-of-the-art protocols for precise quantification of metabolites and metabolic enzymes by metabolomics and proteomics, respectively, with a particular focus on the role of metals such as iron and copper in regulating histone demethylases.

The next part of the book provides frameworks for attaining atomic-resolution understanding of histone methyltransferase structure and function. In Chapter 7, Jon Wilson outlines the general approach for protein purification, crystallization, and X-ray structure determination with a focus on tricks to make this endeavor successful for histone methyltransferases in particular. For their part, Spangler and McGinty detail in Chapter 8 their methods for determining the structures of histone methyltransferases in complex with nucleosomes using cryogenic electron microscopy, including steps from protein complex and grid preparation to model building.

Essential to propelling our comprehension of histone methyltransferases for the last fifteen years, techniques for the genome-wide profiling of site-specific histone methylations occupy a central place in this collection. First, as these methods rely on highly specific antibodies capable of discriminating between closely related differentially methylated species, deriving these precious reagents represents a crucial process in and of itself. Accordingly, in Chapter 9, Zorro Shahidian and Daujat describe a pipeline to develop, purify, and validate antibodies against methylated histone forms that will allow researchers to generate excellent tools for studying histone methyltransferase function. Haruhiko Koseki and colleagues then present in Chapter 10 a protocol for examining genome-wide patterns of histone methylation by CUT&Tag in mouse embryonic tissues, with a battery of useful accompanying procedures for studying histone methyltransferases in the context of development. In Chapter 11, Brind'Amour and Lorincz detail the powerful method they have perfected to profile histone methylation in very low numbers of cells using ultra-low-input native ChIP-seq (ULI-NChIP-seq). Next, we describe a method for automated CUT&RUN using the KingFisher Duo Prime robotic magnetic bead handler in Chapter 12, as an alternative to a similar adaptation already published in the literature using another robot [3]. Finally, this part of the book closes with a vital and comprehensive guide to the bioinformatic analysis of histone methylation profiling data by Nicolas Servant in Chapter 13, from quality control to mapping and differential enrichment analysis, complete with a workflow listing all the relevant computational tools and their applications.

Beyond methylation of histones strictly speaking, histone methyltransferases have also been found to modify nonhistone substrates, including but not limited to the cases of automethylation addressed in Chapter 5, with important biological consequences. Chapter 14 and Chapter 15 propose two alternative strategies for discovering novel methyltransferase substrates. Chapter 14 by Guianvarc'h and colleagues describes a chemical biology approach in which potential substrates are systematically labeled using bioorthogonal moieties to enable enrichment and identification through mass spectrometry. Chapter 15 by Weirich and Jeltsch describes an equally systematic approach using peptide SPOT arrays to identify favorable amino acid sequence contexts for targeting by the histone methyltransferase, thus helping to narrow down candidate proteins for a more focused analysis.

Perhaps the most fundamental biological question regarding histone methyltransferases is how chromatin carrying site-specifically methylated histones comes to be interpreted as a biochemical signal by the transcriptional machinery and other nuclear effectors. To address this very problem, Bartke and colleagues impart in Chapter 16 a complete and authoritative method to identify nuclear proteins that specifically recognize nucleosomes containing methylated histones using affinity pull-downs and quantitative mass spectrometry. By specifically and sensitively uncovering the so-called readers of histone methylations, this approach provides a pathway to understanding the biological meaning of these enzymatic events.

Aside from their roles in influencing transcription and other nuclear processes in an immediate sense, methylated histones are also believed to participate in the epigenetic maintenance of transcription states, and reliable methods to track the mitotic inheritance of histone methylation patterns have consequently grown increasingly important. In Chapter 17, Alabert and colleagues present their ingenious technique, known as nascent chromatin capture, to isolate newly replicated DNA by pulse-chase with a modified nucleotide while using SILAC labeling to discriminate inherited from newly deposited histones and thereby precisely map the transmission of histone methylation marks. Chapter 18 by Ragunathan and colleagues tackles the problem from a different angle, describing an ectopic protein tethering assay first implemented in fission yeast that allows a detailed genetic and mechanistic dissection of the conditions required for histone methylation patterns to be inherited through cell division. To round out these inheritance-focused methods, Menon and Howard deliver a powerful yet accessible approach in Chapter 19 to constructing computational models capable of revealing the scenarios of histone methylation dynamics that are underpinned by a given set of assumptions or experimental observations.

Lastly, the need for pharmacological inhibitors of histone methyltransferases, not only as purely mechanistic research tools but also, given the evidence of causative roles for mutated or misregulated histone methyltransferases in human diseases, as potential therapeutics, has prompted small-molecule drug screens to discover such compounds. In the final chapter of the book, Trojer and colleagues take us through the design and execution of such a screen, using their proven success identifying a selective inhibitor of the histone methyltransferase EZH2 as a representative case study.

The remarkable diversity of methods compiled in this volume speaks to the rapid rise to prominence of histone methyltransferases in the fields of genome regulation and developmental biology. We trust that researchers of all backgrounds and interests will find some of the protocols herein useful for studies they may undertake involving these important and intriguing enzymes. In closing, we offer our sincerest thanks to each of the authors for taking the time to contribute their expertise and wisdom to this collection in the service of their fellow investigators.

Paris, France

Daniel Holoch
Raphaël Margueron

References

1. Chen D, Ma H, Hong H et al (1999) Regulation of transcription by a protein methyltransferase. Science 284:2174–2177. https://doi.org/10.1126/science.284.5423.2174

2. Rea S, Eisenhaber F, O'Carroll D et al (2000) Regulation of chromatin structure by site-specific histone H3 methyltransferases. Nature 406:593–599. https://doi.org/10.1038/35020506

3. Janssens DH, Wu SJ, Sarthy JF et al (2018) Automated in situ chromatin profiling efficiently resolves cell types and gene regulatory programs. Epigenetics Chromatin 11: 74–14. https://doi.org/10.1186/s13072-018-0243-8

Contents

Part V Analysis of Histone Methylation in Cells and Genomes

Part VI Discovery of Histone Methyltransferase Substrates and Methylated Histone Interactors

Part VII Inheritance of Histone Methylation Patterns

Part VIII Finding Inhibitors of Histone Methyltransferases

Contributors

SETAREH AFLAKI • *Institut Curie, Paris Sciences et Lettres Research University, Sorbonne University, Paris, France; INSERM U934/CNRS UMR3215, Paris, France*

CONSTANCE ALABERT • *Genome Regulation and Expression, School of Life Sciences, MSI/WTB Complex, University of Dundee, Dundee, UK*

AMANDA AMES • *Department of Biological Chemistry, University of Michigan, Ann Arbor, MI, USA*

TILL BARTKE • *Institute of Functional Epigenetics, Helmholtz Zentrum München, Neuherberg, Germany*

JULIE BRIND'AMOUR • *Department of Medical Genetics, University of British Columbia, Vancouver, BC, Canada; Département de Biomédecine Vétérinaire, Université de Montréal, Montréal, QC, Canada*

FABIENNE BURLINA • *Sorbonne Université, École normale supérieure, PSL University, CNRS, Laboratoire des Biomolécules, LBM, Paris, France*

TATIANA CAÑEQUE • *Institut Curie, 26 rue d'Ulm, Paris, France; PSL Université, Paris, France; Chemical Biology of Cancer Laboratory, CNRS UMR 3666, INSERM U1143, Paris, France*

NICO CANTONE • *Constellation, A MorphoSys Company, Cambridge, MA, USA*

RICHARD T. CUMMINGS • *Constellation, A MorphoSys Company, Cambridge, MA, USA*

SYLVAIN DAUJAT • *Biotechnology and Cell Signaling, CNRS UMR7242, University of Strasbourg, Illkirch Cedex, France*

ALEXANDRE DÉSERT • *Sorbonne Université, École normale supérieure, PSL University, CNRS, Laboratoire des Biomolécules, LBM, Paris, France*

ALLYSON A. ERLENDSON • *Department of Biochemistry and Biophysics, Oregon State University, Corvallis, OR, USA*

MICHAEL FREITAG • *Department of Biochemistry and Biophysics, Oregon State University, Corvallis, OR, USA*

ANJA GROTH • *Novo Nordisk Foundation Center for Protein Research (CPR), University of Copenhagen, Copenhagen, Denmark; Biotech Research and Innovation Centre (BRIC), University of Copenhagen, Copenhagen, Denmark*

DOMINIQUE GUIANVARC'H • *Université Paris-Saclay, CNRS, Institut de Chimie Moléculaire et des Matériaux d'Orsay (ICMMO), UMR 8182, Orsay, France*

KARINE GUITOT • *Université Paris-Saclay, CNRS, Institut de Chimie Moléculaire et des Maté riaux d'Orsay (ICMMO), UMR 8182, Orsay, France*

DANIEL HOLOCH • *Institut Curie, Paris Sciences et Lettres Research University, Sorbonne University, Paris, France; INSERM U934/CNRS UMR3215, Paris, France*

MARTIN HOWARD • *Department of Computational and Systems Biology, John Innes Centre, Norwich, UK*

NWAMAKA J. IDIGO • *Wellcome Centre for Cell Biology, School of Biological Sciences, University of Edinburgh, Edinburgh, UK*

ALBERT JELTSCH • *Department of Biochemistry, Institute of Biochemistry and Technical Biochemistry, University of Stuttgart, Stuttgart, Germany*

Takashi Kondo • *Laboratory for Developmental Genetics, RIKEN Center for Integrative Medical Sciences (RIKEN-IMS), Yokohama, Kanagawa, Japan*

Haruhiko Koseki • *Laboratory for Developmental Genetics, RIKEN Center for Integrative Medical Sciences (RIKEN-IMS), Yokohama, Kanagawa, Japan; Department of Cellular and Molecular Medicine, Graduate School of Medicine, Chiba University, ChibaChiba, Japan*

Ajay Larkin • *Department of Biological Chemistry, University of Michigan, Ann Arbor, MI, USA*

Chul-Hwan Lee • *Department of Biomedical Sciences, Seoul National University College of Medicine, Seoul, Republic of Korea; Department of Pharmacology, Seoul National University College of Medicine, Seoul, Republic of Korea; BK21 FOUR Biomedical Science Project, Seoul National University College of Medicine, Seoul, Republic of Korea; Ischemic/hypoxic Disease Institute, Seoul National University College of Medicine, Seoul, Republic of Korea*

Guohong Li • *National Laboratory of Biomacromolecules, CAS Center for Excellence in Biomacromolecules, Institute of Biophysics, Chinese Academy of Sciences, Beijing, China; University of Chinese Academy of Sciences, Beijing, China*

Cuifang Liu • *National Laboratory of Biomacromolecules, CAS Center for Excellence in Biomacromolecules, Institute of Biophysics, Chinese Academy of Sciences, Beijing, China*

Matthew C. Lorincz • *Department of Medical Genetics, University of British Columbia, Vancouver, BC, Canada*

Raphaël Margueron • *Institut Curie, Paris Sciences et Lettres Research University, Sorbonne University, Paris, France; INSERM U934/CNRS UMR3215, Paris, France*

Robert K. McGinty • *Department of Biochemistry and Biophysics, School of Medicine, University of North Carolina at Chapel Hill, Chapel Hill, NC, USA; Division of Chemical Biology and Medicinal Chemistry, Eshelman School of Pharmacy, University of North Carolina at Chapel Hill, Chapel Hill, NC, USA; Lineberger Comprehensive Cancer Center, School of Medicine, University of North Carolina at Chapel Hill, Chapel Hill, NC, USA*

Govind Menon • *Department of Computational and Systems Biology, John Innes Centre, Norwich, UK*

Audrey Michaud • *Institut Curie, Paris Sciences et Lettres Research University, Paris, France; INSERM U934/CNRS UMR3215, Paris, France*

Sebastian Müller • *Institut Curie, 26 rue d'Ulm, Paris, France; PSL Université, Paris, France; Chemical Biology of Cancer Laboratory, CNRS UMR 3666, INSERM U1143, Paris, France*

Kyosuke Nakamura • *Novo Nordisk Foundation Center for Protein Research (CPR), University of Copenhagen, Copenhagen, Denmark; Biotech Research and Innovation Centre (BRIC), University of Copenhagen, Copenhagen, Denmark*

Nhuong Nguyen • *Friedrich Miescher Institute for Biomedical Research, Basel, Switzerland*

Yasuhide Ohinata • *Department of Cellular and Molecular Medicine, Graduate School of Medicine, Chiba University, Chiba, Chiba, Japan*

Luis Popoca • *Department of Biochemistry and Molecular Pharmacology, NYU School of Medicine, New York, NY, USA; Howard Hughes Medical Institute, Chevy Chase, MD, USA*

Kaushik Ragunathan • *Department of Biological Chemistry, University of Michigan, Ann Arbor, MI, USA*

Raphaël Rodriguez • *Institut Curie, 26 rue d'Ulm, Paris, France; PSL Université, Paris, France; Chemical Biology of Cancer Laboratory, CNRS UMR 3666, INSERM U1143, Paris, France*

Melissa Seman • *Cellular and Molecular Biology Program, University of Michigan, Ann Arbor, MI, USA*

Nicolas Servant • *Institut Curie, Bioinformatics core facility, Paris, France; INSERM U900, Paris, France; PSL Research University, Paris, France; Mines Paris Tech, Fontainebleau, Paris, France*

Lara Zorro Shahidian • *Institute of Biomedicine and Biotechnology of Cantabria (IBBTEC), University of Cantabria, Santander, Spain*

Fabien Sindikubwabo • *Institut Curie, 26 rue d'Ulm, Paris, France; PSL Université, Paris, France; Chemical Biology of Cancer Laboratory, CNRS UMR 3666, INSERM U1143, Paris, France*

Cathy J. Spangler • *Department of Biochemistry and Biophysics, School of Medicine, University of North Carolina at Chapel Hill, Chapel Hill, NC, USA*

Junichiro Takano • *Laboratory for Developmental Genetics, RIKEN Center for Integrative Medical Sciences (RIKEN-IMS), Yokohama, Kanagawa, Japan*

Patrick Trojer • *Constellation, A MorphoSys Company, Cambridge, MA, USA*

Shengjiang Tu • *Department of Chemistry and Biochemistry, The Ohio State University, Columbus, OH, USA*

Andrey Tvardovskiy • *Institute of Functional Epigenetics, Helmholtz Zentrum München, Neuherberg, Germany*

Philipp Voigt • *Epigenetics Programme, Babraham Institute, Cambridge, UK; Wellcome Centre for Cell Biology, School of Biological Sciences, University of Edinburgh, Edinburgh, UK*

Sara Weirich • *Department of Biochemistry, Institute of Biochemistry and Technical Biochemistry, University of Stuttgart, Stuttgart, Germany*

Jon R. Wilson • *The Francis Crick Institute, London, UK*

Nayuta Yakushiji-Kaminatsui • *Laboratory for Developmental Genetics, RIKEN Center for Integrative Medical Sciences (RIKEN-IMS), Yokohama, Kanagawa, Japan*

Jicheng Zhao • *National Laboratory of Biomacromolecules, CAS Center for Excellence in Biomacromolecules, Institute of Biophysics, Chinese Academy of Sciences, Beijing, China*

Part I

Introduction and Biological Context

Chapter 1

Not all Is SET for Methylation: Evolution of Eukaryotic Protein Methyltransferases

Allyson A. Erlendson and Michael Freitag

Abstract

Dynamic posttranslational modifications to canonical histones that constitute the nucleosome (H2A, H2B, H3, and H4) control all aspects of enzymatic transactions with DNA. Histone methylation has been studied heavily for the past 20 years, and our mechanistic understanding of the control and function of individual methylation events on specific histone arginine and lysine residues has been greatly improved over the past decade, driven by excellent new tools and methods. Here, we will summarize what is known about the distribution and some of the functions of protein methyltransferases from all major eukaryotic supergroups. The main conclusion is that protein, and specifically histone, methylation is an ancient process. Many taxa in all supergroups have lost some subfamilies of both protein arginine methyltransferases (PRMT) and the heavily studied SET domain lysine methyltransferases (KMT). Over time, novel subfamilies, especially of SET domain proteins, arose. We use the interactions between H3K27 and H3K36 methylation as one example for the complex circuitry of histone modifications that make up the "histone code," and we discuss one recent example (*Paramecium* Ezl1) for how extant enzymes that may resemble more ancient SET domain KMTs are able to modify two lysine residues that have divergent functions in plants, fungi, and animals. Complexity of SET domain KMT function in the well-studied plant and animal lineages arose not only by gene duplication but also acquisition of novel DNA- and histone-binding domains in certain subfamilies.

Key words Euchromatin, Heterochromatin, Histone, PRMT, SET, H3K36, H3K27, Protists, Fungi, Plant, Animal

1 Introduction

Chromatin is the key architectural feature organizing most eukaryotic genomes into structurally distinct domains, resulting in varying accessibility to transcriptional machinery [1]. Chromatin is an assembly of proteins and RNA that wrap DNA into repeating units of ~150 bp, called nucleosomes, each of which contains a histone octamer of dimers of H2A, H2B, H3, and H4. In the past, much emphasis has been placed on the idea that chromatin compacts the cell's genetic material and organizes the nucleus into

Raphaël Margueron and Daniel Holoch (eds.), *Histone Methyltransferases: Methods and Protocols*,
Methods in Molecular Biology, vol. 2529, https://doi.org/10.1007/978-1-0716-2481-4_1,

complex hierarchical structures. Yet, to function properly, chromatin must be dynamic and is thus subject to regulation in space and time as development of organisms, differentiation of tissues, and responses to the environment may demand. The local dynamics of chromatin are dictated by interactions of DNA with core histone octamers, with the linker histone H1 and numerous other DNA-binding proteins providing additional structural organization. Histones and other proteins can be chemically altered by posttranslational modifications (PTMs), including methylation, acetylation, phosphorylation, ubiquitination, and more. While these modifications are found along the entire histone sequence, modification in the basic N-terminal tails are most widely studied [1]. Incisive studies by mass spectrometry have uncovered hundreds of PTMs on all histones, though it remains unclear whether all of them carry biological significance [2–6].

Changes to the PTM chromatin landscape are accomplished by histone "writers" and "erasers," while histone "reader" proteins modulate the recruitment of downstream effectors [1]. This modification landscape is complex, with mechanistic understanding still rudimentary; it involves multiple histones, and sometimes neighboring nucleosomes. Histone methylation is just one PTM that is involved in numerous fundamental processes, affecting DNA replication, DNA repair, genome maintenance, and access to the transcriptional start site by transcription factors, and all these processes require alterations in local chromatin, which are achieved by changes in histone methylation status which are coupled to concomitant changes in the status of other PTMs, e.g., acetylation, phosphorylation, and ubiquitination.

Long-range interactions, or the modulation of larger chromatin domains, are defined by characteristic protein and DNA modifications—including nucleosome occupancy, cytosine or adenine DNA methylation, and histone PTMs—resulting in changes to gene expression [1]. Often described as "loosely packed," euchromatin is found in gene-rich, transcriptionally active regions of the genome. These regions have lower nucleosome occupancy, are characterized by histone lysine acetylation, histone H3 lysine 4 di- and trimethylation (H3K4me2/3), H3K36me3, and H3K79me3 (Fig. 1a). In contrast, domains of heterochromatin are transcriptionally silent, existing in two major forms—always condensed "constitutive heterochromatin" and reversibly transcriptionally silent "facultative heterochromatin" (Fig. 1b). Constitutive heterochromatin, characterized by H3K9me3 and cytosine DNA methylation, is found primarily in regions with repetitive DNA, including centromeric and subtelomeric regions, and transposable elements. Facultative heterochromatin is generally transcriptionally silent, but is expressed under appropriate conditions, in response to external or internal stimuli, aiding in proper spatio-temporal gene expression. Chromatin thus exists along a dynamic continuum from

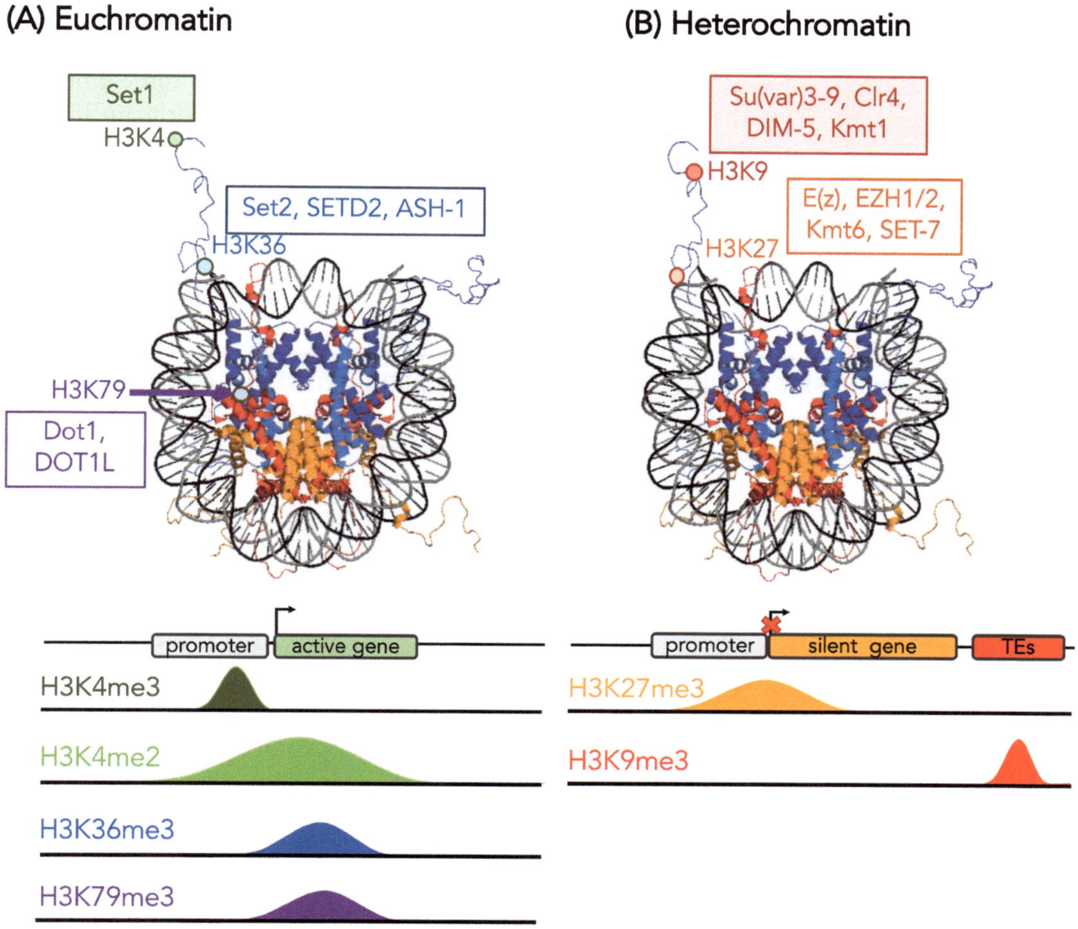

Fig. 1 Selected histone modifications that are correlated with (**a**) euchromatin or (**b**) heterochromatin and their idealized distribution on protein-coding genes. H3K4me2/3 are catalyzed by KMT2/Set1 proteins and are usually found in promoter or 5′ regions of genes. H3K36me2/3 are catalyzed by both Set2 and Ash1-like proteins and in many organisms cover all protein-coding genes, or expressed genes. While early studies showed effects of Ash1 on H3K4me2/3, recent results obtained with filamentous fungi suggest that Ash1-mediated H3K36me2/3 is correlated with subtelomeric facultative heterochromatin and affects H3K27me3. The non-SET KMT, Dot1, methylates a surface-exposed H3K79 residue in the H3 globular domain and is mostly correlated with active transcription. In many eukaryotes, facultative heterochromatin is marked by H3K27 methylation by the KMT6/E(z) subfamily of SET domain proteins, while constitutive heterochromatin is marked by H3K9me2/3, catalyzed by the KMT1/Su(var)3–9 subfamily, first discovered in *Drosophila* and called Clr4 in *S. pombe* and DIM-5 in *N. crassa*. Not shown is H4K20 methylation, which affects cell cycle regulation and DNA repair, and is correlated with gene repression. Also not shown here are the activities of the various protein arginine methyltransferases (PRMTs), as they have not been universally confirmed in most eukaryotes

inaccessible, transcriptionally silent heterochromatin to accessible, transcriptionally active euchromatin [7].

Here we will discuss the occurrence of histone methyltransferases in eukaryotes, and their shared and sometimes divergent functions in arginine and lysine methylation, with some emphasis on the interactions between two conserved SET domain histone methyltransferase complexes: PRC2, which methylates histone H3K27, and ASH1, which methylates histone H3K36. As is true for other families of chromatin-modifying proteins [8], expansion of gene families occurred in all eukaryotic clades, and novelty often arose by addition of expansion of DNA- or histone-binding motifs.

2 Meet the Organisms: Deep Phylogenetic Sequencing Makes New Models

Most scientists working on histone methyltransferases settle early on their "favorite model" organism. Thus, plant biologists are experts on the many histone modification enzymes controlling plant development or host–pathogen interactions, and the same can be said of human geneticists who may be particularly interested in pathologies caused by PTM dysregulation during development or cancer. Some organisms have been essential general models, e.g. among the ciliates *Tetrahymena* and among the fungi budding (*Saccharomyces cerevisiae*) and fission (*Schizosaccharomyces pombe*) yeast, but by now we know that they lack important PTMs present in plants or many animals, and thus filamentous ascomycete fungi (like *Neurospora crassa*, *Fusarium* spp., or *Zymoseptoria tritici*) and basidiomycete yeasts (like the human pathogen *Cryptococcus neoformans*) have become additional models to decipher the general histone methylation landscape and the interactions and dependencies between different methylation states.

Of course, eukaryotic biology is much more diverse than the choice of model organisms reflects, and thus one aim of this chapter is to explore the complement of protein methyltransferases that can affect histones, and thus chromatin structure, from all eukaryotic supergroups [9, 10]. Taxonomy has come a long way since the days of the "Five Kingdoms" hypothesis; instead, we now recognize at least seven supergroups of eukaryotes, still leaving six large clades unassigned, and we are not even quite sure yet how the supergroups form monophyletic clades [9]. Advances in genome sequencing give us access to at least the predicted proteomes of many new taxa, including important human pathogens, organisms important for carbon sequestration and other ecological issues, plants and their pathogens, and a large number of animal taxa. For our analyses, we selected representatives from most supergroups, at least one species each when high-quality genome drafts were available (Fig. 2). This still cannot do justice to the diversity of taxa in supergroups; it is clear from analyses of plants, fungi, and animals, the best-studied eukaryotes, that numerous lineages maintain or expand most known protein methyltransferases but that many lineages within kingdoms lose specific activities.

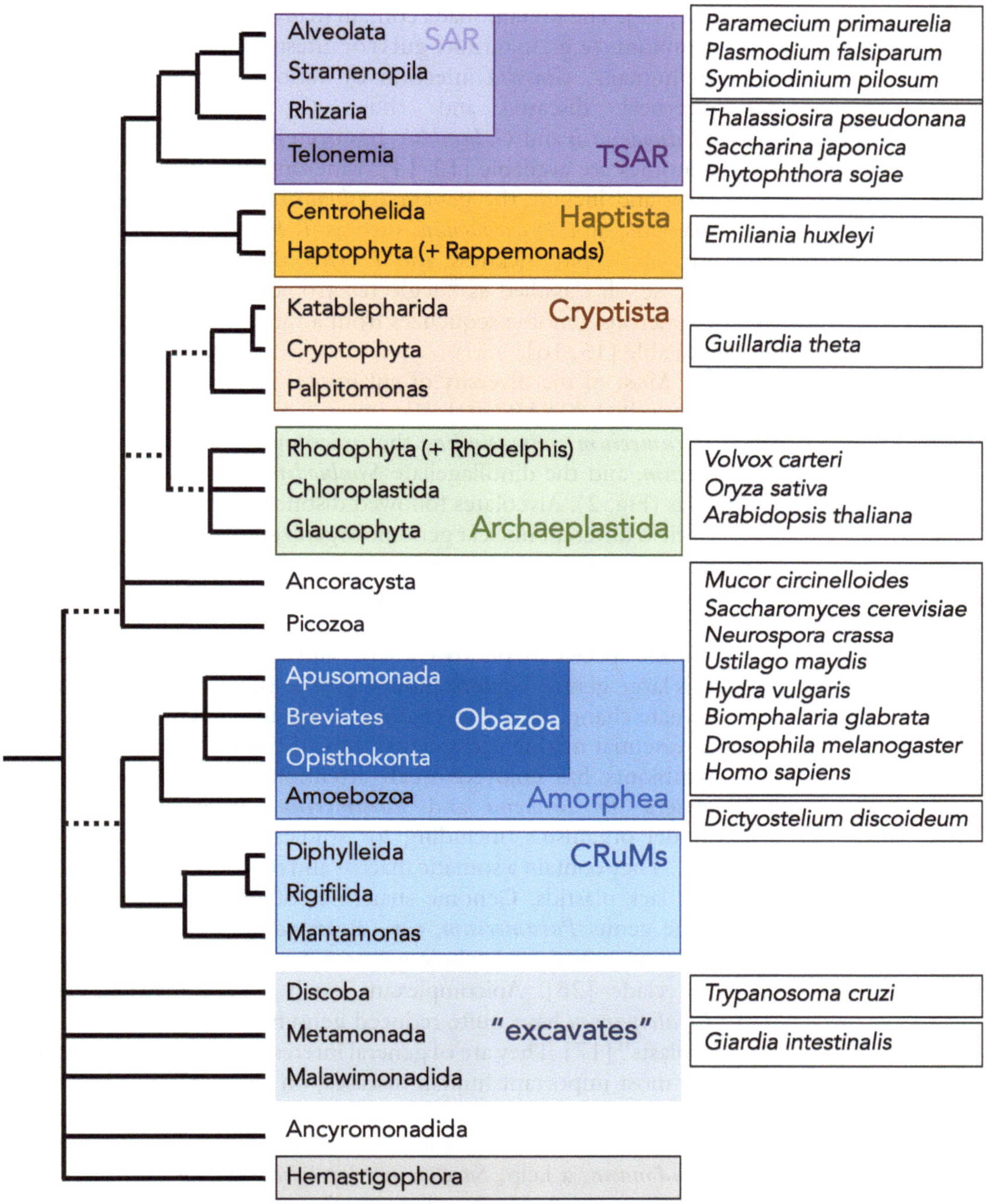

Fig. 2 Proposed supergroups within the eukaryotes (according to [9]) and species selected to investigate distribution and relationships of eukaryotic protein methyltransferases. See text for details on the organisms

The Discoba and Metamonada belong to the former supergroup of "excavates," an early and deeply diverging lineage of eukaryotes (Fig. 2). Many protists that evolved from this group are parasites and thus often have reduced genomes shaped by loss of traits caused by the evolving host–pathogen interactions

[11, 12]. The Metamonada contain many anaerobic species that are symbionts (e.g., in termite guts) or intestinal parasites of mammals. In humans, *Giardia* infection by various species causes serious diarrheal disease, and thus, *G. intestinalis* (synonyms *G. duodenalis* and *G. lamblia*) has been heavily studied and several genomes are available [12–14]. Kinetoplastids belong to the Discoba and include the genera *Leishmania*, causal agents of Leishmaniasis, and *Trypanosoma*, such as *T. brucei*, the causal agent of African sleeping sickness, and *T. cruzi*, the causal agent of Chagas disease, all classified as "neglected tropical diseases" by the NIH. Numerous genome sequences from a variety of pathovars are now available [15, 16].

Most of the diversity of eukaryotes resides in the supergroup now called "TSAR" [9]. We selected three Alveolata, the ciliate *Paramecium primaurelia*, the apicomplexan *Plasmodium falsiparum*, and the dinoflagellate *Symbiodinium pilosum* for our analyses (Fig. 2). Alveolates followed distinct evolutionary trajectories to yield different nuclear genome organization [17, 18]. Dinoflagellates in particular are quite distinct from other eukaryotes in that they do not use histones to organize chromatin; instead, they have condensed liquid-crystalline chromosomes [19–21]. Many *Symbiodinium* species are photosynthetic, and all coral symbionts are from this large genus. Understanding symbiosis, especially in this age of climate change, is of outstanding importance to ensure the survival of essential marine ecosystems; thus, the study of corals and their symbionts has enjoyed much attention [22]. Ciliates, like the genera *Tetrahymena* and *Paramecium*, have been important model organisms, including for studies on gene silencing [23–25]. They contain a somatic macro- and reproductive micronucleus but lack plastids. Genome structures of representatives from the large genus *Paramecium*, e.g., *P. primaurelia* and *P. tetraurelia*, have intensified after whole-genome duplications were detected in the clade [26]. Apicomplexans, such as the malarial parasite, *P. falciparum* have quite reduced genomes, with degenerate "apicoplasts" [17]. They are of general interest because malaria is one of the most important human diseases, on the rise partly because of climate change [27].

From the Stramenopila, we selected a diatom, *Thalassiosira pseudonana*, a kelp, *Saccharina japonica*, and an oomycete, *Phytophthora sojae*. Marine diatoms like *T. pseudonana* are widely distributed throughout all oceans and are models for light absorption and carbon metabolism, including how diatoms may affect global carbon cycling [28]. "Kelps" or "seaweed" belong to a large group of marine brown algae, and *S. japonica* is one of the commercially important species for food production [29–31]. The oomycete genus *Phytophthora* includes some of the most devastating plant pathogens [32, 33]. Some are relative specialists and infect only specific plants, like *P. sojae* on soybeans and *P. infestans* mainly on potatoes, and others are generalists and are able to infect a large group of plants, e.g., *P. ramorum* or *P. cinnamomi* on many diverse woody plants. These species have the potential to be extremely invasive and change the whole ecosystems in a relatively short amount of time.

Within the Haptista, there are many species of marine and freshwater protists. *Emiliania huxleyi* belongs to photosynthetic plankton found in oceans from the equator to subpolar regions that form the basis of marine food webs [34]. It can form extensive blooms in nutrient-depleted waters that impact ocean temperatures and carbon balance but contributions of this species or plankton as a whole are not yet well understood [35–37]. The Cryptista contains species of flagellate algae that have a secondary plastid within a cytoplasm that also contains a vestigial nucleomorph, evidence of eukaryotic endosymbiosis [38]. *Guillardia theta* is the only characterized member of the genus and the first cryptophyte with a sequenced genome [39].

The Archaeplastida includes all land plants and green algae, the photoautotrophic red algae (Rhodophyta), and their non-photosynthetic sister group (Rhodelphis), as well as a distinct group of freshwater algae (Glaucophyta). From this group, we selected a multicellular green alga, *Volvox carteri*, which has become a model organism to study evolution of multicellularity, and two land plants, the monocot and most cultivated grain species, rice (*Oryza sativa*), and the dicot and best understood plant, *Arabidopsis thaliana*.

The Amorphea includes amoebae and the Opisthokonta, which includes the fungi and animals. All of these groups have been very heavily studied, and thus most analyses center on comparisons of animals to each other, or to the fungi and plants. We selected some of the obvious candidates, such as the amoeba ("slime mold") *Dictyostelium discoideum* [40], the fungi *Mucor circinelloides* (an emerging human pathogen belonging to the former "zygomycetes" [41]), the budding yeast *S. cerevisiae* [42], the ascomycete *N. crassa* [43], and the hemibasidiomycete *Ustilago maydis* (a global pathogen on maize [44]). From the animals, we chose *Hydra vulgaris* [45], *Biomphalaria glabrata* (a snail host of schistosome parasites) [46], the fruit fly *Drosophila melanogaster* [47], and human, *Homo sapiens* [48].

The take-home message from our representative sampling is that histone methylation capabilities are of ancient origin and that most arginine and lysine methyltransferase proteins are found in at least some taxa from all extant supergroups. This has been borne out by an in-depth phylogenomics analysis of many taxa that found "punctate retention" of histone methylation genes across eukaryotes [49].

3 Classes of Histone Methyltransferases

Protein methyltransferases (MTases) evolved to specifically methylate arginine or lysine residues of target substrates, and many of them play instrumental roles in regulating the structure and function of chromatin. Universally, histone methyltransferases (HMTs) use *S*-adenosyl-L-methionine (SAM or AdoMet), an intermediate metabolite of methionine metabolism, as a methyl radical donor, yielding methylated arginines or lysines and releasing the cofactor product *S*-adenosyl-L-homocysteine (SAH). HMTs fall into three groups: protein arginine methyltransferases (PRMTs), non-SET domain KMTs, and SET domain lysine methyltransferases (KMTs).

3.1 Protein Arginine Methyltransferases (PRMTs)

Arginine residues can be monomethylated, carried out by Type I, II, and III enzymes, or dimethylated, either asymmetrically (Type I) or symmetrically (Type II; reviewed in [50, 51]). While three types of PRMTs are distinguished by the methylarginine they produce, the active sites of these enzymes are similar though they differ in the substrate binding pocket, thus restricting their activity toward specific target proteins [52–54]. Based on the primary sequence of the AdoMet MTase domain and the presence of additional protein domains or motifs, at least ten distinct eukaryotic PRMTs can be distinguished (Fig. 3a). The current nomenclature for these proteins is challenging to follow; as we will also see for the much better studied SET domain proteins, there is no universally accepted numbering system. Most supergroups have genes encoding representatives of PRMT1, PRMT5, PRMT6, PRMT7, and PRMT10, while PRMT2 and PRMT8 are restricted to animals.

The N-terminus of the conserved methyltransferase region of PRMT3 is composed of a Rossmann-like fold, with five beta strands alternating with four alpha helixes to form an extended beta sheet structure. These folds create a SAM binding pocket [52]. Toward the C-terminal region of the conserved methyltransferase domain a barrel-like structure is observed, comprising the active site of the PRMT, and there has been much progress made on discovering small molecule inhibitors [50, 55].

3.1.1 Distribution of PRMTs in the Eukaryotes

Of the 22 species selected for extensive BLAST searches, only one, the diplomonad *Giardia intestinalis* in the Metamonada supergroup, did not return any reliable "hits" against plant, animal, or fungal PRMTs (Table 1). The fact that *Giardia* does not seem to have any arginine methylation has recently come to light; apparently there are functional equivalents to fulfill this important role, not just for histone but protein methylation in general [56]. Absence of PRMTs had been previously postulated [57], and as diplomonads include important human pathogens, neofunctionalization to substitute for PRMT function immediately suggested novel approaches for pharmacological intervention.

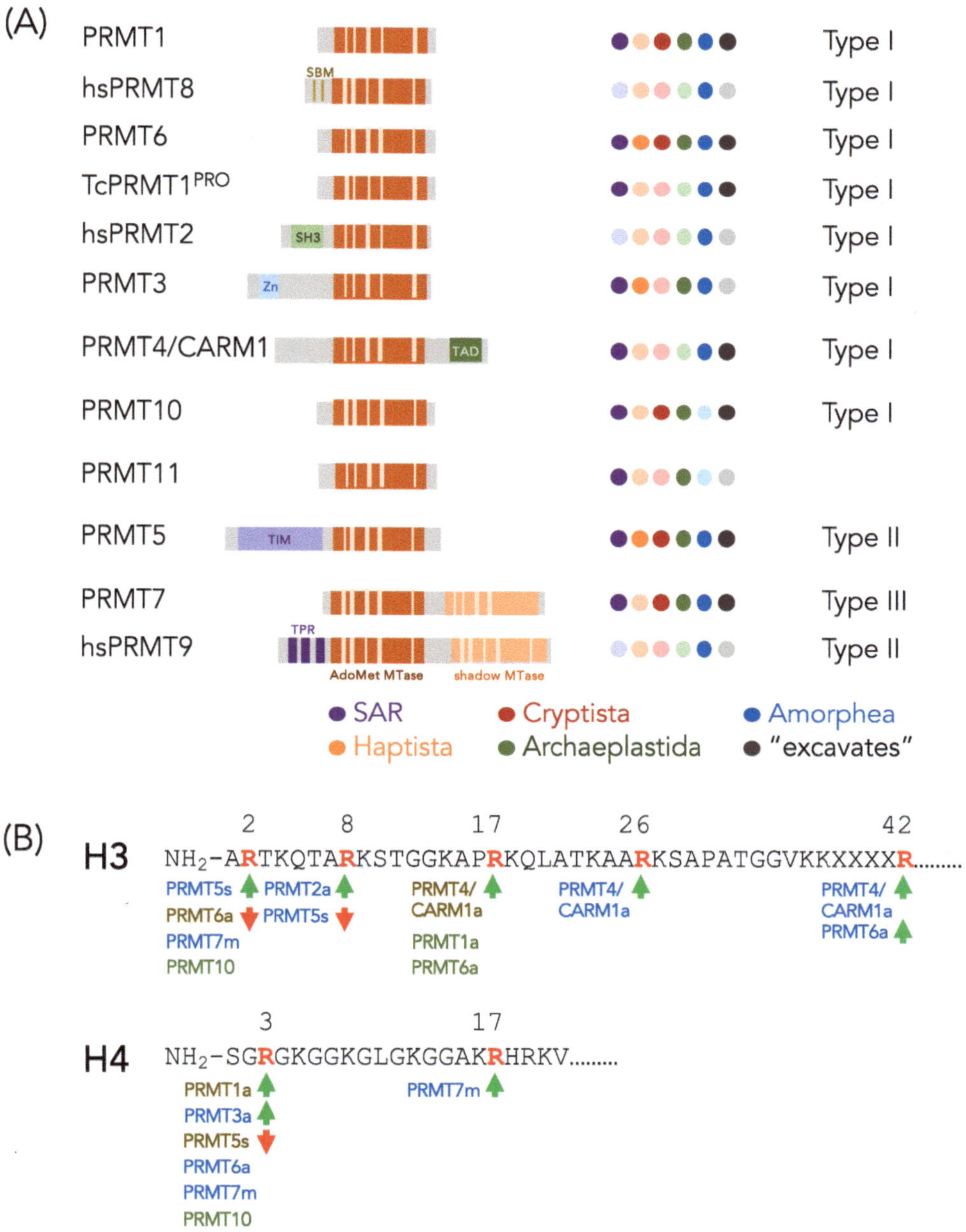

Fig. 3 Eukaryotic protein arginine methyltransferases (PRMTs) and their known histone substrate specificity. (**a**) Classification of PRMTs mostly following the subfamilies described in mammals. The PRMT catalytic domains are shown in brown, with the conserved motifs shaded in light tan. PRMT7 and PRMT9 proteins both contain discernible duplications of the AdoMet MTase domain, called "shadow MTase" here. Within the subfamilies, some novelty is generated by addition of the various functional domains shown (i.e., SBM, SH3, Zn-binding domains in the PRMT1, PRMT6, and PRMT3 groups). Presence of PRMT families in the 22 taxa examined is indicated by the shaded colored circles; solid colors denote presence and pastel colors denote absence in the taxa studied, but the latter does not indicate that other species in the respective supergroups may not encode these PRMTs. (**b**) Documented activities of eukaryotic PRMTs on histone H3 and H4 tails. Blue type denotes activity demonstrated in the Amorphea (mostly animals and fungi), green type denotes activity demonstrated in Archaeplastida, and gold type denotes activity found in both Archaeplastida and Amorphea. Green arrows denote correlation with gene expression, while red arrows denote correlation with transcriptional gene silencing

Table 1
GenBank accession numbers for PRMT homologs discussed and their most likely placement in PRMT subfamilies

Name	PRMT1	PRMT2	PRMT3	PRMT4	PRMT5	PRMT6	PRMT7	PRMT8	PRMT9	PRMT10	PRMT11
Thalassiosira pseudonana	XP_002288454.1		XP_002295201.1		XP_002293139.1	XP_002290528.1				XP_002297359.1	
			XP_002297358.1								
Symbiodinium pilosum				CAE7435993.1	CAE7718648.1	CAE7494238.1	CAE7761882.2			CAE7236492.1	CAE7513706.1
							CAE7331634.1				
Saccharina japonica				5116170.1	5102158.1						
Paramecium primaurelia	CAD8085966.1				CAD8046498.1				CAD8078653.1		
	CAD8100299.1								CAD8088540.1		
	CAD8085964.1								CAD8082635.1		
									CAD8107569.1		
Plasmodium falsiparum	PF3D7_1426200				PF3D7_1361000				PF3D7_0811500		
Phytophthora sojae	XP_009526476.1			XP_009523191.1	XP_009532297.1		XP_009536099.1			XP_009514661.1	
	XP_009526537.1										
Emiliania huxleyi			XP_005757501.1	XP_005766316.1	XP_005764086.1	XP_005765892.1				XP_005792814.1	
			XP_005768836.1		XP_005782814.1					XP_005776194.1	
										XP_005793688.1	
										XP_005764430.1	
Guillardia theta	XP_005825583.1				XP_005837555.1	XP_005836817.1	XP_005835264.1		XP_005822462.1		
	XP_005824608.1					XP_005842057.1	XP_005833971.1				
	XP_005818099.1										
	XP_005834566.1										

Volvox carteri	XP_002946963.1		XP_002953088.1	XP_002958435.1	XP_002946852.1					XP_002946350.1	XP_002958150.1
			XP_002950898.1								
Oryza sativa	XP_015612441.1		XP_015647334.1	XP_015647687.1	XP_015627032.1	XP_015614476.1	XP_015641005.1			XP_015642494.1	
Arabidopsis thaliana	NP_194680.1		NP_187835.2	NP_199713.2	OAO97588.1	NP_188637.2	OAO96767.1			NP_563720.1	
Trypanosoma cruzi	XP_815715.1				XP_808137.1	XP_811231.1	XP_815328.1		XP_822106.1		
	XP_821921.1					XP_806643.1	XP_819617.1		XP_817659.1		
Giardia Intestinalis											
Dictyostelium discoideum	XP_635288.1				XP_637240.1	XP_643270.2					
						XP_636224.1					
Mucor circinelloides	EPB85895.1		EPB81838.1	EPB90571.1	EB87598.1				EPB82839.1		
Saccharomyces cerevisiae	NP_009590.1		NP_010753.1		KZV13214.1						
Neurospora crassa	XP_963910.1		XP_956875.2		XP_011393468.1						
Ustilago maydis	XP_011392245.1			XP_011387335.1	XP_011387515.1	XP_011371078.1					
Hydra vulgaris	XP_002157035.1		XP_002165428.3	XP_002169585.2			XP_012566808.1		XP_002155504.2		
Biomphalaria giabrata	XP_013087638.1		XP_013064251.1	XP_013071413.1	XP_013084844.1	XP_013077111.1	XP_013029096.1		XP_013084436.1		
Drosophila melanogaster	NP_650017.1	NP_001285600.1	NP_731984.1	NP_001262445.1	ADU79249.1	NP_650322.1	NP_611753.4	NP_609478.1	NP_650321.1		
Homo sapiens	NP_001527.3	NP_001526.2	NP_005779.1	NP_001357018.1	NP_006100.2	NP_060607.2	NP_001338072.1	NP_062828.3	NP_612373.2		

Another "deep-branching" group in the former excavates supergroup, the Discoba, includes *Trypanosoma cruzi*, which encodes five different PRMTs, the same as found previously in *T. brucei* [57]. There has been sustained interest in PRMT function in kinetoplastids, resulting in studies on the function and interactions of all PRMTs [58] but especially the novel "PRMT3-like" protein that was found to be necessary for PRMT1 function [59, 60]. This protein has thus been renamed PRMT1^{PRO} (for "prozyme"), and PRMT1 has become PRMT^{ENZ}; a recent review summarizes these and other functional studies [61].

With the exception of *Plasmodium* [61–63], there are no published data on activities or function on PRMTs in the selected species in the Haptista (*E. huxleyi*), Cryptista (*G. theta*), and the diverse TSAR supergroup; these taxa show the most diverse patterns of presence or absence of PRMTs (Table 1). For example, the diatom *T. pseudonana* has two potential PRMT3 homologs, but neither includes a Zn-finger motif, and they are either predicted to be much longer or shorter than the homologs found in fungi and animals. It encodes another four PRMTs, similar to trypanosomes but has PRMT10 rather than PRMT7 in addition to PRMT1, PRMT5, and PRMT6. In contrast, the oomycete *P. sojae* has two PRMT1s, PRMT4, PRMT5, PRMT7, and PRMT10, but lacks PRMT3 and PRMT6. The ciliate *P. primaurelia* and the apicomplexan *P. falsiparum* encode just three PRMTs, clear homologs of PRMT1 and PRMT5, and another PRMT that is similar to *T. brucei* PRMT^{PRO}. Whether this protein has the same function in activating PRMT^{ENZ} is not resolved. Only two predicted proteins with PRMT signatures were detected in the genome of the kelp, *S. japonica*. While dinoflagellates have long been known to lack histones, there are at least six PRMTs predicted from recently published genome sequences, including "PRMT11," which was also detected in *Volvox* in our representative sampling of eukaryotic genomes. All *Symbiodinium* PRMTs are predicted to modify other proteins though it is conceivable that some modify the histones of the hosts, corals.

Compared to the taxa previously mentioned, most Archaeplastida (plants and green, as well as some red algae) have expanded this gene family, encoding eight or nine PRMTs [64], some of which (e.g., PRMT1 and PRMT4) are present in pairs that are at least partially redundant [51]. The green alga *V. carteri* seems to lack clear homologs for PRMT6 and PRMT7 but has two proteins, PRMT11 and PRMT12, that may be the best homologs for these PRMTs (Table 1).

Within the Amorphea, amoebae like *D. discoideum* encode three PRMTs (PRMT1, PRMT5, and two isoforms of PRMT6). Basal lineages of fungi, like the former "zygomycetes," represented here by *M. circinelloides*, encode five PRMTs, including a homolog of the PRMT1^{PRO} protein found in *T. brucei* (Table 1). This lineage and the basidiomycetes also have PRMT4/CARM1 homologs that the ascomycetes (e.g., *S. cerevisiae*, *S. pombe*, and *N. crassa*) lack. Instead, these well-studied model organisms all have a minimal complement of PRMTs, namely PRMT1, PRMT3, and PRMT5 [65, 66]. Animals like *Hydra* and the snail *B. glabrata* have five and seven PRMTs, respectively; there are no functional studies available. *Drosophila* and mammals encode nine different PRMTs [67], with human PRMT2 and PRMT9 arising from the PRMT6 and PRMT7 family, respectively. Similarly, PRMT8 seems to be most closely related to the PRMT1 family. It is important to remember that the numbering system of nonmammalian animal PRMTs does not necessarily match protein similarity based on the human numbering system [67]; some of the animal PRMTs may also have been missed in previous analyses.

Overall, while comparisons of the whole protein sequences or just the MTase domain across the selected taxa resolve some protein phylogenies well, e.g., PRMT5, the PRMT1^{PRO} group, PRMT7, PRMT10, PRMT4, and PRMT1, placement of PRMT2, PRMT3, and PRMT6 is more difficult to resolve, largely because in some taxa the specific motifs found in animal PRMTs are lacking. Compared to some of the other histone modification gene families, PRMTs fall into the group that have been most widely retained since the "Last Common Eukaryotic Ancestor," LECA [49]. Intriguingly, at least one report has also suggested the presence of methylated arginine in bacteria [53].

3.1.2 Histone Methylation Catalyzed by PRMTs

PRMTs are well known to posttranslationally modify many proteins, including transcription factors, co-activators and co-repressors, and signaling factors involved in the cell cycle and oncogenesis; an in-depth review is beyond the purpose of this chapter but is available elsewhere [50]. PRMT activities on core histone tails constitute a rather minor part of their substrate repertoire, but they have been reviewed [50, 51, 67, 68] and are best-studied in mammalian cells (Fig. 3b). Methylated arginine residues of H3 and H4 are correlated with both active and silenced transcription, for example, H4R3me2a, catalyzed by PRMT1 or PRMT3, is a mark for active transcription, but H4R3me2s, catalyzed by PRMT5, is correlated with gene silencing. Similarly, on the H3 tail, H3R2me2s is a mark for active transcriptionm but H3R8me2s is a repressive mark, and both the reactions are catalyzed by PRMT5. PRMT6 catalyzes H2AR29me2a, which results in repression of transcription. Because methylarginines do not just correlate with active or silent transcription but are also involved in crosstalk between other histone modifications it is clear that there are several layers and potentially redundant circuits for specific outcomes of the "histone code." For example, there is evidence for crosstalk between PRMT7 and PRMT5, as H4R17me by PRMT7 may activate PRMT5 to yield H4R3me2s, a mark repressive for transcription [69].

Less is known from plants but early work in both *Arabidopsis* and rice revealed multiple histone methylation sites (Fig. 3b). In rice, PRMT1 generates H3R17me2 and H4R3me2, PRMT4 generates H3R17me2, PRMT5 generates H4R3me2, PRMT6b generates H3R2me2 and H3R17me2, and PRMT10 generates H3R2me2 and H4R3me2 [64]. Much work remains to be done, especially in the protists and filamentous fungi, to uncover the full scope of gene regulation by histone arginine methylation.

3.2 Non-SET Domain Histone Lysine Methyltransferases (KMTs)

Unlike HMTs that target histone tails for methylation, the non-SET-domain-containing methyltransferase Dot1 ("Disruptor of telomeric silencing 1") methylates H3K79, a surface-exposed residue in the H3 globular core. Dot1 is still the sole non-SET domain lysine methyltransferase, first discovered in budding yeast in a genetic screen for proteins involved in position effect variegation [70] and shown to modify gene silencing [71]. The mammalian homolog, DOT1L ("DOT1-like"), is important for transcriptional regulation, cell cycle regulation, and the DNA damage response [72]. In one classification, Dot1 homologs are labeled KMT4 [73].

Most eukaryotes have a single gene encoding Dot1 homologs, though some, like *T. brucei*, have two enzymes, one for H3K79 mono- and dimethylation (Dot1A) and one for H3K79 trimethylation (Dot1B) [72]. In many animals, either several genes (e.g., in *Caenorhabditis elegans*) or splice variants (in mammals) have been detected. Some fungi and most plants do not have genes encoding Dot1 homologs [74].

The overall size and structure of DOT1 homologs vary greatly, with the highest levels of sequence similarity in the N-terminus. Both yeast and human Dot1 have active sites capable of mono-, di-, and trimethylation. Crystal structures of DOT1L in complex with the methyl donor, SAM, showed that the N-terminal HMT domain is comprised of a series of open α/β structures surprisingly similar to that of PRMTs [75]. Differences surrounding the Dot1 active sites confer target specificity. Through incisive studies over the past two decades, Dot1L has emerged as one of the paradigms for histone modification crosstalk. Dot1 is activated by ubiquitination of H2B lysine 120 (H2BK120ub) [76] and structural work on how this is accomplished has recently been reviewed [77]. Additional structural studies showed how not only H2BK120ub but also H4K16 acetylation (H4K16ac) results in allosteric stimulation of Dot1 activity, both in vivo and in vitro [78].

3.3 SET Domain KMTs

SET domain-containing proteins, named after three proteins that were first discovered in *D. melanogaster*, namely suppressor of variegation 3–9 [Su(var)3–9], enhancer of zeste [E(z)], and trithorax (Trx), can be found in the genomes of all eukaryotes and in some bacteria [8, 51, 56, 74, 79–83]. The recent advances in whole-genome and metagenome sequencing have uncovered that histone methylation by SET domain group proteins is an ancient process; phylogenies of selected SET domain proteins show that early diverging eukaryotes carry genes for many of the well-studied subfamilies. Many heavily studied model organisms, like budding or fission yeast, however, have lost genes for specific KMTs, and while the proteomes of budding and fission yeast harbor 12 and 13 SET domain proteins, respectively, many filamentous fungi (like *N. crassa*, *M. circinelloides*, and *Ustilago maydis*) encode as many as 20 SET domain proteins [82]. This protein family is even more expanded in mammalian and plant proteomes, where between 40 and 60 SET domain proteins are found [84]. Especially in the non-model eukaryotes, protein KMT activity has been experimentally attributed to only a subset of these proteins by either in vivo or in vitro methods, and fewer still are histone methyltransferases with known activity on specific lysine residues.

3.3.1 Distribution of SET Domain Proteins in Eukaryotes

Plant and mammalian SET domain proteins are well studied, and most previous work characterized seven or eight only partially overlapping subfamilies (Table 2); phylogenetic analyses of ciliate SET domain proteins uncovered 13 monophyletic eukaryotic clades [80], but based on the uncertain relationship in several subfamilies, there may be as many as 15 SET domain subfamilies. Many of the plant and animal proteins in these subfamilies have homologs and orthologs in the fungi, amoebae, and the SAR clade, though some of the truly well-studied plant, fungal, and mammalian KMTs have no obvious homologs in the SAR group or the early and deeply branching clades Discoba and Metamonada (Fig. 4a). In many of these clades, especially the plant-type SMYD and SETD subfamilies of KMTs, carrying both a SET domain and a zinc-finger MYND domain or a Rubisco LSMT substrate-binding domain, respectively, appear to be expanded to include many more family members than in plants, fungi, and animals. A complete accounting and curation for all SET domain proteins in the SAR clade is beyond this review, but it is curious that *Symbiodinium*, i.e., a genus without histones, seems to encode dozens of proteins with SET domains of the SMYD and SETD type.

Subfamilies in mammalian genomes have been renamed according to a system proposed after numerous model genome sequences had been nearly completed, relying on ordering KMT subfamilies by date of discovery [73]. Subfamily numbering in plants does not adhere to this classification, and even in single species, there are often multiple names for the same gene or isoform, as is common in mammals as well. For our purposes, we grouped subfamilies by known or predicted substrates and followed the mammalian nomenclature (Fig. 4a).

Homologs of *Drosophila* Su(var)3-9 belong to the KMT1 or plant Suv subfamily and methylate H3K9; most studied enzymes are capable of catalyzing mono-, di-, and trimethylation and are essential for gene silencing in constitutive heterochromatin

Table 2

GenBank accession numbers for SET domain KMTs found in selected eukaryotes and their most likely placement in KMT subfamilies

	Supergroup	SAR				Haptista	Cryptista	Archaeaplastida	Amorphea					
	Group	Diatom	Ciliate	Apicomplexa	Oomycete			Chloroplastida	Amoebozoa	Fungi			Animals	
Subfamily	Names	*Tp*	*Pt*	*Pf*	*Ps*	*Eh*	*Gt*	*At*	*Dd*	*Mc*	*Sc*	*Nc*	*Dm*	*Hs*
KMT1	Su(var)3-9, KMT1			PF3D7_0827800_SET3	XP_009518856.1	XP_005778404.1	XP_005825525.1	At5G04940	XP_646062.1	EPB89792.1	none	XP_957479.2	NP_524357.2	
SUV39	SUV39H1, KMT1A							At1G73100						NP_001269095.1
H3K9me	SUV39H2, KMT1B							At1G17770						NP_003164.1
	G9a, KMT1C			PF3D7_0508100_SET9				At2G24740					NP_569834.1	AAD21812.1
	GLP/ EuHMT1, KMT1D							At2G05900						BAB56104.1
	ESET/ SETDB1, KMT1E							At5G13960					NP_611966.3	NP_001380889.1
	SETDB2, KMT1F							At4G13460						NP_114121.2
								At3G03750 At2G35160 At2G22740 At1G04050 At5G43990						
								At3G04380						
								At2G23740						
KMT2	SET1, KMT2	XP_002291362.1	GSPATG00013040001	PF3D7_0629700_SET1	XP_009536801.1	XP_005764528.1	XP_005837587.1	At2G31650	XP_636258.1	EPB84754.1	NP_011987.1	XP_961572.3		
SET1	Trx, MLL1, KMT2A	XP_002296329.1					XP_005839674.1	At1G05830					NP_476769.1	NP_005924.2
H3K4me	Trx, MLL2, KMT2B		GSPATG00014017001					At3G61740						AAD56420.1
	Trr, MLL3, KMT2C		GSPATG00025368001	PF3D7_0910000_SET4				At4G27910						AAK00583.1
	Trr, MLL4, KMT2D		GSPATG00031547001	PF3D7_1355300_SET6				At5G53430						AAH09337.2
	hSET1A, KMT2F		GSPATG00018768001					At4G15180					NP_001015221.1	NP_055527.1

	hSET1B, KMT2G		GSPATG00035094001					At5G42400						NP_001340274.1
	ATXR1-4	XP_002290717.1												
KMT3	SET2, KMT3	XP_002294263.1	GSPATG00013040001	PF3D7_1322100	XP_009537692.1	XP_005794164.1	XP_005832326.1	At2G44150	XP_647576.1	EPB87143.1	NP_012367.2	XP_957740.1	NP_001263029.1	
SET2	SET2, KMT3A	XP_002296152.1	GSPATG00003275001				XP_005818219.1	At3G59960		EPB88931.1			NP_572888.2	NP_054878.5
H3K36me	NSD1, KMT3B	XP_002296152.1	GSPATG00004957001		XP_009539893.1			At4G30860						AAK92049.1
	SMYD2, KMT3C	XP_002294263.1						At1G76710						AAF68983.1
	ASH1, set-3, KMT2H				XP_009518980.1			At1G77300				XP_964116.3	AAB01100.1	AAH11635.1
KMT5	Set9, KMT5	XP_002291638.1			XP_009522567.1	XP_005776989.1		At2G33290		EPB87004.1	none	XP_963033.2		
SUV4-20	Pr-SET7/8, SET8, KMT5A			PF3D7_0403900_SET8						EPB81915.1			NP_001247100.1	Q9NQR1.3
H4K20me	SUV4-20H1, KMT5B													NP_001356355.1
	SUV4-20H2, KMT5C												NP_001245453.1	
KMT6	E(z), KMT6	XP_002290191.1	XP_001436830.1_EZL1		XP_009524339.1		XP_005836861.1	At1G02580		EPB84754.1	none	XP_965043.2	AAC46462.1	
E9z)	EZH1, KMT6A		GSPATG00032888001		XP_009519967.1		XP_005834965.1	At2G23380						XP_011522819.1
H3K27me	EZH2, E(z), KMT6B		GSPATG00012695001		XP_009516860.1			At4G02020						NP_004447.2
			GSPATG00013305001		XP_009521676.1									
			PTETG1700020001		XP_009526799.1									
ATXR	ATXR5, ATXR6	XP_002293206.1	GSPATG00025951001					At5G09790						
H3K27me1			GSPATG00033097001					At5G24330						
			GSPATG00036078001											
			GSPATG00035182001											
			GSPATG00021145001											
KMT7	SET7/9						XP_005842301.1		XP_636856.1					
														NP_085151.1
KMT8	RIZ, PRDM			PF3D7_1214200_SET5										

(continued)

Table 2
(continued)

	Supergroup	SAR				Haptista	Cryptista	Archaeaplastida	Amorphea					
	Group	Diatom	Ciliate	Apicomplexa	Oomycete			Chloroplastida	Amoebozoa	Fungi			Animals	
Subfamily	Names	*Tp*	*Pt*	*Pf*	*Ps*	*Eh*	*Gt*	*At*	*Dd*	*Mc*	*Sc*	*Nc*	*Dm*	*Hs*
H3K9me													NP_001261444.1	NP_001380915.1
SMYD				PF3D7_1115200_SET7			XP_005818844.1	At2G17900	XP_629856.1					Q8NB12.1
		XP_002292103.1			XP_009533210.1		XP_005827049.1	At2G19640	XP_644412.1_SMYD3L			XP_959981.1		NP_064582.2
		XP_002290198.1			XP_009533210.1	XP_005777382.1	XP_005826792.1	At1G26760	XP_002649106.1	EPB91989.1		XP_957968.2	NP_650955.1	NP_001161212.1
								At3G21820	XP_629629.1			XP_965349.2		Q8IYR2
								At5G06620						NP_006053.2
SETD		XP_002287161.1	GSPATG00003961001		XP_009533919.1	XP_005783180.1	XP_001713614.1	At1G01920				XP_964786.2		
		XP_002286075.1	GSPATG00027281001				XP_005828987.1	At1G14030				XP_963144.2		
		XP_002295030.1	GSPATG00020579001					At1G24610			NP_015160.1	XP_963594.3		NP_001153777.1
		XP_002296912.1	GSPATG00031631001					At2G18850						
		XP_002294212.1	GSPATG00024000001					At3G07670						
			GSPATG00015535001					At3G55080						
			GSPATG00012103001					At3G56570						
			GSPATG00016863001					At4G20130						
			GSPATG00037631001					At5G14260						
			GSPATG00010387001					At5G17240						
			GSPATG00038239001											
			GSPATG00007235001											
			GSPATG00012599001											
			GSPATG00013397001											
			GSPATG00012308001											
SETMAR														NP_006506
H3K4, H3K36														

Tt, *Thalassiosira pseudonana*, *Pt*, *Paramecium tetraurelia*, *Pf*, *Plasmodium falsiparum*, *Ps*, *Phytophthora sojae*, *Eh*, *Emiliania huxleyi*, *Gt*, *Guillardia theta*, *At*, *Arabidopsis thaliana*, *Dd*, *Dictyostelium discoideum*, *Mc*, *Mucor circinelloides*, *Sc*, *S. cerevisiae*, *Nc*, *Neurospora crassa*, *Dm*, *Drosophila melanogaster*, *Hs*, *Homo sapiens*

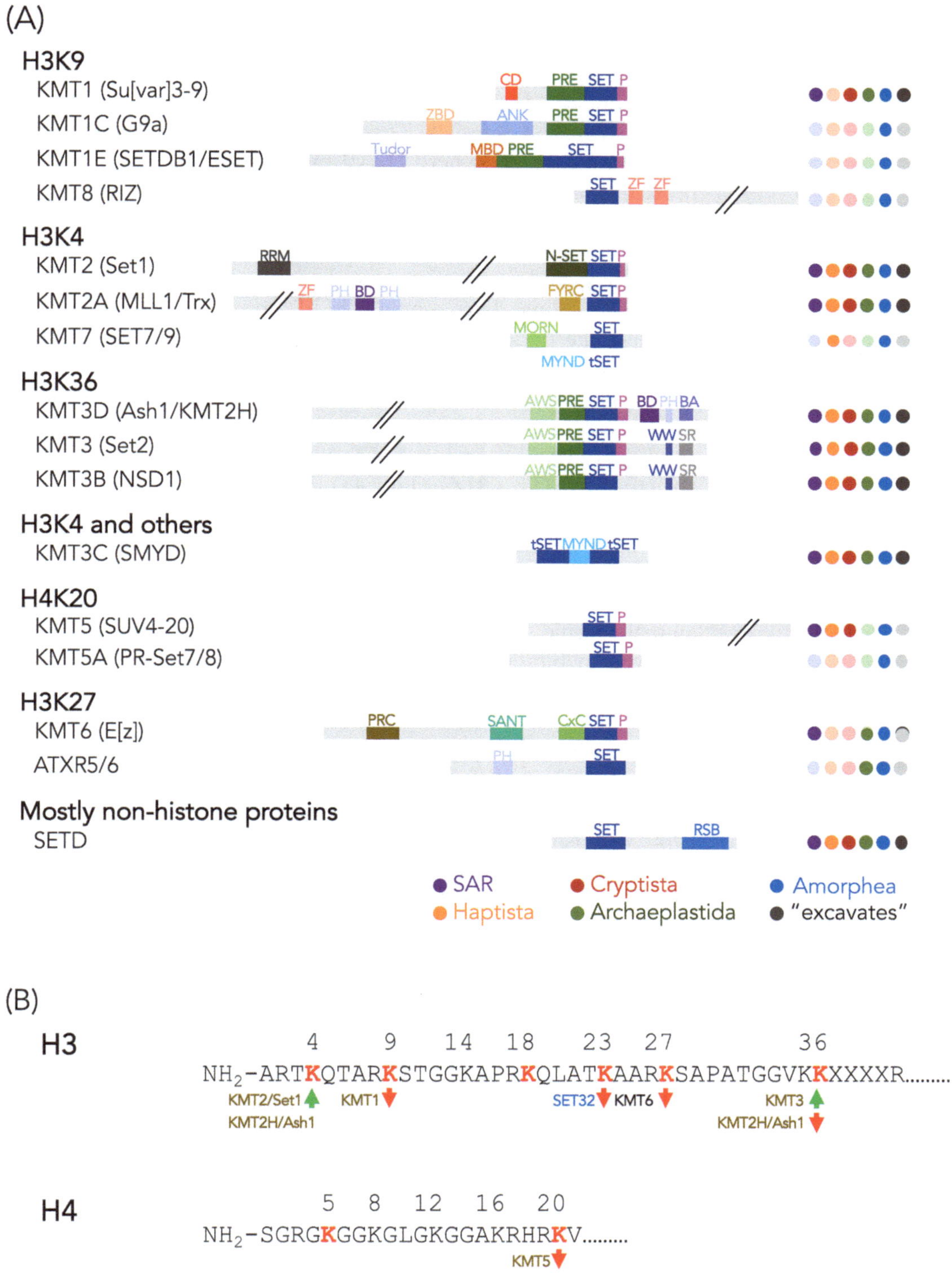

Fig. 4 Eukaryotic SET domain lysine methyltransferases (KMTs) and their known histone substrate specificity. (**a**) Classification of KMTs grouped by known substrate specificity and mostly following subfamilies described in mammals. See text for details. Presence of PRMT families in the 22 taxa examined is indicated by the shaded colored circles; solid colors denote presence and pastel colors denotes absence in the taxa studied, but the latter does not indicate that other species in the respective supergroups may not encode these KMTs. (**b**) Documented activities of eukaryotic KMTs on histone H3 and H4 tails. Blue type denotes activity

(Fig. 4b). Within the SAR clade, there are potential homologs in *Plasmodium* (SET3) and *P. sojae*, and some of the other taxa studied here, like *Giardia*, also have putative KMT1 homologs [56], none of which have been studied in great detail. Ciliates like *Paramecium* and especially *Tetrahymena* have been models for gene silencing for decades; here H3K9 methylation is carried out by a protein more similar to KMT6 (E[z]) [85], while true Su(var)3–9 homologs are absent [80]. Within the Amorphea, *Dictyostelium* has one potential homolog that may be mis-annotated as the protein is quite long and is predicted to include a NimA kinase motif (Table 2). Fungi have single KMT1 homologs, though whole families, like budding yeast and its relatives, have lost the ability to methylate H3K9. Only the animals have additional members of H3K9-specific KMTs that catalyze H3K9me in euchromatin or under certain conditions, such as KMT1C (G9a), KMT1E (SETDB1/ESET), and a KMT with a quite different primary structure, KMT8 (RIZ) (Fig. 4a). With between eight and ten proteins, the KMT1 family is expanded in most plant species that have been studied [74, 83, 86]. Shared features among all the KMT1s include pre- and post-SET domains, but the N-terminal chromo domain (CD) is lacking in many family members; the animal-specific KMT1s have additional motifs that are often involved in binding chromatin proteins or histones, such as the Tudor, zinc-binding (ZBD), ankyrin, or methyl-binding domains (MBD).

Homologs for *Saccharomyces* Set1 and *Drosophila* Trx belong to the KMT2 or plant Trx subfamily and catalyze H3K4 methylation, a histone modification associated with active transcription (Fig. 1a); KMT2 homologs were found in most taxa examined (Table 2). Again, while many taxa have single KMT2 homologs, this family is greatly expanded in plants and animals with at least seven different proteins. Pre- and post-SET domains are present, as are PWWP, PHD, ring finger, and FYRC motifs (Fig. 4a). Two additional subfamilies of KMTs are known to act on H3K4, namely KMT7 (SET7/9), found in animals, and KMT3C (SMYD) that in animals includes ~400–450 aa proteins with a large SET domain that is interrupted by a MYND domain. These proteins may act not only on H3K4 but also on other histone residues and indeed non-histone substrates, as many of the protist- or plant-type SMYD proteins may do.

Fig. 4 (continued) demonstrated in the Amorphea (mostly animals and fungi; SET32 is a novel H3K23 MTase from *C. elegans*), green type denotes activity demonstrated in Archaeplastida, gold type denotes activity found in both Archaeplastida and Amorphea, and black type denotes activity found in Archaeplastida, Amorphea, and "excavates." Green arrows denote correlation with gene expression, while red arrows denote correlation with transcriptional gene silencing

Saccharomyces Set2, *Drosophila* Ash1, and mammalian NSD KMTs belong to the KMT3 or plant Ash subfamily, which is well conserved across eukaryotes and is recognized to methylate H3K36. Traditionally, this histone mark has been associated with active transcription, because its appearance is correlated with transcript elongation in *S. cerevisiae*; however, subsequent studies showed that its intrinsic function is to interfere with transcription efficiency [87]. Thus, it is perhaps not surprising that H3K36me2/3 are also involved in the generation and maintenance of facultative heterochromatin, as will be discussed below. Again, plants and animals have more family members than the other taxa examined. One group of proteins originally grouped with KMT2 or Set1 homologs are ASH1 proteins [73]; however, they, as all other KMT3 proteins, contain an AWS domain, and they have by now been shown to catalyze H3K36 methylation, even though earlier studies showed that ASH1 affected H3K4me. In KMT3 proteins, the C-terminal motifs vary widely between proteins among the various taxa examined (Fig. 4a).

Schizosaccharomyces Set9 and animal SUV4-20 proteins constitute the KMT5 subfamily, known to methylate H4K20 and also correlated with the maintenance or generation of heterochromatin; there are overall fewer homologs in the eukaryotes studied here. In fungi, several taxa lack this protein, and in *Arabidopsis*, a Suv subfamily protein, SUVH2, is capable of methylating H4K20. Animals have additional, shorter SET domain proteins (PR-Set7/8) that carry out H4K20 mono-methylation.

Homologs of *Drosophila* E(z) belong to the KMT6 or plant E(z) subfamily and carry out H3K27 methylation, the canonical histone mark for facultative heterochromatin. This family is expanded in plants and animals but not in fungi; *S. cerevisiae* and *S. pombe*, and the industrially or medically important taxa *Aspergillus* and *Penicillium,* lack this protein. Many protists have potential KMT6 homologs (Table 2). Belonging to a different KMT subfamily, the *Arabidopsis* ATXR5 and ATXR6 proteins (plant subfamily IV) carry out H3K27me1 and are involved in regulation of re-replication of heterochromatin [88, 89]; no obvious homologs for these proteins exist in fungi or animals.

As mentioned above, the SMYD (plant subfamily VI) and SETD (plant subfamily VII) are still poorly defined groups in terms of sequence and function and appear greatly expanded in the SAR and deeply branching clades (Table 2, Fig. 4a); many of these KMTs have nonhistone substrates but activity on specific histone residues has been observed both in vivo and in vitro. In conclusion, the distribution and relationships between the extant SET domain subfamilies allows the assertion that histone methylation by these KMTs is an ancient process that was lost in many lineages over evolutionary time. There is strong support for an ancient origin of the KMT2 (Set1), KMT3 (Set2/Ash1), KMT6 (E[z]), ATXR, SMYD, and SETD subfamilies [80]. In contrast, the KMT1 (Su[var]3-9), KMT5 (Su[var]4-20), KMT7, and KMT8 subfamilies appear to be more recent additions to the ensemble of SET domain KMTs.

3.3.2 Function of Selected SET Domain-Containing Protein Complexes

While discussing the distributions of SET domain KMTs, we already mentioned their preferred histone substrates; there is insufficient space to discuss functional studies that contributed to this general understanding for all KMT function in plants, fungi, and animals. Instead, we will focus on selected aspects of the relationships between KMT2 (Set1), KMT3 (Ash1), and KMT6 (E[z]) complexes. This quickly expanding subject of chromatin biology aims to decipher regulation of opposing chromatin features, for example, how bivalent chromatin promoters influence gene expression, and how the balance of PcG (Polycomb Group)-mediated gene silencing and TrxG (Trithorax Group)-dependent expression affects development and disease [90, 91].

The H3K4 methyltransferase complex, COMPASS (Complex Proteins Associated with Set1), is a highly conserved family of proteins functioning—in combination with other complexes—to maintain developmentally appropriate patterns of gene expression. The subunits of the COMPASS of yeast are comprised of the KMT2 subfamily member Set1, as well as Bre2 (Cps60), Swd1 (Cps50), Spp1 (Cps40), Swd2 (Cps35), Swd3 (Cps30), Sdc1 (Cps25), and Shg1 (Cps15) [92]. While a KMT2 homolog is always present, the subunits can vary greatly among the eukaryotes. Fungi have one COMPASS complex, containing a single homolog of *S. cerevisiae* Set1, but the number of complexes is greatly expanded in *Drosophila* and humans, containing at least three or six COMPASS families, and each capable of H3K4 methylation with non-redundant functions in the cell [93, 94]. Responsible for the "bulk" H3K4me2/3 at promoters and gene bodies of actively transcribed genes, Set1A (KMT2F) COMPASS is important for the regulation of stem cell differentiation [95, 96]. MLL (KMT2A) COMPASS is primarily responsible for the deposition of H3K4me3 marks specifically regulating Hox genes clusters [97], while MLL2 (KMT2B) COMPASS has a role in maintaining bivalent chromatin [98]. Monomethylase activity has been primarily attributed to the MLL3/MLL4 (KMT2C/D) COMPASS—deletion of MLL3 and 4 resulted in substantial losses of monomethylation, particularly at enhancer regions [98]. H3K4 methylation has long been associated with actively transcribed regions as early studies found a correlation between levels of H3K4 methylation and transcriptional activation in *Tetrahymena* macronuclei [99]. Later work established a connection between MLL (KMT2A) activity and Hox gene expression [100]. Chromatin patterns of H3K4 methylation are dependent not only on COMPASS but also on

RNA polymerase II (RNAPII)-mediated transcription. An association between COMPASS and Paf1 (polymerase associated factor 1) has been shown integral for the recruitment of COMPASS to RNAPII and therefore actively transcribed chromatin [101].

In contrast to the TrxG COMPASS complex, PcG proteins form complexes that promote and maintain the formation of repressive facultative heterochromatin. These proteins thus act in direct opposition to COMPASS and H3K4 methylation. Their proper regulation is essential for multicellular development and differentiation, X-chromosome inactivation, and the repression of cancer development. In humans, there are at least two such complexes, Polycomb Repressive Complex 1 and 2 (PRC1 and PRC2). PRC2 catalyzes the deposition of H3K27me3, and PRC1 is believed to either maintain this heterochromatic mark and directly interfere with transcription [102] or act as a guide to bring PRC2 to the appropriate regions [103]; these two options are not mutually exclusive. Subunits of PRC1 vary greatly, but in animals include Polycomb (Pc) a chromatin "reader" protein that binds H3K27me3, suggesting a mutualistic function between PRC1 and PRC2 [104], a catalytic RING protein, which is known to ubiquitinate lysine 119 of histone H2A, and homologs of *Drosophila* polyhomeotic protein (Phc). Even though PRC1 is required in animals, plants and fungi lack clear homologs for most PRC1 subunits [82, 105], although in plants Pc is replaced by a version of HP1, a protein that binds H3K9me3 in other organisms [106]. In fungi, functional homologs for Pc remain to be discovered; most likely there is a completely different group of protein complexes involved.

PRC2 is conserved and likely an ancient protein complex. Three subunits are essential: E(z)/EZH1/2, EED (Early Ectoderm Development), and SUZ12 (Suppressor of Zeste). E(z) homologs are KMT6 lysine methyltransferases, catalyzing H3K27 mono-, di-, and trimethylation. The WD40 beta propeller domain of EED recognizes H3K27me3 and is believed to aid in propagation of the repressive mark. The function of SUZ12 is still not completely understood, though its presence is required for the establishment and maintenance of H3K27me3 [107, 108], and it is likely serving as a "recruitment platform" for additional PRC2 subunits, such as p55/RbAp46/48 and others. In plants, with three E(z) homologs, and in mammals, with two EZH proteins, multiple PRC2 complexes are formed. Targeting of PRC2 in mammals seems to involve CG-rich DNA [108], targeting in *Drosophila* is accomplished by binding to "Polycomb response elements" (PREs) [109, 110], and specific binding motifs may also be important for PRC2 targeting in plants [111]. There are also examples in mammals (ES cells), where non-coding RNAs have been shown as another means to target PRC2 to specific genes [112]. No such elements have been conclusively identified in fungi, suggesting other mechanisms for PRC2 targeting. Because of their relative simplicity, PRC2 complexes of the basidiomycete yeast, *Cryptococcus*, and the ascomycetes *Neurospora*, *Fusarium*, and *Zymoseptoria* have become models to aid in the general understanding of how PRC2 interacts with other histone marks [113–115].

3.3.3 The Relationship Between H3K36 and H3K27 Methylation

Studies in animals and fungi have suggested antagonism not just between TrxG (H3K4 methylation) and PcG (H3K27 methylation) proteins but also between H3K36 methylation by KMT3/Ash1 and H3K27 methylation by KMT6/E(z). Members of the original KMT3 (ScSet2) subfamily bind to the elongating RNAPII and mono-, di-, or trimethylate H3K36. Distribution of H3K36me3, catalyzed by the single Set2 enzyme in budding yeast, is correlated with active transcription; in other fungi, H3K36me3 covers most annotated genes, though it is more pronounced near the 3′ end of genes. The true function for Set2-catalyzed H3K36me3 is repression of transcription [87], and ScSet2 interacts with the two largest RNAPII subunits by binding to phosphorylated serine 2 of the C-terminal domain [116]. In other fungi, plants, and animals, a second group of KMT3s, the Ash1 homologs, are also capable of H3K36 methylation. All KMT3 enzymes have AWS (associated with SET), WW, and SRI (Set2-Rpb1-interacting) domains but plant and fungal Ash1-like KMT3s lack the C-terminal PHD, BAH, or Bromo domains found in animal Ash1 (Fig. 4a; [117]).

In vitro studies have shown that H3K4me and H3K36me peptides or nucleosomes inhibit reconstituted PRC2 [118, 119] and mutations that change or eliminate H3K36me3 result in mislocalized PcG proteins in animals [120, 121]. Deletion of KMT3/Ash1/SET-3 in *Fusarium fujikuroi* [122] resulted in region-specific increases of H3K27me3 in subtelomeric regions but H3K36 methylation catalyzed by KMT3/Set2 on active genes appeared largely undisturbed; this was similar to findings in *Drosophila* where H3K36me2 inhibited H3K27me3 [123]. Deletion of *kmt6/set-7* in *N. crassa* had no effect on KMT3/Ash1-mediated H3K36me2 in a *kmt3/set-2* strain, while genome-wide loss of Ash1-mediated H3K36me2 resulted in loss (180 regions) or gain (128 regions) of H3K27 methylation and upregulation or transcriptional silencing of genes [124]. Deleting the SRI domain of *Neurospora* KMT3/SET-2 removed most H3K36me3 yet was slightly additive in combination with a catalytically inactive *ash-1^{Y888F}* mutation, suggesting that *Neurospora* KMT3/Ash1 catalyzes not just H3K36me2 but also at least some H3K36me3; Ash1 deletion was shown to be lethal in this fungus [124]. Overall, H3K36me2-methylated regions depending on KMT3/Ash1 are associated with poorly transcribed and usually transcriptionally silent genes mostly in subtelomeric regions. Thus, H3K36 methylation on subtelomeric transcriptionally silenced genes is necessary for the proper accumulation and maintenance of H3K27 methylation in the same regions in both *Neurospora* and *Fusarium*, but sometimes with opposite consequences for transcription. In *F. fujikuroi*, the absence of KMT3/Ash1-catalyzed H3K36me3 resulted in enhanced chromosome instability, measured by the frequency of loss of an "accessory," i.e., conditionally dispensable, chromosome [122]. An increase of H3K27me3 in inappropriate regions, e.g., after loss of H3K9me3, also resulted in increased genome instability in *Z. tritici* [125].

Earlier studies showed that *Drosophila* Ash1 inhibits H3K27me3 accumulation [123] and that the presence of H3K36me histone tails inhibits PRC2 activity [119]. Recent structural studies by cryoelectron microscopy revealed how H3K36me-modified and -unmodified histone tails affect KMT6 directly [126]. PRC2 contacts two nucleosomes: the substrate nucleosome is bound by the EZH2 CXC domain, and the allosteric nucleosome is contacted by EED and the EZH2 SBD and SANT1 domains. In this configuration, H3K36 lies directly opposite to the EZH2-CXC-DNA interaction surface. Positioning of the H3K27 residue in the catalytic center is sensitive to the chemistry of the H3K36 side chain; mutations of H3K36A or H3K36R do not provide a correct fit. Methylation of H3K36 appears to directly interfere with PRC2 catalysis.

Previous studies showed that H3K36 methylation can repress PRC2 activity by PRC2-associated Polycomb-like proteins via their H3K36me3-binding Tudor domains [127–129]. Comparing activity of a full-length PHF1-PRC2 on unmodified and H3Kc36me3 (an H3K36 analog) on mononucleosomes showed that H3K27 methylation was inhibited on H3Kc36me3 mononucleosomes [126], even though one function of PHF1 is to increase PRC2 residence time on nucleosomes [130]. Genetic studies with H3K36R and H3K36A mutant larvae confirmed reduction of H3K27me3 levels, including on HOX genes [126]. Although all of these results suggest a direct influence of KMT3/Ash1-catalyzed H3K36 methylation on PRC2 activity, genetic experiments support the idea that Ash1 catalytic activity may also contribute indirectly, for example, by methylation of non-histone substrates like KMT2/Trx [131, 132].

3.3.4 A Single Enzyme that Methylates both H3K9 and H3K27

The ciliate *Paramecium tetraurelia* has been shown to methylate both H3K9 and H3K27, both in vivo and in vitro, by use of a single enzyme, Ezl1 [85]. The main overlapping function of both histone modifications is transcriptional silencing of transposable elements (TEs), as loss of Ezl1 results in transcription from TEs but not from core genes. This contrasts with the well-studied function of H3K27 methylation in plants, fungi, and animals, where presence or absence of H3K27 methylation controls development and differentiation.

This observation suggested that E(z) proteins are the more ancient subfamily as they are conserved in the SAR clade as well as the Archaeplastida and Amorphea [80, 85]. This "double-marking" is found in several fungi naturally, e.g., in the subtelomeric regions of the ascomycete fungi *N. crassa* [133, 134], *F. graminearum* [135], and *Zymoseptoria* [125] and the basidiomycete yeast *Cryptococcus neoformans* [136]. Marking with both H3K9me3 and H3K27me2/3 is enhanced to cover usually H3K9-methylated regions when HP1 is lacking [134], suggesting that the *Neurospora* E(z) homolog, SET-7, or PRC2 as a whole have intrinsic abilities to be directed toward constitutive heterochromatin. This has also been found in bryophytes like *Marchantia* [137], mammals, and *C. elegans*, at least in certain regions [138–140]. Similarly, when H3K9 methylation is abolished by mutation of the KMT1 homologs of *N. crassa*, *C. neoformans*, or *Z. tritici*, H3K27me3 moves from its usual locations into formerly H3K9-methylated regions again, revealing intrinsic abilities to be directed toward constitutive heterochromatin; this phenomenon has also been observed in mammalian H3K9me3 mutants [141].

The results obtained with *Paramecium* also suggest that H3K9me3 modification predates the evolution of a dedicated enzyme, like the members of the KMT1 subfamily, and this is borne out by results in other ciliates, diatoms, and *Chlamydomonas* where KMT1 homologs are lacking [80, 85, 105, 142, 143]. How then is recognition of the correct target sequence accomplished? Three-dimensional structures of the C-terminal SET domains of HMTs reveal much of what we know about the domain's function. In protein databases one will find many structures of SET domains and SET domain-containing proteins, both in *apo* form or complexed with ligands and cofactors, such as the methyl donor SAM, the cofactor product SAH, substrate peptides, and zinc. While structural variations exist between SET domains of different KMT subfamilies, key structural features apply to all SET domains. There are two distinct architectural features, a conserved β-barrel and a pseudoknot structure comprising the enzyme's active site. These anti-parallel β-sheets position the catalytic residues of the SET domain (e.g., N688, H689, and Y726 in human EZH2), separated by approximately 36 amino acids in the primary sequence, into the active site fold. While there is debate in the field as to the exact mechanism of the methylation of a substrate peptide of a SET domain-containing protein, catalysis by KMT requires these conserved residues and a protonated amino group on the substrate lysine. As the substrate enters the active site, a hydrogen bond between the catalytic tyrosine and the amide proton is enough to change the electronic chemistry of the nitrogen at the N-terminus of the lysine, promoting its nucleophilic attack on the sulphonium methyl group of the SAM cofactor. These catalytic intermediates are further stabilized by other active site residues (N688 and H689), and they, in combination with the binding pocket, promote the release of the cofactor product SAH [144].

What dictates the precise chemical specificity against substrates of SET domain KMTs is heavily studied, especially with an eye to pharmacological interventions [50, 55]. It results from a combination of amino acids comprising the enzyme's substrate binding site as well as the consensus amino acid motifs in biological targets. Histone tails are quite basic; the interactions that drive substrate specificity are largely polar in nature [145, 146]. For the two well-studied targets of methylation by KMT1 and KMT6, H3K9 and H3K27, the sequences directly flanking the target lysines in the histone H3 tail are identical: A-R-K-S, and yet the two enzymes that catalyze these reactions, KMT1/Su(var)3-9 and KMT6/EZH2, specifically recognize which lysine to methylate. Adjacent to the conserved ARKS region, the consensus is different, T-A-R-K-S-T for Su(var)3–9, compared to A-A-R-K-S-A for EZH2, and thus the two enzymes may achieve specificity by recognizing a different binding pocket or tunnel, and by their different catalytic sites. The intriguing study on *Paramecium* Ezl1 compared the SET domain features of *Neurospora* KMT1 (DIM-5), human EZH2, and Ezl1, and modeled potential catalytic site interactions that would allow dual specificity [85]. Based on this, and additional studies of protist and fungal enzymes, one goal for the future is to take advantage of the ancient origin of KMT2, KMT3, and KMT6 enzymes to study the catalytic characteristics of extant or "reverse evolved" enzymes, as has been done successfully with families of extant and ancient transcription factors [147, 148].

4 Summary

Reports from the literature and our representative sampling of eukaryotic supergroups allow the conclusion that protein methyltransferases are ancient. There is evidence that PRMTs arose with the last common eukaryotic ancestor but that SET domain KMTs are even more ancient. Phylogenies of PRMTs for PRMT5, the PRMT1^{PRO} group, PRMT7, PRMT10, PRMT4, and PRMT1 suggest monophyletic placement but the relationships between PRMT2, PRMT3, and PRMT6 are more difficult to resolve. Dot1-like non-SET KMTs occur most often as single proteins in eukaryotes, although genome or gene duplications can result in species with two specialized homologs (e.g., in *T. brucei*), which allows for functional specialization (e.g., for mono-, di, or tri-methylation). In animals, several Dot1L genes or splice variants are found. However, Dot1 is not universally conserved in eukaryotes, as some fungi and most plants do not have homologs. Phylogenetic analyses of SET domain proteins suggest the existence of more than a dozen different subfamilies, based on the uncertain relationship in several subfamilies, and much work remains to establish the substrates and functions for many of these proteins, especially in the less well understood taxa.

Overall, the distribution between extant PRMT and SET domain subfamilies allows the assertion that histone methylation by these proteins is an ancient process that was lost in many lineages over evolutionary time. For example, there is much evidence to support the ancient origin of the KMT2 (Set1), KMT3 (Set2/Ash1), KMT6 (E[z]), ATXR, SMYD, and SETD subfamilies, while the KMT1 (Su[var]3-9), KMT5 (Su[var]4-20), KMT7 (SET7/9), and KMT8 (RIZ) subfamilies are of more recent origin. Studies of *Paramecium* Ezl1, a dual-specificity KMT6-type protein that catalyzes both H3K9me and H3K27me, open the door to interesting new mechanistic studies that may allow the "reverse evolution" of some extant KMTs to uncover their evolutionary origin and capabilities.

Acknowledgments

We thank the members of the Freitag Lab for discussions. We apologize to our colleagues whose original research is not cited because of space constraints. Accession numbers and sequences of proteins discussed are available upon request. Work in the Freitag lab is supported by grants from the NSF (MCB1818006), the NIH (R01GM132644), and the US-Israel BSF (#2019034).

References

1. Allis CD (2015) Epigenetics, 2nd edn. Cold Spring Harbor Laboratory Press, Cold Spring Harbor, NY
2. Johnson L, Mollah S, Garcia BA, Muratore TL, Shabanowitz J, Hunt DF, Jacobsen SE (2004) Mass spectrometry analysis of *Arabidopsis* histone H3 reveals distinct combinations of post-translational modifications. Nucleic Acids Res 32:6511–6518
3. Wu T, Yuan T, Tsai SN, Wang C, Sun SM, Lam HM, Ngai SM (2009) Mass spectrometry analysis of the variants of histone H3 and H4 of soybean and their post-translational modifications. BMC Plant Biol 9:98
4. Lu CC, Coradin M, Porter EG, Garcia BA (2021) Accelerating the field of epigenetic histone modification through mass spectrometry-based approaches. Mol Cell Proteomics 20:100006
5. Sidoli S, Kori Y, Lopes M, Yuan ZF, Kim HJ, Kulej K, Janssen KA, Agosto LM, da Cunha JPC, Andrews AJ, Garcia BA (2019) One minute analysis of 200 histone posttranslational modifications by direct injection mass spectrometry. Genome Res 29:978–987
6. Garcia B (2016) Quantitative proteomics for understanding the histone code. J Biomol Tech 24(supplement):S10
7. Allshire RC, Madhani HD (2018) Ten principles of heterochromatin formation and function. Nat Rev Mol Cell Bio 19:229–244
8. Iyer LM, Anantharaman V, Wolf MY, Aravind L (2008) Comparative genomics of transcription factors and chromatin proteins in parasitic protists and other eukaryotes. Int J Parasitol 38:1–31
9. Burki F, Roger AJ, Brown MW, Simpson AGB (2020) The new tree of eukaryotes. Trends Ecol Evol 35:43–55
10. Burki F (2014) The eukaryotic tree of life from a global Phylogenomic perspective. CSH Perspect Biol 6:a016147

11. Leger MM, Kolisko M, Kamikawa R, Stairs CW, Kume K, Cepicka I, Silberman JD, Andersson JO, Xu FF, Yabuki A, Eme L, Zhang QQ, Takishita K, Inagaki Y, Simpson AGB, Hashimoto T, Roger AJ (2017) Organelles that illuminate the origins of *trichomonas* hydrogenosomes and *giardia* mitosomes. Nat Ecol Evol 1:0092
12. Morrison HG, McArthur AG, Gillin FD, Aley SB, Adam RD, Olsen GJ, Best AA, Cande WZ, Chen F, Cipriano MJ, Davids BJ, Dawson SC, Elmendorf HG, Hehl AB, Holder ME, Huse SM, Kim UU, Lasek-Nesselquist E, Manning G, Nigam A, Nixon JEJ, Palm D, Passamaneck NE, Prabhu A, Reich CI, Reiner DS, Samuelson J, Svard SG, Sogin ML (2007) Genomic minimalism in the early diverging intestinal parasite *Giardia lamblia*. Science 317:1921–1926
13. Xu FF, Jex A, Svard SG (2020) A chromosome-scale reference genome for *Giardia intestinalis* WB. Sci Data 7:38
14. Adam RD (2000) The *Giardia lamblia* genome. Int J Parasitol 30:475–484
15. Sodre CL, Chapeaurouge AD, Kalume DE, Lima LD, Perales J, Fernandes O (2009) Proteomic map of *Trypanosoma cruzi* CL Brener: the reference strain of the genome project. Arch Microbiol 191:177–184
16. El-Sayed NM, Myler PJ, Bartholomeu DC, Nilsson D, Aggarwal G, Tran AN, Ghedin E, Worthey EA, Delcher AL, Blandin G, Westenberger SJ, Caler E, Cerqueira GC, Branche C, Haas B, Anupama A, Arner E, Aslund L, Attipoe P, Bontempi E, Bringaud F, Burton P, Cadag E, Campbell DA, Carrington M, Crabtree J, Darban H, da Silveira JF, de Jong P, Edwards K, Englund PT, Fazelina G, Feldblyum T, Ferella M, Frasch AC, Gull K, Horn D, Hou LH, Huang YT, Kindlund E, Ktingbeil M, Kluge S, Koo H, Lacerda D, Levin MJ, Lorenzi H, Louie T, Machado CR, McCulloch R, McKenna A, Mizuno Y, Mottram JC, Nelson S, Ochaya S, Osoegawa K, Pai G, Parsons M, Pentony M, Pettersson U, Pop M, Ramirez JL, Rinta J, Robertson L, Salzberg SL, Sanchez DO, Seyler A, Sharma R, Shetty J, Simpson AJ, Sisk E, Tammi MT, Tarteton R, Teixeira S, Van Aken S, Vogt C, Ward PN, Wickstead B, Wortman J, White O, Fraser CM, Stuart KD, Andersson B (2005) The genome sequence of *Trypanosoma cruzi*, etiologic agent of Chagas disease. Science 309:409–415
17. Gardner MJ, Hall N, Fung E, White O, Berriman M, Hyman RW, Carlton JM, Pain A, Nelson KE, Bowman S, Paulsen IT, James K, Eisen JA, Rutherford K, Salzberg SL, Craig A, Kyes S, Chan MS, Nene V, Shallom SJ, Suh B, Peterson J, Angiuoli S, Pertea M, Allen J, Selengut J, Haft D, Mather MW, Vaidya AB, Martin DMA, Fairlamb AH, Fraunholz MJ, Roos DS, Ralph SA, McFadden GI, Cummings LM, Subramanian GM, Mungall C, Venter JC, Carucci DJ, Hoffman SL, Newbold C, Davis RW, Fraser CM, Barrell B (2002) Genome sequence of the human malaria parasite *plasmodium falciparum*. Nature 419:498–511
18. Aranda M, Li Y, Liew YJ, Baumgarten S, Simakov O, Wilson MC, Piel J, Ashoor H, Bougouffa S, Bajic VB, Ryu T, Ravasi T, Bayer T, Micklem G, Kim H, Bhak J, LaJeunesse TC, Voolstra CR (2016) Genomes of coral dinoflagellate symbionts highlight evolutionary adaptations conducive to a symbiotic lifestyle. Sci Rep 6:39734
19. Bouligand Y, Norris V (2001) Chromosome separation and segregation in dinoflagellates and bacteria may depend on liquid crystalline states. Biochimie 83:187–192
20. Rizzo PJ (1991) The enigma of the dinoflagellate chromosome. J Protozool 38:246–252
21. Rizzo PJ (2003) Those amazing dinoflagellate chromosomes. Cell Res 13:215–217
22. Gonzalez-Pech RA, Stephens TG, Chen YB, Mohamed AR, Cheng YY, Shah S, Dougan KE, Fortuin MDA, Lagorce R, Burt DW, Bhattacharya D, Ragan MA, Chan CX (2021) Comparison of 15 dinoflagellate genomes reveals extensive sequence and structural divergence in family Symbiodiniaceae and genus Symbiodinium. BMC Biol 19:73
23. Preer JR Jr (2000) Epigenetic mechanisms affecting macronuclear development in *paramecium* and *Tetrahymena*. J Eukaryot Microbiol 47:515–524
24. Ruiz F, Vayssie L, Klotz C, Sperling L, Madeddu L (1998) Homology-dependent gene silencing in *paramecium*. Mol Biol Cell 9:931–943
25. Cummings DJ, Tait A, Goddard JM (1974) Methylated bases in DNA from *Paramecium aurelia*. Biochim Biophys Acta 374:1–11
26. Aury JM, Jaillon O, Duret L, Noel B, Jubin C, Porcel BM, Segurens B, Daubin V, Anthouard V, Aiach N, Arnaiz O, Billaut A, Beisson J, Blanc I, Bouhouche K, Camara F, Duharcourt S, Guigo R, Gogendeau D, Katinka M, Keller AM, Kissmehl R, Klotz C, Koll F, Le Mouel A, Lepere G, Malinsky S, Nowacki M, Nowak JK, Plattner H, Poulain J, Ruiz F, Serrano V, Zagulski M, Dessen P, Betermier M, Weissenbach J, Scarpelli C, Schachter V, Sperling L, Meyer E, Cohen J,

Wincker P (2006) Global trends of whole-genome duplications revealed by the ciliate *paramecium tetraurelia*. Nature 444:171–178

27. Agyekum TP, Botwe PK, Arko-Mensah J, Issah I, Acquah AA, Hogarh JN, Dwomoh D, Robins TG, Fobil JN (2021) A systematic review of the effects of temperature on *anopheles* mosquito development and survival: implications for malaria control in a future warmer climate. Int J Environ Res Public Health 18:7255
28. Clement R, Dimnet L, Maberly SC, Gontero B (2016) The nature of the CO2-concentrating mechanisms in a marine diatom, *Thalassiosira pseudonana*. New Phytol 209:1417–1427
29. Duan DL, Wang XL, Yao JT, Liu DY, Hong WG, Hu ZM, Xia YH, Shao ZR, Li QY, Zhang J (2013) Draft genome sequences of *Saccharina japonica*. Phycologia 52:27–27
30. Liu T, Wang XM, Wang GL, Jia SG, Liu GM, Shan GL, Chi S, Zhang J, Yu YH, Xue T, Yu J (2019) Evolution of complex thallus alga: genome sequencing of *Saccharina japonica*. Front Genet 10:378
31. Ye NH, Zhang XW, Miao M, Fan X, Zheng Y, Xu D, Wang JF, Zhou L, Wang DS, Gao Y, Wang YT, Shi WY, Ji PF, Li DM, Guan Z, Shao CW, Zhuang ZM, Gao ZW, Qi J, Zhao FQ (2015) *Saccharina* genomes provide novel insight into kelp biology. Nat Commun 6:6986
32. Mao YX, Tyler BM (1996) The *Phytophthora sojae* genome contains tandem repeat sequences which vary from strain to strain. Fungal Genet Biol 20:43–51
33. Zhang X, Liu B, Zou F, Shen DY, Yin ZY, Wang RB, He F, Wang YC, Tyler BM, Fan W, Qian WQ, Dou DL (2019) Whole genome re-sequencing reveals natural variation and adaptive evolution of Phytophthora sojae. Front Microbiol 10:2792
34. Read BA, Kegel J, Klute MJ, Kuo A, Lefebvre SC, Maumus F, Mayer C, Miller J, Monier A, Salamov A, Young J, Aguilar M, Claverie JM, Frickenhaus S, Gonzalez K, Herman EK, Lin YC, Napier J, Ogata H, Sarno AF, Shmutz J, Schroeder D, de Vargas C, Verret F, von Dassow P, Valentin K, Van de Peer Y, Wheeler G, Dacks JB, Delwiche CF, Dyhrman ST, Glockner G, John U, Richards T, Worden AZ, Zhang XY, Grigoriev IV, Allen AE, Bidle K, Borodovsky M, Bowler C, Brownlee C, Cock JM, Elias M, Gladyshev VN, Groth M, Guda C, Hadaegh A, Iglesias-Rodriguez MD, Jenkins J, Jones BM, Lawson T, Leese F, Lindquist E, Lobanov A, Lomsadze A, Malik SB, Marsh ME, Mackinder L, Mock T, Mueller-Roeber B, Pagarete A, Parker M, Probert I, Quesneville H, Raines C, Rensing SA, Riano-Pachon DM, Richier S, Rokitta S, Shiraiwa Y, Soanes DM, van der Giezen M, Wahlund TM, Williams B, Wilson W, Wolfe G, Wurch LL, Annotation EH (2013) Pan genome of the phytoplankton *Emiliania* underpins its global distribution. Nature 499: 209–213
35. Westbroek P, Brown CW, Vanbleijswijk J, Brownlee C, Brummer GJ, Conte M, Egge J, Fernandez E, Jordan R, Knappertsbusch M, Stefels J, Veldhuis M, Vanderwal P, Young J (1993) A model system approach to biological climate forcing - the example of *Emiliania huxleyi*. Glob Planet Chang 8:27–46
36. Seitz R (2011) Bright water: hydrosols, water conservation and climate change. Clim Chang 105:365–381
37. Lefebvre SC, Benner I, Stillman JH, Parker AE, Drake MK, Rossignol PE, Okimura KM, Komada T, Carpenter EJ (2012) Nitrogen source and pCO(2) synergistically affect carbon allocation, growth and morphology of the coccolithophore *Emiliania huxleyi*: potential implications of ocean acidification for the carbon cycle. Glob Chang Biol 18: 493–503
38. Archibald JM, Lane CE (2009) Going, going, not quite gone: Nucleomorphs as a case study in nuclear genome reduction. J Hered 100: 582–590
39. Curtis BA, Tanifuji G, Burki F, Gruber A, Irimia M, Maruyama S, Arias MC, Ball SG, Gile GH, Hirakawa Y, Hopkins JF, Kuo A, Rensing SA, Schmutz J, Symeonidi A, Elias M, Eveleigh RJM, Herman EK, Klute MJ, Nakayama T, Obornik M, Reyes-Prieto A, Armbrust EV, Aves SJ, Beiko RG, Coutinho P, Dacks JB, Durnford DG, Fast NM, Green BR, Grisdale CJ, Hempel F, Henrissat B, Hoppner MP, Ishida KI, Kim E, Koreny LK, Kroth PG, Liu Y, Malik SB, Maier UG, McRose D, Mock T, Neilson JAD, Onodera NT, Poole AM, Pritham EJ, Richards TA, Rocap G, Roy SW, Sarai C, Schaack S, Shirato S, Slamovits CH, Spencer DF, Suzuki S, Worden AZ, Zauner S, Barry K, Bell C, Bharti AK, Crow JA, Grimwood J, Kramer R, Lindquist E, Lucas S, Salamov A, McFadden GI, Lane CE, Keeling PJ, Gray MW, Grigoriev IV, Archibald JM (2012)

Algal genomes reveal evolutionary mosaicism and the fate of nucleomorphs. Nature 492: 59–65

40. Eichinger L, Pachebat JA, Glockner G, Rajandream MA, Sucgang R, Berriman M, Song J, Olsen R, Szafranski K, Xu Q, Tunggal B, Kummerfeld S, Madera M, Konfortov BA, Rivero F, Bankier AT, Lehmann R, Hamlin N, Davies R, Gaudet P, Fey P, Pilcher K, Chen G, Saunders D, Sodergren E, Davis P, Kerhornou A, Nie X, Hall N, Anjard C, Hemphill L, Bason N, Farbrother P, Desany B, Just E, Morio T, Rost R, Churcher C, Cooper J, Haydock S, van Driessche N, Cronin A, Goodhead I, Muzny D, Mourier T, Pain A, Lu M, Harper D, Lindsay R, Hauser H, James K, Quiles M, Babu MM, Saito T, Buchrieser C, Wardroper A, Felder M, Thangavelu M, Johnson D, Knights A, Loulseged H, Mungall K, Oliver K, Price C, Quail MA, Urushihara H, Hernandez J, Rabbinowitsch E, Steffen D, Sanders M, Ma J, Kohara Y, Sharp S, Simmonds M, Spiegler S, Tivey A, Sugano S, White B, Walker D, Woodward J, Winckler T, Tanaka Y, Shaulsky G, Schleicher M, Weinstock G, Rosenthal A, Cox EC, Chisholm RL, Gibbs R, Loomis WF, Platzer M, Kay RR, Williams J, Dear PH, Noegel AA, Barrell B, Kuspa A (2005) The genome of the social amoeba *Dictyostelium discoideum*. Nature 435:43–57

41. Tang X, Zhao L, Chen H, Chen YQ, Chen W, Song Y, Ratledge C (2015) Complete genome sequence of a high lipid-producing strain of *Mucor circinelloides* WJ11 and comparative genome analysis with a low lipid-producing strain CBS 277.49. PLoS One 10: e0137543

42. Goffeau A, Barrell BG, Bussey H, Davis RW, Dujon B, Feldmann H, Galibert F, Hoheisel JD, Jacq C, Johnston M, Louis EJ, Mewes HW, Murakami Y, Philippsen P, Tettelin H, Oliver SG (1996) Life with 6000 genes. Science 274:546–567

43. Galagan JE, Calvo SE, Borkovich KA, Selker EU, Read ND, Jaffe D, FitzHugh W, Ma LJ, Smirnov S, Purcell S, Rehman B, Elkins T, Engels R, Wang S, Nielsen CB, Butler J, Endrizzi M, Qui D, Ianakiev P, Bell-Pedersen D, Nelson MA, Werner-Washburne-M, Selitrennikoff CP, Kinsey JA, Braun EL, Zelter A, Schulte U, Kothe GO, Jedd G, Mewes W, Staben C, Marcotte E, Greenberg D, Roy A, Foley K, Naylor J, Stange-Thomann N, Barrett R, Gnerre S, Kamal M, Kamvysselis M, Mauceli E, Bielke C, Rudd S, Frishman D, Krystofova S, Rasmussen C, Metzenberg RL, Perkins DD, Kroken S, Cogoni C, Macino G, Catcheside D, Li W, Pratt RJ, Osmani SA, DeSouza CP, Glass L, Orbach MJ, Berglund JA, Voelker R, Yarden O, Plamann M, Seiler S, Dunlap J, Radford A, Aramayo R, Natvig DO, Alex LA, Mannhaupt G, Ebbole DJ, Freitag M, Paulsen I, Sachs MS, Lander ES, Nusbaum C, Birren B (2003) The genome sequence of the filamentous fungus *Neurospora crassa*. Nature 422:859–868

44. Kämper J, Kahmann R, Bolker M, Ma LJ, Brefort T, Saville BJ, Banuett F, Kronstad JW, Gold SE, Muller O, Perlin MH, Wosten HA, de Vries R, Ruiz-Herrera J, Reynaga-Pena CG, Snetselaar K, McCann M, Perez-Martin J, Feldbrügge M, Basse CW, Steinberg G, Ibeas JI, Holloman W, Guzman P, Farman M, Stajich JE, Sentandreu R, Gonzalez-Prieto JM, Kennell JC, Molina L, Schirawski J, Mendoza-Mendoza A, Greilinger D, Munch K, Rossel N, Scherer M, Vranes M, Ladendorf O, Vincon V, Fuchs U, Sandrock B, Meng S, Ho EC, Cahill MJ, Boyce KJ, Klose J, Klosterman SJ, Deelstra HJ, Ortiz-Castellanos L, Li W, Sanchez-Alonso P, Schreier PH, Hauser-Hahn I, Vaupel M, Koopmann E, Friedrich G, Voss H, Schluter T, Margolis J, Platt D, Swimmer C, Gnirke A, Chen F, Vysotskaia V, Mannhaupt G, Guldener U, Munsterkotter M, Haase D, Oesterheld M, Mewes HW, Mauceli EW, DeCaprio D, Wade CM, Butler J, Young S, Jaffe DB, Calvo S, Nusbaum C, Galagan J, Birren BW (2006) Insights from the genome of the biotrophic fungal plant pathogen *Ustilago maydis*. Nature 444:97–101

45. Chapman JA, Kirkness EF, Simakov O, Hampson SE, Mitros T, Weinmaier T, Rattei T, Balasubramanian PG, Borman J, Busam D, Disbennett K, Pfannkoch C, Sumin N, Sutton GG, Viswanathan LD, Walenz B, Goodstein DM, Hellsten U, Kawashima T, Prochnik SE, Putnam NH, Shu SQ, Blumberg B, Dana CE, Gee L, Kibler DF, Law L, Lindgens D, Martinez DE, Peng JS, Wigge PA, Bertulat B, Guder C, Nakamura Y, Ozbek S, Watanabe H, Khalturin K, Hemmrich G, Franke A, Augustin R, Fraune S, Hayakawa E, Hayakawa S, Hirose M, Hwang JS, Ikeo K, Nishimiya-Fujisawa C, Ogura A, Takahashi T, Steinmetz PRH, Zhang XM, Aufschnaiter R, Eder MK, Gorny AK, Salvenmoser W, Heimberg AM, Wheeler BM, Peterson KJ, Boettger A, Tischler P, Wolf A, Gojobori T, Remington KA, Strausberg RL, Venter JC, Technau U,

Hobmayer B, Bosch TCG, Holstein TW, Fujisawa T, Bode HR, David CN, Rokhsar DS, Steele RE (2010) The dynamic genome of *hydra*. Nature 464:592–596

46. Adema CM, Hillier LW, Jones CS, Loker ES, Knight M, Minx P, Oliveira G, Raghavan N, Shedlock A, do Amaral LR, Arican-Goktas HD, Assis JG, Baba EH, Baron OL, Bayne CJ, Bickham-Wright U, Biggar KK, Blouin M, Bonning BC, Botka C, Bridger JM, Buckley KM, Buddenborg SK, Caldeira RL, Carleton J, Carvalho OS, Castillo MG, Chalmers IW, Christensens M, Clifton S, Cosseau C, Coustau C, Cripps RM, Cuesta-Astroz Y, Cummins SF, Di Stephano L, Dinguirard N, Duval D, Emrich S, Feschotte C, Feyereisen R, FitzGerald P, Fronick C, Fulton L, Galinier R, Gava SG, Geusz M, Geyer KK, Giraldo-Calderon GI, Gomes MD, Gordy MA, Gourbal B, Grunau C, Hanington PC, Hoffmann KF, Hughes D, Humphries J, Jackson DJ, Jannotti-Passos LK, Jeremias WD, Jobling S, Kamel B, Kapusta A, Kaur S, Koene JM, Kohn AB, Lawson D, Lawton SP, Liang D, Limpanont Y, Liu SJ, Lockyer AE, Lovato TL, Ludolf F, Magrini V, McManus DP, Medina M, Misra M, Mitta G, Mkoji GM, Montague MJ, Montelongo C, Moroz LL, Munoz-Torres MC, Niazi U, Noble LR, Oliveira FS, Pais FS, Papenfuss AT, Peace R, Pena JJ, Pila EA, Quelais T, Raney BJ, Rast JP, Rollinson D, Rosse IC, Rotgans B, Routledge EJ, Ryan KM, Scholte LLS, Storey KB, Swain M, Tennessen JA, Tomlinson C, Trujillo DL, Volpi EV, Walker AJ, Wang T, Wannaporn I, Warren WC, Wu XJ, Yoshino TP, Yusuf M, Zhang SM, Zhao M, Wilson RK (2017) Whole genome analysis of a schistosomiasis-transmitting freshwater snail. Nat Commun 8:15451

47. Adams MD, Celniker SE, Holt RA, Evans CA, Gocayne JD, Amanatides PG, Scherer SE, Li PW, Hoskins RA, Galle RF, George RA, Lewis SE, Richards S, Ashburner M, Henderson SN, Sutton GG, Wortman JR, Yandell MD, Zhang Q, Chen LX, Brandon RC, Rogers YHC, Blazej RG, Champe M, Pfeiffer BD, Wan KH, Doyle C, Baxter EG, Helt G, Nelson CR, Miklos GLG, Abril JF, Agbayani A, An HJ, Andrews-Pfannkoch C, Baldwin D, Ballew RM, Basu A, Baxendale J, Bayraktaroglu L, Beasley EM, Beeson KY, Benos PV, Berman BP, Bhandari D, Bolshakov S, Borkova D, Botchan MR, Bouck J, Brokstein P, Brottier P, Burtis KC, Busam DA, Butler H, Cadieu E, Center A, Chandra I, Cherry JM, Cawley S, Dahlke C, Davenport LB, Davies A, de Pablos B, Delcher A, Deng ZM, Mays AD, Dew I, Dietz SM, Dodson K, Doup LE, Downes M, Dugan-Rocha S, Dunkov BC, Dunn P, Durbin KJ, Evangelista CC, Ferraz C, Ferriera S, Fleischmann W, Fosler C, Gabrielian AE, Garg NS, Gelbart WM, Glasser K, Glodek A, Gong FC, Gorrell JH, Gu ZP, Guan P, Harris M, Harris NL, Harvey D, Heiman TJ, Hernandez JR, Houck J, Hostin D, Houston DA, Howland TJ, Wei MH, Ibegwam C, Jalali M, Kalush F, Karpen GH, Ke ZX, Kennison JA, Ketchum KA, Kimmel BE, Kodira CD, Kraft C, Kravitz S, Kulp D, Lai ZW, Lasko P, Lei YD, Levitsky AA, Li JY, Li ZY, Liang Y, Lin XY, Liu XJ, Mattei B, McIntosh TC, McLeod MP, McPherson D, Merkulov G, Milshina NV, Mobarry C, Morris J, Moshrefi A, Mount SM, Moy M, Murphy B, Murphy L, Muzny DM, Nelson DL, Nelson DR, Nelson KA, Nixon K, Nusskern DR, Pacleb JM, Palazzolo M, Pittman GS, Pan S, Pollard J, Puri V, Reese MG, Reinert K, Remington K, Saunders RDC, Scheeler F, Shen H, Shue BC, Siden-Kiamos I, Simpson M, Skupski MP, Smith T, Spier E, Spradling AC, Stapleton M, Strong R, Sun E, Svirskas R, Tector C, Turner R, Venter E, Wang AHH, Wang X, Wang ZY, Wassarman DA, Weinstock GM, Weissenbach J, Williams SM, Woodage T, Worley KC, Wu D, Yang S, Yao QA, Ye J, Yeh RF, Zaveri JS, Zhan M, Zhang GG, Zhao Q, Zheng LS, Zheng XQH, Zhong FN, Zhong WY, Zhou XJ, Zhu SP, Zhu XH, Smith HO, Gibbs RA, Myers EW, Rubin GM, Venter JC (2000) The genome sequence of *Drosophila melanogaster*. Science 287:2185–2195

48. Venter JC, Adams MD, Myers EW, Li PW, Mural RJ, Sutton GG, Smith HO, Yandell M, Evans CA, Holt RA, Gocayne JD, Amanatides P, Ballew RM, Huson DH, Wortman JR, Zhang Q, Kodira CD, Zheng XQH, Chen L, Skupski M, Subramanian G, Thomas PD, Zhang JH, Miklos GLG, Nelson C, Broder S, Clark AG, Nadeau C, McKusick VA, Zinder N, Levine AJ, Roberts RJ, Simon M, Slayman C, Hunkapiller M, Bolanos R, Delcher A, Dew I, Fasulo D, Flanigan M, Florea L, Halpern A, Hannenhalli S, Kravitz S, Levy S, Mobarry C, Reinert K, Remington K, Abu-Threideh J, Beasley E, Biddick K, Bonazzi V, Brandon R, Cargill M, Chandramouliswaran I, Charlab R, Chaturvedi K, Deng ZM, Di Francesco V, Dunn P, Eilbeck K, Evangelista C, Gabrielian AE, Gan W, Ge WM, Gong FC, Gu ZP, Guan P, Heiman TJ, Higgins ME, Ji RR, Ke ZX, Ketchum KA, Lai ZW, Lei YD, Li ZY, Li

JY, Liang Y, Lin XY, Lu F, Merkulov GV, Milshina N, Moore HM, Naik AK, Narayan VA, Neelam B, Nusskern D, Rusch DB, Salzberg S, Shao W, Shue BX, Sun JT, Wang ZY, Wang AH, Wang X, Wang J, Wei MH, Wides R, Xiao CL, Yan CH, Yao A, Ye J, Zhan M, Zhang WQ, Zhang HY, Zhao Q, Zheng LS, Zhong F, Zhong WY, Zhu SPC, Zhao SY, Gilbert D, Baumhueter S, Spier G, Carter C, Cravchik A, Woodage T, Ali F, An HJ, Awe A, Baldwin D, Baden H, Barnstead M, Barrow I, Beeson K, Busam D, Carver A, Center A, Cheng ML, Curry L, Danaher S, Davenport L, Desilets R, Dietz S, Dodson K, Doup L, Ferriera S, Garg N, Gluecksmann A, Hart B, Haynes J, Haynes C, Heiner C, Hladun S, Hostin D, Houck J, Howland T, Ibegwam C, Johnson J, Kalush F, Kline L, Koduru S, Love A, Mann F, May D, McCawley S, McIntosh T, McMullen I, Moy M, Moy L, Murphy B, Nelson K, Pfannkoch C, Pratts E, Puri V, Qureshi H, Reardon M, Rodriguez R, Rogers YH, Romblad D, Ruhfel B, Scott R, Sitter C, Smallwood M, Stewart E, Strong R, Suh E, Thomas R, Tint NN, Tse S, Vech C, Wang G, Wetter J, Williams S, Williams M, Windsor S, Winn-Deen E, Wolfe K, Zaveri J, Zaveri K, Abril JF, Guigo R, Campbell MJ, Sjolander KV, Karlak B, Kejariwal A, Mi HY, Lazareva B, Hatton T, Narechania A, Diemer K, Muruganujan A, Guo N, Sato S, Bafna V, Istrail S, Lippert R, Schwartz R, Walenz B, Yooseph S, Allen D, Basu A, Baxendale J, Blick L, Caminha M, Carnes-Stine J, Caulk P, Chiang YH, Coyne M, Dahlke C, Mays AD, Dombroski M, Donnelly M, Ely D, Esparham S, Fosler C, Gire H, Glanowski S, Glasser K, Glodek A, Gorokhov M, Graham K, Gropman B, Harris M, Heil J, Henderson S, Hoover J, Jennings D, Jordan C, Jordan J, Kasha J, Kagan L, Kraft C, Levitsky A, Lewis M, Liu XJ, Lopez J, Ma D, Majoros W, McDaniel J, Murphy S, Newman M, Nguyen T, Nguyen N, Nodell M, Pan S, Peck J, Peterson M, Rowe W, Sanders R, Scott J, Simpson M, Smith T, Sprague A, Stockwell T, Turner R, Venter E, Wang M, Wen MY, Wu D, Wu M, Xia A, Zandieh A, Zhu XH (2001) The sequence of the human genome. Science 291:1304–1351

49. Weiner AKM, Ceron-Romero MA, Yan Y, Katz LA (2020) Phylogenomics of the epigenetic toolkit reveals punctate retention of genes across eukaryotes. Genome Biol Evol 12:2196–2210

50. Hwang JW, Cho Y, Bae GU, Kim SN, Kim YK (2021) Protein arginine methyltransferases: promising targets for cancer therapy. Exp Mol Med 53:788–808

51. Liu CY, Lu FL, Cui X, Cao XF (2010) Histone methylation in higher plants. Annu Rev Plant Biol 61:395–420

52. Zhang X, Zhou L, Cheng XD (2000) Crystal structure of the conserved core of protein arginine methyltransferase PRMT3. EMBO J 19:3509–3519

53. Zhang M, Xu JY, Hu H, Ye BC, Tan MJ (2018) Systematic proteomic analysis of protein methylation in prokaryotes and eukaryotes revealed distinct substrate specificity. Proteomics 18:1700300

54. Tewary SK, Zheng YG, Ho MC (2019) Protein arginine methyltransferases: insights into the enzyme structure and mechanism at the atomic level. Cell Mol Life Sci 76:2917–2932

55. Kaniskan HU, Martini ML, Jin J (2018) Inhibitors of protein methyltransferases and demethylases. Chem Rev 118:989–1068

56. Emery-Corbin SJ, Hamey JJ, Ansell BRE, Balan B, Tichkule S, Stroehlein AJ, Cooper C, McInerney BV, Hediyeh-Zadeh S, Vuong D, Crombie A, Lacey E, Davis MJ, Wilkins MR, Bahlo M, Svard SG, Gasser RB, Jex AR (2020) Eukaryote-ConservedMethylarginine is absent in diplomonads and functionally compensated in *giardia*. Mol Biol Evol 37:3525–3549

57. Fisk JC, Read LK (2011) Protein arginine methylation in parasitic protozoa. Eukaryot Cell 10:1013–1022

58. Lott K, Zhu L, Fisk JC, Tomasello DL, Read LK (2014) Functional interplay between protein arginine methyltransferases in *Trypanosoma brucei*. Microbiol Open 3:595–609

59. Hashimoto H, Kafkova L, Raczkowski A, Jordan KD, Read LK, Debler EW (2020) Structural basis of protein arginine methyltransferase activation by a catalytically dead homolog (Prozyme). J Mol Biol 432: 410–426

60. Kafkova L, Debler EW, Fisk JC, Jain K, Clarke SG, Read LK (2017) The major protein arginine methyltransferase in *Trypanosoma brucei* functions as an enzyme-Prozyme complex. J Biol Chem 292:2089–2100

61. Campagnaro GD, Nay E, Plevin MJ, Cruz AK, Walrad PB (2021) Arginine methyltransferases as regulators of RNA-binding protein activities in pathogenic Kinetoplastids. Front Mol Biosci 8:692668

62. Fan Q, Miao J, Cui L, Cui LW (2009) Characterization of PRMT1 from *plasmodium falciparum*. Biochem J 421:107–118

63. Zeeshan M, Kaur I, Joy J, Saini E, Paul G, Kaushik A, Dabral S, Mohmmed A, Gupta D, Malhotra P (2017) Proteomic identification and analysis of arginine-methylated proteins of *plasmodium falciparum* at asexual blood stages. J Proteome Res 16:368–383
64. Ahmad A, Dong YZ, Cao XF (2011) Characterization of the PRMT gene family in Rice reveals conservation of arginine methylation. PLoS One 6:e22664
65. Borkovich KA, Alex LA, Yarden O, Freitag M, Turner GE, Read ND, Seiler S, Bell-Pedersen-D, Paietta J, Plesofsky N, Plamann M, Goodrich-Tanrikulu M, Schulte U, Mannhaupt G, Nargang FE, Radford A, Selitrennikoff C, Galagan JE, Dunlap JC, Loros JJ, Catcheside D, Inoue H, Aramayo R, Polymenis M, Selker EU, Sachs MS, Marzluf GA, Paulsen I, Davis R, Ebbole DJ, Zelter A, Kalkman ER, O'Rourke R, Bowring F, Yeadon J, Ishii C, Suzuki K, Sakai W, Pratt R (2004) Lessons from the genome sequence of *Neurospora crassa*: tracing the path from genomic blueprint to multicellular organism. Microbiol Mol Biol Rev 68:1–108
66. Schmoll M, Dattenbock C, Carreras-Villasenor N, Mendoza-Mendoza A, Tisch D, Aleman MI, Baker SE, Brown C, Cervantes-Badillo MG, Cetz-Chel J, Cristobal-Mondragon GR, Delaye L, Esquivel-Naranjo EU, Frischmann A, Gallardo-Negrete Jde J, Garcia-Esquivel M, Gomez-Rodriguez EY, Greenwood DR, Hernandez-Onate M, Kruszewska JS, Lawry R, Mora-Montes HM, Munoz-Centeno T, Nieto-Jacobo MF, Nogueira Lopez G, Olmedo-Monfil V, Osorio-Concepcion M, Pilsyk S, Pomraning KR, Rodriguez-Iglesias A, Rosales-Saavedra MT, Sanchez-Arreguin JA, Seidl-Seiboth V, Stewart A, Uresti-Rivera EE, Wang CL, Wang TF, Zeilinger S, Casas-Flores S, Herrera-Estrella A (2016) The genomes of three uneven siblings: footprints of the lifestyles of three *Trichoderma* species. Microbiol Mol Biol Rev 80:205–327
67. Wang YC, Li C (2012) Evolutionarily conserved protein arginine methyltransferases in non-mammalian animal systems. FEBS J 279: 932–945
68. Jahan S, Xu WN, Ilic A, Davie JR (2017) Protein arginine methyltransferases (PRMTs) and transcriptionally active chromatin domains. Biochem Cell Biol 95:182–182
69. Jain K, Clarke SG (2019) PRMT7 as a unique member of the protein arginine methyltransferase family: a review. Arch Biochem Biophys 665:36–45
70. Singer MS, Kahana A, Wolf AJ, Meisinger LL, Peterson SE, Goggin C, Mahowald M, Gottschling DE (1998) Identification of high-copy disruptors of telomeric silencing in *Saccharomyces cerevisiae*. Genetics 150: 613–632
71. van Leeuwen F, Gafken PR, Gottschling DE (2002) Dot1p modulates silencing in yeast by methylation of the nucleosome core. Cell 109:745–756
72. Kim W, Choi M, Kim JE (2014) The histone methyltransferase Dot1/DOT1L as a critical regulator of the cell cycle. Cell Cycle 13:726–738
73. Allis CD, Berger SL, Cote J, Dent S, Jenuwein T, Kouzarides T, Pillus L, Reinberg D, Shi Y, Shiekhattar R, Shilatifard A, Workman J, Zhang Y (2007) New nomenclature for chromatin-modifying enzymes. Cell 131:633–636
74. Pontvianne F, Blevins T, Pikaard CS (2010) *Arabidopsis* histone lysine methyltransferases. Adv Bot Res 53:1–22
75. Min J, Feng Q, Li Z, Zhang Y, Xu RM (2003) Structure of the catalytic domain of human DOT1L, a non-SET domain Nucleosomal histone methyltransferase. Cell 112:711–723
76. Ng HH, Xu RM, Zhang Y, Struhl K (2002) Ubiquitination of histone H2B by Rad6 is required for efficient Dot1-mediated methylation of histone H3 lysine 79. J Biol Chem 277:34655–34657
77. Worden EJ, Wolberger C (2019) Activation and regulation of H2B-ubiquitin-dependent histone methyltransferases. Curr Opin Struc Biol 59:98–106
78. Valencia-Sanchez MI, De Ioannes P, Wang M, Truong DM, Lee R, Armache JP, Boeke JD, Armache KJ (2021) Regulation of the Dot1 histone H3K79 methyltransferase by histone H4K16 acetylation. Science 371(363): eabc6663
79. Cui LW, Fan Q, Cui L, Miao J (2008) Histone lysine methyltransferases and demethylases in *plasmodium falciparum*. Int J Parasitol 38:1083–1097
80. Lhuillier-Akakpo M, Frapporti A, Denby Wilkes C, Matelot M, Vervoort M, Sperling L, Duharcourt S (2014) Local effect of enhancer of Zeste-like reveals cooperation of epigenetic and cis-acting determinants for zygotic genome rearrangements. PLoS Genet 10:e1004665

81. Alvarez-Venegas R (2014) Bacterial SET domain proteins and their role in eukaryotic chromatin modification. Front Genet 5:65
82. Freitag M (2017) Histone methylation by SET domain proteins in fungi. Annu Rev Microbiol 71:413–439
83. Zhou HY, Liu YH, Liang YW, Zhou D, Li SF, Lin S, Dong H, Huang L (2020) The function of histone lysine methylation related SET domain group proteins in plants. Protein Sci 29:1120–1137
84. Herz HM, Garruss A, Shilatifard A (2013) SET for life: biochemical activities and biological functions of SET domain-containing proteins. Trends Biochem Sci 38: 621–639
85. Frapporti A, Pina CM, Arnaiz O, Holoch D, Kawaguchi T, Humbert A, Eleftheriou E, Lombard B, Loew D, Sperling L, Guitot K, Margueron R, Duharcourt S (2019) The Polycomb protein Ezl1 mediates H3K9 and H3K27 methylation to repress transposable elements in paramecium. Nat Commun 10: 2710
86. Chen DH, Qiu HL, Huang Y, Zhang L, Si JP (2020) Genome-wide identification and expression profiling of SET DOMAIN GROUP family in *Dendrobium catenatum*. BMC Plant Biol 20:40
87. Strahl BD, Grant PA, Briggs SD, Sun ZW, Bone JR, Caldwell JA, Mollah S, Cook RG, Shabanowitz J, Hunt DF, Allis CD (2002) Set2 is a nucleosomal histone H3-selective methyltransferase that mediates transcriptional repression. Mol Cell Biol 22:1298–1306
88. Jacob Y, Stroud H, Leblanc C, Feng S, Zhuo L, Caro E, Hassel C, Gutierrez C, Michaels SD, Jacobsen SE (2010) Regulation of heterochromatic DNA replication by histone H3 lysine 27 methyltransferases. Nature 466:987–991
89. Jacob Y, Feng S, LeBlanc CA, Bernatavichute YV, Stroud H, Cokus S, Johnson LM, Pellegrini M, Jacobsen SE, Michaels SD (2009) ATXR5 and ATXR6 are H3K27 monomethyltransferases required for chromatin structure and gene silencing. Nat Struct Mol Biol 16:763–768
90. Bernstein BE, Mikkelsen TS, Xie X, Kamal M, Huebert DJ, Cuff J, Fry B, Meissner A, Wernig M, Plath K, Jaenisch R, Wagschal A, Feil R, Schreiber SL, Lander ES (2006) A bivalent chromatin structure marks key developmental genes in embryonic stem cells. Cell 125:315–326
91. Kuroda MI, Kang H, De S, Kassis JA (2020) Dynamic competition of Polycomb and Trithorax in transcriptional programming. Annu Rev Biochem 89:235–253
92. Miller T, Krogan NJ, Dover J, Erdjument-Bromage H, Tempst P, Johnston M, Greenblatt JF, Shilatifard A (2001) COMPASS: a complex of proteins associated with a trithorax-related SET domain protein. Proc Natl Acad Sci U S A 98:12902–12907
93. Thornton JL, Westfield GH, Takahashi YH, Cook M, Gao X, Woodfin AR, Lee JS, Morgan MA, Jackson J, Smith ER, Couture JF, Skiniotis G, Shilatifard A (2014) Context dependency of Set1/COMPASS-mediated histone H3 Lys4 trimethylation. Genes Dev 28:115–120
94. Eissenberg JC, Shilatifard A (2010) Histone H3 lysine 4 (H3K4) methylation in development and differentiation. Dev Biol 339:240–249
95. Sze CC, Cao KX, Collings CK, Marshall SA, Rendleman EJ, Ozark PA, Chen FX, Morgan MA, Wang L, Shilatifard A (2017) Histone H3K4 methylation-dependent and -independent functions of Set1A/COMPASS in embryonic stem cell self-renewal and differentiation. Genes Dev 31:1732–1737
96. Wu M, Wang PF, Lee JS, Martin-Brown S, Florens L, Washburn M, Shilatifard A (2008) Molecular regulation of H3K4 Trimethylation by Wdr82, a component of human Set1/COMPASS. Mol Cell Biol 28:7337–7344
97. Wang PF, Lin CQ, Smith ER, Guo H, Sanderson BW, Wu M, Gogol M, Alexander T, Seidel C, Wiedemann LM, Ge K, Krumlauf R, Shilatifard A (2009) Global analysis of H3K4 methylation defines MLL family member targets and points to a role for MLL1-mediated H3K4 methylation in the regulation of transcriptional initiation by RNA polymerase II. Mol Cell Biol 29:6074–6085
98. Hu DQ, Garruss AS, Gao X, Morgan MA, Cook M, Smith ER, Shilatifard A (2013) The Mll2 branch of the COMPASS family regulates bivalent promoters in mouse embryonic stem cells. Nat Struc Mol Biol 20:1093–1097
99. Strahl BD, Ohba R, Cook RG, Allis CD (1999) Methylation of histone H3 at lysine 4 is highly conserved and correlates with transcriptionally active nuclei in *Tetrahymena*. Proc Natl Acad Sci U S A 96:14967–14972
100. Milne TA, Briggs SD, Brock HW, Martin ME, Gibbs D, Allis CD, Hess JL (2002) MLL targets SET domain methyltransferase activity

to Hox gene promoters. Mol Cell 10:1107–1117

101. Krogan NJ, Dover J, Wood A, Schneider J, Heidt J, Boateng MA, Dean K, Ryan OW, Golshani A, Johnston M, Greenblatt JF, Shilatifard A (2003) The Paf1 complex is required for histone H3 methylation by COMPASS and Dot1p: linking transcriptional elongation to histone methylation. Mol Cell 11:721–729

102. Aranda S, Mas G, Di Croce L (2015) Regulation of gene transcription by Polycomb proteins. Science Adv 1:e1500737

103. Gal-Yam EN, Egger G, Iniguez L, Holster H, Einarsson S, Zhang XM, Lin JC, Liang GN, Jones PA, Tanay A (2008) Frequent switching of Polycomb repressive marks and DNA hypermethylation in the PC3 prostate cancer cell line. Proc Natl Acad Sci U S A 105: 12979–12984

104. Margueron R, Reinberg D (2011) The Polycomb complex PRC2 and its mark in life. Nature 469:343–349

105. Shaver S, Casas-Mollano JA, Cerny RL, Cerutti H (2010) Origin of the polycomb repressive complex 2 and gene silencing by an E (z) homolog in the unicellular alga *Chlamydomonas*. Epigenetics 5:301–312

106. Merini W, Calonje M (2015) PRC1 is taking the lead in PcG repression. Plant J 83:110–120

107. Pasini D, Bracken AP, Jensen MR, Lazzerini Denchi E, Helin K (2004) Suz12 is essential for mouse development and for EZH2 histone methyltransferase activity. EMBO J 23: 4061–4071

108. Laugesen A, Hojfeldt JW, Helin K (2019) Molecular mechanisms directing PRC2 recruitment and H3K27 methylation. Mol Cell 74:8–18

109. Kahn TG, Dorafshan E, Schultheis D, Zare A, Stenberg P, Reim I, Pirrotta V, Schwartz YB (2016) Interdependence of PRC1 and PRC2 for recruitment to Polycomb response elements. Nucleic Acids Res 44:10132–10149

110. Horard B, Tatout C, Poux S, Pirrotta V (2000) Structure of a polycomb response element and in vitro binding of polycomb group complexes containing GAGA factor. Mol Cell Biol 20:3187–3197

111. Xiao J, Jin R, Yu X, Shen M, Wagner JD, Pai A, Song C, Zhuang M, Klasfeld S, He C, Santos AM, Helliwell C, Pruneda-Paz JL, Kay SA, Lin X, Cui S, Garcia MF, Clarenz O, Goodrich J, Zhang X, Austin RS, Bonasio R, Wagner D (2017) Cis and trans determinants of epigenetic silencing by Polycomb repressive complex 2 in *Arabidopsis*. Nat Genet 49:1546–1552

112. Rinn JL, Kertesz M, Wang JK, Squazzo SL, Xu X, Brugmann SA, Goodnough LH, Helms JA, Farnham PJ, Segal E, Chang HY (2007) Functional demarcation of active and silent chromatin domains in human HOX loci by noncoding RNAs. Cell 129:1311–1323

113. Lewis ZA (2017) Polycomb Group Systems in Fungi: new models for understanding Polycomb repressive complex 2. Trends Genet 33: 220–231

114. Wiles ET, Selker EU (2017) H3K27 methylation: a promiscuous repressive chromatin mark. Curr Opin Genet Dev 43:31–37

115. Ridenour JB, Möller M, Freitag M (2020) Polycomb repression without bristles: facultative heterochromatin and genome stability in fungi. Genes 11:638

116. Smolle M, Workman JL, Venkatesh S (2013) reSETting chromatin during transcription elongation. Epigenetics 8:10–15

117. Veerappan CS, Avramova Z, Moriyama EN (2008) Evolution of SET domain protein families in the unicellular and multicellular Ascomycota fungi. BMC Evol Biol 8:190

118. Schmitges FW, Prusty AB, Faty M, Stützer A, Lingaraju GM, Aiwazian J, Sach R, Hess D, Li L, Bunker RD, Wirth U, Bouwmeester T, Bauer A, Ly-Hartig N, Zhao K, Chan H, Gu J, Gut H, Fischle W, Müller J, Thomä N (2011) Histone methylation by PRC2 is inhibited by active chromatin marks. Mol Cell 42:330–341

119. Yuan W, Xu M, Huang C, Liu N, Chen S, Zhu B (2011) H3K36 methylation antagonizes PRC2-mediated H3K27 methylation. J Biol Chem 286:7983–7989

120. Lu C, Jain SU, Hoelper D, Bechet D, Molden RC, Ran LL, Murphy D, Venneti S, Hameed M, Pawel BR, Wunder JS, Dickson BC, Lundgren SM, Jani KS, De Jay N, Papillon-Cavanagh S, Andrulis IL, Sawyer SL, Grynspan D, Turcotte RE, Nadaf J, Fahiminiyah S, Muir TW, Majewski J, Thompson CB, Chi P, Garcia BA, Allis CD, Jabado N, Lewis PW (2016) Histone H3K36 mutations promote sarcomagenesis through altered histone methylation landscape. Science 352:844–849

121. Gaydos LJ, Rechtsteiner A, Egelhofer TA, Carroll CR, Strome S (2012) Antagonism between MES-4 and Polycomb repressive complex 2 promotes appropriate gene expression in C. elegans germ cells. Cell Rep 2: 1169–1177

122. Janevska S, Baumann L, Sieber CMK, Münsterkotter M, Ulrich J, Kämper J, Güldener U, Tudzynski B (2018) Elucidation of the two H3K36me3 histone methyltransferases Set2 and Ash1 in *fusarium fujikuroi* unravels their different chromosomal targets and a major impact of Ash1 on genome stability. Genetics 208:153–171
123. Papp B, Muller J (2006) Histone trimethylation and the maintenance of transcriptional ON and OFF states by trxG and PcG proteins. Genes Dev 20:2041–2054
124. Bicocca VT, Ormsby T, Adhvaryu KK, Honda S, Selker EU (2018) ASH1-catalyzed H3K36 methylation drives gene repression and marks H3K27me2/3-competent chromatin. elife 7:e41497
125. Möller M, Schotanus K, Soyer JL, Haueisen J, Happ K, Stralucke M, Happel P, Smith KM, Connolly LR, Freitag M, Stukenbrock EH (2019) Destabilization of chromosome structure by histone H3 lysine 27 methylation. PLoS Genet 15:e1008093
126. Finogenova K, Bonnet J, Poepsel S, Schafer IB, Finkl K, Schmid K, Litz C, Strauss M, Benda C, Muller J (2020) Structural basis for PRC2 decoding of active histone methylation marks H3K36me2/3. elife 9:e61964
127. Musselman CA, Avvakumov N, Watanabe R, Abraham CG, Lalonde ME, Hong Z, Allen C, Roy S, Nunez JK, Nickoloff J, Kulesza CA, Yasui A, Cote J, Kutateladze TG (2012) Molecular basis for H3K36me3 recognition by the Tudor domain of PHF1. Nat Struct Mol Biol 19:1266–1272
128. Qin S, Guo Y, Xu C, Bian C, Fu M, Gong S, Min J (2013) Tudor domains of the PRC2 components PHF1 and PHF19 selectively bind to histone H3K36me3. Biochem Biophys Res Commun 430:547–553
129. Cai L, Rothbart SB, Lu R, Xu B, Chen WY, Tripathy A, Rockowitz S, Zheng D, Patel DJ, Allis CD, Strahl BD, Song J, Wang GG (2013) An H3K36 methylation-engaging Tudor motif of polycomb-like proteins mediates PRC2 complex targeting. Mol Cell 49: 571–582
130. Choi J, Bachmann AL, Tauscher K, Benda C, Fierz B, Muller J (2017) DNA binding by PHF1 prolongs PRC2 residence time on chromatin and thereby promotes H3K27 methylation. Nat Struct Mol Biol 24:1039–1047
131. Dorafshan E, Kahn TG, Glotov A, Savitsky M, Walther M, Reuter G, Schwartz YB (2019) Ash1 counteracts Polycomb repression independent of histone H3 lysine 36 methylation. EMBO Rep 20:e46762
132. Dorafshan E, Kahn TG, Glotov A, Savitsky M, Schwartz YB (2019) Genetic dissection reveals the role of Ash1 domains in counteracting polycomb repression. G3 (Bethesda) 9:3801–3812
133. Basenko EY, Sasaki T, Ji L, Prybol CJ, Burckhardt RM, Schmitz RJ, Lewis ZA (2015) Genome-wide redistribution of H3K27me3 is linked to genotoxic stress and defective growth. Proc Natl Acad Sci U S A 112: E6339–E6348
134. Jamieson K, Wiles ET, McNaught KJ, Sidoli S, Leggett N, Shao Y, Garcia BA, Selker EU (2016) Loss of HP1 causes depletion of H3K27me3 from facultative heterochromatin and gain of H3K27me2 at constitutive heterochromatin. Genome Res 26:97–107
135. Connolly LR, Smith KM, Freitag M (2013) The *fusarium graminearum* histone H3 K27 methyltransferase KMT6 regulates development and expression of secondary metabolite gene clusters. PLoS Genet 9:e1003916
136. Dumesic PA, Homer CM, Moresco JJ, Pack LR, Shanle EK, Coyle SM, Strahl BD, Fujimori DG, Yates JR 3rd, Madhani HD (2015) Product binding enforces the genomic specificity of a yeast polycomb repressive complex. Cell 160:204–218
137. Montgomery SA, Tanizawa Y, Galik B, Wang N, Ito T, Mochizuki T, Akimcheva S, Bowman JL, Cognat V, Marechal-Drouard L, Ekker H, Hong SF, Kohchi T, Lin SS, Liu LD, Nakamura Y, Valeeva LR, Shakirov EV, Shippen DE, Wei WL, Yagura M, Yamaoka S, Yamato KT, Liu C, Berger F (2020) Chromatin Organization in Early Land Plants Reveals an ancestral association between H3K27me3, transposons, and constitutive heterochromatin. Curr Biol 30:573–588 e577
138. Ho JWK, June YL, Liu T, Alver BH, Lee S, Ikegami K, Sohn KA, Minoda A, Tolstorukov MY, Appert A, Parker SCJ, Gu TT, Kundaje A, Riddle NC, Bishop E, Egelhofer TA, Hu SS, Alekseyenko AA, Rechtsteiner A, Asker D, Belsky JA, Bowmanm SK, Chens QB, Chen RAJ, Day DS, Dong Y, Dose AC, Duan XK, Epstein CB, Ercan S, Feingold EA, Ferrari F, Garrigues JM, Gehlenborg N, Good PJ, Haseley P, He D, Herrmann M, Hoffman MM, Jeffers TE, Kharchenko PV, Kolasinska-Zwierz P, Kotwaliwale CV, Kumar N, Langley SA, Larschan EN, Latorre I, Libbrecht MW, Lin XQ, Park R, Pazin MJ, Pham HN, Plachetka A, Qin B, Schwartz YB, Shoresh N, Stempor P, Vielle A, Wang CY, Whittle CM, Xue HL, Kingstonm RE, Kim JH, Bernstein BE, Dernburg AF, Pirrotta V, Kuroda MI, Noble WS,

Tullius TD, Kellis M, MacAlpine DM, Strome S, Elgin SCR, Liu XS, Lieb JD, Ahringer J, Karpen GH, Park PJ (2014) Comparative analysis of metazoan chromatin organization. Nature 512:449–452

139. Hawkins RD, Hon GC, Lee LK, Ngo Q, Lister R, Pelizzola M, Edsall LE, Kuan S, Luu Y, Klugman S, Antosiewicz-Bourget J, Ye Z, Espinoza C, Agarwahl S, Shen L, Ruotti V, Wang W, Stewart R, Thomson JA, Ecker JR, Ren B (2010) Distinct Epigenomic landscapes of pluripotent and lineage-committed human cells. Cell Stem Cell 6: 479–491

140. Mozzetta C, Pontis J, Fritsch L, Robin P, Portoso M, Proux C, Margueron R, Ait-Si-Ali S (2014) The histone H3 lysine 9 methyltransferases G9a and GLP regulate polycomb repressive complex 2-mediated gene silencing. Mol Cell 53:277–289

141. Peters AH, Kubicek S, Mechtler K, O'Sullivan RJ, Derijck AA, Perez-Burgos L, Kohlmaier A, Opravil S, Tachibana M, Shinkai Y, Martens JH, Jenuwein T (2003) Partitioning and plasticity of repressive histone methylation states in mammalian chromatin. Mol Cell 12:1577–1589

142. Taverna SD, Li H, Ruthenburg AJ, Allis CD, Patel DJ (2007) How chromatin-binding modules interpret histone modifications: lessons from professional pocket pickers. Nat Struct Mol Biol 14:1025–1040

143. Veluchamy A, Rastogi A, Lin X, Lombard B, Murik O, Thomas Y, Dingli F, Rivarola M, Ott S, Liu XY, Sun YZ, Rabinowicz PD, McCarthy J, Allen AE, Loew D, Bowler C, Tirichine L (2015) An integrative analysis of post-translational histone modifications in the marine diatom *Phaeodactylum tricornutum*. Genome Biol 16:102

144. Qian C, Zhou MM (2006) SET domain protein lysine methyltransferases: structure, specificity and catalysis. Cell Mol Life Sci 63: 2755–2763

145. Xiao B, Jing C, Kelly G, Walker PA, Muskett FW, Frenkiel TA, Martin SR, Sarma K, Reinberg D, Gamblin SJ, Wilson JR (2005) Specificity and mechanism of the histone methyltransferase Pr-Set7. Genes Dev 19: 1444–1454

146. Couture JF, Dirk LMA, Brunzelle JS, Houtz RL, Trievel RC (2008) Structural origins for the product specificity of SET domain protein methyltransferases. Proc Natl Acad Sci U S A 105:20659–20664

147. Starr TN, Thornton JW (2017) Exploring protein sequence-function landscapes. Nat Biotechnol 35:125–126

148. Anderson DW, McKeown AN, Thornton JW (2015) Intermolecular epistasis shaped the function and evolution of an ancient transcription factor and its DNA binding sites. elife 4:e07864

Part II

In Vitro Enzymatic Assays

Chapter 2

Detection and Quantification of Histone Methyltransferase Activity In Vitro

Nwamaka J. Idigo and Philipp Voigt

Abstract

Histone methyltransferases (HMTs) catalyze the methylation of lysine and arginine residues in histone as well as nonhistone substrates. In vitro histone methyltransferase assays have been instrumental in identifying HMTs, and they continue to be invaluable tools for the study of these important enzymes, revealing novel substrates and modes of regulation.

Here we describe a universal protocol to examine HMT activity in vitro that can be adapted to a range of HMTs, substrates, and experimental objectives. We provide protocols for the detection of activity based on incorporation of ^{3}H-labeled methyl groups from S-adenosylmethionine (SAM), methylation-specific antibodies, and quantification of the reaction product S-adenosylhomocysteine (SAH).

Key words Chromatin, Transcription, Histone methylation, Histone posttranslational modification, Nucleosomes, *S*-adenosylmethionine, *S*-adenosylhomocysteine

1 Introduction

Histone methyltransferases (HMTs) comprise three families of enzymes that methylate specific lysine and arginine residues, most prominently within histones H3 and H4 [1]. With the exception of the DOT1 family, lysine-specific HMTs feature a characteristic SET domain that catalyzes the mono-, di-, or trimethylation of the ε-amino group of lysine. Arginine residues are mono- or dimethylated at their guanidino group by enzymes of the PRMT family. Histone methylation marks are key regulators of transcription, setting up chromatin environments that reflect active, poised, or repressed transcriptional states at promoters, gene bodies, enhancers, and other genomic regions [2, 3]. Histone methyl marks are thought to act through the recruitment of specific binding or "reader" proteins that activate or repress transcription [4–6]. Several HMT complexes themselves feature reader domains [7], establishing feedback and crosstalk between different methyl marks and

Raphaël Margueron and Daniel Holoch (eds.), *Histone Methyltransferases: Methods and Protocols*,
Methods in Molecular Biology, vol. 2529, https://doi.org/10.1007/978-1-0716-2481-4_2,

other histone modifications such as ubiquitination, acetylation, and phosphorylation.

Histone methylation [8] and HMT activity were first observed in the 1960s [9, 10]. However, individual HMTs were not identified until the late 1990s [11, 12]. Assays to examine the enzymatic activity of HMTs in vitro were instrumental in these discoveries. Since then, HMT assays have continued to generate novel insight into these important regulators of gene expression, revealing novel histone and nonhistone substrates, automethylation, auxiliary subunits that modulate activity, and allosteric regulation through histone mark, RNA, or DNA binding. Several HMTs such as Polycomb Repressive Complex 2 (PRC2) are mutated or misregulated in cancer and other disease states, making them attractive targets for pharmacological intervention [13–15]. HMT assays are therefore also relevant to drug discovery, enabling identification and characterization of HMT inhibitors.

HMT assays are based on the same principles that underlie activity assays for other transferases such as protein kinases and acetyltransferases. Especially early versions of these assays shared very similar workflows based on the detection of radiolabel transfer from radioactive cofactors onto substrates. HMTs catalyze the transfer of a methyl group from the methyl donor *S*-adenosylmethionine (SAM) to substrate lysine or arginine residues, resulting in a methylated product and S-adenosylhomocysteine (SAH) (Fig. 1). In HMT assays, enzymatic activity is assessed via detection of either reaction product (summarized in Fig. 1). Use of SAM containing a ^{3}H- or ^{14}C-labeled methyl group allows for radiometric detection of methylated products. Methylated peptides or proteins are separated from unspent SAM via gel electrophoresis or—in early versions of this assay—binding to cellulose phosphate filter paper disks. This assay format not only is highly sensitive but also provides information about the identity of the substrate(s), allowing the identification of novel substrates and target residues. However, radiometric assay formats cannot easily distinguish between mono-, di-, and trimethylation. They also require access to facilities suitable for work with radioactive material.

If the substrate is known and a suitable antibody for its methylated form is available, reactions can also be carried out with nonradioactive SAM and methylated products detected via Western blot or other antibody-based approaches. While avoiding the need for radioactivity and allowing the resolution of different methylation states, this assay format is limited to reactions with known products, and antibody avidity might compromise sensitivity. Mass spectrometry (MS)-based approaches can also be used to identify and quantify methylated proteins or peptides after HMT reactions (*see*, e.g., [16]). While very powerful, the requirement for MS instrumentation limits throughput compared to other approaches and may

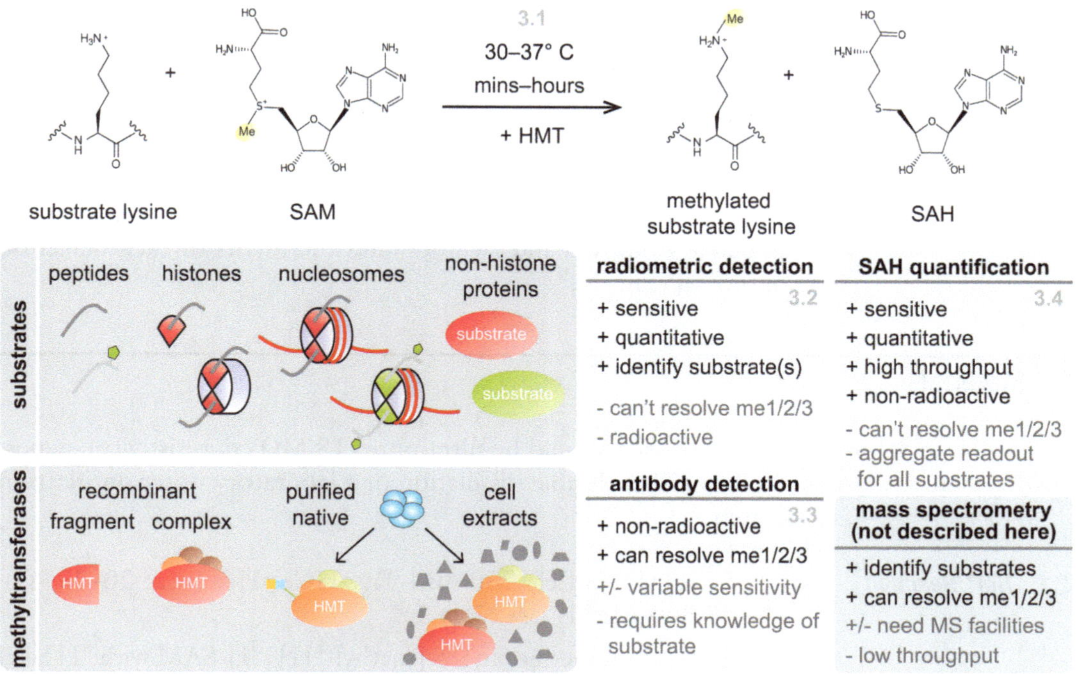

Fig. 1 Schematic overview of HMT reaction and guide for the selection of detection methods, HMT preparations, and substrates. HMTs catalyze the transfer of a methyl group from SAM to substrate lysine (shown here) or arginine residues. Detection of HMT activity is based on detecting either the methylated lysine or SAH, the demethylated form of SAM. The corresponding subheading numbers are given for each detection method described in this chapter. Boxes show different options for histone and nonhistone substrates as well as sources of HMT preparations

render this approach less suited for routine experiments, especially if the required facilities are not readily available.

Recently, HMT assays based on the detection of the reaction product SAH have become increasingly popular due to their quantitative nature and ease of use [17]. Many SAH detection methods are based on coupled enzyme reactions (*see*, e.g., [18–20]). Luminescence-based assays have been developed that involve a series of reactions, converting SAH to adenosine, which is progressively phosphorylated to ATP. ATP is then quantified via luciferase activity [20]. More widespread adoption of these assays has been facilitated by recent availability of commercial kits, abolishing the need to purify SAH-converting and luciferase enzymes. SAH conversion-based assays are sensitive and directly provide quantitative data. However, because they provide an aggregate readout of methylation activity toward potentially several different substrates present in the reaction (e.g., multiple histone residues as well as automethylation sites), their interpretation may not always be straightforward. Nonetheless, SAH-based assays are powerful tools for obtaining quantitative data on well-defined reactions and for high-throughput applications such as drug screening.

Here, we provide a universal protocol for HMT assays (Subheading 3.1). We focus on commonly used and readily accessible detection approaches, including radiometric detection (Subheading 3.2), antibody-based detection (Subheading 3.3), and SAH quantification via the methyltransferase-Glo assay (Subheading 3.4). A suitable assay strategy can be tailored to individual experimental needs, based on the specific question to be addressed, available resources and facilities, and the merits of each detection approach (Fig. 1).

2 Materials

All water used should be ultra-pure (18 MΩ cm resistivity), generated by either double distillation or a laboratory water purification system.

2.1 HMT Reaction

1. 5× HMT buffer: 250 mM Tris–HCl pH 8.5 (at 20–25 °C), 25 mM $MgCl_2$ (*see* **Note 1**).
2. Adenosyl-L-methionine, *S*-[methyl-^{3}H] (^{3}H-SAM with ^{3}H label present in the methyl group, available, e.g., from Perkin Elmer or Hartmann Analytic). Aliquot, store at −20 °C, and avoid repeated freeze–thaw cycles (radiometric assays only, *see* **Note 2**).
3. *S*-Adenosylmethionine (SAM) solution. Available pre-made from various suppliers, e.g., NEB. Alternatively, make stock solution by dissolving SAM in 5 mM H_2SO_4, 10% ethanol, to a final concentration of 2.5–20 mM. Aliquot and store at −20 °C (*see* **Note 2**).
4. 0.2 M DTT solution. Aliquot and store at −20 °C. Avoid repeated freeze–thaw cycles and discard working aliquot after 1 month of use.
5. 3× SDS-PAGE sample buffer: 190 mM Tris–HCl (pH 6.8), 150 mM DTT, 6% (w/v) SDS, 0.3% (w/v) bromophenol blue, 30% (v/v) glycerol. Aliquot and store at −20 °C.
6. 0.5% (v/v) Trifluoroacetic acid (TFA) (SAH quantification only).
7. Histone methyltransferase preparation in suitable buffer (*see* **Note 3**).
8. Suitable methylation substrate such as histone peptides (*see* **Note 4**), recombinant or native histones (*see* **Note 5**), or nucleosome preparations (*see* **Note 6**).
9. Heat block (ideally with heated lid to prevent evaporation of reactions and condensation on the lids of tubes) or incubator set to reaction temperature, usually 30 °C or 37 °C (*see* **Note 7**).

10. Laboratory equipped and certified for the use of ^{3}H radioactive material (radiometric assays only, *see* **Note 8**).

2.2 Radiometric Detection of HMT Activity Based on Incorporation of Tritiated Methyl Groups

1. SDS-PAGE reagents and equipment to run standard protein mini gels. Use of prestained protein marker is recommended to monitor transfer.
2. Reagents and equipment for tank or semi-dry transfer.
3. PVDF membrane (0.2 μm pore size recommended for histones or peptides).
4. Laboratory platform rocker.
5. Coomassie stain solution: 45% (v/v) methanol, 10% (v/v) acetic acid, 0.25% (w/v) Coomassie brilliant blue R.
6. Destain solution: 45% (v/v) methanol, 10% (v/v) acetic acid.
7. Camera or imaging system capable of imaging Coomassie-stained membranes. Alternatively, a scanner can be used.
8. Intensifying screen for low-energy beta emitters (e.g., Carestream BioMax TranScreen LE).
9. Autoradiography film with high sensitivity for ^{3}H (e.g., Carestream BioMax MS).
10. Adhesive luminescent marker or ruler (e.g., Stratagene Glogos II Autorad Markers).
11. Pre-exposed sheet of autoradiography film of same size as in **step 9**.
12. Autoradiography cassette and bag.
13. −80 °C freezer.
14. Darkroom with red safelight and X-ray film processor.
15. Optional: scintillation vials and liquid scintillation cocktail suitable for ^{3}H.
16. Optional: scintillation counter suitable for ^{3}H detection.
17. For peptide substrates: ENLIGHTNING Rapid Autoradiography Enhancer (Perkin Elmer) or Amersham Amplify Fluorographic Reagent (GE).
18. For peptide substrates: Whatman 3MM filter paper.
19. For peptide substrates: gel drying equipment.

2.3 Antibody-Based Detection of HMT Activity

1. SDS-PAGE reagents and equipment to run standard mini gels.
2. Reagents and equipment for tank or semi-dry transfer.
3. Nitrocellulose or PVDF membrane.
4. Laboratory platform rocker.

5. Antibody suitable for Western blot against the methyl mark to be analyzed.
6. Secondary antibody coupled to horseradish peroxidase or fluorescent moiety.
7. TBS-T: 20 mM Tris pH 7.6, 137 mM NaCl, 0.1% (v/v) Tween.
8. Blocking solution: 5% (w/v) nonfat dry milk in TBS-T.
9. Antibody dilution solution: 2% (w/v) BSA in TBS-T.
10. Reagents and equipment for chemiluminescence or fluorescence-based detection.

2.4 Detection of HMT Activity Via SAH Quantification

1. MTase-Glo methyltransferase kit (Promega).
2. *S*-Adenosylhomocysteine (SAH) solution. Included as 15 μM stock solution with the MTase-Glo methyltransferase kit (Promega). Alternatively, make stock solution by dissolving SAH in DMSO to a final concentration of 1.5–15 mM, aliquot and store at −20 °C.
3. Single tube or microplate luminometer or multimode detection system for the detection of luminescence and required plasticware.

3 Methods

3.1 HMT Reaction

The protocol outlined below is suitable for a broad range of HMTs and substrates (Fig. 1); however, some optimization might be required in order to reliably detect the activity for the specific HMT–substrate combination to be studied.

The assay can accommodate HMTs from a wide range of sources and of varying levels of purity, from whole-cell or nuclear extract to highly purified proteins (Fig. 1). Some HMTs are amenable to expression in *E. coli*, whereas others require eukaryotic expression systems such as Sf9 insect cells, especially in the case of multi-subunit HMT complexes. HMTs can also be purified in their native state from mammalian cells using conventional fractionation and chromatography approaches (*see*, e.g., [21, 22]), or using affinity purification combined with ectopic expression (*see*, e.g., [23]) or introduction of affinity tags into endogenous loci using CRISPR/Cas9 genome editing [24]. Of note, many HMTs such as PRC2 require auxiliary subunits for activity (*see*, e.g., [11, 21, 23]). Therefore, isolated catalytic subunits or domains may fail to be enzymatically active. Moreover, HMTs with C-terminal SET domains usually require a free C terminus for activity. Therefore, C-terminal fusions of SET domains with affinity tags are often enzymatically inactive.

Several considerations guide selection of the substrate. Histone peptides are often good models of native substrates (reviewed in [25, 26]). They are readily available from several suppliers and thus a straightforward way to provide a readout for HMT activity. However, they require knowledge of the substrate residue and also lack the native nucleosomal context. Therefore, histone peptides are ill-suited for the study of HMT regulation by nucleosomal features such as DNA, interaction motifs such as the H2A–H2B acidic patch, or other histone marks. The native substrate of most HMTs are nucleosomes within chromatin. Mononucleosomes and nucleosomal arrays are therefore good approximations for native HMT substrates. They can be reconstituted from recombinant or native core histones and DNA containing nucleosome positioning sequences such as the synthetic 601 sequence [27–30]. Alternatively, nucleosomes can be purified from native chromatin via MNase digestion. Recombinant histones can either be provided in unmodified form or with site-specific modifications, which can be introduced, e.g., via cysteine alkylation or native chemical ligation [25, 26]. Recombinant histones are well-suited to identify novel target residues, which can be confirmed using arginine- or lysine-to-alanine point mutations. Native nucleosomes comprise an ensemble of differentially modified histones, allowing the detection of HMT activity even if that activity requires pre-existing histone marks. Use of native histones, however, may also be associated with reduced activity, as the substrate residue may already be extensively modified or other pre-existing histone marks may inhibit activity. Site-specifically modified recombinant histones can help define the mechanisms underlying such histone mark crosstalk (*see*, e.g., [31–33]). Non-histone substrates can be identified using cellular extracts or peptide arrays, or, if their identity is known, studied using recombinant proteins or domains.

Choice of detection methods is similarly guided by the experimental requirements and substrates chosen, but also by available instrumentation and facilities. As detailed above and summarized in Fig. 1, radiometric detection (Subheading 3.2) is unrivalled for the identification of novel substrates, whereas antibody-based assays (Subheading 3.3) provide convenient means to study reactions with known substrates in a semiquantitative way. Detection based on SAH generation (Subheading 3.4) allows quantitative, high-throughput analysis of reactions with known substrates.

Many applications for HMT assays involve comparison between a control reaction and reactions that feature, for example, differentially modified substrates, auxiliary subunits, or inhibitors in order to study regulation of the HMT in focus. In such cases, preparation of a master mix including all common components is advisable when setting up the assay.

1. For each reaction to be performed, prepare a tube containing the substrate and other reaction-specific components as applicable (e.g., inhibitors, proteins, or nucleic acids whose effect on HMT activity is to be tested). If using peptides as substrates (*see* **Note 4**), add peptide to a final concentration of 0.1–1 mM (2.5–25 nmol per 25-μl reaction). If using core histones (*see* **Note 5**) or nucleosomes (*see* **Note 6**) as substrates, add the equivalent of 0.1–4 μg of total histone protein (~36 nM to 1.47 μM or ~0.9–36.7 pmol if using recombinant *Xenopus laevis* histone octamers). Add corresponding buffer to achieve equal volume for each substrate–effector mix (*see* **Note 9**). Keep on ice while preparing master mix.
2. Prepare sufficient master mix for all reactions to be performed. For each 25-μl reaction, combine in the following order: distilled water up to a total reaction volume of 25 μl (including substrate and other variable components from **step 1**), 5 μl 5× methylation buffer, 0.5 μl 0.2 M DTT (4 mM final), 20 μM unlabeled SAM (e.g., 0.5 μl of a 1 mM pre-dilution of the stock solution) or 25–75 kBq of ^{3}H-SAM (if performing radiometric detection, *see* **Note 2**), and HMT preparation (*see* **Note 3**). 10–40 nM (0.25–1 pmol, 12.5–50 ng for a 50 kDa protein) of enzyme is usually sufficient to achieve detectable activity, but higher amounts may be required for some HMTs. If larger reaction volumes are required to accommodate dilute enzyme or substrate/effector preparations, scale up 5× methylation buffer, DTT, and SAM accordingly to maintain their final concentrations.
3. To start reaction, add master mix to substrate. Mix by flicking the tubes and centrifuge briefly to collect liquid.
4. Incubate at temperature of choice (e.g., 30 °C, *see* **Note 7**) for 0.5–4 h for end-point assays or for shorter incubation times if performing time course experiments or determining enzymatic parameters such as Michaelis–Menten kinetics (*see* **Note 10**).
5. Unless performing quantification of SAH production, stop reactions by adding 12.5 μl of 3× SDS sample buffer (or appropriate volume to reach 1× final concentration) and boil at 95 °C for 5 min. Centrifuge for 10 s at full speed to collect liquid. Proceed to detection steps or store at −20 °C until ready to perform detection. If performing quantification of SAH production (Subheading 3.4), stop reactions by adding 6.25 μl 0.5% TFA. Mix well, centrifuge to collect liquid, and incubate at room temperature for 5 min before proceeding to the detection reaction or freeze at −20 °C until ready to perform detection.

3.2 Radiometric Detection of HMT Activity Based on Incorporation of Tritiated Methyl Groups

This protocol outlines the steps required to detect methylation radiometrically in reactions using ^{3}H-SAM. Incorporation of tritiated methyl groups is detected via autoradiography after gel electrophoresis and transfer to PVDF membranes (**steps 1–12**). To obtain quantitative data, liquid scintillation counting of signal from individual bands can be performed after or instead of exposure to film (optional **steps 13–15**). Older versions of radiometric protocols based on absorption of methylated proteins onto filter paper disks do not provide meaningful advantages over the protocol outlined below. They are therefore not covered here but described elsewhere [34]. Dispose of any radioactive waste from this protocol according to local regulations (*see* **Note 8**).

1. Perform SDS polyacrylamide gel electrophoresis (SDS-PAGE) using gels of suitable percentage for the substrate used. For histone substrates, we find 15% gels to provide good resolution for core histones. If using peptides, please follow the adapted protocol outlined below.
2. Perform semi-dry or tank blot transfer onto a PVDF membrane (*see* **Note 11**).
3. Stain PVDF membrane with Coomassie stain solution for 2–5 min with constant agitation on a platform rocker.
4. Remove the Coomassie stain solution and wash the membrane with destain solution for 5 min under agitation on a platform rocker.
5. Repeat **step 4** twice or until substrate bands are clearly visible.
6. Air-dry membrane until completely dry (*see* **Note 12**).
7. Document membrane with a camera, imaging system, or scanner.
8. Using adhesive tape or microtube labels, attach the dried membrane to an old, already exposed film of identical size to the one used for autoradiography detection. Attach a small piece of adhesive luminescent marker or ruler to the old film, e.g., next to the membrane's top right corner. This will facilitate identification of substrate bands by allowing the matching of signal on the developed film (**step 12**) to bands on the membrane.
9. Insert intensifying screen into an autoradiography cassette and place the film-attached membrane face down on the intensifying screen. Follow the pictograms on the screen. Align screen and film-attached membrane to lower left corner of the cassette.
10. In the dark, insert an unexposed autoradiography film in between the two sleeves of the intensifying screen. Use the film to which the membrane is attached as a guide to position the unexposed film. Close the cassette and put into an autoradiography bag.

11. Expose for 12–96 h in a −80 °C freezer.
12. Develop film in an X-ray film processor.
13. Optional: To obtain quantitative data, carefully cut individual substrate bands from the membrane with a scalpel or razorblade.
14. Optional: Place individual bands into scintillation vials and add liquid scintillation cocktail.
15. Optional: Perform scintillation counting of sample and background (containing only scintillation cocktail) vials in a scintillation counter suitable for ^{3}H.

If using peptides as substrates, the following modified protocol should be followed:

1. Perform SDS-PAGE using 10% SDS polyacrylamide gels. Carry out electrophoresis until dye front has traveled about halfway into the resolving gel. This should concentrate peptides into sharp bands close to the dye front.
2. Stain the gel in Coomassie staining solution for 30–60 min with agitation.
3. Destain gel with several washes of destain solution (10–15 min each) until the bands of interest are clearly visible and background is devoid of Coomassie stain.
4. Document the gel with a camera, imaging system, or scanner.
5. Cover the gel in ENLIGHTNING solution and incubate for 30 min under constant agitation on a platform rocker.
6. Place gel onto water-presoaked Whatman paper cut to be slightly bigger than the gel.
7. Cover with clear wrap and dry on a gel drying setup.
8. Remove clear wrap and expose autoradiography film for 12–96 h in an autoradiography cassette in a −80 °C freezer. Use a piece of adhesive luminescent marker as described above in **step 8** to mark orientation of the gel.
9. Develop film in an X-ray film processor.

3.3 Detection of Methylated Peptides or Histones Using Antibodies

If using methylation-specific antibodies to detect HMT activity, HMT reactions should be performed with nonradioactive SAM. Use of native histones or nucleosomes as substrates may prevent use of antibodies for detection, as preexisting methylation would be picked up by the antibody, potentially masking signal generated in the HMT reaction, especially for lowly active HMTs. The protocol below provides an outline for a Western blot protocol with methylation-specific antibodies, but any established Western blot protocol should provide similar results. To verify equal loading of substrate across reactions to be compared, it is recommended to reprobe the membrane or to carry out a parallel Western blot with part of the reaction (e.g., 10%) using a methylation-insensitive antibody against the substrate (e.g., H3 antibody if analyzing methylation of sites on H3). All incubation and wash steps are carried out on a laboratory platform rocker, at room temperature or 4 °C as indicated.

1. Resolve protein or peptide samples by SDS-PAGE as described in Subheading 3.2.
2. Transfer to nitrocellulose or PVDF membrane by wet or semi-dry transfer. If using peptide substrates, use membrane with 0.2 μm or similar pore size and adjust transfer duration to avoid transferring peptides through the membrane.
3. Briefly wash membrane in TBS-T for 2–3 min.
4. Remove TBS-T and incubate membrane in blocking solution for at least 20 min at room temperature or overnight at 4 °C.
5. Wash with TBS-T for 5 min.
6. Dilute methylation-specific primary antibody in antibody dilution solution using the dilution recommended by the manufacturer (e.g., 1:1000). Incubate for 1–2 h at room temperature or overnight at 4 °C.
7. Remove antibody solution and wash membrane three times with TBS-T for 5 min each.
8. Dilute appropriate secondary antibody in antibody dilution solution using the dilution recommended by the manufacturer (e.g., 1:5000). Incubate for 30–60 min at room temperature.
9. Remove secondary antibody solution and wash membrane three times with TBS-T for 5 min each.
10. Carry out chemiluminescence detection if using horseradish peroxidase-coupled secondary antibody. Document with imaging system capable of detecting chemiluminescence or using X-ray film and X-ray film processor. Alternatively, if using fluorescent secondary antibody, document with appropriate imaging system.

3.4 Detection of HMT Activity Via SAH Quantification

This protocol provides an outline for SAH quantification using the MTase-Glo methyltransferase assay kit (Promega, *see* also [20]). The manufacturer's manual provides detailed information for various assay formats and can be consulted for additional information as needed.

1. Set up a SAH standard curve in 1× HMT buffer (1:5 dilution of 5× HMT buffer in water). Prepare 50 μl of a 1 μM SAH solution. Combine 25 μl 1 μM SAH with 25 μl of 1× HMT

buffer to generate 500 nM SAH standard. Continue analogous serial twofold dilutions down to 7.8125 nM (8 standards in total, 7.8125–1000 nM). For very high activities, consider expanding the standard curve to 10 μM SAH. Also prepare a blank of 1× HMT buffer without SAH.

2. Add 6.25 μl of 0.5% TFA to each 25-μl SAH standard solution and blank.
3. Thaw 10× MTase-Glo reagent on ice and mix gently but thoroughly by inverting the tube. Do not vortex. Aliquot into single-use portions upon first thawing and store at −80 °C.
4. Immediately before use, prepare a 1:2 dilution of 10× MTase-Glo reagent in water to obtain sufficient 5× MTase-Glo reagent for all samples to be analyzed and two replicates of each standard and blank.
5. Add 2 μl 5× MTase-Glo reagent to 8 μl of TFA-treated SAH standard or HMT reaction stopped by the addition of TFA. This and the following steps can be carried out in tubes or directly in 96- or 384-well plates if using a plate reader to detect luminescence in **step 9**. Mix by shaking plates or flicking tubes, spin briefly to collect liquid at the bottom of wells or tubes.
6. Incubate at room temperature for 30 min.
7. Thaw MTase-Glo Detection Solution on ice and mix thoroughly by inverting the tube. Do not vortex. Aliquot into single-use portions upon first thawing and store at −20 °C.
8. Add 10 μl MTase-Glo Detection Solution to each well or tube.
9. Incubate at room temperature for 30 min. Mix by shaking plates or flicking tubes, spin briefly to collect liquid at the bottom of wells or tubes.
10. Measure luminescence in a tube or plate luminometer.
11. To calculate SAH production, perform linear regression of the blank-subtracted luminescence values (in RLU, relative light units) of the SAH standards to generate a linearly fitted standard curve. Use the slope and offset of the resulting graph to calculate SAH concentration in the HMT reaction samples. As SAH standards and HMT reactions underwent the same dilution steps, a HMT reaction sample with a blank-corrected RLU value equivalent to the RLU of, for example, a blank-corrected 0.5 μM SAH standard would have an SAH concentration of 0.5 μM.

4 Notes

1. We find this buffer to support activity of many HMTs from different sources. However, buffer composition can be adapted to meet individual requirements. It is advisable to have robust buffer capacity at a basic pH, as both lysine- and arginine-specific HMTs have pH optima that fall in the basic range, often around 9–10 (*see*, e.g., [35–37]). Increasing the pH above 8.5 can therefore increase activity. Salt concentration should be kept around or below 50 mM, in order to minimize potential interference with substrate binding and to prevent unwrapping of nucleosomes (*see* **Note 9**). However, some HMTs may require higher salt concentrations for stability. Carrier protein such as BSA may be added to enhance stability of HMTs and substrates and to prevent loss of protein due to adsorption to tube walls. Addition of DTT curbs oxidative damage during the reaction.
2. We find ^{3}H-SAM with high specific activity (ideally ~3 TBq/mmol or higher) to provide the highest sensitivity in HMT reactions. If using nonradiolabeled SAM, concentrations should be above the K_m for SAM for the HMT. Concentrations of 10–20 μM SAM in the reaction fulfil this criterion for many HMTs [38] and can be used as a starting point if the K_m is not known for the HMT used. Note that both unlabeled and radiolabeled SAM are prone to decay, and stock solutions may lose significant activity within 6 months or less of storage. Check age of SAM stocks if observing a loss of activity in HMT reactions and consider replacing them with fresh stocks. To maximize stability, stock solutions of SAM should be kept in a strongly acidic buffer (e.g., 5 mM sulfuric acid pH 2, 10% ethanol), as SAM is unstable at neutral or alkaline pH. Aliquot SAM stocks to avoid repeated freeze/thaw cycles. SAM aliquots should be stored at −20 °C. Storage at −80 °C accelerates decay. To account for appreciable decay at elevated temperature and pH, consider replenishing SAM in reactions longer than 2 h to maintain activity.
3. The optimal amount of HMT to be used per reaction depends on the identity of the HMT and the specific activity of the preparation. If not carried out carefully, complex purification protocols may yield enzyme preparations of poor quality, not only due to experimental issues such as repeated freeze–thaw cycles of purification intermediates or failure to work on ice at all times, but also due to introduction of compounds that can negatively affect activity of some HMTs, such as some detergents. If no activity is observed with the suggested amounts of HMT, higher amounts of enzyme can be used. However, the volume of enzyme preparation per HMT reaction should be

kept to a minimum in order to minimize the introduction of potential inhibitors of activity that may be present in the enzyme preparation. Moreover, excessive amounts of salt introduced with the HMT preparation can inhibit activity (*see* **Note 9**). We find that dialyzing HMTs into a low-salt buffer such as BC100 (20 mM Tris–HCl pH 8.0, 100 mM KCl, 0.2 mM EDTA, 20% glycerol, 1 mM DTT; DTT added fresh before use) for storage and subsequent use works well for HMT reactions. If using affinity purification to obtain HMTs, such a dialysis step will also remove eluents such as FLAG peptide. However, we have not found the latter to affect activity. Aliquot enzyme preparations to avoid repeated freeze/thaw cycles and store at −80 °C to preserve activity. Flash-freeze in liquid nitrogen, thaw rapidly but with minimal heat (e.g., by rolling tubes between your fingers), and place on ice as soon as thawed to minimize time spent in transition between frozen and liquid states.

4. A wide range of modified and unmodified histone peptides is readily available from several commercial sources. However, if resorting to custom peptide synthesis to generate peptide substrates, a few points should be considered. Peptides should ideally be at least 20 residues in length, with the substrate residue in the center of the sequence, as neighboring residues are often required for substrate recognition. Unless peptides end at the native N or C termini of the protein or histone, modify peptides with N-terminal acetyl and C-terminal amide groups to avoid extraneous charges at the termini of what would be internal peptides within the protein. For both commercial and custom-made peptides, the overall mass of lyophilized material is often a poor measure for actual peptide amounts due to varying amounts of residual salt lyophilized along with the peptide after synthesis and purification. It is therefore highly advisable to add a C-terminal tyrosine, allowing quantification by UV spectroscopy, or C-termial amino biotinylated lysine to enable relative quantification by Western blotting with a biotin antibody. When comparing HMT activity on different peptides, it is crucial to obtain accurate relative quantification to ensure equal amounts of peptide across different reactions. Peptides should be resuspended to a final concentration of 10 mM in water (very small amounts of TFA or ammonium hydroxide can be added to help solubilization of peptides if needed). Aliquot in protein low-bind tubes (to prevent adsorption of peptide to tube walls) and store at −20 °C. To verify that residual TFA or other compounds in peptide solutions do not affect HMT activity by altering HMT reaction pH, measure reaction pH by spotting 0.5–1 μl onto a suitable pH strip when first using a specific peptide–buffer combination.

5. Recombinant, unmodified histones are available from commercial sources or can be expressed and purified from *E. coli* using well-established protocols (see, e.g., [27, 28]). An ever-increasing selection of site-specifically modified histones generated via cysteine alkylation or native chemical ligation approaches is also available commercially as individual histones, histone octamers, or fully assembled nucleosomes. Several protocols for the generation of site-specifically modified histones are available (*see*, e.g., [5, 39, 40]). Native histones can be isolated from native sources such as chicken erythrocytes [30] or HeLa cells but are also available commercially. When using histones as substrates, they can be provided as individual histones or assembled into H2A-H2B dimers, H3-H4 tetramers, or histone octamers. To maintain their solubility in the absence of DNA, histones and histone complexes are often provided in buffers containing 2 M NaCl. Such preparations may introduce significant amounts of salt into HMT reactions, potentially leading to inhibition (*see* **Note 9**). Of note, low concentrations of salt in HMT reactions promote dissociation of histone octamers into H2A–H2B dimers and H3–H4 tetramers [41], so that histone substrates will be mostly present as dimers and tetramers instead of octamers.
6. For most HMTs active on chromatin, it is advisable to use nucleosomes rather than histone octamers as substrates. However, some HMTs are more active on histones than nucleosomes. Recombinant and native histones can be assembled into nucleosomes and nucleosome arrays using well-established salt dialysis-based protocols (see, e.g., [27, 28]). DNA templates for assembly are often based on the 601 sequence [29], which can be generated by PCR for mono- and dinucleosomes or provided as plasmids containing 601 repeats separated by linkers of varying length. Arrays can also be excised from these plasmids in order to obtain a fully positioned nucleosome array. Care should be taken not to over- or underassemble nucleosomes, which can be monitored by native gel electrophoresis for mono- and dinucleosomes or by restriction digest for arrays. Undersaturation with histone octamers is usually well tolerated by many HMTs, but may preclude analysis of specific modes of regulation that require the presence of neighboring nucleosomes to read nucleosome spacing [42]. Reconstituted chromatin is usually assembled into TE buffer (10 mM Tris–HCl pH 8.0, 1 mM EDTA), which is advantageous for HMT reactions due to its low osmolarity.
7. Most mammalian HMTs should be optimally active at temperatures between 30 and 37 °C. Lower temperatures can be used, e.g., if contaminating protease activity or the thermal stability of HMT or substrate are an issue. HMTs from other

sources such as plants may have lower optimal reaction temperatures. For example, *Neurospora crassa* Dim-5 is optimally active at 10 °C, but only 50% active at 30 °C and essentially inactive at 37 °C [36].

8. When performing HMT assays with ^{3}H-SAM and using radiometric detection protocols, make sure to handle radioactive materials with care and in line with local rules and regulations. While performing reactions, briefly centrifuge tubes before opening to collect samples at the bottom of tubes, minimizing the risk of radioactive contamination. Note that all tubes, pipette tips, blotting paper, and other consumables that have been in contact with ^{3}H-SAM need to be treated as radioactive waste. In addition, buffers from SDS-PAGE, blotting transfer, and gel or membrane staining protocols need to be disposed of properly as radioactive liquid waste. Equipment such as gel tanks and blotting setups should be monitored regularly for contamination.
9. As mentioned above, salt concentration can affect HMT activity (see, e.g., [43]). It is therefore crucial to match the buffer composition as closely as possible between samples. When preparing the substrate pre-mix (**step 1** of the HMT reaction protocol), use the appropriate substrate buffer to adjust volumes between samples, e.g., TE if comparing activity between differentially modified nucleosomes prepared in TE. Similarly, use the corresponding solvent or buffer to balance volumes between samples if adding small-molecule inhibitors or additional proteins such as auxiliary subunits, or when comparing different HMTs or different HMT preparations. When analyzing effects of additional proteins on activity, obtaining exact buffer matches can be difficult. In these cases, boiling the protein preparation for 5 min at 95 °C to denature the proteins prior to addition to the HMT reaction can provide a better control. However, some proteins may resist heat denaturation.
10. If the primary goal of the assay is to demonstrate HMT activity, incubation times can be extended to as much as overnight if dealing with poorly active enzymes or preparations with low activity. However, we find 30 min to 4 h to be sufficient to observe activity for most HMTs, if providing a suitable substrate. When aiming to perform kinetic analysis to determine Michaelis–Menten or other enzyme kinetic parameters, shorter time points within the linear range of the reaction will be required to accurately determine initial velocities. Such assays may also require adapting or titrating substrate and SAM concentrations, e.g., if determining K_m values.

11. PVDF membranes need to be used here, as nitrocellulose membranes are incompatible with the Coomassie stain and destain solutions. Pre-soak PVDF in methanol before use. Methanol for this purpose can be reused.

12. Drying the PVDF membrane completely after destaining is important to reduce background and ensure optimal signal over background for Coomassie-stained bands. The membrane should therefore be dry before documenting it. Drying can be accelerated using a hair dryer or laboratory heat gun. However, these must be run in cool mode to not heat up the membrane, as this may compromise detection sensitivity. When using a hair dryer, carefully hold the membrane using forceps and use a low fan speed to prevent the membrane from breaking, ripping, or becoming airborne.

Acknowledgments

We thank Pablo De Ioannes and Karim-Jean Armache for helpful discussions and advice on SAH quantification methods. Work in the Voigt lab is supported by the Wellcome Trust ([104175/Z/14/Z], Sir Henry Dale Fellowship to P.V.) and through funding from the European Research Council (ERC) under the European Union's Horizon 2020 research and innovation programme (ERC-STG grant agreement No. 639253 to P.V.). The Wellcome Centre for Cell Biology is supported by core funding from the Wellcome Trust [203149].

References

1. Greer EL, Shi Y (2012) Histone methylation: a dynamic mark in health, disease and inheritance. Nat Rev Genet 13(5):343–357
2. Li B, Carey M, Workman JL (2007) The role of chromatin during transcription. Cell 128(4): 707–719
3. Zhou VW, Goren A, Bernstein BE (2011) Charting histone modifications and the functional organization of mammalian genomes. Nat Rev Genet 12(1):7–18
4. Musselman CA, Lalonde M-E, Côté J, Kutateladze TG (2012) Perceiving the epigenetic landscape through histone readers. Nat Struct Mol Biol 19(12):1218–1227
5. Bartke T, Vermeulen M, Xhemalce B, Robson SC, Mann M, Kouzarides T (2010) Nucleosome-interacting proteins regulated by DNA and histone methylation. Cell 143(3): 470–484
6. Vermeulen M, Eberl HC, Matarese F, Marks H, Denissov S, Butter F, Lee KK, Olsen JV, Hyman AA, Stunnenberg HG, Mann M (2010) Quantitative interaction proteomics and genome-wide profiling of epigenetic histone marks and their readers. Cell 142(6):967–980
7. Zhang T, Cooper S, Brockdorff N (2015) The interplay of histone modifications - writers that read. EMBO Rep 16(11):1467–1481
8. Murray K (1964) The occurrence of epsilon-N-methyl lysine in histones. Biochemistry 3: 10–15
9. Kim S, Paik WK (1965) Studies on the origin of epsilon-*N*-methyl-*L*-lysine in protein. J Biol Chem 240(12):4629–4634
10. Murn J, Shi Y (2017) The winding path of protein methylation research: milestones and new frontiers. Nat Rev Mol Cell Biol 18(8): 517–527

11. Rea S, Eisenhaber F, O'Carroll D, Strahl BD, Sun ZW, Schmid M, Opravil S, Mechtler K, Ponting CP, Allis CD, Jenuwein T (2000) Regulation of chromatin structure by site-specific histone H3 methyltransferases. Nature 406(6796):593–599
12. Chen D, Ma H, Hong H, Koh SS, Huang SM, Schurter BT, Aswad DW, Stallcup MR (1999) Regulation of transcription by a protein methyltransferase. Science 284(5423): 2174–2177
13. Helin K, Dhanak D (2013) Chromatin proteins and modifications as drug targets. Nature 502(7472):480–488
14. Richart L, Margueron R (2020) Drugging histone methyltransferases in cancer. Curr Opin Chem Biol 56:51–62
15. Piunti A, Shilatifard A (2021) The roles of Polycomb repressive complexes in mammalian development and cancer. Nat Rev Mol Cell Biol 22(5):326–345
16. Li Y, Trojer P, Xu C-F, Cheung P, Kuo A, Drury WJ, Qiao Q, Neubert TA, Xu R-M, Gozani O, Reinberg D (2009) The target of the NSD family of histone lysine methyltransferases depends on the nature of the substrate. J Biol Chem 284(49):34283–34295
17. Luo M (2012) Current chemical biology approaches to interrogate protein methyltransferases. ACS Chem Biol 7(3):443–463
18. Drake KM, Watson VG, Kisielewski A, Glynn R, Napper AD (2014) A sensitive luminescent assay for the histone methyltransferase NSD1 and other SAM-dependent enzymes. Assay Drug Dev Technol 12(5):258–271
19. Duchin S, Vershinin Z, Levy D, Aharoni A (2015) A continuous kinetic assay for protein and DNA methyltransferase enzymatic activities. Epigenetics Chromatin 8(1):56–59
20. Hsiao K, Zegzouti H, Goueli SA (2016) Methyltransferase-Glo: a universal, bioluminescent and homogenous assay for monitoring all classes of methyltransferases. Epigenomics 8(3):321–339
21. Cao R, Wang L, Wang H, Xia L, Erdjument-Bromage H, Tempst P, Jones RS, Zhang Y (2002) Role of histone H3 lysine 27 methylation in Polycomb-group silencing. Science 298(5595):1039–1043
22. Nishioka K, Rice JC, Sarma K, Erdjument-Bromage H, Werner J, Wang Y, Chuikov S, Valenzuela P, Tempst P, Steward R, Lis JT, Allis CD, Reinberg D (2002) PR-Set7 is a nucleosome-specific methyltransferase that modifies lysine 20 of histone H4 and is associated with silent chromatin. Mol Cell 9(6): 1201–1213
23. Kuzmichev A, Nishioka K, Erdjument-Bromage H, Tempst P, Reinberg D (2002) Histone methyltransferase activity associated with a human multiprotein complex containing the enhancer of Zeste protein. Genes Dev 16(22):2893–2905
24. Doyon Y, Côté J (2016) Preparation and analysis of native chromatin-modifying complexes. Methods Enzymol 573:303–318
25. Voigt P, Reinberg D (2011) Histone tails: ideal motifs for probing epigenetics through chemical biology approaches. Chembiochem 12(2): 236–252
26. Müller MM, Muir TW (2015) Histones: at the crossroads of peptide and protein chemistry. Chem Rev 115(6):2296–2349
27. Dyer PN, Edayathumangalam RS, White CL, Bao Y, Chakravarthy S, Muthurajan UM, Luger K (2004) Reconstitution of nucleosome core particles from recombinant histones and DNA. Methods Enzymol 375:23–44
28. Luger K, Rechsteiner TJ, Flaus AJ, Waye MM, Richmond TJ (1997) Characterization of nucleosome core particles containing histone proteins made in bacteria. J Mol Biol 272(3): 301–311
29. Thåström A, Lowary PT, Widlund HR, Cao H, Kubista M, Widom J (1999) Sequence motifs and free energies of selected natural and non-natural nucleosome positioning DNA sequences. J Mol Biol 288(2):213–229
30. Simon RH, Felsenfeld G (1979) A new procedure for purifying histone pairs H2A + H2B and H3 + H4 from chromatin using hydroxylapatite. Nucleic Acids Res 6(2):689–696
31. Voigt P, LeRoy G, Drury WJ, Zee BM, Son J, Beck DB, Young NL, Garcia BA, Reinberg D (2012) Asymmetrically modified nucleosomes. Cell 151(1):181–193
32. Schmitges FW, Prusty AB, Faty M, Stützer A, Lingaraju GM, Aiwazian J, Sack R, Hess D, Li L, Zhou S, Bunker RD, Wirth U, Bouwmeester T, Bauer A, Ly-Hartig N, Zhao K, Chan H, Gu J, Gut H, Fischle W, Müller J, Thomä NH (2011) Histone methylation by PRC2 is inhibited by active chromatin marks. Mol Cell 42(3):330–341
33. Yuan W, Xu M, Huang C, Liu N, Chen S, Zhu B (2011) H3K36 methylation antagonizes PRC2-mediated H3K27 methylation. J Biol Chem 286(10):7983–7989
34. Jacob Y, Voigt P (2018) In vitro assays to measure histone methyltransferase activity using different chromatin substrates. Methods Mol Biol 1675:345–360
35. Trievel RC, Beach BM, Dirk LMA, Houtz RL, Hurley JH (2002) Structure and catalytic

mechanism of a SET domain protein methyltransferase. Cell 111(1):91–103
36. Zhang X, Tamaru H, Khan SI, Horton JR, Keefe LJ, Selker EU, Cheng X (2002) Structure of the Neurospora SET domain protein DIM-5, a histone H3 lysine methyltransferase. Cell 111(1):117–127
37. Kipp DR, Quinn CM, Fortin PD (2013) Enzyme-dependent lysine deprotonation in EZH2 catalysis. Biochemistry 52(39): 6866–6878
38. Richon VM, Johnston D, Sneeringer CJ, Jin L, Majer CR, Elliston K, Jerva LF, Scott MP, Copeland RA (2011) Chemogenetic analysis of human protein methyltransferases. Chem Biol Drug Des 78(2):199–210
39. Simon MD, Chu F, Racki LR, de la Cruz CC, Burlingame AL, Panning B, Narlikar GJ, Shokat KM (2007) The site-specific installation of methyl-lysine analogs into recombinant histones. Cell 128(5):1003–1012
40. Chen Z, Grzybowski AT, Ruthenburg AJ (2014) Traceless semisynthesis of a set of histone 3 species bearing specific lysine methylation marks. Chembiochem 15(14):2071–2075
41. Eickbush TH, Moudrianakis EN (1978) The histone core complex: an octamer assembled by two sets of protein-protein interactions. Biochemistry 17(23):4955–4964
42. Yuan W, Wu T, Fu H, Dai C, Wu H, Liu N, Li X, Xu M, Zhang Z, Niu T, Han Z, Chai J, Zhou XJ, Gao S, Zhu B (2012) Dense chromatin activates Polycomb repressive complex 2 to regulate H3 lysine 27 methylation. Science 337(6097):971–975
43. Stützer A, Liokatis S, Kiesel A, Schwarzer D, Sprangers R, Söding J, Selenko P, Fischle W (2016) Modulations of DNA contacts by linker histones and post-translational modifications determine the mobility and modifiability of Nucleosomal H3 tails. Mol Cell 61(2): 247–259

Chapter 3

In Vitro Histone Demethylase Assays

Shengjiang Tu

Abstract

Dynamic histone methylation regulates gene activation and repression. It is involved in proliferation, differentiation, lineage specification, and development. Histone demethylase assays are invaluable in studying histone demethylation substrate recognition, kinetics, regulation, and inhibition by small molecules, many of which are potential therapeutics. Here we describe general procedures to purify recombinant enzymes from different expression hosts, and to prepare a broad range of substrates, as well as to set up a variety of in vitro histone demethylase assays. These assays provide useful tools for discoveries from enzymes to drugs.

Key words Gene transcription, Chromatin, Nucleosome, Histones, Histone modifications, Histone methylation, Histone demethylases, Demethylase assays

1 Introduction

Gene transcription occurs in the context of chromatin, where octamers of histone proteins H3, H4, H2A, H2B, and 146 bp of DNA form the basic unit, the nucleosome [1]. Histone modifications, along with DNA methylation and chromatin structure, are key subjects of transcription regulation [2]. The tails of histone proteins are extensively modified by many families of enzymes, including histone acetyltransferases (HATs), deacetylases (HDACs), kinases, methyltransferases (HMTs), ubiquitin ligases, and more [3–5]. Histone modifications impact transcription, as modified histone tails can be recognized by chromatin-associated proteins. Many of them are transcription factor interacting proteins and are important components of co-activators and co-repressors [6]. As a result, histone modifications are involved in cell lineage specification, development, cell cycle control, DNA damage responses, and tumorigenesis [7].

Histone methylation exemplifies the above complex cellular regulation [8, 9]. Lysine and arginine residues on histones can be methylated. Histone lysine methylation is extensively studied due

Raphaël Margueron and Daniel Holoch (eds.), *Histone Methyltransferases: Methods and Protocols*,
Methods in Molecular Biology, vol. 2529, https://doi.org/10.1007/978-1-0716-2481-4_3,

to its complexity and diverse functions. For example, in histone tails, multiple lysine sites can be methylated, mainly including H3K4, H3K9, H3K27, H3K36, H3K79, and H4K20. In addition, each lysine can have 4 different methylation states (me0, me1, me2, and me3), which can be recognized by PHD, Chromo, Tudor, WD40, and other related domains [2, 10]. Adding additional complexity, histone methylation was discovered to be reversible. Two families of demethylases, flavin-dependent mono−/di-methyl demethylases (LSD1/LSD2) and Fe^{2+}/α-ketoglutarate dependent JmjC-domain-containing mono−/di−/tri-methyl demethylases, contribute to the dynamic regulation of histone methylation [11–13]. Since the discovery of these demethylases, progress has been made to identify their substrates, substrate recognition mechanisms, associated complex and enzyme activity regulation, and their involvement in diseases, particularly in oncology [14, 15]. Mis-regulation and somatic mutations of histone demethylases have been implicated in multiple cancers [16]. Small-molecule inhibitors are an active research area in oncology [17].

At a mechanistic level, several themes emerge. Many demethylases have DNA-binding domains and chromatin-binding domains, which have targeting or regulatory roles. Second, posttranslational modifications such as phosphorylation might further modulate histone demethylase activities [18]. Third, demethylases might form protein complexes, which can contribute to demethylase activity and substrate specificity. The above insights into histone demethylases are profound. A primary consideration is the enzyme source. In cases where other domains are necessary for demethylase activity or targeting, full-length or constructs with neighboring domains might be needed. In addition, different expression hosts can be tested. This includes *E. coli*, yeast, baculovirus, and mammalian cells. For the latter two hosts, it is advantageous to co-express and co-purify demethylase complexes. Another critical issue is testing substrates, for example, histone peptides, octamers, and nucleosomes in their methylated forms (me1-3). Sometimes, combinatorially modified histone substrates are needed. They can be prepared with chemical synthesis (peptides) or histone methyltransferases. In addition, histones with methyl-lysine analogs can be installed via a chemical approach [19, 20]. If resources are available, native chemical ligation (NCL) or expressed protein ligation (EPL) with pre-methylated and other modifications on histones could be utilized to test precisely defined methylated histone substrates [21]. Lastly, we need to select the assay format. For peptide substrates, mass spectrometry readout is the most robust. As an indirect readout, formaldehyde dehydrogenase can be deployed in some cases to detect demethylation by-product formaldehyde (Fig. 1). For histone octamer or nucleosome substrates, radioactivity monitoring of histone-methyltransferase-treated substrates is a

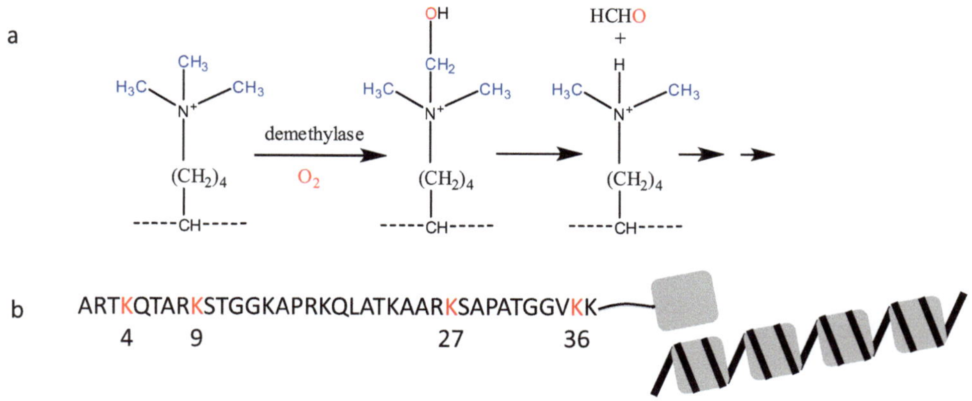

c 1. Peptide Mass Spectrometry assay
2. Formaldehyde dehydrogenase assay
3. Radioactivity monitoring assay
4. Demethylation assay with Western Blots

Fig. 1 (**a**) Demethylation mechanism. Fe^{2+}/α-ketoglutarate dependent dioxygenases, catalyze the hydroxylation step via an oxygen radical mechanism. Subsequent spontaneous elimination of formaldehyde results in the first demethylation cycle. After three cycles, trimethylated lysine will be converted to unmethylated lysine. (**b**) Demethylase substrates. As an example, the histone H3 N-terminal tail is listed, with major demethylation sites highlighted. Histone octamers and nucleosomes are also potential substrates. (**c**) A list of demethylase assays

sensitive assay. Alternatively, Western blots can be used as a convenient readout assay (Fig. 1).

In this chapter, for in vitro histone lysine demethylation assays, we focus on the Fe^{2+}/α-ketoglutarate-dependent JmjC family of demethylases as they form the majority of histone demethylases. All of the assay formats described below can also be easily adapted to flavin-dependent mono-/di-methyl demethylase assays [11, 22]. We will start with general procedures of recombinant enzyme preparation. We then describe histone substrate preparation steps. In the end, we will introduce a few common demethylation assays. This is not a comprehensive assay review; our primary focus is to give researchers options which might lead to novel enzyme discovery and robust assay results.

2 Materials

2.1 Enzyme Purification

2.1.1 Purification of Recombinant GST Fusion Proteins from *E. coli*

1. Sonicator.
2. Refrigerated shaker.
3. SDS-PAGE apparatus.
4. 100 mg/mL (1000×) Ampicillin.
5. pGEX6p2 vector.

6. TOP10 chemically competent cells.
7. BL21-CodonPlus (DE3)-RIL competent cells.
8. DNA polymerase.
9. Restriction enzymes.
10. T4 DNA ligase.
11. Site-directed mutagenesis kit.
12. LB medium: 10 g/L tryptone, 10 g/L NaCl, 5 g/L yeast extract, pH 7.0 (sterilized by autoclaving).
13. LB agar plates: autoclave 1.5% agar in LB (w/v), cool to 55°, and pour 20 mL into each 10 cm plate.
14. 1 M IPTG.
15. UV-Vis spectrophotometer with temperature control capacity.
16. PBS buffer: 140 mM NaCl, 2.7 mM KCl, 10 mM Na_2HPO_4, 1.8 mM KH_2PO_4, pH 7.3.
17. PBS lysis buffer: PBS with 1 mM PMSF, 2 mM DTT, 1% Triton X-100.
18. PBS wash buffer: 490 mM NaCl, 2.7 mM KCl, 10 mM Na_2HPO_4, 1.8 mM KH_2PO_4, pH 7.3, 1 mM PMSF, 2 mM DTT.
19. Glutathione-agarose beads.
20. GST elution buffer: 50 mM Tris–HCl, 20 mM reduced glutathione, pH 8.0.
21. Amicon Ultra centrifugal units, MWCO 30 kDa.
22. Liquid nitrogen.
23. HEPES protein storage buffer: 30 mM HEPES, pH 7.5, 1 mM β-mercaptoethanol.
24. 10% SDS-PAGE gel (or 4–15% gradient gel).
25. 4× SDS loading buffer: 200 mM Tris–HCl, pH 6.8, 200 mM DTT, 8% SDS, 40% glycerol, 0.4% bromophenol blue.
26. SDS-PAGE running buffer: 25 mM Tris base, 0.19 M glycine, 0.1% SDS (no pH adjustment).
27. Coomassie staining buffer: 0.1% Coomassie Brilliant Blue, 50% methanol (v/v), 10% glacial acetic acid (v/v).
28. Coomassie destaining buffer: 50% methanol (v/v), 10% glacial acetic acid (v/v).

2.1.2 Purification of Recombinant Enzymes from Yeast

1. SD-ura: 6.7 g/L yeast nitrogen base without amino acids, 0.77 g/L -ura dropout amino acid mix + 2% glucose (w/v, added after autocleaving other components).
2. SC-ura: 6.7 g/L yeast nitrogen base without amino acids, 0.77 g/L -ura dropout amino acid mix + 2% sucrose (w/v, added after autocleaving other components) (for protein induction, sucrose is replaced with 2% galactose).

3. SD-ura plates: pour 20 mL warm (55 °C) 2% agar in SD-ura after autoclaving into each 10 cm plate.
4. UV-Vis spectrophotometer with temperature control capacity.
5. Liquid nitrogen.
6. Coffee grinder.
7. Mortar and pestle.
8. Ni-NTA resin.
9. Lysis buffer A: 20 mM HEPES, pH 7.5, 1 M NaCl, 5% Glycerol, 2 mM β-mercaptoethanol, 2.5 μg/mL pepstatin, 2.5 μg/mL leupeptin, 2 mM PMSF.
10. Bradford reagent.
11. Imidazole wash buffer I: Lysis buffer A + 20 mM imidazole, pH 7.5.
12. Imidazole wash buffer II: Lysis buffer A + 50 mM imidazole, pH 7.5.
13. Imidazole elution buffer: Lysis buffer A + 250 mM imidazole, pH 7.5.
14. Amicon Ultra centrifugal units, MWCO 30 kDa.
15. HEPES protein storage buffer: 30 mM HEPES, pH 7.5, 1 mM β-mercaptoethanol.
16. SDS-PAGE apparatus.
17. 10% SDS-PAGE gel (or 4–15% gradient gel).
18. 4× SDS loading buffer: 200 mM Tris–HCl, pH 6.8, 200 mM DTT, 8% SDS, 40% glycerol, 0.4% bromophenol blue.
19. SDS-PAGE running buffer: 25 mM Tris base, 0.19 M glycine, 0.1% SDS (no pH adjustment).
20. Coomassie staining buffer: 0.1% Coomassie Brilliant Blue, 50% methanol (v/v), 10% glacial acetic acid (v/v).
21. Coomassie destaining buffer: 50% methanol (v/v), 10% glacial acetic acid (v/v).
22. Filter paper.
23. Transfer apparatus.
24. Nitrocellulose transfer membrane.
25. SDS-PAGE transfer buffer: 25 mM Tris base, 0.19 M glycine, 0.1% SDS, 20% methanol.
26. TBST: 20 mM Tris–HCl, pH 7.5, 140 mM NaCl, 0.05% Tween-20.
27. Anti-His6 antibody: 1:1000 dilution in 5% (w/v) BSA in TBST.
28. HRP-conjugated second antibodies: 1:5000 dilution in 5% (w/v) milk in TBST.

29. Luminol solution: 50 mg Sodium Luminol, 62 μL 30% H_2O_2, 200 mL 0.1 M Tris–HCl, pH 8.6–8.7 (avoid light).
30. Enhancer: 11 mg p-coumaric acid in 10 mL DMSO (avoid light).
31. ChemiDoc imaging system.

2.1.3 Recombinant Enzymes and Complexes Purification from Sf9 Cells and Mammalian Cells

1. 500 mL,1 L spinner flasks.
2. Magnetic stir plate or shaker apparatus.
3. Incubator set to 27 °C.
4. Six-well cell culture plates.
5. pFastBac HT vector.
6. TOP10 chemically competent *E. coli* cells.
7. DH10Bac chemically competent *E. coli* cells.
8. DNA polymerase.
9. Restriction enzymes.
10. T4 DNA ligase.
11. Site-directed mutagenesis kit.
12. M13 Forward(−40) primer: GTTTTCCCAGTCACGAC.
13. M13 Reverse primer: CAGGAAACAGCTATGAC.
14. Internal primer.
15. Taq DNA polymerase.
16. SOC medium: 20 g/L tryptone, 5 g/L yeast extract, 0.5 g/L NaCl, 10 mM $MgCl_2$, 20 mM glucose.
17. SOC plate: 1.5% (1.5 g/100 mL) agar in SOC.
18. Kanamycin: 50 mg/mL (1000× stock solution).
19. Gentamicin: 7 mg/mL (1000× stock solution).
20. Tetracycline: 10 mg/mL (1000× stock solution).
21. X-Gal: 100 mg/mL (1000× stock solution).
22. IPTG: 40 mg/mL (1000× stock solution).
23. SF-900 III serum-free media (SFM).
24. ExpiFectamine Sf transfection Reagent.
25. Opti-MEM I Reduced Serum Medium.
26. 0.2 μm low-protein-binding filter.
27. Ni-NTA resin.
28. Lysis buffer A: 20 mM HEPES, pH 7.5, 1 M NaCl, 5% Glycerol, 2 mM β-mercaptoethanol, 2.5 μg/mL pepstain, 2.5 μg/mL leupeptin, 2 mM PMSF.
29. Modified lysis buffer A: 20 mM HEPES, 350 mM NaCl, 10 mM imidazole, pH 7.5, 5% Glycerol, 2 mM β-mercaptoethanol, 2.5 μg/mL pepstatin, 2.5 μg/mL leupeptin, 2 mM PMSF.

30. Sonicator.
31. Modified Imidazole wash buffer I: 20 mM HEPES, 350 mM NaCl, 20 mM imidazole, pH 7.5, 5% Glycerol, 2 mM β-mercaptoethanol, 2.5 μg/mL pepstatin, 2.5 μg/mL leupeptin, 2 mM PMSF.
32. Modified Imidazole wash buffer II: 20 mM HEPES, 350 mM NaCl, 50 mM imidazole, pH 7.5, 5% Glycerol, 2 mM β-mercaptoethanol, 2.5 μg/mL pepstatin, 2.5 μg/mL leupeptin, 2 mM PMSF.
33. Modified Imidazole elution buffer: 20 mM HEPES, 350 mM NaCl, 250 mM imidazole, pH 7.5, 5% Glycerol, 2 mM β-mercaptoethanol, 2.5 μg/mL pepstatin, 2.5 μg/mL leupeptin, 2 mM PMSF.
34. Amicon Ultra centrifugal units, MWCO 30 kDa.
35. HEPES protein storage buffer: 30 mM HEPES, pH 7.5, 1 mM β-mercaptoethanol.
36. Bradford reagent.
37. Liquid nitrogen.
38. pFLAG-CMV-4 vector.
39. 293-F cells.
40. 10-cm cell culture plates.
41. 15-cm cell culture plates.
42. PBS buffer: 140 mM NaCl, 2.7 mM KCl, 10 mM Na_2HPO_4, 1.8 mM KH_2PO_4, pH 7.3.
43. BC350 buffer: 20 mM Tris–HCl, pH 7.9, 350 mM NaCl, 10% glycerol, 0.1% NP40, 0.2 mM PMSF, 0.5 mM DTT, 1 μg/mL pepstatin 1 μg/mL leupeptin.
44. BC100 buffer: 20 mM Tris–HCl, pH 7.9, 100 mM NaCl, 10% glycerol, 0.2 mM PMSF, 0.5 mM DTT, 1 μg/mL pepstatin, 1 μg/mL leupeptin.
45. M2 beads.
46. Flag peptide: 0.2 mg/mL in BC100.
47. Lipofectamine 2000.
48. G418: 100 mg/mL stock solution.
49. Dounce homogenizer.
50. Buffer A: 20 mM Tris, pH 7.9, 0.5 mM DTT, 0.2 mM PMSF, 1 μg/mL pepstatin 1 μg/mL leupeptin.
51. Buffer C: 20 mM Tris, pH 7.9, 1.5 mM $MgCl_2$, 0.42 M NaCl, 0.2 mM EDTA, 0.5 mM DTT, 0.2 mM PMSF, 1 μg/mL pepstatin, 1 μg/mL leupeptin.
52. SDS-PAGE apparatus.
53. 10% SDS-PAGE gel (or 4–15% gradient gel).

54. 4× SDS loading buffer: 200 mM Tris–HCl, pH 6.8, 200 mM DTT, 8% SDS, 40% glycerol, 0.4% bromophenol blue.
55. SDS-PAGE running buffer: 25 mM Tris base, 0.19 M glycine, 0.1% SDS (no pH adjustment).
56. Coomassie staining buffer: 0.1% Coomassie Brilliant Blue, 50% methanol (v/v), 10% glacial acetic acid (v/v).
57. Coomassie destaining buffer: 50% Methanol (v/v), 10% glacial acetic acid (v/v).
58. Filter paper.
59. Transfer apparatus.
60. Nitrocellulose transfer membrane.
61. SDS-PAGE transfer buffer: 25 mM Tris base, 0.19 M glycine, 0.1% SDS, 20% methanol.
62. TBST: 20 mM Tris–HCl, pH 7.5, 140 mM NaCl, 0.05% Tween-20.
63. Anti-His6 antibody: 1:1000 dilution in 5% (w/v) BSA in TBST.
64. Anti-Flag antibody: 1:1000 dilution in 5% (w/v) BSA in TBST.
65. HRP-conjugated second antibodies: 1:5000 dilution in 5% (w/v) milk in TBST.
66. Luminol solution: 50 mg Sodium Luminol, 62 μL 30% H_2O_2, 200 mL 0.1 M Tris–HCl, pH 8.6–8.7 (avoid light).
67. Enhancer: 11 mg p-coumaric acid in 10 mL DMSO (avoid light).
68. ChemiDoc imaging system.

2.2 Substrate Preparation

1. Yeast Peptides (12mer-13mer) at 5–10 mg scale. Confirm peptide purity and methylation status by HPLC-MS (*see* **Note 1**):
   (a) H3K4me1/2/3: ARTK(me1/2/3)KQTARKST 12mer.
   (b) H3K36me1/2/3: STGGVK(me1/2/3)KKPHRY 12mer.
   (c) H3K79me1/2/3: IAQDFK(me1/2/3)KTDLRFQ 13mer.
2. Human peptides (19-21mer) at 5–10 mg scale. Confirm peptide purity and methylation status by HPLC-MS:
   (a) H3K4me1/2/3: ARTK(me1/2/3)QTARKSTGGKAPRKQL 20mer.
   (b) H3K9me1/2/3: ARTKQTARK(me1/2/3)STGGKAPRKQL 20mer.
   (c) H3K27me1/2/3: RKQLATKAARK(me1/2/3)SAPATGGVKK 21mer.
   (d) H3K36me1/2/3: RKSAPATGGVK(me1/2/3)KPHRYQPGT 20mer.

(e) H3K79me1/2/3: RLVREIAQDFK(me1/2/3)TDLRFQSSA 20mer.

(f) H4K20me1/2/3: KGGAKRHRK(me1/2/3)VLRDNIQGIT 19mer.

3. S200 gel filtration column and fraction collector (or FPLC protein purification system).
4. pET3-H2A plasmid.
5. pET3-H2B plasmid.
6. pET3-H3 plasmid.
7. pET3-H4 plasmid.
8. pET3-H3K4C-C110A plasmid.
9. Site-directed mutagenesis kit.
10. TOP10 chemically competent cells.
11. BL21 (DE3) pLys competent cells.
12. pG5E4 plasmid.
13. 100 mg/mL (1000×) Ampicillin.
14. 25 mg/mL (1000×) Chloramphenicol.
15. 2x YT medium: 16 g/L tryptone, 10 g/L NaCl, 5 g/L yeast extract, pH 7.0 (sterilized by autoclaving).
16. 1 M IPTG.
17. UV-Vis spectrophotometer.
18. Sonicator.
19. Glass Dounce homogenizer.
20. Histone wash buffer: 50 mM Tris–HCl, pH 7.5, 100 mM NaCl, 1 mM β-mercaptoethanol, 0.2 mM PMSF.
21. Triton X-100.
22. Histone unfolding buffer: 6 M fresh guanidine–HCl, 20 mM Tris–HCl, pH 7.5, 5 mM DTT.
23. Histone refolding buffer: 2 M NaCl, 10 M Tris–HCl, pH 7.5, 1 mM EDTA, 5 mM β-mercaptoethanol.
24. Dialysis tubing, MWCO 6–8 kDa.
25. Stirring bar and plate.
26. SDS-PAGE apparatus.
27. 15% SDS-PAGE gels.
28. 4× SDS loading buffer: 200 mM Tris–HCl, pH 6.8, 200 mM DTT, 8% SDS, 40% glycerol, 0.4% bromophenol blue.
29. SDS-PAGE running buffer: 25 mM Tris base, 0.19 M glycine, 0.1% SDS (no pH adjustment).
30. Coomassie staining buffer: 0.1% Coomassie Brilliant Blue, 50% methanol (v/v), 10% glacial acetic acid (v/v).

31. Coomassie destaining buffer: 50% methanol (v/v), 10% glacial acetic acid (v/v).
32. Amicon Ultra centrifugal units, MWCO 30 kDa.
33. 4 M KCl.
34. TE: 10 mM Tris–HCl, pH 7.5, 1 mM EDTA.
35. Filter paper.
36. Transfer apparatus.
37. Nitrocellulose transfer membrane.
38. SDS-PAGE transfer buffer: 25 mM Tris base, 0.19 M glycine, 0.1% SDS, 20% methanol.
39. TBST: 20 mM Tris–HCl, pH 7.5, 140 mM NaCl, 0.05% Tween-20.
40. Primary antibodies against methylated histones (site- and methylation-status-, i.e., me3/me2/me1/me0, –specific).
41. HRP-conjugated second antibodies: 1:5000 dilution in 5% (w/v) milk in TBST.
42. Luminol solution: 50 mg Sodium Luminol, 62 μL 30% H_2O_2, 200 mL 0.1 M Tris–HCl, pH 8.6–8.7 (avoid light).
43. Enhancer: 11 mg p-coumaric acid in 10 mL DMSO (avoid light).
44. ChemiDoc imaging system.
45. HEPES protein storage buffer: 30 mM HEPES, pH 7.5, 1 mM β-mercaptoethanol.
46. PVDF membrane.
47. 3H-Enhancer.
48. BioMax XAR film.
49. Peristaltic pump.
50. Reconstitution buffer (High salt): 2 M KCl, 10 mM Tris–HCl, pH 7.5, 1 mM EDTA, 1 mM DTT.
51. TE buffer: 10 mM Tris–HCl, pH 7.5, 1 mM EDTA, 1 mM DTT.
52. Slide-A-Lyzer mini dialysis device, 10 K MWCO.
53. 5× HMT buffer: 250 mM Tris–HCl, pH 8.5, 20 mM DTT, 25 mM $MgCl_2$.
54. SAM: 400 μM in water.
55. SAM[3H]: 50–85 Ci/mmol, 0.55 mCi/mL.
56. Amicon Ultra centrifugal units, MWCO 3 kDa.
57. Liquid scintillation cocktail.
58. Liquid scintillation counter.
59. Sonicator bath.

60. Alkylation buffer: 1 M HEPES, pH 7.8, 4 M guanidine–HCl, 10 mM methionine.
61. β-Mercaptoethanol.
62. 1 M DTT, freshly made.
63. (2-Bromoethyl)-trimethylammonium bromide.
64. (2-Chloroethyl)-dimethylammonium chloride.
65. (2-Chloroethyl)-methylammonium chloride.
66. (2-Bromoethyl)-ammonium bromide.
67. PD-10 columns.

2.3 Demethylation Assays

1. Histone peptides (*see* Subheading 2.2).
2. Demethylase proteins (purified from Subheading 2.1).
3. 2× HEPES buffer: 60 mM HEPES, pH 7.5.
4. 60 mM α-ketoglutarate.
5. 200 mM ascorbic acid (fresh preparation).
6. 60 mM DTT.
7. 10 mM $(NH_4)_2Fe(SO_4)_2 \cdot 6H_2O$ in 200 mM ascorbic acid (fresh preparation).
8. 100 mM EDTA.
9. MALDI-TOF instrument (accessed via a Mass Spectrometry/Proteomics facility).
10. Saturated matrix solution: α-cyano-4-hydroxy cinnamic acid in 50% acetonitrile/0.1% TFA/50% H_2O.
11. MALDI target plate.
12. FlexAnalysis software.
13. 20 mg/mL NAD^+.
14. 0.01 U/μL FDH, reconstituted in water.
15. UV spectrophotometer with temperature control capacity.
16. ^3H-radiolabeled octamer and nucleosomes (from Subheading 3.2.3).
17. NASH reagent: 0.2% (v/v) 2,4-pentanedione in 0.1 M acetic acid, 3.89 M ammonium acetate.
18. 100% trichloroacetic acid (TCA), stored at 4 °C.
19. 1-Pentanol.
20. Liquid scintillation cocktail.
21. Liquid scintillation counter.
22. SDS-PAGE apparatus.
23. 15% SDS-PAGE gels.
24. 4× SDS loading buffer: 200 mM Tris–HCl, pH 6.8, 200 mM DTT, 8% SDS, 40% glycerol, 0.4% bromophenol blue.
25. SDS-PAGE running buffer: 25 mM Tris base, 0.19 M glycine, 0.1% SDS (no pH adjustment).

26. Filter paper.
27. Transfer apparatus.
28. Nitrocellulose transfer membrane.
29. TBST: 20 mM Tris–HCl, pH 7.5, 140 mM NaCl, 0.05% Tween-20.
30. Primary antibodies against methylated histones (site- and methylation-status-, i.e., me3/me2/me1/me0, –specific).
31. HRP-conjugated second antibodies: 1:5000 dilution in 5% (w/v) milk in TBST.
32. Luminol solution: 50 mg Sodium Luminol, 62 μL 30% H_2O_2, 200 mL 0.1 M Tris–HCl, pH 8.6–8.7 (avoid light).
33. Enhancer: 11 mg p-coumaric acid in 10 mL DMSO (avoid light).
34. ChemiDoc imaging system.

3 Methods

3.1 Enzyme Purification

Histone demethylase preparation could arguably be the most critical component of successful enzyme discovery or assay outcomes. Many factors can impact the assay outcome, such as substrates used, other domains within the demethylases, enzyme folding, Fe^{2+}/-α-ketoglutarate co-factors, solubility, posttranslational modifications, and associated proteins. We provide a variety of options for demethylase purification to improve the chance to obtain potent enzymes. In addition, the histone demethylase assay is an oxidase assay. Free radical intermediates are generated, which might inactivate the demethylases themselves, and many oxidases might have low turnover numbers [23]. To avoid uncoupled radical oxidative damage at active sites, demethylases are purified as apo-enzymes and devoid of Fe^{2+}/α-ketoglutarate cofactors.

3.1.1 Purification of Recombinant GST Fusion Proteins from E. coli

Here we use *E. coli* recombinant GST fusion protein of yeast histone H3K36me3 demethylase RPH1 (KDM4) as an example. In addition, many histone methyltransferases (for octamer/nucleosome methylation in Subheading 3.2.3) can be purified as GST fusion proteins in a similar manner.

1. Amplify yeast genomic DNA with RPH1 primers and subclone the fragment into pGEX6p2 vector using TOP10 cells to give the construct GST-RPH1(1–373), with the JmjC domain at the C-terminus. For all recombinant proteins, their Fe^{2+} ion binding mutants (His to Ala mutation in the HxD/E motif) should be cloned with a site-directed mutagenesis kit and purified in parallel as negative controls [13] (*see* **Note 2**).
2. Transform the plasmids into BL21-CodonPlus (DE3)-RIL cells for overexpression.

3. Inoculate one single colony in 25 mL LB with antibiotics (Ampicillin). Grow the culture at 37 °C and 250 rpm overnight.
4. Next morning, transfer 10 mL overnight culture into 1 L LB culture with antibiotics. Continue growing at 37 °C. When OD600 reaches 0.6 in 2–3 h, cells are chilled in ice-water for 20 min and induced afterwards at 25 °C and 0.5 mM IPTG for 8 h. For new proteins, induction conditions can be optimized with varying temperatures (16–30 °C) and IPTG concentrations (0.1–1 mM).
5. Spin the cells at 7000 × *g* for 15 min. Freeze the pellets at −20 °C.
6. All downstream protein purification steps are performed at 4 °C. Thaw the cell pellet and resuspend it in 60–100 mL PBS lysis buffer.
7. Sonicate three rounds on ice. Set the sonicator amplitude/ power at a medium to high level, 60 cycles, 1 s on, 3 s off. Rest 5 min on ice before the next round. Spin down the lysate at 18,000 × *g* for 1–2 h. Verify protein expression and solubility on SDS-PAGE gels.
8. While waiting, wash 1–3 mL glutathione-agarose beads with water three times. Load the beads onto a glass column for batch purification. Wash the column twice with cold PBS. Use a stopcock to adjust the flow rate. Alternatively, an FPLC system and prepacked columns can be used for the GST fusion protein purification.
9. Load the supernatant from **step** 7 onto the column. Adjust the stopcock to let the solution flow very slowly (1 h for 100 mL). Load the flow-through back. Repeat twice. Wash the column with 100 mL PBS wash buffer.
10. Elute the proteins with 20 mL reduced glutathione (GST elution buffer). Run SDS-PAGE gel and perform Coomassie staining to check protein quality and purity.
11. Use Amicon Ultra centrifugal filter units (MWCO 30 kDa) to concentrate and perform buffer exchange with HEPES protein storage buffer. Use UV280 to quantify protein concentration.
12. Make small 50 μL aliquots and use liquid N_2 to snap freeze them for later single usage.

3.1.2 Purification of Recombinant Enzymes from Yeast

Recombinant yeast demethylases can be overexpressed from budding yeast and purified using an affinity column. Budding yeast has relatively simpler histone methylation patterns. Yeast overexpression and knockout strains can be used to predict histone demethylase substrates. The above yeast genetics approach can facilitate demethylase discovery [13]. In addition, as a eukaryotic organism, yeast preserves some protein posttranslational modifications. So recombinant proteins purified from yeast can be valuable enzyme sources. We adapt a yeast protein overexpression approach from a moveable ORF library [24]. The JHD1 (KDM2), RPH1 (KDM4), and JHD2 (KDM5) overexpression strains are constructed by transforming BY4743 with a GAL1-promoter-derived expression vector (BG1805) carrying the coding sequence of each gene and a His_6 tag [13]. Maintain yeast cultures at 30 °C and 200 rpm. Ni-NTA resin is used to purify the yeast proteins.

1. Day 1. In the afternoon, streak a SD-ura plate.
2. Day 3. Inoculate one single colony in 10 mL SD-ura (2% glucose) medium, 30 °C overnight.
3. Day 4. Transfer 2–3 mL overnight culture into 50 mL SC-ura medium (2% sucrose), grow 8–10 h. Dilute the culture into 300 mL SC-ura with 2% sucrose to give an initial OD600 of 0.02. Grow the culture overnight.
4. Day 5. When OD600 reaches 1.2, spin down the cells at 4000 × *g* for 5 min. Resuspend the cells in 300 mL of SC-ura with 2% galactose. Grow the cells under the induction condition for 4–10 h. Harvest the cells by centrifugation at 4000 × *g* for 10 min, weigh and freeze the cell pellet at −80 °C (*see* **Note 3**).
5. Day 6. Add 1/3 volume (3.3 mL/10 g pellet) lysis buffer A. Drip the slurry into liquid N_2 with a pipette to produce cell popcorns. Place the cell popcorns in a coffee grinder (pre-cooled at 4 °C). Grind for 3 min to yield a fine powder. Transfer the powder into a mortar which is sitting in a bath of liquid nitrogen. Grind the powder for 1 h. Store the powder at −20 °C or continue directly to the next step.
6. Mix powder with 1/3 volume Imidazole wash buffer I. Stir at 4 °C for 5 min, till it thaws. Centrifuge at 16,000 × *g* for 1 h.
7. Transfer the supernatant to pre-equilibrated Ni-NTA resin in a glass column in batch purification mode. Collect the flow through. Repeat once.
8. Wash with 50 mL each of Imidazole wash buffer I and II (20 mM and 50 mM Imidazole, respectively).
9. Elute with 10 mL Imidazole elution buffer. Collect fractions (1–2 mL each). Use Bradford reagent to monitor the elution process. Run SDS-PAGE gel, pool peak fractions.
10. Concentrate the protein and perform buffer exchange with HEPES protein storage buffer, using Amicon Ultra centrifugal filter units (MWCO 30 kDa). Quantify protein concentration by Bradford assay. Verify proteins by Western blots with anti-His_6 antibody or gene-specific antibodies.
11. Make small 10 μL aliquots and use liquid N_2 to snap freeze it for later single usage.

3.1.3 Purification of Recombinant Enzymes and Complexes from Sf9 Cells

For human and mouse histone-modifying enzymes, transfecting insect cells (Sf9 or Sf21) with recombinant baculovirus is a good option to purify active enzymes with good yield. In some cases, recombinant proteins purified from *E. coli* might not be active. Recombinant baculovirus protein purification from insect cells might be one of the few options to obtain active enzymes [25, 26]. Here we describe the generation of recombinant baculovirus from the pFastBac HT B vector and purification of His_6 or Flag-tagged enzymes from Sf9 cells. All insect culture work is performed in a 27 °C humidified incubator.

1. Clone the gene of interest into the pFastBac HT B plasmid, using TOP10 chemically competent cells. Catalytically inactive mutant plasmids are prepared with a site-directed mutagenesis kit.
2. Transform the pFastBac plasmids (*see* **Note 4**) containing the gene of interest and mutant counterparts into DH10Bac chemically competent *E. coli* cells. Grow them 2–3 days on SOC agar plates with kanamycin (50 μg/mL), gentamicin (7 μg/mL), tetracycline (10 μg/mL), X-gal (100 μg/mL), and IPTG (40 μg/mL).
3. Pick up large white colonies for plasmid purification. Use PCR with M13 Forward (−40) and M13 Reverse primer pair as well as M13 Forward and an internal primer pair to confirm the correct bacmid. A Taq DNA polymerase can be used for PCR. Store them at 4 °C.
4. Transfect 1 μg bacmid DNA into one million Sf9 cells with 10 μL ExpiFectamine Sf transfection Reagent in Opti-MEM in a well of a six-well plate. Incubate the cells for 72–96 h.
5. Check for signs of virus infection (*see* **Note 5**). Collect the medium and centrifuge at 500 × *g* for 5 min. Remove cells. Take the supernatant and filter it through a 0.2-μm filter. Store the P0 virus stock at 4 °C. Cover it with aluminum foil and avoid light.
6. Amplify the virus stock till P2 or P3 by successively adding preceding supernatants to new 72- to 96-h cultures and recovering new supernatants. Check expression in Sf9 whole-cell extracts by Western blot and Coomassie gel and choose desired titer (*see* **Note 6**).
7. Infect 500–1000 mL Sf9 cells in a few spinner flasks with SF-900 III SFM at two million cells/mL. 72 h Postinfection, harvest the cells by centrifugation. For Ni-NTA purification, resuspend the cells in 50–100 mL modified Lysis buffer A (adjust NaCl to 350 mM, add 10 mM Imidazole). Alternatively, the Ni-NTA step can be skipped and Flag purification can be performed directly; in this case, resuspend in BC350. Use a sonicator to lyse the cells. Sonicate the cells three rounds. Set the amplitude/power at a medium to high level as in Subheading 3.1.2.

8. If applicable, proceed with Ni-NTA purification as in Subheading 3.1.2 with modified lysis buffer A based solutions.
9. For Flag purification, incubate the Ni-NTA eluates or the lysate with 50–300 μL M2 beads for 4 h or overnight.
10. Wash with BC350 six times. Elute with 200 μL 0.2 mg/mL Flag peptide in BC100. Collect 5–7 fractions.
11. Run Coomassie gel and Western blot. Pool peak fractions. Concentrate the protein and perform buffer exchange with HEPES protein storage buffer with Amicon Ultra centrifugal filter units.
12. Quantify protein concentration by Bradford assay.
13. Snap freeze the protein in 10 μL aliquots in liquid N_2 and store at −80 °C for single usage.

3.1.4 Purification of Histone Demethylases and Enzyme Complexes from Mammalian Cells

In cases where enzyme complexes are involved in histone demethylation regulation, purification of Flag-tagged histone demethylase complexes might be a valuable approach. However, low yield and potential contamination could be major issues. Sensitive radioactivity assays and mutant controls will be needed. The following nuclear extraction step is adapted from Dignam and Roeder [27]. If necessary, recombinant proteins can be purified from nuclear pellet fractions [28].

1. Clone the enzyme of interest into a mammalian cell expression vector, for example, pFLAG-CMV-4 (*see* **Note 7**).
2. Transfect the plasmid into 293-F cells in a 10-cm plate with Lipofectamine 2000. In parallel, a catalytically inactive mutant plasmid should also be transfected (*see* **Note 8**).
3. 72 h later, split the cells at 1:100, 1:400, and 1:1000 into a 15-cm plate, and treat with 1 mg/mL G418.
4. Two weeks later, pick single colonies, expand the colonies with 200 μg/mL G418, and check protein expression level by Western blot. Select a colony with a high level of expression for further expansion.
5. Culture the cells in 15 cm plates. Scrape the cells off 20–40 plates and wash with five volumes of PBS.
6. Resuspend the cells in five volumes of buffer A. After 10 min on ice, lyse the cells with a glass Dounce homogenizer.
7. Spin down the cells at 1000 × *g* for 10 min. The supernatant is the S100 fraction (cytosol fraction).
8. Wash the pellet with five volumes of buffer A and spin down at 14,000 × *g* for 5 min. The pellet can be stored at −80 °C.
9. Resuspend the pellet in three volumes of buffer C with a glass Dounce homogenizer. Let sit for 20 min. Stir or pipet the mix occasionally.

10. Spin down at 14,000 × *g* for 30 min. Keep the supernatant as nuclear extract.
11. Quantify the nuclear extract protein concentration by Bradford assay.
12. Use liquid N_2 to snap freeze the extract at −80 °C.
13. Purify the protein with anti-Flag M2 beads (all washes should be done at 3500 × *g* for 3 min): Wash 100 μL M2 Flag beads with BC350 three times. Thaw the nuclear extract on ice. Spin at 14,000 × *g* for 10 min. Keep the supernatant as nuclear extract. Quantify protein concentration by Bradford assay. Mix 10 mg nuclear extract with the above processed beads. Rock the mix at 4 °C overnight.
14. Wash six times with BC350. Wash once with BC100. Transfer the beads into fresh tubes during the last wash. Proteins are eluted with 100 μL 0.2 mg/mL Flag peptide in BC100 buffer. Collect five elutions.
15. Analyze fractions by Western blot and pool peak fractions.
16. Concentrate the protein and perform buffer exchange against HEPES protein storage buffer with Amicon Ultra centrifugal units (MWCO 30 kDa). Snap-freeze the purified proteins from the nuclear extract fraction in small aliquots in liquid N_2 and store at −80 °C for single usage.

3.2 Substrate Preparation

3.2.1 Peptide Substrates

1. Dissolve the methylated peptides in water at 2.5 mM each. Store the peptide solutions at −20 °C (*see* **Note 9**).

3.2.2 Prepare Recombinant Octamer and Nucleosome

Recombinant *Xenous laevis* core histones H2A, H2B, H3, and H4 are purified following published procedures with minor modifications [29, 30].

1. Transform pET3-based plasmids of full-length H2A, H2B, H3, and H4 or H3C110A mutant into BL21 (DE3) pLys *E. coli* strain for overexpression.
2. Grow 1 L 2x YT culture each. At OD600 of 0.40, induce protein expression with 0.2–0.4 mM IPTG at 37 °C for 2 h.
3. Purify histone proteins via unfolding from the inclusion body. Resuspend cell pellets in 35 mL histone wash buffer, and sonicate three rounds at medium to high amplitude/power till the lysis solution is not sticky. Spin down at 12,000 × *g* for 30 min at 4 °C. Remove the supernatant. Resuspend the pellet in 75 mL histone wash buffer with extra 1% Triton X-100. Homogenize the mix with a glass Dounce homogenizer for 20 strokes. Spin down at 12,000 × *g* for 20 min at 4 °C. Remove the supernatant. Repeat the 1% Triton X-100 wash step once. Then repeat the wash step without Triton X-100. Weigh the pellet, resuspend at 2 mg/mL in histone unfolding buffer. Measure the absorbance at 276 nm and calculate histone concentrations.

4. Mix the four histones at equal molar ratio, at a total final concentration of 1 mg/mL and total volume of 6 mL. With a dialysis tubing, dialyze three times (>6 h each step) against 600 mL of histone refolding buffer at 4 °C.
5. Load proteins onto a S200 gel filtration column. Purify octamer from aggregate and dimer/tetramer. Run 15% SDS-PAGE gels to monitor the purity. Pool octamer fractions and concentrate them with Amicon Ultra centrifugal units (MWCO 30 kDa). Store octamer at 4 °C.
6. pG5E4 plasmid or a purified fragment [31], which contains multiple tandem repeats of 5S rDNA nucleosome positioning sequences is used for nucleosome assembling, at 0.9:1 ratio (DNA: Histones, or 1:1, 1.1:1 ratio). Add 4 M KCl into DNA to a final 2 M KCl concentration, then add histone octamers.
7. Transfer the DNA-protein mix into a small mini dialysis device on top of a floater. The floater is in 400 mL Reconstitution buffer (high salt) in a 2-L breaker with a stirring bar. Use a peristaltic pump to pump in 1600 mL TE buffer (low salt) at 1 mL/min for 1 h, then 2 mL/min overnight.
8. At the end of the gradient, dialyze against 400 mL TE buffer for 3 h. Store nucleosomes at 4 °C.

3.2.3 Methylation of Histone Octamers by In Vitro HMT Reactions

Histone methyltransferases can be purified recombinantly as described in Subheading 3.1. Detailed procedures of specific methylase purification can be found in related references [28, 32]. The enzymatic approach to prepare methylated histones is necessary for radioactivity-based histone demethylation assays.

1. In a 20 μL reaction, add 4 μL 5× HMT buffer, water, 2 μg histone octamer (nucleosomes), 1 μL 400 μM unlabeled SAM and 0.1–10 μg purified histone methyltransferase (*see* **Note 10**). Incubate at 30 °C for 1 h.
2. Confirm methylation status by Western blotting.
3. Following positive confirmation of methylation, scale up the reactions. Dialyze the reaction mixture with HEPES protein storage buffer.
4. For radiolabeled octamer or nucleosomes, in the reaction described in **step 1**, add additional 1 μL SAM [3H] (50–85 Ci/mmol, 0.55 mCi/mL). Confirm methylation by Western blots (*see* **Note 11**) and fluorography. For fluorography, separate the histones by 15% SDS-PAGE and transfer them into a PVDF membrane. Spray the membrane with 3H-ENHANCE and dry it in air. Use BioMax XAR film for exposure at −80 °C overnight. In addition, histone bands can be excised from the membrane and quantified with a liquid scintillation counter.

3.2.4 Histones with Methyl-Lysine Analog

Recently, Simon and Shokat developed a chemical method to site specifically install methyl-lysine analogs (MLAs) [19]. The four MLAs, Kc(me1), Kc(me2), Kc(me3), and Kc(me0) can be used to mimic me1/2/3/0, four different methylation states, with the only difference being a sulfur atom replacing the CH_2 group at the gamma position. The MLA histones have been reported to maintain methyl-histone binding and demethylation capacities. They can be quite useful, particularly when trimethylated lysine is not easily produced with enzymatic methylation steps. However, in certain cases, quantitative differences have been reported, when compared with natural methylated lysine [20]. Here we adapt the Simon and Shokat method. As an example, the following steps describe installation of analogs of trimethyl lysine, Kc(me3) at the histone H3 lysine 4 site.

1. H3K4C-C110A double mutant (or Lys to Cys mutants at other methylation sites) is generated from the pET3 plasmids with site-directed mutagenesis. The mutant histone is purified from the inclusion body as described in Subheading 3.2.2 (*see* **Note 12**).
2. After the S200 gel filtration step, the mutant histone fractions are pooled.

   Mutant histones are then dialyzed against water with 5 mM β-Mercaptoethanol and lyophilized.
3. Dissolve 10 mg lyophilized histone mutant protein in 950 μL alkylation buffer. Sonicate the protein mixture for 10–15 min if solubility is a problem. Add 20 μL fresh 1 M DTT, incubate at 37 °C for 1 h. Agitate occasionally.
4. Mix 100 mg of (2-bromoethyl)-trimethylammonium bromide solid with the above histone protein. Incubate the reactions at dark at 50 °C for 2.5 h. Flick every 30 min.
5. Add 10 μL freshly prepared 1 M DTT. Agitate the tube and let the reaction continue for another 2.5 h.
6. Stop the reaction by adding 50 μL β-mercaptoethanol. Dilute the reaction to 2.5 mL with alkylation buffer.
7. Purify the histone protein with a PD-10 column. Elute the protein with 2 mM β-mercaptoethanol in water.
8. Quantify protein concentration by measuring absorbance at 276 nm. Use reported extinction coefficients (H3 at 4040 M^{-1} cm^{-1} and H4 at 5040 M^{-1} cm^{-1}). Lyophilize the MLA histone.
9. Use the MLA histone to replace the parental histone, and assemble octamer and nucleosomes as described in Subheading 3.2.2.

3.3 Demethylation Assays

3.3.1 Peptide Mass Spectrometry Demethylase Assay

This is the preferred assay if enzymatic reaction is positive when peptide substrates are used. Demethylation extent can be assayed and quantified. Kinetic data can be obtained if time points are collected. Set up 60 μL reactions.

1. Prepare 200 mM ascorbic acid freshly. To prepare $(NH_4)_2Fe(SO_4)_2{\cdot}6H_2O$ solution, first make 100 mM $(NH_4)_2Fe(SO_4)_2{\cdot}6H_2O$ solution stock in 200 mM ascorbic acid solution by freshly dissolving the $(NH_4)_2Fe(SO_4)_2{\cdot}6H_2O$ salt into 200 mM fresh ascorbic acid solution. Afterwards, dilute the 100 mM $(NH_4)_2Fe(SO_4)_2{\cdot}6H_2O$ solution into 200 mM ascorbic acid solution by ten-fold. The resulting 10 mM $(NH_4)_2Fe(SO_4)_2{\cdot}6H_2O$ solution will be used as a 100× working solution (*see* **Note 13**).
2. Add reagents in the following order, 2× HEPES buffer, water, 0.8 μL 2.5 mM peptide, 1.5 μL 200 mM ascorbic acid, 1 μL 60 mM α-ketoglutarate, 1 μL DTT, 1–20 μg demethylase protein, 0.6 μL 10 mM $(NH_4)_2Fe(SO_4)_2{\cdot}6H_2O$. In parallel, set up controls without $(NH_4)_2Fe(SO_4)_2{\cdot}6H_2O$, enzyme, and include catalytically inactive mutants.
3. Incubate the reactions at 37 °C for 30–120 min. Stop the reactions by adding 1 μL of 100 mM EDTA. Freeze the samples at −20 °C till further analysis.
4. Dilute 1 μL of the above reaction mixture with 5 μL of saturated matrix solution. Spot the samples on MALDI target plates with triplicates for MALDI-TOF MS analysis. Use Flex-Analysis software to quantify methylation level changes (*see* **Notes 14** and **15**).

3.3.2 FDH (Formaldehyde Dehydrogenase) Assay

In histone demethylation reactions, formaldehyde is generated from the methyl group after removal. When formaldehyde dehydrogenase and NAD^+ are presented, formaldehyde will be oxidized into formic acid, and NADH will be produced. Because NADH has a UV absorption peak at 340 nm, the UV340 signal can be used to indirectly monitor the demethylation reaction [33]. This is particularly valuable in enzymatic kinetics and inhibitor studies.

1. In a 240 μL reaction, add 2× HEPES buffer, water, 6 μL 200 mM ascorbic acid, 4 μL 60 mM α-ketoglutarate, 10 μL 2.5 mM peptide, 11 μL 20 mg/mL NAD^+, 10 μL 0.01 U/μL FDH, 1–20 μg demethylase protein, 2.4 μL 10 mM $(NH_4)_2Fe(SO_4)_2{\cdot}6H_2O$ (*see* **Note 16**).
2. Incubate at 37 °C for 20 min; use a UV spectrometry to monitor absorption at 340 nm (*see* **Note 17**). Calculate the production of NADH and formaldehyde based on the extinction coefficient of NADH (6220 M^{-1} cm^{-1}).

3.3.3 Demethylation with Radioactivity Assay

At histone demethylase discovery stage, radioactivity monitoring of the methyl group is particularly valuable due to its sensitivity and fast readout. Since radioactive materials are involved, safety measures must be taken at all times. Formaldehyde produced in the demethylation reactions is separated from proteins and salts and reacted with NASH reagent [12, 34]. The product containing the radio label is enriched in the final product 3,5-diacetyl-1,4-dihydrolutidine (DDL) after 1-pentanol extraction. Radioactivity signals are recorded on a liquid scintillation counter.

1. Use 2 μg [3H]-radiolabeled methylated histone octamer or nucleosomes (Subheading 3.2.3) as the substrate. Follow a similar procedure to that used above in the peptide demethylation assay (Subheading 3.3.1, **steps 1** and **2**). Seal the reaction tubes with parafilm.
2. After 1–2 h at 37 °C, centrifuge the tubes at 1000 × *g* for 1 min at 4 °C. Stop the reactions by adding 25% volume of cold 100% (w/v) TCA to the reaction mixture to bring TCA concentration to 20%. Cool the samples at −20 °C for 5 min, then 4 °C for 30 min or longer. Spin at 20,000 × *g* or max speed for 10 min at 4 °C. Take the supernatant.
3. Add 1 volume of NASH reagent to the supernatant. Incubate the mix at 37 °C for 50 min. Add 1 volume of 1-pentanol and vortex. Spin down at 20,000 × *g* or max speed for 5 min.
4. Transfer the upper layer into a liquid scintillation cocktail. Record radioactivity signal on a liquid scintillation counter.

3.3.4 Demethylation with Western Blots

1. Use 2 μg methylated histone octamer or nucleosomes as the substrate. Follow similar procedure to that used above in the peptide demethylation assay (Subheading 3.3.1, **steps 1** and **2**) (*see* **Note 18**).
2. Stop the reactions by adding 2× SDS-PAGE loading buffer. Histones are separated on a 15% SDS-PAGE gel. Standard Western blots with modified histones are used to evaluate the demethylation process. Briefly, after transfer, the membrane is blocked with 5% milk in TBST for 30 min. After brief wash with water, the membranes are mixed with diluted first antibodies overnight on a rocker. Wash three times with TBST, each time 5–10 min. Rock the membranes with diluted second antibodies for 45 min. After similar three washes with TBST, wash with water for 1 min. Add freshly mixed substrate (5 mL luminol and 50 μL enhancer). Take images with a ChemiDoc imaging system with 5 s to 3 min exposure time.

4 Notes

1. When choosing substrates for demethylase assays, methylated histone peptides are preferred. Mass spectrometry-based peptide demethylase assay is a direct readout of demethylation and is quantitative. In addition, peptides with combinatorial modifications can be readily synthesized. Furthermore, substrate contamination is not a concern for peptide substrates. We suggest testing methylated histone peptides (12–21mer) with the methylated lysine site in the middle. Short peptides such as 12mers have simpler mass spectra and lower cost. For organisms with simpler methylation patterns, like yeast, 12mers might be sufficient. When testing human and mouse enzymes, longer peptides can be tested.
2. GST fusion proteins are preferred, as GST fusion protein purification can achieve higher purity with potential benefits of improved solubility and stability. Unlike Ni-NTA purification, where the metal ion Ni^{2+} is involved, GST fusion demethylases can be purified in apo form, which avoids potential metal interference. As for construct selection, we recommend testing both full-length proteins and the JmjC domain alone. For the few yeast JmjC proteins, we had solubility issues when full-length proteins were expressed even after multiple rounds of optimization. When designing catalytically inactive mutants, we recommend mutating Fe^{2+} binding sites, such as the HxD/E motif right after the second β-sheet [13].
3. At the enzyme discovery stage, it will be advantageous to use one set of cultures including WT controls to quantify histone methylation changes in vivo. Histones can be purified and quantified with mass spectrometry [13] or a panel of anti-methylated histone antibodies with Western blotting. Unlike in vitro enzymatic assays where the right substrates and purified active enzymes have to be combined to have positive readouts, properly controlled in vivo experiments can be used to rapidly discover histone demethylases and their substrate sites [13].
4. If histone demethylase or histone methyltransferase protein complexes are involved, co-purification the complex from Sf9 might be a good approach. We recommend transforming individual plasmids and obtaining virus P2 or P3 stocks separately. We suggest that the enzymatically active subunit has Flag-his tags. For protein expression, optimize ratio of individual virus stock for best yield and activity.
5. Include a control well without bacmid transfection for comparison. Examine the cells daily under microscope. Infected cells are larger with 20–50% increase in cell diameter. Other signs of infection include granular appearance, slow or no cell growth, as well as cell detachment and lysis.

6. It is critical to obtain a high titer of baculovirus for recombinant expression. Follow the Bac-to-Bac Baculovirus Expression system user guide. Maintain the Sf9 cells well. Adjust spinner setup and track cell doubling time. Keep the cell density at 0.5–4 million cells/mL and avoid higher density. Sterilize spinners well. P1 or P2 virus stock can also be obtained from Baculovirus CROs, including Kinnakeet Biotechnology.
7. Similar to GTS-fusion recombinants, catalytically inactive mutants are essential when assaying candidate enzymes purified from mammalian cells as the yield is much lower and the potential of co-purifying contaminant enzymes is much higher.
8. We strongly recommend performing separate transient transfection (including mutants) experiments in a proper mammalian cell line to test in vivo demethylase activity. A panel of anti-methylated histone antibodies can be used to evaluate histone methylation changes via immunofluorescence assays [25].
9. When dissolving peptides, make sure the peptides are completely dissolved. Consult with peptide synthesis vendors on solubility beforehand. Sonication or pH change might help in some cases.
10. Histone methyltransferases can be purified recombinantly as described in Subheading 3.1. Some enzymes from commercial sources can be used for in vitro methylation. Typical enzymes are the following: GST-SET7/9, MLL1 complex, GST-G9a, GST-SET2, GST-DOT1L, GST-SUV4-20H1, GST-PR-SET7, and PRC2 complex from Sf9 purification. When selecting the methyltransferases, it is important to consider lysine site, methylation extent, octamer/nucleosome preference, and enzyme activity. In most of the reactions, the products are octamer/nucleosomes with mixed states of methylation. Western blots can be used to evaluate the product methylation pattern. In some cases, purified octamers or nucleosomes from mammalian cells can be used [28], but potential contamination in the octamer/nucleosome preparations might complicate data interpretation. In those situations, mutant enzymes need to be tested as a negative control.
11. If Western blots are used in subsequent demethylation assays, it will be advantageous to achieve higher levels of methylation at this step. If resources are available, histone semi-synthesis via NCL and EPL can be deployed to achieve precise and high levels of methylation. The chemical strategies can also introduce combinatorial modifications, which is quite useful if demethylation is dependent on modification status near the methylation site.

12. For the other three MLAs, refer to the published procedure for details [35]. It is necessary to check if there are Cys residues present other than the methylation site to avoid unintended installation.
13. Fe^{2+} salt will be easily oxidized in water. Dissolve the iron salt directly in a fresh ascorbic acid solution to maintain the active Fe^{2+} form. Use a freshly prepared iron solution each time. α-Ketoglutarate and DTT can be stored at −20 °C for repeated usage.
14. Standard desalting with C18 ZipTips is not necessary. Some peptides might not retain well on C18 ZipTips. In this assay, to have good quality of mass spectrometry data, we minimize introducing Na^{+}, K^{+}, or other unnecessary metal ions in the final assay buffer.
15. Peptide mass spectrometry demethylation assay is quantitative for positive results. However, a negative result does not mean the potential demethylases under test are inactive. We suggest that a full panel of peptides (listed in Subheading 2.2) be used for testing. Furthermore, additional modifications at neighboring sites might have regulatory roles, particularly if the demethylase has other chromatin binding domains or belongs to a chromatin-binding protein complex. In those cases, combinatorially modified peptides can be tested. It is also possible that octamer or nucleosome substrates are needed for the demethylase activities.
16. If there is enough ascorbic acid in the buffer, Fe^{2+} will not interfere with FDH assay; otherwise, oxidation of Fe^{2+} will make the data noisy.
17. FDH assay can be used to check reaction kinetics; however, caution must be taken to include negative controls, and UV 340 nm detection requires robust enzymes for this assay.

    The reaction is performed at 37 °C by using a UV spectrometer with temperature control capacity. At lower temperatures, the reaction is slow and NADH signal might be weak. Before testing demethylases, it will be helpful to set up standard HCHO FDH reactions with a series of HCHO dilutions to figure out the detection limit and signal ranges. Increase the substrate and enzyme concentrations or reduce reaction volume, if the signal is not high enough. It is advised to check UV signal stability during incubation.
18. This assay might be valuable when methyl-lysine analogs (MLAs) are used as the substrates. In addition, purified octamers or nucleosomes from mammalian cells can be used, if potential contamination issues are properly addressed. Since no specialized instruments or radioactivity are involved, this assay can easily be implemented in most labs. We suggest this assay when the histone demethylases involved are relatively

potent. Most methylated histone antibodies (site and methylation state specific) are readily available. Ideally for each lysine site of interest, four antibodies (against me0–3) are used. For quantification purposes, it is strongly recommended to use antihistones (e.g., H3) as loading controls. Antibody specificity can be pre-validated by peptide blotting.

Acknowledgments

The author is very grateful to Dr. Ming-Daw Tsai and all Tsai lab members, OSU Campus Chemical Instrument Center, and Dr. Danny Reinberg lab.

References

1. Luger K, Mäder AW, Richmond RK, Sargent DF, Richmond TJ (1997) Crystal structure of the nucleosome core particle at 2.8 Å resolution. Nature 389:251–260. https://doi.org/10.1038/38444
2. Suganuma T, Workman JL (2011) Signals and combinatorial functions of histone modifications. Annu Rev Biochem 80:473–499. https://doi.org/10.1146/annurev-biochem-061809-175347
3. Brownell JE, Zhou J, Ranalli T, Kobayashi R, Edmondson DG, Roth SY, Allis CD (1996) Tetrahymena histone acetyltransferase a: a homolog to yeast Gcn5p linking histone acetylation to gene activation. Cell 84:843–851. https://doi.org/10.1016/S0092-8674(00)81063-6
4. Taunton J, Hassig CA, Schreiber SL (1996) A mammalian histone deacetylase related to the yeast transcriptional regulator Rpd3p. Science 272:408–411. https://doi.org/10.1126/science.272.5260.408
5. Kouzarides T (2007) Chromatin modifications and their function. Cell 128:693–705
6. Li B, Carey M, Workman JL (2007) The role of chromatin during transcription. Cell 128:707–719
7. Margueron R, Trojer P, Reinberg D (2005) The key to development: interpreting the histone code? Curr Opin Genet Dev 15:163–176
8. Rea S, Eisenhaber F, O'Carroll D, Strahl BD, Sun ZW, Schmid M, Opravil S, Mechtier K, Ponting CP, Allis CD, Jenuwein T (2000) Regulation of chromatin structure by site-specific histone H3 methyltransferases. Nature 406:593–599. https://doi.org/10.1038/35020506
9. Zhang Y, Reinberg D (2001) Transcription regulation by histone methylation: interplay between different covalent modifications of the core histone tails. Genes Dev 15:2343–2360
10. Cheng X, Blumenthal RM (2010) Coordinated chromatin control: structural and functional linkage of DNA and histone methylation. Biochemistry 49:2999–3008
11. Shi Y, Lan F, Matson C, Mulligan P, Whetstine JR, Cole PA, Casero RA, Shi Y (2004) Histone demethylation mediated by the nuclear amine oxidase homolog LSD1. Cell 119:941–953. https://doi.org/10.1016/j.cell.2004.12.012
12. Tsukada YI, Fang J, Erdjument-Bromage H, Warren ME, Borchers CH, Tempst P, Zhang Y (2006) Histone demethylation by a family of JmjC domain-containing proteins. Nature 439:811–816. https://doi.org/10.1038/nature04433
13. Tu S, Bulloch EMM, Yang L, Ren C, Huang WC, Hsu PH, Chen CH, Liao CL, Yu HM, Lo WS, Freitas MA, Tsai MD (2007) Identification of histone demethylases in Saccharomyces cerevisiae. J Biol Chem 282:14262–14271. https://doi.org/10.1074/jbc.M609900200
14. Helin K, Dhanak D (2013) Chromatin proteins and modifications as drug targets. Nature 502:480–488
15. Mosammaparast N, Shi Y (2010) Reversal of histone methylation: biochemical and molecular mechanisms of histone demethylases. Annu Rev Biochem 79:155–179
16. Højfeldt JW, Agger K, Helin K (2013) Histone lysine demethylases as targets for anticancer therapy. Nat Rev Drug Discov 12:917–930

17. Kaniskan HÜ, Martini ML, Jin J (2018) Inhibitors of protein methyltransferases and demethylases. Chem Rev 118:989–1068
18. Dimitrova E, Turberfield AH, Klose RJ (2015) Histone demethylases in chromatin biology and beyond. EMBO Rep 16:1620–1639. https://doi.org/10.15252/embr.201541113
19. Simon MD, Chu F, Racki LR, de la Cruz CC, Burlingame AL, Panning B, Narlikar GJ, Shokat KM (2007) The site-specific installation of methyl-lysine analogs into recombinant histones. Cell 128:1003–1012. https://doi.org/10.1016/j.cell.2006.12.041
20. Simon MD, Shokat KM (2012) A method to site-specifically incorporate methyl-lysine analogues into recombinant proteins. In: Methods in enzymology. Academic Press Inc., Cambridge, pp 57–69
21. Holt M, Muir T (2015) Application of the protein semisynthesis strategy to the generation of modified chromatin. Annu Rev Biochem 84:265–290
22. Forneris F, Binda C, Vanoni MA, Mattevi A, Battaglioli E (2005) Histone demethylation catalysed by LSD1 is a flavin-dependent oxidative process. FEBS Lett 579:2203–2207. https://doi.org/10.1016/j.febslet.2005.03.015
23. Vaillancourt FH, Yin J, Walsh CT (2005) SyrB2 in syringomycin E biosynthesis is a nonheme FeII α-ketoglutarate- and O2-dependent halogenase. Proc Natl Acad Sci U S A 102:10111–10116. https://doi.org/10.1073/pnas.0504412102
24. Gelperin DM, White MA, Wilkinson ML, Kon Y, Kung LA, Wise KJ, Lopez-Hoyo N, Jiang L, Piccirillo S, Yu H, Gerstein M, Dumont ME, Phizicky EM, Snyder M, Grayhack EJ (2005) Biochemical and genetic analysis of the yeast proteome with a movable ORF collection. Genes Dev 19:2816–2826. https://doi.org/10.1101/gad.1362105
25. Klose RJ, Yan Q, Tothova Z, Yamane K, Erdjument-Bromage H, Tempst P, Gilliland DG, Zhang Y, Kaelin WG (2007) The retinoblastoma binding protein RBP2 is an H3K4 demethylase. Cell 128:889–900. https://doi.org/10.1016/j.cell.2007.02.013
26. Cao R, Zhang Y (2004) SUZ12 is required for both the histone methyltransferase activity and the silencing function of the EED-EZH2 complex. Mol Cell 15:57–67. https://doi.org/10.1016/j.molcel.2004.06.020
27. Dignam JD, Lebovitz RM, Roeder RG (1983) Accurate transcription initiation by RNA polymerase II in a soluble extract from isolated mammalian nuclei. Nucleic Acids Res 11:1475–1489. https://doi.org/10.1093/nar/11.5.1475
28. Fang J, Wang H, Zhang Y (2003) Purification of histone methyltransferases from HeLa cells. Methods Enzymol 377:213–226. https://doi.org/10.1016/S0076-6879(03)77012-8
29. Luger K, Rechsteiner TJ, Richmond TJ (1999) Preparation of nucleosome core particle from recombinant histones. Methods Enzymol 304:3–19. https://doi.org/10.1016/S0076-6879(99)04003-3
30. Dyer PN, Edayathumangalam RS, White CL, Bao Y, Chakravarthy S, Muthurajan UM, Luger K (2003) Reconstitution of nucleosome Core particles from recombinant histones and DNA. Methods Enzymol 375:23–44. https://doi.org/10.1016/S0076-6879(03)75002-2
31. Neely KE, Hassan AH, Wallberg AE, Steger DJ, Cairns BR, Wright APH, Workman JL (1999) Activation domain-mediated targeting of the SWI/SNF complex to promoters stimulates transcription from nucleosome arrays. Mol Cell 4:649–655. https://doi.org/10.1016/S1097-2765(00)80216-6
32. Nishioka K, Reinberg D (2003) Methods and tips for the purification of human histone methyltransferases. Methods 31:49–58. https://doi.org/10.1016/S1046-2023(03)00087-2
33. Lizcano JM, Unzeta M, Tipton KF (2000) A spectrophotometric method for determining the oxidative deamination of methylamine by the amine oxidases. Anal Biochem 286:75–79. https://doi.org/10.1006/abio.2000.4782
34. Kleeberg U, Klinger W (1982) Sensitive formaldehyde determination with NASH's reagent and a "tryptophan reaction". J Pharmacol Methods 8:19–31. https://doi.org/10.1016/0160-5402(82)90004-3
35. Simon MD (2010) Installation of site-specific methylation into histones using methyl lysine analogs. Curr Protoc Mol Biol, Chapter 21: Unit 21.18.1-10

Part III

Exploring the Regulation of Histone Methyltransferase Enzymatic Activity

Chapter 4

Preparation and Characterization of Chromatin Templates for Histone Methylation Assays

Cuifang Liu, Jicheng Zhao, and Guohong Li

Abstract

In eukaryotic cells, chromatin plays an important role in gene regulation by controlling the access of the transcription machinery to DNA. In this chapter, we will describe methods for generating different chromatin templates to investigate the impact of histone variants and chromatin structure on histone methyltransferase activities. For this purpose, we take Polycomb Repressive Complex 2 (PRC2) as an example and investigate how its activity on H3K27me3 is regulated by the histone variants H3.3 and H2A.Z and higher-order chromatin structure.

Key words Chromatin structure, Histone variant, H3.3, H2A.Z, PRC2

1 Introduction

Eukaryotic chromatin is hierarchically organized from its fundamental repeating unit, the nucleosome, into a more condensed higher-order chromatin. Chromatin structure must be precisely and dynamically regulated to modulate transcription, DNA replication, and other DNA-related processes by different epigenetic mechanisms [1]. For instance, histone variant deposition/replacement has been shown to play critical roles in gene transcription and DNA replication through modulating higher-order chromatin structure and histone methyltransferase activities [2–6]. In contrast to canonical histones, histone variants are synthesized and incorporated into chromatin in a replication-uncoupled manner. Histone variant H2A.Z has been reported to stabilize the nucleosome, but H3.3 can modulate the effect of H2A.Z on nucleosome stability, which leads to reduced stability for nucleosomes containing both H2A.Z and H3.3 [6–10]. In addition, H3.3 and H2A.Z have been reported to influence PRC2 activity by modulating chromatin compaction in vitro and in mES cells [5, 11, 12]. Moreover, it was reported that PRC2 and Polycomb Repressive Complex

Raphaël Margueron and Daniel Holoch (eds.), *Histone Methyltransferases: Methods and Protocols*,
Methods in Molecular Biology, vol. 2529, https://doi.org/10.1007/978-1-0716-2481-4_4,

1 (PRC1) activities are regulated by nucleosome density and H1-dependent chromatin compaction, respectively [13, 14]. Therefore, in this chapter, we describe in vitro methods for generating different chromatin templates to investigate the impact of histone variants (H3.3 and H2A.Z) and higher-order chromatin structure on PRC2 activity.

2 Materials

2.1 Expression and Purification of Recombinant Histones

1. pET expression plasmids: pET3a-H2A, pET3a-H2B, pET3a-H3, pET3a-H4, pET3a-H2A.Z, pET3a-H3.3, pET3a-HA-H2A, pET3a-HA-3 × FLAG-H2A, pET28a-H1e.
2. BL21 (DE3) pLys cells.
3. LB-agar plates and LB medium with ampicillin (100 μg/mL) and chloramphenicol (25 μg/mL).
4. 37 °C shaking incubator.
5. Spectrophotometer.
6. IPTG (Isopropyl-β-D-thiogalactoside).
7. Wash buffer: 10 mM Tris–HCl, pH 7.5, 100 mM NaCl, 1 mM EDTA, 1 mM benzamidine.
8. Wash buffer containing 1% Triton X-100.
9. Sonicator.
10. Unfolding buffer: 7 M guanidine–HCl, 20 mM Tris–HCl, pH 7.5, 10 mM DTT.
11. High-salt buffer: 1 M NaCl, 50 mM Tris–HCl, pH 7.5, 5 mM 2-mercaptoethanol.
12. Materials for dialysis: dialysis bag (intercept molecular weight 8 kDa).
13. Buffer A: 50 mM Tris–HCl, pH 8.0, 0.5 M NaCl.
14. Buffer B: 50 mM Tris–HCl, pH 8.0, 1.0 M NaCl.
15. AKTA purification system with fraction collector.
16. HiTrap Heparin HP column.
17. Materials for SDS-PAGE and Coomassie staining:

    For SDS-PAGE gel: 30% acrylamide, 1.5 M Tris–HCl (pH 6.8), 10% SDS, 10% APS, 1 M Tris (pH 8.8), TEMED.

    For Coomassie staining: 10% acetic acid, 50% methanol, 1.25 g/500 mL Coomassie R250 dye, filtered.
18. BC100 buffer: 100 mM NaCl, 20 mM Tris–HCl, pH 7.5, 20% glycerol, 1 mM PMSF.

2.2 Purification of Different DNA Templates

1. Plasmids: pWM530-12× 177-601, pWM530-4× 177-601.
2. LB medium and TB medium with ampicillin (100 μg/mL).
3. DH5α cells.
4. Buffer S1: 25 mM Tris–HCl, pH 8.0, 50 mM glucose, 10 mM EDTA.
5. Buffer S2: 0.2 M NaOH, 1% SDS at room temperature.
6. Buffer S3: 3 M KOAc, 2 M HOAc, pH 5.2.
7. Gauze for filtration.
8. Isopropyl alcohol.
9. Ethanol.
10. TE buffer: 10 mM Tris–HCl, pH 8.0, 1 mM EDTA.
11. RNase A.
12. 4 M NaCl.
13. 40% PEG 6000.
14. Phenol:chloroform (1:1).
15. Chloroform:isoamyl alcohol (24:1).
16. 3 M NaOAc (pH 5.2).
17. 1% agarose gel.
18. Restriction enzymes: EcoRV, AlwNI, BsaI, DraIII, AvaI.

2.3 Reconstitutions and Purifications of Histone Octamers

Histone octamers include canonical octamers (H2A/H2B/H3/H4, HA-H2A/H2B/H3/H4, HA-3 × FLAG-H2A/H2B/H3/H4), H2A.Z-octamers (H2A.Z/H2B/H3/H4), H3.3-octamers (H2A /H2B /H3.3/H4), H2AK119ub1-octamers (H2AK119ub1/H2B/H3/H4).

1. Refolding buffer: 2 M NaCl, 10 mM Tris–HCl, pH 7.5, 1 mM EDTA, 5 mM 2-mercaptoethanol.
2. Dialysis bag (intercept molecular weight 8 kDa).
3. Sizing column (Superdex 200, GE).
4. AKTA purification system with fraction collector.
5. Materials for SDS-PAGE and Coomassie staining:

   For SDS-PAGE gel: 30% acrylamide, 1.5 M Tris–HCl (pH 6.8), 10% SDS, 10% APS, 1 M Tris (pH 8.8), TEMED.

   For Coomassie staining: 10% acetic acid, 50% methanol, 1.25 g/500 mL Coomassie R250 dye, filtered.

2.4 Nucleosome and Chromatin Fiber Assembly

1. TEN buffer: 10 mM Tris–HCl, pH 8.0, 2 M NaCl, 1 mM EDTA.
2. TE buffer: 10 mM Tris–HCl, pH 8.0, 1 mM EDTA.
3. Dialysis tube (Millipore, D-TubeTM Dialyzer Mini-MWCO 6–8 kDa).

4. Peristaltic pump.
5. HE buffer: 10 mM HEPES, pH 8.0, 0.1 mM EDTA.
6. T4 DNA ligase (NEB).
7. Ligation Buffer: 70 mM Tris–HCl, pH 7.5, 5 mM $MgCl_2$, 1 mM ATP, and 10 mM DTT.
8. TE buffer supplemented with 0.6 M NaCl.

2.5 Characterizing the Chromatin Structure Using Electron Microscopy (EM)

1. 0.4% glutaraldehyde (Sigma).
2. HE buffer: 10 mM HEPES, pH 8.0, 0.1 mM EDTA.
3. 40 mM spermidine (20×).
4. Grids.
5. Ethanol.
6. Tungsten filament wire.
7. Filter papers.
8. 2% Uranylacetate.
9. FEI Tecnai G2 Spirit 120 kV transmission electron microscope.

2.6 Analyzing Chromatin Structure Using Sedimentation Velocity of Analytical Ultracentrifugation (AUC)

1. Beckman Coulter ProteomeLab XL-I.
2. Ultrascan II software.

2.7 HMT Assay

1. 5× Methylation buffer: 0.25 M Tris–HCl, pH 8.5, 25 mM $MgCl_2$.
2. DTT.
3. ^{3}H-SAM.
4. Enzyme.
5. 30 °C incubator.
6. 5× SDS loading buffer: 10% SDS, 500 mM DTT, 50% Glycerol, 250 mM Tris–HCl, and 0.5% bromophenol blue dye, pH 6.8.
7. Materials for SDS-PAGE gel: 30% acrylamide, 1.5 M Tris–HCl (pH 6.8), 10% SDS, 10% APS, 1 M Tris (pH 8.8), TEMED.
8. PVDF membrane.
9. Methanol.
10. Coomassie solution: 10% acetic acid, 50% methanol, 1 g/500 mL Coomassie R250 dye, filtered.
11. Destaining solution: 10% acetic acid, 50% methanol.
12. EN3HANCE spray (Perkin Elmer).
13. ^{3}H sensitive film (GE).

3 Methods

3.1 Expression and Purification of Recombinant Histones

The following protocol is used as previously described with some modifications [15]. Both canonical histones (H2A, H2B, H3, H4, HA-H2A and HA-3 × FLAG-H2A) and H2A/H3 variants (H2A. Z, H3.3) have been subcloned into the pET vector series, and histone H1e has been subcloned into pET28a plasmid with 6 × His tag. All the proteins form inclusion bodies when they are expressed at high levels in bacteria. Both canonical histones and H2A/H3 variants are purified under denaturing conditions, but histone H1e is purified under native conditions.

3.1.1 For the Preparation of Canonical Histones and H2A/H3 Variants

1. Transform BL21 (DE3) pLys cells with the pET-histone expression plasmids and plate on LB-agar plates with ampicillin and chloramphenicol. Incubate at 37 °C overnight.
2. Inoculate single colonies in 5 mL of LB medium containing ampicillin and chloramphenicol, at 37 °C overnight. Take about five different colonies to test for induction (*see* **Note 1**). Induction condition is described in Luger et al. [15].
3. Select the best colony and inoculate in 4.5 L LB medium containing ampicillin and chloramphenicol. Shake at 37 °C until the OD600 is 0.5–0.7.
4. Induce with 0.5 mM IPTG for 4 h at 37 °C.
5. Harvest the cells by centrifugation at room temperature. At this step, the cells can be stored at −20 °C.
6. Resuspend the cell pellets with 100 mL Wash buffer.
7. Sonicate cell suspension on ice, 5 s on/8 s off, 400 W, 99 times.
8. Centrifuge at 39191 × *g* for 20 min at 4 °C. The pellet contains inclusion bodies of the corresponding histone proteins.
9. Wash the pellet by completely resuspending in 100 mL Wash buffer plus 1% Triton X-100 followed by sonication (5 s on/8 s off, 400 W, 99 times) on ice. Spin for 10 min at 4 °C and 39191 × *g*.
10. Repeat this step once with Wash buffer plus 1% Triton X-100 and twice with Wash buffer without Triton X-100. After the last wash, the drained pellet can be stored at −20 °C.
11. Resuspend the pellet with 30 mL Unfolding buffer and stir gently for 1 h at room temperature.
12. Collect the dissolved material by centrifugation at 39191 × *g* for 30 min at 20 °C. The purified histones (Fig. 1) can be kept at−80 °C for long-term storage.

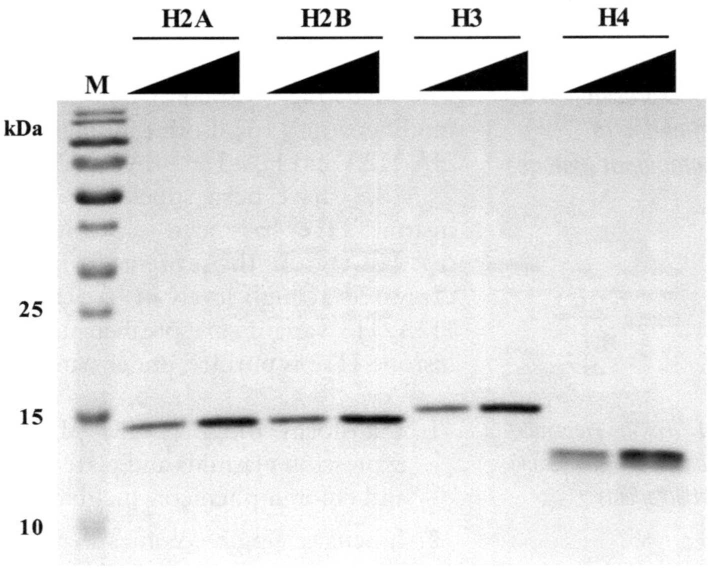

Fig. 1 Four canonical histones (H2A, H2B, H3, H4) analyzed by 15% SDS-PAGE

*3.1.2 For the Preparation of H1e, the Expression Condition Is the Same as Canonical Histones (Subheading 3.1.1, **steps 1–5**): The Purification Protocol of H1e*

1. Resuspend the cell pellets with 100 mL Wash buffer.
2. Sonicate cell suspension on ice, 5 s on/8 s off, 400 W, 99 times.
3. Centrifuge at 39191 × *g* for 20 min at 4 °C. The pellet contains inclusion bodies of H1e protein.
4. Resuspend the pellet with 30 mL High-salt buffer and stir gently for 1 h at room temperature.
5. Collect the dissolved material by centrifugation at 39191 × *g* for 30 min at 4 °C.
6. Collect the supernatant and dialyze in Buffer A for 3 h.
7. Load onto a Heparin column equilibrated in Buffer A. Wash the column with Buffer A until the absorbance reaches the baseline, then elute H1e with Buffer B through the AKTA purification system.
8. Analyze the fractions of H1e by 15% SDS-PAGE and keep fractions with higher protein concentration and higher purity.
9. Dialyze selected fractions against BC100 buffer and store at −80 °C.

   Of note, the site-specifically modified histones (methylation, ubiquitination, etc.) are synthesized by native chemical ligation [16–18].

3.2 Purification of Different DNA Templates

3.2.1 12× 177 bp 601 DNA

1. Inoculate a single colony of DH5α transformed with plasmid pWM530-12× 177-601 (Fig. 2) in 50 mL LB medium for overnight culture from a freshly transformed plate (*see* **Note 2**).
2. Amplify the culture by transferring 30 mL of the primary culture into 1.5 L TB medium with ampicillin at 37 °C for 4–5 h, and then incubate the culture at 42 °C, for about 12–13 h.

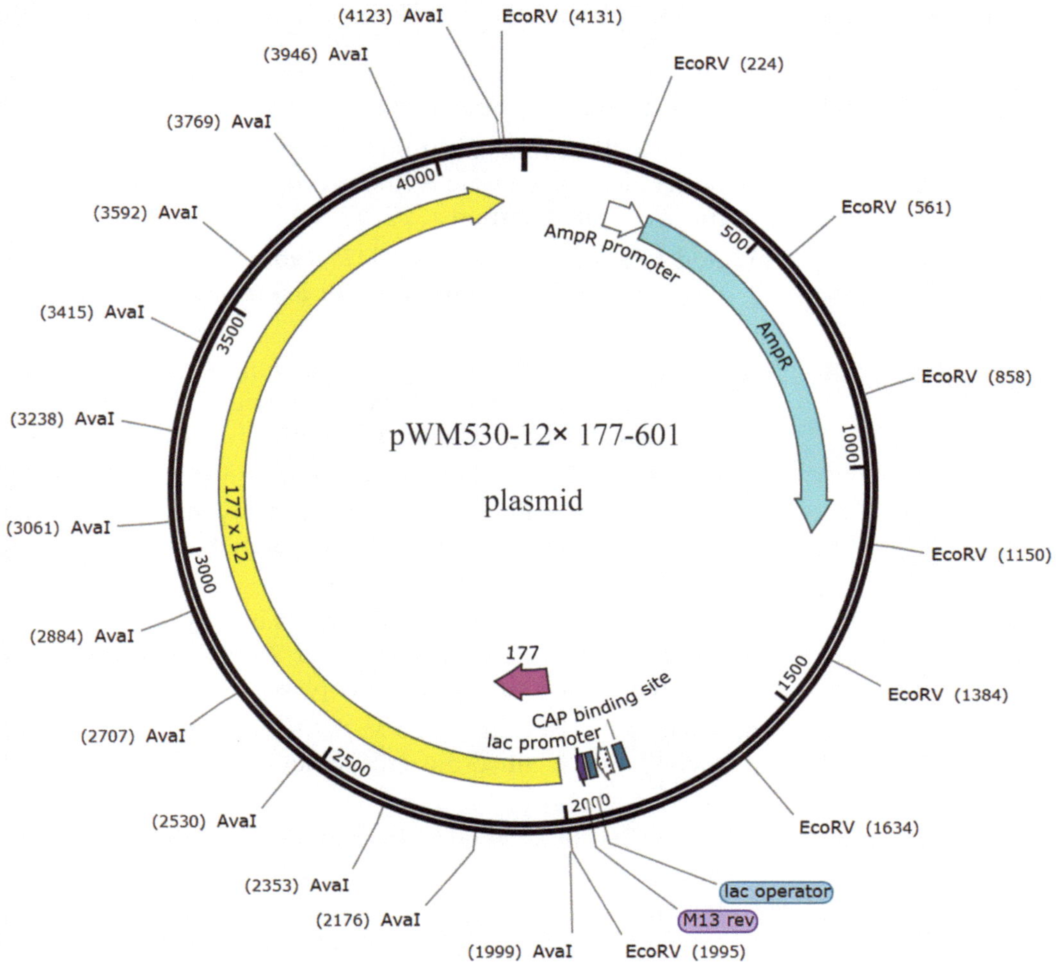

Fig. 2 pWM530-12× 177-601 plasmid map. Unique restriction sites are indicated

3. Harvest the cells by centrifugation at 4670 × *g* for 30 min at room temperature.
4. Resuspend the pellet with 60 mL cold Buffer S1 and vortex to form a uniform suspension.
5. Add 120 mL fresh Buffer S2 (*see* **Note 3**), turn gently upside down to fully mix, leave at room temperature for 5 min (not more than 10 min).
6. Then carefully add 210 mL cold Buffer S3, and mix it in a direction to make a homogeneous suspension, and incubate on ice for 10 min.
7. Centrifuge at 4 °C 4670 × *g* for 30 min, carefully aspirate supernatant, remove suspended impurities by four layers of gauze filtration, transfer to 2 L beaker.
8. To recover plasmid DNA, add 0.52 times volume of isopropyl alcohol into the sample, fully mix and keep at room temperature for 15 min.

9. Centrifuge at 15,000 × *g* 15 min to pellet the DNA, then rinse tube wall and DNA pellet with 70% ethanol, and spin at 15,000 × *g* 10 min at 4 °C. Discard the supernatant so that residual ethanol is evaporated.
10. Remove RNA contamination by dissolving DNA in 30 mL TE buffer plus 3 mg RNase A (to final concentration 100 μg/mL) for overnight digestion at 37 °C.
11. Add 1/5 volume of 4 M NaCl, 2/5 volume of 40% PEG 6000, mix evenly in 37 °C water bath for 5 min and then keep on ice for 30 min.
12. Centrifuge at 20,000 × *g* for 15 min (4 °C), discard the supernatant (containing RNA), wash with 70% ethanol once, then dissolve the precipitate in 20 mL TE buffer.
13. In order to remove protein contamination and PEG, add 1/5 volume of phenol:chloroform (1:1), vortex and mix and spin at 20 °C and 20,000 × *g* for 10 min, carefully collect the upper water phase, repeat this step twice.
14. Then add 1/5 volume of chloroform:isoamyl alcohol (24:1), vortex to mix and spin at 20 °C at 20,000 × *g* for 10 min. Carefully collect the upper water phase.
15. Then add 1/10 volume of 3 M NaOAc (pH 5.2) and 2.5 times the volume of anhydrous ethanol and keep at −20 °C for at least 7–8 h.
16. Spin at 20,000 × *g* and 4° for 15 min. The DNA precipitate is then washed with 70% ethanol, dried and dissolved in the TE buffer.
17. Check the sample by gel electrophoresis with 1% agarose gel. The plasmid DNA obtained from this step can be kept at −20 °C for long-term storage.
18. The plasmids are digested by EcoRV, and then the 12× 177 bp 601 DNA fragments are purified through the PEG precipitation (Fig. 3) or gel extraction, which can be kept at −20 °C for long-term storage.

3.2.2 4× 177 bp 601 DNA for Sequential Ligation

1. 4× 177 bp 601 DNA with AlwNI or BsaI or DraIII sites are designed as described previously [19], which are synthesized and ligated into pWM530 vectors.
2. Individual DNA plasmids are purified as described above, then they are digested with corresponding enzymes (AlwNI, BsaI and DraIII) to release the 4× 177 bp 601 DNA fragments from plasmids.
3. The 4× 177 bp 601 DNA fragments are purified through PEG precipitation.

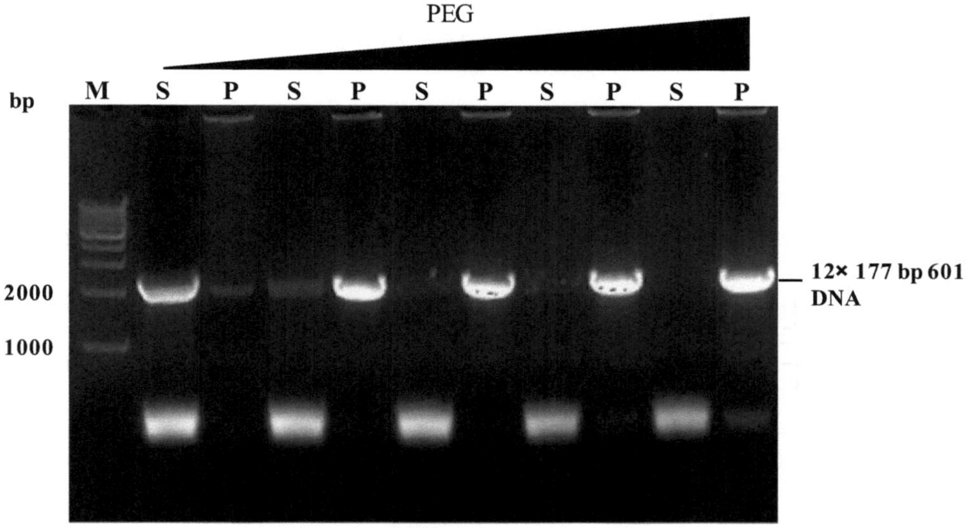

Fig. 3 PEG precipitation to isolate 12× 177 bp DNA. *S* supernatant, *P* precipitate

3.2.3 177 bp DNA with 601 Sequence

There are two ways to get the 177 bp DNA with 601 sequence.

1. AvaI enzyme sites are located between every neighbor 177 bp DNA, so the DNA can be obtained by ScaI digestion of the purified 12× 177 bp 601 DNA fragments.
2. The single copy of the 177 bp DNA can also be purified by AvaI digestion of the pWM530-12× 177-601 plasmid, followed by PEG precipitation or gel extraction.

3.3 Reconstitution and Purification of Histone Octamers

1. Mix the four histone proteins (H2A, H2B, H3, H4) in equimolar ratios (*see* **Note 4**), then dialyze at 4° against at least 2 changes of 2 L Refolding buffer. The second dialysis step should be performed overnight.
2. Centrifuge to remove precipitation. Concentrate the sample according to the volume necessary for the sizing column.
3. Purify the histone octamers from aggregates and dimers/tetramers on a sizing column (Superdex 200, GE) and analyze the fractions by 15% SDS-PAGE (*see* **Note 5**) (Fig. 4).
4. After pooling, determine the concentration. Octamers can be kept at −80 °C for long-term storage or at 4 °C for short-term storage.

3.4 Nucleosome and Chromatin Fiber Assembly

Nucleosomes and chromatin fibers are assembled using the salt-dialysis method as previously described [6]. The reconstitution samples are obtained by mixing histone octamers and DNA in high-salt buffer (2 M NaCl) and slowly lowering the concentration of NaCl to 0.6 M by pump dialysis (Fig. 5).

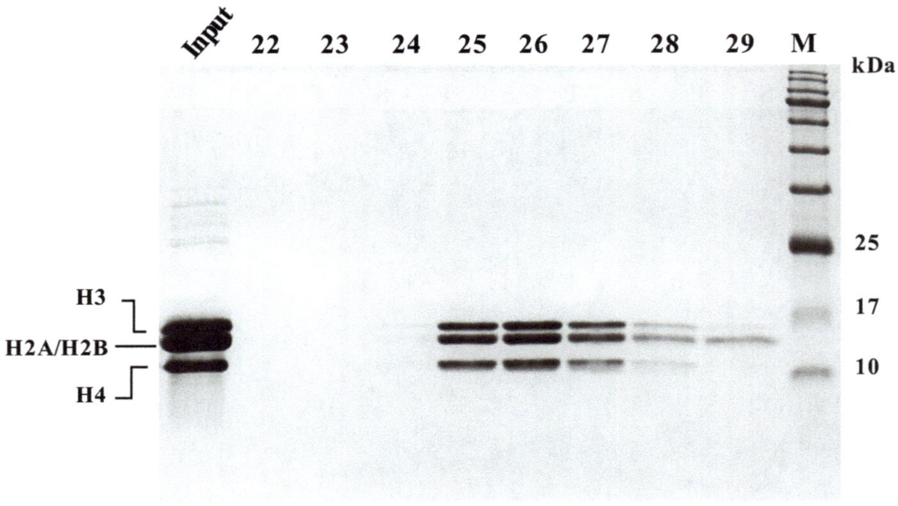

Fig. 4 Purification of histone octamers through Superdex 200

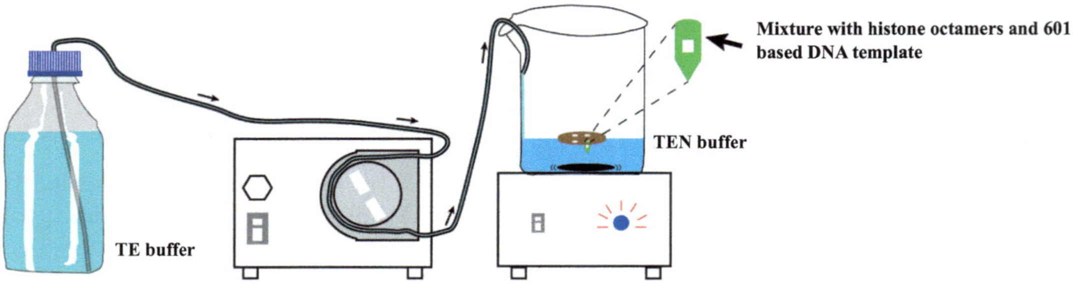

Fig. 5 Nucleosome and chromatin fiber reconstitution system

1. Mix the histone octamers and mono-177 bp DNA or 4× 177 bp 601 DNA or 12× 177 bp 601 DNA templates at the ratio about 1:1 in TEN buffer. Put the mix into dialysis tube.
2. Dialyze the mix in TEN buffer at 4 °C.
3. Then slowly pump TE buffer into the dialysis buffer to decrease the salt concentration to 0.6 M NaCl after 16–19 h. For linker histone H1e incorporation (*see* **Note 6**), H1e with different molar ratios relative to the mono-nucleosomes is added at this step and further dialyzed in TE buffer with 0.6 M NaCl for 3 h.
4. The mixture is further dialyzed in HE buffer for 4 h. The reconstituted nucleosomes and chromatin fibers are characterized by Electron Microscopy (EM) images and Analytical ultracentrifugation (AUC) investigation (Subheadings 3.5 and 3.6).
5. Heterotypic nucleosome arrays are generated as previously reported [19]:
   (a) 3 sequential 4× polynucleosome chromatin blocks (Fig. 6a) [14] are ligated by T4 DNA ligase in the ligation buffer and incubated 4 h at 16 °C.

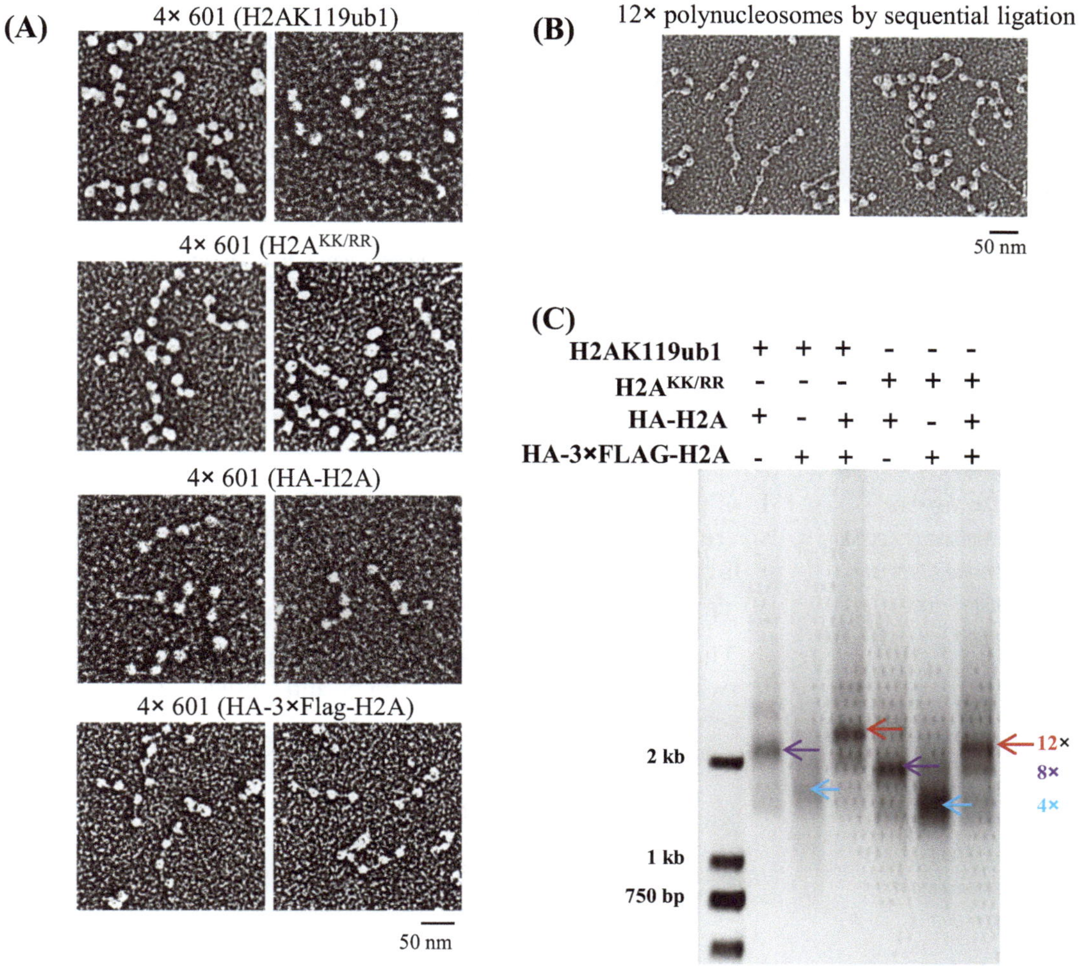

Fig. 6 Chromatin sequential ligation (adapted from [14]). (**a**) EM images of 4× nucleosome arrays. (**b**) EM images of 12× nucleosome arrays after chromatin ligation. (**c**) Ligated nucleosome arrays analyzed by 1% agarose gel

(b) Ligations are then centrifuged at 17,000 × *g* at 4 °C for 10 min to collect 12× polynucleosomes.

(c) The chromatin precipitation is dissolved in TE buffer and dialyzed with 2 L TE buffer overnight.

(d) The ligation products are analyzed by EM and 1% agarose gel (Fig. 6b, c) [14].

(e) For H1e-compacted chromatin, ligated chromatin is dialyzed against 0.5 L TE + 0.6 M NaCl, then H1e is added and further dialyzed for 3 h, and finally the chromatin is dialyzed in 500 mL HE buffer for 4 h.

3.5 Characterizing the Chromatin Structure Using Electron Microscopy (EM)

3.5.1 Metal Shadowing with Tungsten to Analyze the Mono-Nucleosomes and Nucleosome Array

1. Reconstituted nucleosome samples are fixed with 0.4% glutaraldehyde (Fluka) in HE buffer on ice for 30 min. The DNA concentrations of nucleosome samples are 1 μg/mL (*see* **Note 7**).
2. After fixation, 2 mM spermidine is added into the sample solution to enhance the absorption of the nucleosomes to the grids.
3. Then samples are loaded to the glow-discharged carbon-coated EM grids, incubated for 2 min and then blotted.
4. The grids are washed stepwise in 20 mL baths of 0%, 25%, 50%, 75%, and 100% ethanol solution for 4 min, each at room temperature, air-dried, and then shadowed with tungsten at an angle of 10° with rotation.

3.5.2 The Negative Staining Method Is Applied to Analyze the Chromatin Fibers with Histone H1e Incorporation

1. Chromatin samples in fixative solution (0.4% glutaraldehyde) are incubated on glow-discharged carbon-coated EM grids for 1 min (*see* **Note 8**).
2. The excess sample solution is removed using filter papers.
3. The grid is incubated in 2% uranylacetate for staining for 30 s twice, blotted with filter papers and allowed to air-dry for several minutes.

3.5.3 The Samples Are Evaluated Under a FEI Tecnai G2 Spirit 120 kV Transmission Electron Microscope (Fig. 7)

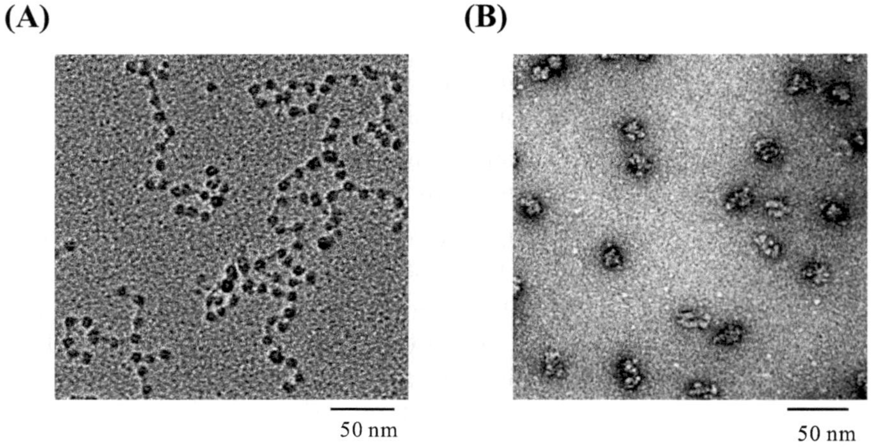

Fig. 7 EM images of 12× nucleosome arrays by metal shadowing (**a**) and 30-nm chromatin fiber by negative staining (**b**)

3.6 Analyzing Chromatin Structure Using Sedimentation Velocity of Analytical Ultracentrifugation (AUC)

AUC is a powerful technique for in vitro structural investigation. As the rate of sedimentation of a molecule depends on both its shape and its size, AUC can be used to analyze chromatin of any size.

1. Prepare nucleosome arrays or chromatin fibers compacted by H1e.
2. Sedimentation experiments are performed on a Beckman Coulter ProteomeLab XL-I using a four-hole An-60Ti rotor. The samples with the initial absorbance at 260 nm of about 0.5–0.8 are equilibrated for 2 h at 20 °C under vacuum in the centrifuge before sedimentation (*see* **Note 9**).
3. Absorbance at 260 nm is measured in continuous scan mode during sedimentation at 20,000 rpm in 12-mm double-sector cells.
4. The data are analyzed by enhanced van Holde–Weischet analysis using Ultrascan II software (*see* **Note 10**) (Fig. 8).

3.7 HMT Assay

1. Mix water, 5× Methylation buffer and DTT in EP tubes for each reaction according to Table 1.
2. Add nucleosome substrate and enzyme into the upper mixture and finally add ^{3}H-SAM for each reaction according to Table 1 (*see* **Notes 11** and **12**).
3. Pipet well up and down to uniformly mix all reactions. Then incubate at 30 °C for 1 h.
4. Stop reaction by adding 5× SDS loading buffer and heat at 100 °C for 5 min.

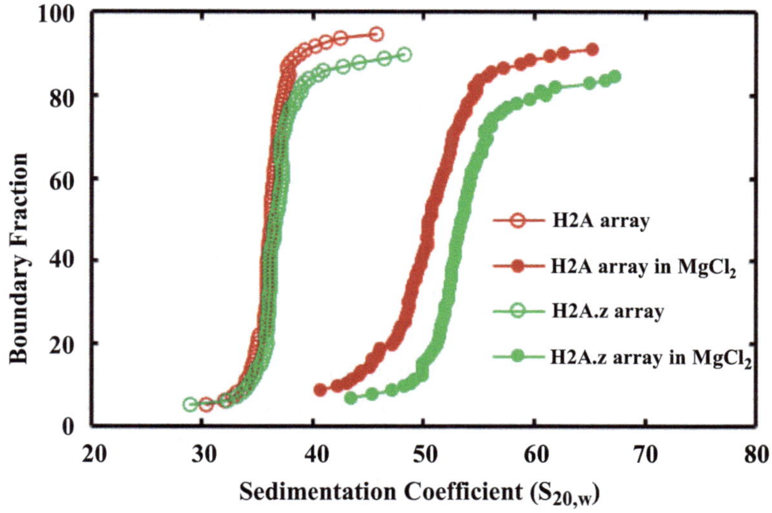

Fig. 8 AUC to monitor the effect of H2A.Z on chromatin compaction

Table 1
Methylation reaction

Agents	Volume
5× methylation buffer	5 μL
^{3}H-SAM	1.5 μL (0.5 μM final)
Nucleosome/Core histone (1–3 μg final)	*X* μL
Enzyme (10–200 ng)	*Y* μL
0.2 M DTT	0.5 μL
Water	*Z* μL
Total	25 μL

5. The reaction products are loaded on a 15% SDS-PAGE and transferred onto PVDF membrane (activate the membrane in methanol before use).
6. Stain the membrane with Coomassie solution for 5 min.
7. Destain with Destaining solution, 10–20 min.
8. Dry the membrane. Scan or take a picture at this step.
9. Spray the membrane with EN3HANCE spray (Perkin Elmer) under the hood, and let the membrane dry (you can use a hair dryer to save some time).
10. Expose 12–24 h at −80 °C with ^{3}H sensitive film (GE) (*see* **Notes 13** and **14**).

4 Notes

1. A small-scale expression test should be done before large-scale expression and purification of proteins. The entire process of protein purification should be carried out under low temperatures, except for unfolding with 7 M guanidine-HCl.
2. Before large-scale expression and purification, check whether the number of repetitive sequences is correct by restriction digestion using EcoRV.
3. Buffer S2 should be prepared at room temperature to prevent SDS precipitation.
4. When mixing the four histones, you can add more H2A and H2B to reduce the formation of tetramers and hexamers, which cannot be separated from octamers.
5. Octamers should be kept at 0–4 °C to avoid complex dissociation. Check the peak fractions by 15% SDS-PAGE, choose the fractions according to the stoichiometry of octamers. High-

quality octamers are generally within the second peak. Clean the column with NaOH or acetic acid as recommended by the supplier.

6. H1e must be added at 600 mM NaCl, so it can be assembled to the correct position under this condition.
7. The amount of 12× nucleosome array used for metal shadowing should be 20–40 ng.
8. At least 300 ng chromatin fiber is used for negative staining, and the staining time can be adjusted according to the effect.
9. For the best data quality, the initial absorbance of the samples at 260 nm is about 0.5–0.8 OD.
10. $S_{20,w}$ values (sedimentation coefficient corrected for water at 20 °C) are calculated with a partial specific volume of 0.622 mL/g for chromatin and buffer density and viscosity adjusted. The average sedimentation coefficients are determined at the boundary midpoint.
11. Final salt concentration is critical, so if your sample is in high-salt buffer, dialyze it prior to the enzymatic assay. Final salt concentration should not exceed 75 mM for maximum efficiency.
12. pH is also important, and HMT activity is higher when pH increased.
13. To re-expose the membrane, change the cassette or let it dry well.
14. We use different chromatin substrates to perform the HMT assay to investigate the impact of histone variants or chromatin structure on methyltransferase activity. For example, at the mono-nucleosome level histone variant H2A.Z enhances Suv4-20h1 activity [4] (Fig. 9a). However, H2A.Z-mediated stimulation of PRC2 activity is observed for oligo-nucleosomal substrates [5] (Fig. 9b). Histone variant H3.3 does not enhance PRC2 activity, but it can counteract the H2A.Z-mediated stimulation of PRC2 activity [5] (Fig. 9b). In principle, H2A.Z can promote PRC2 enzymatic activity by facilitating chromatin folding to generate a preferred substrate for PRC2; however, the stimulation is alleviated by H3.3 which counteracts H2A.Z-mediated chromatin compaction [5]. These investigations suggest that histone variants regulate methyltransferase activity by modulating higher-order chromatin structure. Indeed, several studies have shown that PRC2 activity is enhanced by dense chromatin or H1-containing chromatin [13, 20]. Besides, H1 can also stimulate Suv39h1 activity [21]. Previous studies have shown that H2AK119ub1 can recruit Jarid2-PRC2 and RYBP/YAF2-PRC1 to facilitate the deposition of H3K27me3 and H2AK119ub1 on histone

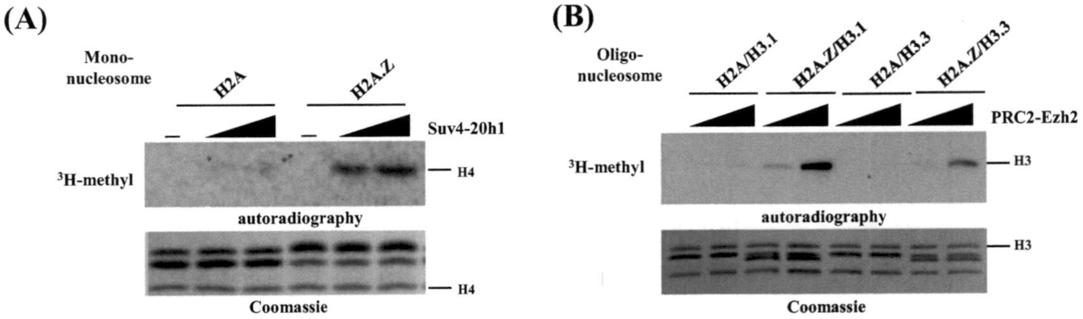

Fig. 9 Effect of H2A.Z on Suv4-20h1 and PRC2 activity. (**a**) H2A.Z enhances Suv4-20h1 activity at mono-nucleosome level. (**b**) H2A.Z enhances PRC2 activity at oligo-nucleosome level (adapted from [5])

tails [14, 22]. Here, we can use sequentially ligated 12× Nucleosome arrays or H1-compacted chromatin fibers to test whether H2AK119ub1 can drive a proximal-distal spreading of H3K27me3 and whether H1-dependent chromatin compaction could speed up the spreading of H3K27me3.

Acknowledgments

This work was supported by grants from the Ministry of Science and Technology of China (2017YFA0504202 to G.L.; 2019YFA0508903 to J.Z.), the National Natural Science Foundation of China (31991161,31521002 to G.L.; 32070604 to J.C.) and the Beijing Municipal Science and Technology Committee (Z201100005320013 to G.L.). The work was also supported by the CAS Key Research Program on Frontier Science (QYZDY-SSW-SMC020 to G.L.), the Chinese Academy of Sciences (CAS) Strategic Priority Research Program (XDB19040202), an HHMI International Research Scholar grant (55008737) to G.L.

References

1. Chen P, Li G (2010) Dynamics of the higher-order structure of chromatin. Protein Cell 1: 967–971
2. Jin C, Zang C, Wei G, Cui K, Peng W, Zhao K, Felsenfeld G (2009) H3.3/H2A.Z double variant-containing nucleosomes mark nucleosome-free regions' of active promoters and other regulatory regions. Nat Genet 41: 941–945
3. Talbert PB, Henikoff S (2010) Histone variants — ancient wrap artists of the epigenome. Nat Rev Mol Cell Biol 11:264–275
4. Long H, Zhang L, Lv M, Wen Z, Zhang W, Chen X, Zhang P, Li T, Chang L, Jin C (2020) H2A.Z facilitates licensing and activation of early replication origins. Nature 577:576–581
5. Wang Y, Long H, Yu J, Dong L, Wassef M, Zhuo B, Li X, Zhao J, Wang M, Liu C, Wen Z, Chang L, Chen P, Wang Q, Xu X, Margueron R, Li G (2018) Histone variants H2A.Z and H3.3 coordinately regulate PRC2-dependent H3K27me3 deposition and gene expression regulation in mES cells. BMC Biol 16(107)
6. Chen P, Zhao J, Wang Y, Wang M, Long H, Liang D, Huang L, Wen Z, Li W, Li X (2013) H3.3 actively marks enhancers and primes gene transcription via opening higher-ordered chromatin. Genes Dev 27:2109–2124
7. Park YJ, Dyer PN, Tremethick DJ, Luger K (2004) A new fluorescence resonance energy transfer approach demonstrates that the

histone variant H2AZ stabilizes the histone octamer within the nucleosome. J Biol Chem 279:24274–24282

8. Hoch DA, Stratton JJ, Gloss LM (2007) Protein–protein frster resonance energy transfer analysis of nucleosome core particles containing H2A and H2A.Z. J Mol Biol 371:971–988
9. Jin C, Felsenfeld G (2007) Nucleosome stability mediated by histone variants H3.3 and H2A.Z. Genes Dev 21:1519–1529
10. Thakar A, Gupta P, Ishibashi T, Finn R, Silva-Moreno B, Uchiyama S, Fukui K, Tomschik M, Ausio J, Zlatanova J (2009) H2A.Z and H3.3 histone variants affect nucleosome structure: biochemical and biophysical studies. Biochemistry 48:10852–10857
11. Creyghton MP, Markoulaki S, Levine SS, Hanna J, Lodato MA, Sha K, Young RA, Jaenisch R, Boyer LA (2008) H2AZ is enriched at Polycomb complex target genes in ES cells and is necessary for lineage commitment. Cell 135:649–661
12. Banaszynski L, Wen D, Dewell S, Whitcomb S, Lin M, Diaz N, Elssser S, Chapgier A, Goldberg A, Canaani E, Rafii S, Zheng D, Allis CD (2013) Hira-dependent histone H3.3 deposition facilitates PRC2 recruitment at developmental loci in ES cells. Cell 155: 107–120
13. Yuan W, Wu T, Fu H, Dai C, Wu H, Liu N, Li X, Xu M, Zhang Z, Niu T (2012) Dense chromatin activates Polycomb repressive complex 2 to regulate H3 lysine 27 methylation. Science 337:971–975
14. Zhao J, Wang M, Chang L, Yu J, Song A, Liu C, Huang W, Zhang T, Wu X, Shen X et al (2020) RYBP/YAF2-PRC1 complexes and histone H1-dependent chromatin compaction mediate propagation of H2AK119ub1 during cell division. Nat Cell Biol 22:439–452
15. Luger K, Mader AW, Richmond RK, Sargent DF, Richmond TJ (1997) Crystal structure of the nucleosome core particle at 2.8 a resolution. Nature 389:251–260
16. He S, Bauman D, Davis JS, Loyola A, Nishioka K, Gronlund JL, Reinberg D, Meng F, Kelleher N, Mccafferty DG (2003) Facile synthesis of site-specifically acetylated and methylated histone proteins: reagents for evaluation of the histone code hypothesis. Proc Natl Acad Sci U S A 100:12033–12038
17. Blanco-Canosa J, Dawson P (2007) An efficient Fmoc-SPPS approach for the generation of thioester peptide precursors for use in native chemical ligation. Angew Chem 47:6851–6855
18. Fierz B, Kilic S, Hieb AR, Luger K, Muir TW (2012) Stability of nucleosomes containing homogenously Ubiquitylated H2A and H2B prepared using Semisynthesis. J Am Chem Soc 134:19548–19551
19. Muller MM, Fierz B, Bittova L, Liszczak G, Muir TW (2016) A two-state activation mechanism controls the histone methyltransferase Suv39h1. Nat Chem Biol 12:188–193
20. Martin CC, Cao R, Zhang Y (2006) Substrate preferences of the EZH2 histone methyltransferase complex. J Biol Chem 281:8365–8370
21. Lu X, Wontakal SN, Kavi H, Kim BJ, Guzzardo PM, Emelyanov AV, Xu N, Hannon GJ, Zavadil J, Fyodorov DV (2013) Drosophila H1 regulates the genetic activity of heterochromatin by recruitment of Su(var)3-9. Science 340:78–81
22. Kalb R, Latwiel S, Baymaz HI, Jansen PW, Muller CW, Vermeulen M, Muller J (2014) Histone H2A monoubiquitination promotes histone H3 methylation in Polycomb repression. Nat Struct Mol Biol 21:569–571

Chapter 5

Techniques to Study Automethylation of Histone Methyltransferases and its Functional Impact

Luis Popoca and Chul-Hwan Lee

Abstract

The catalytic activity of histone methyltransferases is not restricted to histones but also includes noncanonical substrates. Increasing evidence shows that histone methyltransferases methylate themselves, and automethylation has emerged as a self-regulatory mechanism. Here, we introduce experimental procedures to identify automethylation sites of histone methyltransferases and to investigate the function of automethylation in a reconstituted biochemical system and in cellular contexts.

Key words Histone methyltransferase assay, Automethylation, Self-regulatory mechanism, Regulation of catalytic activity, Active conformation

1 Introduction

The catalytic activity of histone methyltransferase (HMT) is regulated by many factors including self-methylation, so-called automethylation. Several histone lysine methyltransferases methylate themselves (Table 1). These include PRC2 (Polycomb repressive complex 2), SUV39H2 (Suppressor of variegation 3–9 homolog 2), CLR4 (the fission yeast homolog of the mammalian SUV39H1 and SUV39H2 enzymes), and G9a [1–6]. Interestingly, these enzymes have a characteristic in common in that they catalyze repressive histone modifications. PRC2 catalyzes H3K27 methylation, and H3K27me3 is a hallmark of facultative heterochromatin. Other methyltransferases mentioned above methylate H3K9, and this facilitates the formation of constitutive heterochromatin. Recent studies reveal that automethylation positively regulates the catalytic activity of histone methyltransferases on chromatin. Therefore, automethylation of methyltransferases is considered as a self-regulatory mechanism [1, 2, 4]. When automethylation sites are unmethylated or hypo-methylated, the lysine residue occupies the substrate pocket and interferes with histone substrate accessibility. On the other

Raphaël Margueron and Daniel Holoch (eds.), *Histone Methyltransferases: Methods and Protocols*,
Methods in Molecular Biology, vol. 2529, https://doi.org/10.1007/978-1-0716-2481-4_5,

Table 1
Automethylation of histone lysine methyltransferases

	Histone methylation site (s)	Automethylation site(s)	Function of automethylation	References
PRC2	H3K27	EZH2 K510, K514, K515 (EZH1 K511, K515, K516)	Self-activation	[1, 2]
SUV39H2	H3K9	K392	Self-activation	[3]
CLR4 (yeast SUV39H)	H3K9	K455	Self-activation	[4]
G9a	H3K9	K239	Facilitate HP1 binding	[5, 6]

hand, high levels of automethylation promote histone substrate accessibility as automethylation sites are expelled from the substrate pocket.

In this chapter, the method is focused on the automethylation of histone lysine methyltransferases (Fig. 1). However, these procedures can be applied to other posttranslational modifications (PTMs) of histone methyltransferases.

2 Materials

2.1 Histone Methyltransferase Assay

1. 5× histone methyltransferase (HMT) buffer: 250 mM Tris–HCl at pH 8.5, 25 mM $MgCl_2$.
2. *S*-Adenosylmethionine (SAM) (*see* **Note 1**).
3. Radiolabeled SAM (^{3}H-SAM or ^{14}C-SAM) (*see* **Note 2**).
4. Recombinant histone methyltransferases (*see* details in Ref. 7).
5. Reconstituted nucleosomes (*see* details in Ref. 8).
6. 4× STOP buffer: 0.2 M Tris–HCl at pH 6.8, 20% glycerol, 10% SDS, 10 mM β-mercaptoethanol, 0.05% Bromophenol blue.
7. EDTA.
8. Metal block.
9. 0.6-mL microcentrifuge tubes.
10. Heat block.
11. Buffer compatible with liquid chromatography-mass spectrometry (LC-MS).
12. Access to LC-MS equipment or facility.

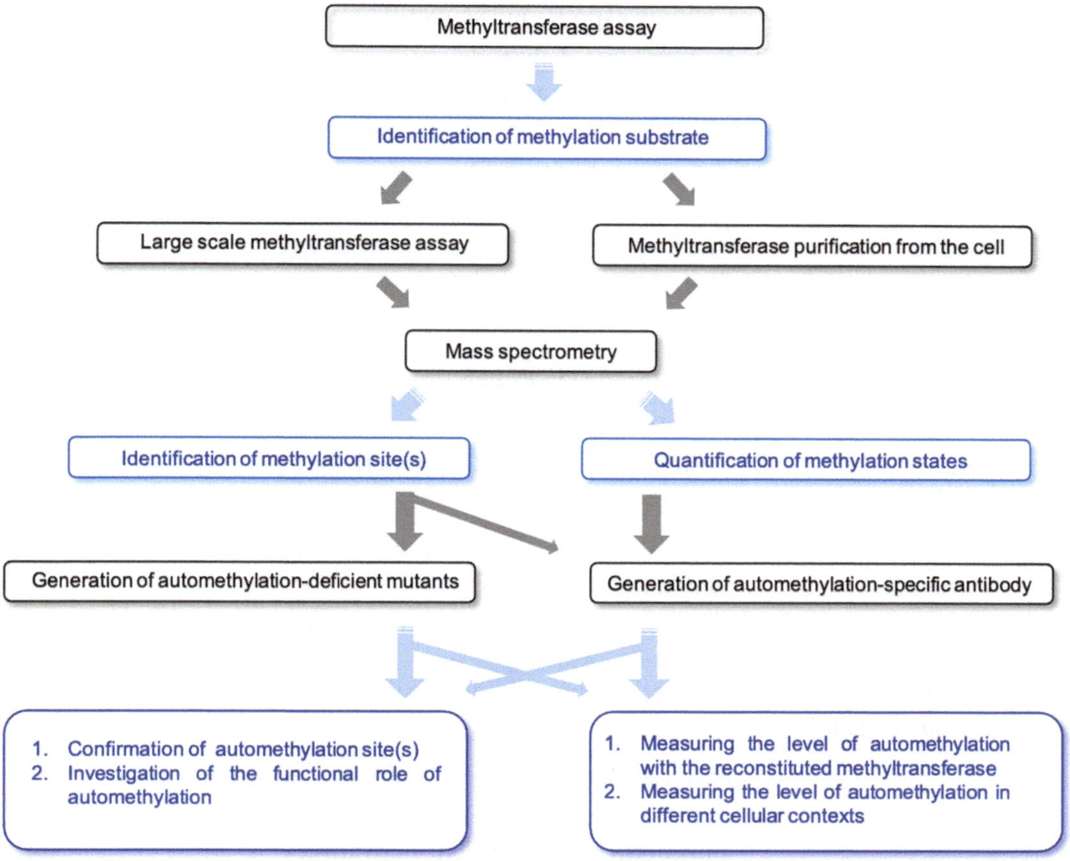

Fig. 1 The flow of experiments to investigate the role of automethylation

2.2 SDS-PAGE Gel Running, Western Blot, and Autoradiography

1. SDS-polyacrylamide gel (*see* details in Ref. 9).
2. SDS-PAGE running buffer: SDS-PAGE running buffer: 0.025 M Tris–HCl at pH 8.3, 0.192 M glycine, 0.1% SDS.
3. Coomassie Staining buffer: 0.025% (w/v) Brilliant blue R, 50% Methanol, 40% Glacial acetic acid, 10% distilled water.
4. Destaining buffer: 50% Methanol, 40% Glacial acetic acid, 10% distilled water.
5. PVDF membrane.
6. Methanol.
7. Western blot transfer buffer: 0.025 M Tris–HCl, 0.192 M glycine, 20% methanol.
8. X-ray film.
9. Biomax Transcreen intensifying screener (*see* **Note 3**).
10. Liquid scintillation cocktail solution.
11. Liquid scintillation vials.
12. Scintillation counter.

2.3 Immuno-precipitation

1. Cell line expressing the methyltransferase of interest.
2. Low-volume dialysis device.
3. Buffer compatible with liquid chromatography-mass spectrometry (LC-MS).
4. Access to LC-MS equipment or facility.

If using epitope tag:

5. Anti-epitope agarose beads.
6. Epitope peptide.

If not using epitope tag:

7. Antibodies.
8. Protein A or Protein G magnetic beads.
9. Magnet apparatus.
10. Ag/Ab (Antigen/Antibody) elution buffer: 0.1 M glycine–HCl, pH 2.5–3.0.
11. Neutralization Buffer: 1 M Tris–HCl; pH 8–9.

2.4 Antibody Generation and Testing

1. Synthesized peptides for production and testing of antibodies.
2. Access to antibody production company.
3. Materials for Enzyme-Linked Immunosorbent Assay (ELISA).
4. Cell line expressing the methyltransferase of interest, and automethylation-deficient derivative.

3 Methods

3.1 Identification of Methylation Substrate Using Methyltransferase Assay

1. Prechill 0.6-mL microcentrifuge tubes by putting them on the ice-chilled metal block.
2. Prepare all components on the ice metal block (*see* **Note 4**).
3. Make a spreadsheet for the assay. The example of a spreadsheet is shown in Table 2.
4. Make a master mix for components that are distributed in all samples (*see* **Note 5**).
5. Serial-dilute the methyltransferase (*see* **Note 6**).
6. Add the desired amount of recombinant methyltransferases to each tube (*see* **Note 7**).
7. (Optional) Add reconstituted nucleosomes if measuring automethylation in the presence of nucleosomes (*see* **Note 8**).
8. Add master mix to wall or cap of the tube and centrifuge the droplet to the bottom of the tube to incorporate the components simultaneously.
9. Centrifuge tubes briefly and tap several times to mix the reaction. Centrifuge again.
10. Incubate samples at 30 °C for 1 h (*see* **Note 9**).

Table 2
An example of a spreadsheet for methyltransferase assay

				Reaction					
		Stock concentration	Final concentration	#1	#2	#3	...	#14	Master Mix[b]
HMT buffer		5×	1×	3 μL	3 μL	3 μL		3 μL	=3 × 15 = 45 μL
DTT		100 mM	4 mM	0.6 μL	0.6 μL	0.6 μL		0.6 μL	=0.6 × 15 = 9 μL
SAM	Cold Radiolabeled (hot)	150 μM[a]	10 μM	1 μL	1 μL	1 μL		1 μL	=1 × 15 = 15 μL
Recombinant methyltransferase			Increasing concentration of HMT	1 μL	1 μL	1 μL		1 μL	
Reconstituted nucleosome[c]			Desired concentration	1 μL	1 μL	1 μL		1 μL	(optional) =1 × 15 = 15 μl
H_2O				6.6 μL	6.6 μL	6.6 μL		6.6 μL	=6.6 × 15 = 99 μL
Total				15 μL	15 μL	15 μL		15 μL	

[a] Making 150 μM Hot/Cold SAM: Mix 60 μL of 7 μM Radiolabeled SAM and 0.88 μL of 10 mM Cold SAM
[b] Make a master mix to reduce sample-to-sample variations and pipet errors
[c] Reconstituted nucleosome can be excluded if you want to measure automethylation without nucleosomes

11. Add 4× STOP buffer (final 1× STOP) to stop the reaction. Mix well and centrifuge. Boil samples for 5 min at 95 °C. Cool down to room temperature and briefly centrifuge.
12. Load samples into the wells of the SDS-PAGE gel for separation of proteins. Run the gel for 1–2 h at 100–150 V.
13. Transfer the protein from the gel to an activated PVDF membrane (*see* **Note 10**).
14. Stain the membrane with Coomassie Blue for 1 min. If you plan to perform Western blot using pan-methyl lysine antibody, do not stain the membrane and proceed with Western blot (*see* **Note 11**).
15. Destain the membrane and dry it at room temperature. Scan the image (Fig. 2, left and *see* **Note 12**).
16. Expose the membrane to the X-ray film for autoradiography (Fig. 2, right). Develop the film using Biomax Transcreen intensifying screener. Depending on enzyme activity and the concentration of radiolabeled SAM, the signal can be detected from 2 to 7 days of exposure.
17. Quantify the band intensity using ImageJ or other programs. To quantify the level of methylation precisely, cut the protein band shown from the Coomassie Blue-stained membrane. Put the cut membrane into 3 mL of scintillation cocktail solution. Count radioactivity using a scintillation counter (*see* **Note 13**).

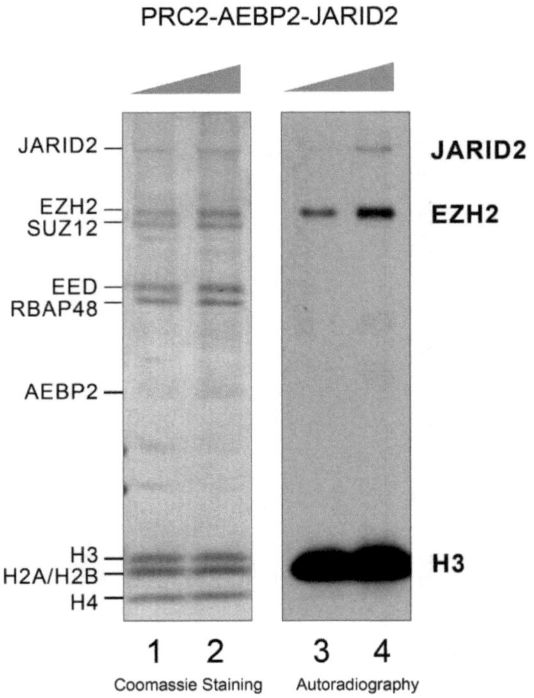

Fig. 2 The result of methyltransferase assay on PVDF membrane. Adapted from [1]

3.2 Identification of Automethylation Site

3.2.1 In Vitro Methylation and Mass Spectrometry

1. Perform methylation assay as described in Subheading 3.1 with minor modification. Mix 50 μg of recombinant methyltransferase,1 mM of cold SAM (non-radioactive), 4 mM DTT in 1× HMT buffer (*see* **Note 14**).
2. Incubate samples at 30 °C for 2 h.
3. Add EDTA (final concentration, 5 mM) to quench the reaction.
4. Dialyze against the buffer that is compatible with LC-MS (*see* **Note 15**).
5. Analyze the potential methylation sites by LC-MS (*see* **Note 16**).

3.2.2 In Vivo Methylation and Mass Spectrometry

1. Choose an adequate cell line that highly expresses the methyltransferase. To obtain high yield and high purity of the methyltransferase from the cell, it is recommended to ectopically express epitope-tagged methyltransferase (*see* **Note 17**).

   If using a cell line with epitope-tagged methyltransferase:

2. Immunoprecipitate epitope-tagged methyltransferase using anti-epitope agarose beads (*see* **Note 18**).
3. Elute the methyltransferase using the epitope peptide.
4. Remove epitope peptide and adjust to the buffer for LC-MS by dialysis or by additional purification (*see* **Note 15**).

   If it is not suitable to use a cell line with epitope-tagged methyltransferase:

5. Immunoprecipitate the methyltransferase using an antibody against the methyltransferase and protein A or G beads (*see* **Note 19**).
6. Elute the methyltransferase using Ag/Ab Elution Buffer. Immediately adjust eluted fractions to physiologic pH by adding Neutralization Buffer (1 M Tris–HCl; pH 8–9).
7. Dialyze to exchange the purified methyltransferase into a buffer suitable for LC-MS or purify it using another protein purification column (*see* **Note 15**).
8. Analyze the potential methylation sites by LC-MS.

3.3 Generation and Confirmation of Automethylation Antibody

1. **Design the antigen**. Determine the peptide sequence containing lysine methylation. Check whether the sequence is specific. To do so, use NCBI BLAST program to search if there is any other identical peptide sequence from other unknown proteins (*see* **Note 20**).
2. Determine the methylation state (mono-, di-, or tri-methylation) of the antigen (*see* **Note 21**). Synthesize methylated peptide for the antigen and non-methylated for later verification of antibody specificity.

3. **Antibody production**. Inject the antigen into the animal (e.g., mouse, rabbit, or rat) to produce an automethylation-specific antibody.
4. Once you obtain the serum from the animal, check whether the serum is specific to the automethylated methyltransferase before the purification of the antibody (*see* **Note 22**).
5. **Antibody purification**. If necessary, purify the antibody using automethylated peptide or automethylated recombinant methyltransferase (*see* **Note 23**).
6. **Antibody verification**. The specificity of the antibody can be validated by several different procedures as described below.
   (a) Apply the automethylation antibody and compare automethylation levels of methylated peptide versus unmethylated peptide using ELISA.
   (b) Apply the automethylation antibody and compare automethylation levels of WT versus catalytically inactive methyltransferase using ELISA or Western blot.
   (c) Apply the automethylation antibody and compare automethylation levels of cell extracts containing WT versus cell extracts containing catalytically inactive methyltransferase.

3.4 Investigate the Function of Automethylation

3.4.1 Compare Automethylated Recombinant Methyltransferase Versus Recombinant Methyltransferase (Fig. 3a)

1. Perform methylation assay as described in Subheading 3.1 with minor modification. Mix 50 μg of recombinant methyltransferase with 4 mM DTT, 1× HMT buffer in the absence or presence of 1 mM of cold SAM. The reaction with cold SAM will hyper-automethylate the methyltransferase while the reaction without cold SAM will only preserve the existing automethylation on the recombinant methyltransferase (hypo-automethylated methyltransferase) (*see* **Note 24**).
2. Compare the level of automethylation by autoradiography as well as Western blot using an automethylation-specific antibody. The former will detect the automethylation that occurred during the methyltransferase assay while the latter will also detect basal automethylation that is present from the recombinant methyltransferase.
3. After the assay, remove residual SAM and SAH (*S*-adenosylhomocysteine) as well as other reaction components by dialysis (*see* **Note 25**).
4. Perform histone methyltransferase assay as described in Subheading 3.1 with oligonucleosomes and either hyper-automethylated methyltransferase or hypo-automethylated methyltransferase.

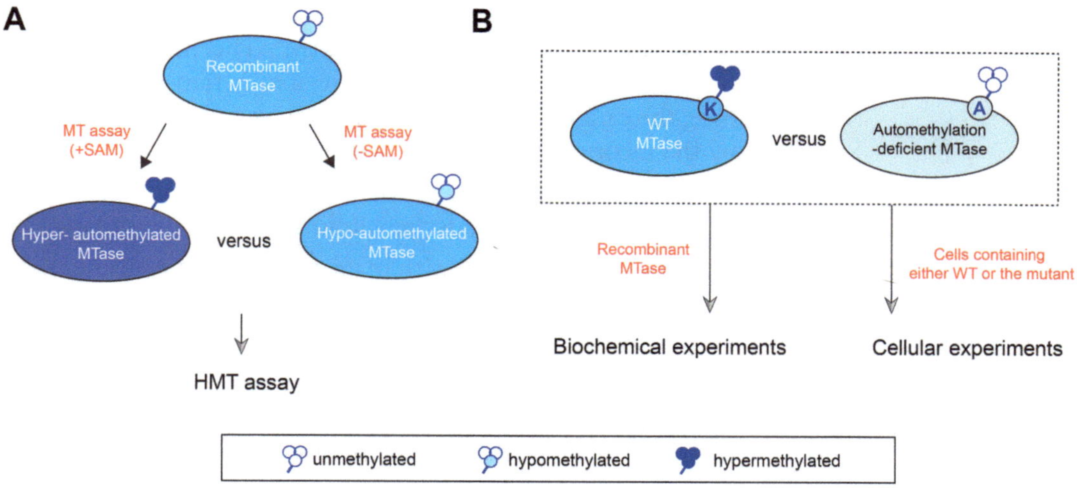

Fig. 3 Experiments to investigate the role of automethylation. (**a**) HMT assay with hypermethylated recombinant methyltransferase (MTase) or hypomethylated MTase. (**b**) Biochemical experiments with recombinant WT or automethylation-deficient MTase (*left*). Cellular experiments with cells containing either WT or automethylation-deficient MTase (*right*). It is recommended to mutate the lysine residue to alanine for automethylation deficiency

3.4.2 Wild-Type Versus Automethylation-Deficient Methyltransferase (Fig. 3b)

1. Generate a mutation on the automethylation site using site-directed mutagenesis (*see* the detailed procedure in Ref. 10 and *see* **Note 26**).
2. To check whether automethylation affects the catalytic activity of methyltransferase on chromatin in vitro, perform histone methyltransferase assay with recombinant WT or automethylation-deficient methyltransferases using oligonucleosomes as the substrate. Check the levels of histone methylation by either autoradiography or Western blot using corresponding antibodies (*see* Subheading 3.1).
3. To investigate the function of automethylation in the cellular context, prepare extracts from cells containing either WT or automethylation-deficient methyltransferases. Check the level of histone methylation by Western blot using corresponding antibodies.
4. To investigate the relevance of automethylation in some disease conditions or at certain developmental stages, prepare cellular extracts and detect the level of automethylation using an automethylation-specific antibody.

4 Notes

1. *S*-adenosyl L-methionine (SAM) has to be dissolved in the manufacturer's recommendation. SAM readily decomposes into S-adenosyl homocysteine (SAH) and 5′-methylthioadenosine very quickly in the aqueous solution.

2. Commercially available ^{3}H-SAM and ^{14}C-SAM are radiolabeled on the methyl group of SAM. The methyl group is transferred to the substrate by the methyltransferase. Therefore, the catalytic activity of methyltransferase can be measured by the incorporation of radioactivity on the substrate.
3. Using intensifying screen will increase the sensitivity for all isotopes and offers greater resolution.
4. It is much easier to deal with many tubes on ice by using metal blocks.
5. Use proper safety equipment such as a lab coat and an acrylic protection shield when handling radioactive materials.
6. When diluting methyltransferase, use the enzyme storage buffer so that all diluted methyltransferase are in the same buffer condition.
7. It is recommended to put methyltransferase samples at the bottom of the tube to prevent any loss.
8. It is recommended to measure the level of automethylation in the absence as well as in the presence of nucleosome substrate and compare the outcome.
9. Optimize the reaction time by measuring the activity at each time point (e.g., 30, 45, 60, 75, 90 min). Then, use the condition where the activity has almost reached the stationary phase.
10. PVDF membrane should be activated by a short rinse (1 min) in methanol before Western transfer. This will hydrate the membrane and improve protein transfer.
11. Pan-methyl lysine antibody will detect the methylation on lysine residues. Therefore, it will not only recognize the methylation from the assay but also the methylation already present in the recombinant methyltransferase. We recommend not relying too much on the result of Western blot using pan-methyl lysine antibody, as the antibody does not recognize all possible methylations.
12. The amount of methyltransferase and nucleosome is shown by Coomassie Blue staining.
13. The methylation signal from X-ray film is not linear. To precisely measure the level of methylation, it is recommended to use a scintillation counter to measure the incorporation of radioactive tritium on methyltransferase.
14. The reaction condition can be modified as each methyltransferase behaves differently.
15. It is extremely critical to remove any types of detergent before the mass spec analysis. We recommend removing detergent at the last step of methyltransferase purification. Make sure to discuss with the Proteomics (or LC-MS) facility personnel before sending samples for LC-MS analysis.

16. It is recommended to use trypsin for the identification of methylation sites as trypsin is commonly used for MS analysis. However, it is not recommended to use trypsin for quantifying the percentage of each methylation state (unmodified-, mono-methylated-, dimethylated-, trimethylated-lysine). Using trypsin will cleave the carboxy terminus of unmodified lysine; therefore, the percentage of each methylation state is not accurate. In this case, it is recommended to use other proteases such as Arg-C which cleaves after the carboxy terminus of Arginine.
17. Make sure to check whether the epitope-tagged methyltransferase does not affect the activity of methyltransferase by comparing the activity of WT versus epitope-tagged methyltransferase.
18. We recommend tagging either FLAG or HA on the methyltransferase. By using these epitopes, you can purify the enzyme with anti-FLAG agarose or anti-HA agarose beads and elute it with FLAG or HA peptide, respectively.
19. Choose either Protein A beads or Protein G depending on the subclass and the biological source of the antibody.
20. Check whether there are any other known posttranslational modifications (PTMs) within the antigen peptide sequence. If these PTMs are at the end of the peptide sequence, exclude the residue from the peptide.
21. It is recommended to choose the most abundant methylation state in the cellular context. Therefore, choose it based on the mass spectrometry result of methyltransferase purified from the cell.
22. It is important to check whether the serum is specific to the automethylated methyltransferase before purifying the antibody. In some cases, the specificity and the antigen detection level of the antibody can be lost after the purification.
23. When purifying antibodies, you can also consider using recombinant methyltransferase as a bait as it maintains 3D structure.
24. As long as recombinant methyltransferases are purified from eukaryotes, the endogenous level of automethylation is present in the purified recombinant methyltransferases. Keep in mind that the level of automethylation is already high from the purified recombinant methyltransferase.
25. After the methyltransferase assay, the methyl group of *S*-adenosylmethionine (SAM) is transferred to the substrate and yields *S*-adenosylhomocysteine (SAH). SAH can act as a methylation inhibitor; therefore, it is necessary to remove both SAM and SAH before the following experiments. If the volume of the sample is low, use a low-volume sample dialysis device that is commercially available.

26. It is typical to mutate the lysine residue to alanine to mimic automethylation-deficient methyltransferase. But keep in mind that the mutant could cause other secondary defects that are not due to the loss of automethylation but K-to-A mutation.

Acknowledgments

We thank Dr. D. Reinberg for critical reading of the manuscript. This work was supported by Research Resettlement Fund for the new faculty of Seoul National University, Creative-Pioneering Researchers Program through Seoul National University, grants from Seoul National University College of Medicine, National Research Foundation of Korea (NRF) grant funded by the Korea government (MSIT) (No. 2021R1C1C1013220), and BK21 Four Biomedical Science Program. L. Popoca is supported by National Institutes of Health (NIH) grant R01CA199652, the Howard Hughes Medical Institute (HHMI), and the Making Headway Foundation St. Baldrick's Research Grant (189290).

References

1. Lee C-H, Yu J-R, Granat J et al (2019) Automethylation of PRC2 promotes H3K27 methylation and is impaired in H3K27M pediatric glioma. Genes Dev 33:1428–1440
2. Wang X, Long Y, Paucek RD et al (2019) Regulation of histone methylation by automethylation of PRC2. Genes Dev 33: 1416–1427
3. Piao L, Nakakido M, Suzuki T, Dohmae N, Nakamura Y, Hamamoto R (2016) Automethylation of SUV39H2, an oncogenic histone lysine methyltransferase, regulates its binding affinity to substrate proteins. Oncotarget 7:22846–22856
4. Iglesias N, Currie MA, Jih G et al (2018) Automethylation-induced conformational switch in Clr4 (Suv39h) maintains epigenetic stability. Nature 560:504–508
5. Chin HG, Estève P-O, Pradhan M et al (2007) Automethylation of G9a and its implication in wider substrate specificity and HP1 binding. Nucleic Acids Res 35:7313–7323
6. Poulard C, Baulu E, Lee BH, Pufall MA, Stallcup MR (2018) Increasing G9a automethylation sensitizes B acute lymphoblastic leukemia cells to glucocorticoid-induced death. Cell Death Dis 9:1038
7. Nishioka K, Reinberg D (2003) Methods and tips for the purification of human histone methyltransferases. Methods 31:49–58
8. Yun M, Ruan C, Huh JW, Li B (2012) Reconstitution of modified chromatin templates for in vitro functional assays. Methods Mol Biol 833:237–253
9. Kurien BT, Scofield RH (2015) Multiple immunoblots by passive diffusion of proteins from a single SDS-PAGE gel. Methods Mol Biol 1312:77–86
10. Li F, Mullins JI (2002) Site-directed mutagenesis facilitated by DpnI selection on hemimethylated DNA. Methods Mol Biol 182: 19–27

Chapter 6

Profiling the Regulation of Histone Methylation and Demethylation by Metabolites and Metals

Sebastian Müller, Fabien Sindikubwabo, Tatiana Cañeque, and Raphaël Rodriguez

Abstract

Here we describe how to profile the contribution of metabolism and implication of metals to histone methylation and demethylation. The techniques described with the adequate protocols are metabolomics, quantitative proteomics, inductively coupled mass spectrometry and nanoscale secondary ion mass spectrometry.

Key words Histone methylation, Histone demethylation, ICP-MS, NanoSIMS, Metabolomics, Metals, Proteomics

1 Introduction

Histone methylation is a chemical reaction by which a methyl group is transferred to a specific amino acid residue of a histone tail by histone methyltransferases. Histone demethylation is the removal of a methyl group from an amino acid residue of a histone tail, mediated by histone demethylases [1]. The underlying reactions require specific metabolites and catalysts for their functioning [2]. Due to their roles in these reactions, the metabolites involved exert direct control over histone methylation, placing metabolism as a key determinant in controlling gene expression [3, 4] and cell plasticity [5, 6]. Although distinct histone methyltranferases and demethylases are involved in these reactions and give specificity in terms of what histone amino acid residues are (de)methylated at what specific genomic loci [7, 8], their abundance is not necessarily rate-limiting for these reactions. The reagents and catalysts of these reactions, however, are rate-limiting. Thus, to understand epigenetic control, it is crucial to measure changes in metal catalyst, metabolites, and substrates involved in these reactions, such as iron [6] and oxygen [9]. Histone methyltransferases transfer a

Raphaël Margueron and Daniel Holoch (eds.), *Histone Methyltransferases: Methods and Protocols*, Methods in Molecular Biology, vol. 2529, https://doi.org/10.1007/978-1-0716-2481-4_6,

methyl group from the metabolite *S*-Adenosyl methionine onto the lysine or arginine residue of histones tails [10, 11] (Fig. 1a). There are two different classes of histone demethylases: Flavin adenine dinucleotide (FAD)-dependent amine oxidases and Fe^{2+} and α-ketoglutarate-dependent hydroxylases (Jumonji-domain containing demethylases) (Fig. 1b,c).

In order to understand the impact of metabolomics to histone methylation in a given biological system or disease [12], one has to analyze the involved metabolites, which can be achieved using quantitative metabolomics, as well as the contribution of the metabolic enzymes involved in these processes, whose expression levels can be profiled using quantitative proteomics. Interestingly, histone demethylation was discovered while studying the contribution of metabolism to epigenetic gene regulation [13, 14]. Since iron is the catalyst in Jumonji-containing demethylases and a master regulator of cell plasticity [6], measuring iron levels is crucial to understanding the dynamics of histone demethylation in a given biological system. Competition between metals, such as the replacement of catalytic iron in Jumonji proteins by nickel [15] may play important roles in disease etiology, given that nickel is not a competent metal catalyst for demethylation. Other metals such as copper [16] also play roles in metabolic control during epigenetic reprogramming, and future work is required to unravel these processes. A reliable method to assess the contribution of metals to histone (de)-methylation is the quantification of elemental iron, which requires a technique termed inductively coupled plasma mass spectrometry (ICP-MS) for quantitative analyses. There are element-specific fluorescent probes, which allow for imaging of metals in specific cellular organelles such as lysosomes [17, 18]. However, such probes do not directly monitor metals, but rather their reactivity with a fluorescent probe. Thus, depending on the chemical reaction, the photon yield and specificity of these probes can be low. Metals can also be imaged directly using nanoscale secondary ion mass spectrometry (NanoSIMS), a technique sparsely employed in cell biology to date [19, 20].

This chapter describes the protocols to perform quantitative metabolomics and proteomics to assess the contribution of metabolites and their converting enzymes to histone (de)methylation. For the quantification of some metabolites, there are commercial kits available that can help confirm the findings of metabolomic studies. Quantitative proteomics is very dependent on the abundance of the protein input and hence the biological material used, as well as the digestion protocol employed when bottom-up proteomics is applied, as described in this chapter. We recommend proteomics over RNA-seq to identify expression levels of metabolic enzymes, as RNA expression does not always accurately reflect protein expression. Since metals play key roles in metabolic control, and iron is a master regulator of histone demethylation, we also

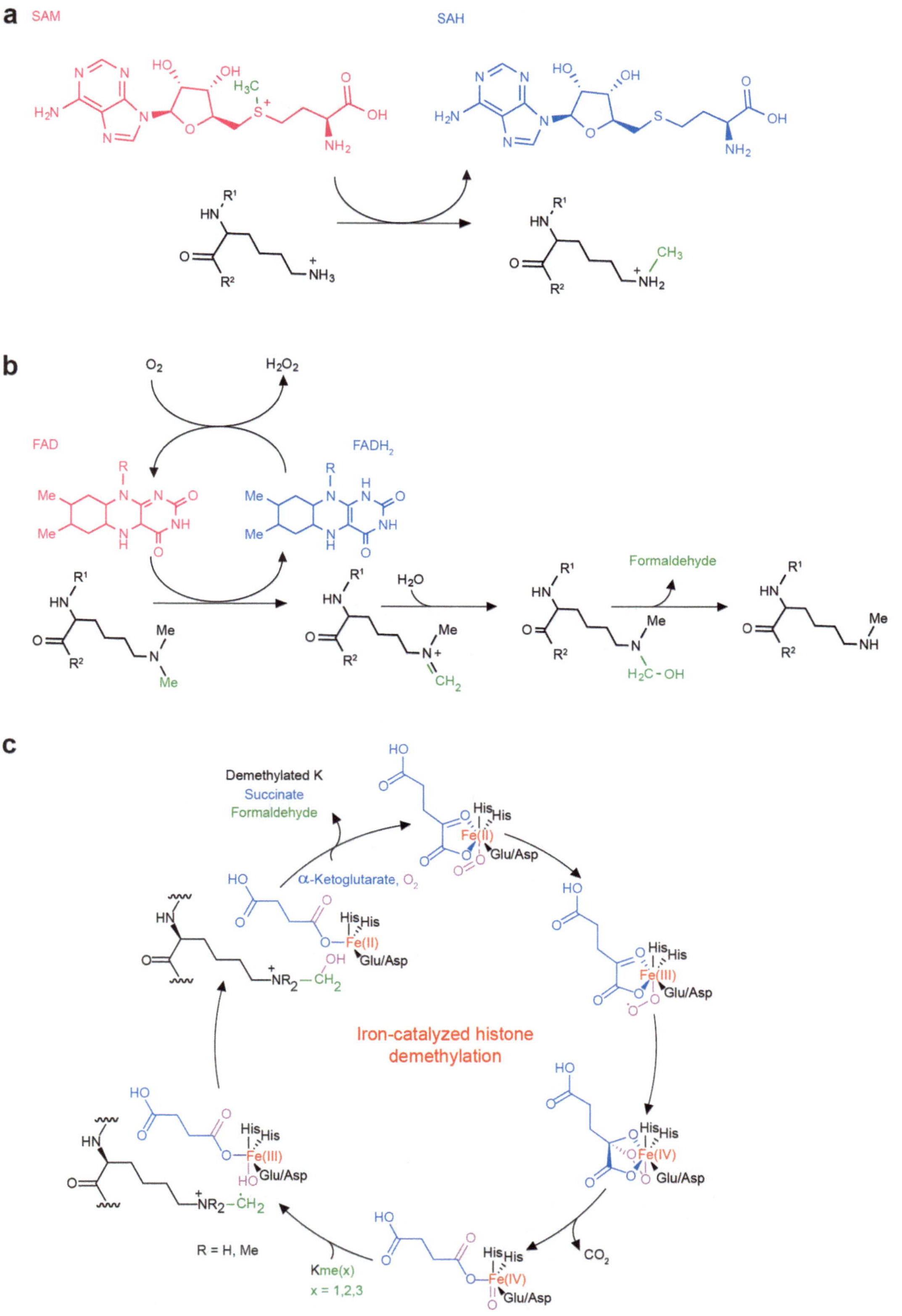

Fig. 1 Reaction schemes of histone methylation and demethylation. (**a**) Histone methylation involves the metabolite *S*-adenosyl methionine (SAM), which donates a methyl group, and gets converted to *S*-adenosyl homocysteine (SAH). (**b**) Mechanism of histone demethylation involving the metabolite flavin adenine

describe in this chapter how to quantitatively measure metal concentrations using ICP-MS and how subcellular localization of metals can be imaged using NanoSIMS. Depending on the biological question, quantitative metabolomics, proteomics, and ICP-MS can be applied to fractionated samples, such as nuclear or mitochondrial extracts, which require an adequate scale-up of material.

2 Materials

2.1 Metabolomics

1. Cells cultured in adequate cell culture medium.
2. 1× Phosphate buffer saline (PBS).
3. Methanol.
4. Lysis buffer: 9:1 methanol: water (v:v).
5. Eppendorf tubes.
6. Cell scrapers.
7. Centrifuge with cooling module.
8. 20 mg/mL methoxyamine in pyridine.
9. *N*-Methyl-*N*-(trimethylsilyl) trifluoroacetamide (MSTFA).
10. Gas Chromatograph (e.g., 7890B GC).
11. Triple quadrupole mass spectrometer (QQQ) (e.g., 7000C).
12. Analysis software (e.g., Agilent Mass Hunter quantitative software).

2.2 Quantitative Proteomics

1. Cells cultured in adequate cell culture medium.
2. Centrifuge with cooling module.
3. 1× PBS.
4. Eppendorf tubes.
5. Lysis buffer:8 M urea, 50 mM NH_4HCO_3 and protease inhibitors cocktail.
6. Dounce Homogenizer.
7. NH_4HCO_3.
8. Dithiothreitol (DTT).

Fig. 1 (continued) dinucleotide (FAD), which gets reduced to $FADH_2$ in a reaction involving molecular oxygen. The methyl group of the histone tail gets converted into an imine, that is then hydroxylated using water. In the final step formaldehyde is eliminated. The only known histone demethylase using this mechanism is lysine-specific histone demethylase 1A (LSD1). (**c**) Catalytic cycle of Fe^{2+} and α-ketoglutarate-dependent hydroxylases, which encompass all known histone demethylases apart from LSD1. Iron is a catalyst in this reaction cycle. This reaction utilizes the metabolite α-ketoglutarate, which is a key metabolite of the Krebs cycle. The final step also involves an elimination of formaldehyde

9. Iodoacetamide.
10. Trypsin/LysC.
11. C18 StageTips.
12. Buffer A: 40:60 MeCN/H_2O with 0.1% formic acid.
13. Speedvac.
14. Buffer B: 2:98 MeCN/H_2O with 0.3% trifluoroacetic acid (TFA).
15. Buffer C: 2:98 MeCN/H_2O with 0.1% formic acid (="buffer A").
16. Buffer D: 100% MeCN and 0.1% formic acid (="buffer B").
17. RSLCnano system (e.g., Ultimate 3000) coupled online to a Q Exactive HF-X with a Nanospray Flex ion source.
18. C18-reversed phase precolumn (75 μm inner diameter × 2 cm; nanoViper Acclaim PepMap 100).
19. Analytical column (75 μm inner diameter × 50 cm; nanoViper C18, 2 μm, 100 Å, Acclaim PepMap RSLC).
20. Mass spectrometer (ultrahigh-field Orbitrap mass analyzer).
21. Mass spectrometry analysis software (Proteome Discoverer version 2.0, myProMS, MassChroQ version 2.2.2).

2.3 Inductively Coupled Mass Spectrometry

1. Cells cultured in adequate cell culture medium.
2. Glass vials equipped with Teflon septa.
3. Metal-free ultrapure 65% nitric acid (VWR, Suprapur, 1.00441.0250).
4. Metal-free ultrapure water (Sigma-Aldrich, 1012620500).
5. 1× PBS.
6. Freeze dryer.
7. Quadrupole-based mass spectrometer (e.g., Agilent 7900 ICP-QMS).
8. Micro-nebulizer (MicroMist) and Scott spray chamber.

2.4 Nano SIMS Imaging

1. Cells cultured on cover slips in adequate cell culture medium.
2. 1× PBS.
3. 2% glutaraldehyde in 0.1 M cacodylate buffer pH 7.4.
4. 1% OsO_4 in 0.1 M cacodylate buffer pH 7.4.
5. EtOH solutions in water: 50%, 70%, 90%.
6. 100% EtOH (dried over molecular sieves).
7. Resin (dodecenylsuccinic anhydride, methyl nadic anhydride, DMP-30: LX112 resin).
8. Embedding capsules.

9. Oven (at 56 °C).
10. Microtome.
11. Silicon chips.
12. NanoSIMS-50 ion microprobe.

3 Methods

3.1 Metabolomics

An example of data obtained with metabolomics is depicted in Fig. 2a.

1. Grow cells at around 70% confluence in growth media in adequate technical replicates (*see* **Note 1**).
2. Place cells in culture plates on ice, remove carefully the media, wash cells quickly (<2 s) three times with 1 mL of ice-cold 1× PBS (*see* **Note 2**).
3. Lyse cells with 1 mL of cold lysis buffer (−20 °C) by scraping and transfer the material into Eppendorf tubes placed on ice.
4. Mix cells by vortexing for 1 min, centrifuge at 15,000 × *g* for 10 min at 4 °C. Transfer 750 μL of supernatant in centrifugation microtubes and store at −80 °C.
5. Evaporate the whole supernatant at 40 °C to obtain dried extracts and add 500 μL of methanol.
6. Transfer 270 μL in injection vial, evaporate at 40 °C to obtain dried extracts.
7. Add 50 μL of methoxyamine and store at room temperature in the dark for 16 h.
8. The following day, add 80 μL of MSTFA and final derivatization occurs at 40 °C for 30 min.
9. To analyze the widely targeted intracellular metabolites, use a gas chromatography coupled to a triple quadrupole mass spectrometer, equipped with a high sensitivity electronic impact source operating in positive mode. The front inlet temperature should be set to 250 °C.
10. Perform the injection in splitless mode. The transfer line and the ion-source temperature should be set to 250 °C and 230 °C, respectively.
11. Fix the septum purge flow at 3 mL/min and the purge flow to split vent operates at 80 mL/min during 1 min.
12. Set the gas saver mode to 15 mL/min after 5 min. The helium gas should flow through the column at 1 mL/min.

13. Hold the column temperature at 60 °C for 1 min, raise to 210 °C (10 °C/min), followed by a step at 230 °C (5 °C/min), reach 325 °C (15 °C/min), and maintain this temperature for 5 min. The collision gas should be nitrogen. As a scan mode, use the MRM for biological samples.
14. Perform the peak detection and integration of each analyte by using a specific software.

3.2 Quantitative Proteomics

An example of data obtained with quantitative proteomics is depicted in Fig. 2b.

1. Grow cells at around 70% confluence in growth media in biological replicates (*see* **Notes 3** and **5**).
2. Collect the whole cell extracts, centrifuge at 500 × *g* for 5 min at 4 °C and discard the supernatant.
3. Wash twice in ice-cold 1× PBS by centrifuging 500 × *g* for 5 min at 4 °C (*see* **Note 4**).
4. Lyse the cells with 200 μL of lysis buffer by douncing ten times with a Dounce Homogenizer.
5. Transfer the lysate into an Eppendorf and mix for 1 h on a rotating wheel at 4 °C (*see* **Note 6**).
6. To analyze the global proteome, use a Q Exactive HF-X quadrupole-Orbitrap mass spectrometer with a label-free approach.
7. Dilute about 30 μg of total protein cell lysate with 90 μL of 25 mM NH_4HCO_3.
8. Reduce proteins by incubation with 5 mM dithiothreitol (DTT) at 57 °C for 1 h.
9. Alkylate with 9 mM iodoacetamide for 30 min at room temperature in the dark.
10. Add trypsin/LysC 1:100 (wt/wt) enzyme: substrate and performed the digestion overnight at 37 °C.
11. Load samples onto a homemade C18 StageTips for desalting.
12. Elute peptides from beads by incubation with buffer A.
13. Dry peptides in a Speedvac and reconstitute them in 15 μL of buffer B prior to liquid chromatography-tandem mass spectrometry (LC-MS/MS) analysis.
14. Separate chromatographically 4 μL of samples by using an RSLCnano system coupled online to a Q Exactive HF-X with a Nanospray Flex ion source.
15. Load peptides onto a C18-reversed phase precolumn, with buffer C at a flow rate of 2.5 μL/min over 4 min.

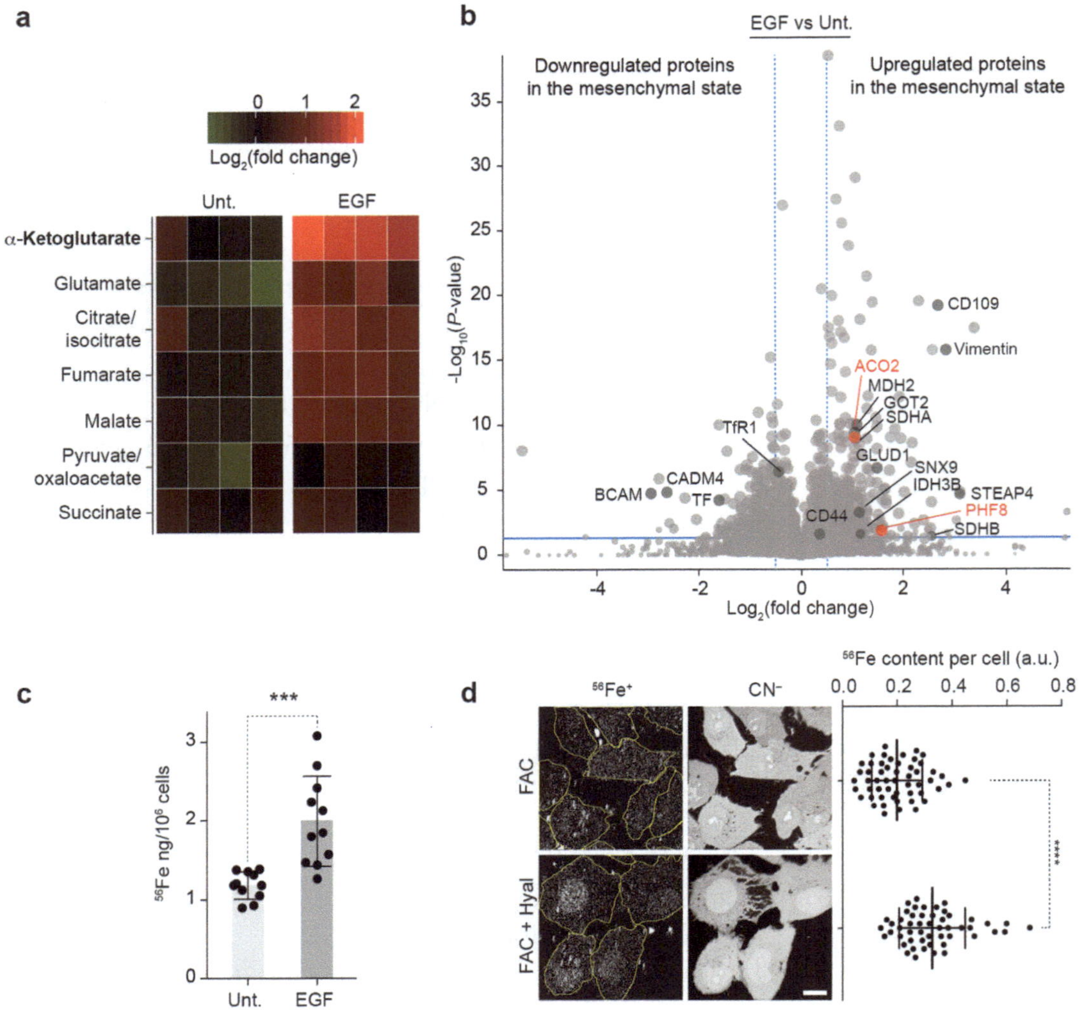

Fig. 2 Examples of obtained data with the described techniques. (The figures have been adapted from Müller S et al. [6]). *EGF* epidermal growth factor, *Unt.* untreated, *FAC* ferrous ammonium citrate, *Hyal* hyaluronic acid. Data were obtained using the triple negative breast cancer cell line MDA-MB-468. (**a**) Analysis of Krebs-cycle metabolites using quantitative proteomics. Several metabolites are upregulated in EGF-treated (more mesenchymal) cells, in particular α-ketoglutarate. (**b**) Quantitative proteomics identifies changes in protein expression during cell plasticity. In the mesenchymal state key metabolic enzymes are upregulated, for instance aconitase 2 (ACO2), and the Fe^{2+} and α-ketoglutarate-dependent histone demethylase PHF8, which demethylates dimethylation of histone H3 at position Lysine 9 (H3K9me2). (**c**) ICP-MS data of iron, quantified in cells treated with EGF. The mesenchymal state has more iron. (**d**) Images obtained using NanoSIMS technology. There is an increase of cellular iron (in particular the nuclei) in cells treated with hyaluronic acid

16. Switch peptides for separation to an analytical column regulated to a temperature of 50 °C with a linear gradient of 2–35% buffer D at a flow rate of 300 nL/min over 211 min.
17. Perform the mass spectrometry full scans in the ultrahigh-field Orbitrap mass analyzer in ranges *m/z* 375–1500 with a resolution of 120,000 at *m/z* 200, the maximum injection time (MIT) should be 50 ms.

18. Set the automatic gain control (AGC) to 3×10^6. The top 20 intense ions are subjected to Orbitrap for further fragmentation via high energy collision dissociation (HCD) activation and a resolution of 15,000 with the intensity threshold kept at 3.2×10^5.
19. Select ions with charge state from 2+ to 6+ for screening and set the normalized collision energy (NCE) at 27. For each scan, set the AGC at 1×10^5, the MIT at 25 ms and a dynamic exclusion of 40 s.
20. Establish the identity of proteins from UniProt human canonical database (downloaded on 22/08/2017) using Sequest HF through proteome discoverer.
21. Set the enzyme specificity to trypsin and a maximum of two missed cleavage sites are allowed.
22. 22.Set oxidized methionine, N-terminal acetylation, and carbamidomethyl cysteine as variable modifications. Set maximum allowed mass deviation to 10 ppm for monoisotopic precursor ions and 0.02 Da for MS/MS peaks.
23. Process the resulting files using software (e.g., myProMS). Use percolator for the False Discovery Rate (FDR) calculation and set it to 1% at the peptide level for the whole study.
24. Perform the label-free quantification by peptide Extracted Ion Chromatograms (XICs) computed with MassChroQ.
25. For protein quantification, use XICs from proteotypic peptides shared between compared conditions (TopN matching) with up to two missed cleavages and carbamidomethyl modifications.
26. Apply median and scale normalization on the total signal to correct the XICs for each biological replicate.
27. Perform a linear model (adjusted on peptides and biological replicates) to estimate the significance of the change in protein abundance. Adjust the *P*-values with a Benjamini–Hochberg FDR procedure with a control threshold set to 0.05.
28. Analyze the fold change-based GO enrichment analysis.

3.3 Inductively Coupled Plasma Mass Spectrometry (ICP-MS)

An example of data obtained with ICP-MS is depicted in Fig. 2c.

1. Clean glass vials equipped with Teflon septa with nitric acid 65% and wash them with ultrapure water (*see* **Notes** 7 and **8**).
2. Plate cells at least 24 h prior to the experiments.
3. Harvest cells with an adaptive protocol (e.g., trypsinization for adherent cells) followed by two washes with 1× PBS.
4. Count cells accurately and transfer them in 1× PBS to the cleaned glass vials.

5. Lyophilize samples for instance by using a freeze dryer.
6. Determine the dry weight of your samples (optional if normalization against dry weight is preferred to cell number) (*see* **Note 9**).
7. Mix samples with ultrapure metal-free nitric acid 65% overnight followed by heating at 80 °C overnight (*see* **Note 10**).
8. Let the samples cool and dilute them with ultrapure metal-free water to a final concentration of 0.475 N nitric acid (*see* **Note 11**).
9. To measure metal concentrations use a quadrupole-based mass spectrometer in low-resolution mode. Sample introduction can be achieved with a micro-nebulizer (0.2 mL/min) through a Scott spray chamber.
10. Isotopes can be measured using a collision-reaction interface with helium gas (5 mL/min) to remove polyatomic interferences.
11. Scandium and indium internal standards need to be injected after inline mixing with the samples to control the absence of signal drift and matrix effects.
12. A mix of certified standards has to be measured at concentrations spanning those of the samples to convert count measurements to concentrations in the solution.
13. Uncertainties on sample concentrations can be calculated using algebraic propagation of ICP-MS blank and sample counts uncertainties.

3.4 NanoSIMS

An example of data obtained with NanoSIMS is depicted in Fig. 2d.

1. Grow your cells on glass cover slips.
2. Wash cells on cover slips twice with 1× PBS.
3. Fix the cells in 2% glutaraldehyde in 0.1 M cacodylate buffer for 24 h.
4. Treat the cells with 1% OsO_4 in 0.1 M cacodylate buffer for 1 h (*see* **Note 12**).
5. Dehydrate cells with sequential EtOH solutions each for 10 min each: 50%, 70%, 2 × 90%, 3 × 100% (dried over molecular sieves).
6. Coated the samples with a 1:1 mixture of resin (dodecenyl succinic anhydride, methylnadic anhydride, DMP-30: LX112 resin) and dry EtOH for 1 h.
7. Embed the samples in pure resin for 1 h.
8. Fill embedding capsules with resin and invert them onto the cover slides and place them in an oven at 56 °C for 24 h (*see* **Note 13**).

9. Cut 0.2 μm sections using a microtome (*see* **Note 14**).
10. Deposit the sections onto a clean silicon chip and introduce them into a NanoSIMS imager.
11. Measure and image your metal of interest.

4 Notes

1. Several technical and biological replicates are recommended for the robustness and confidence of the data.
2. Metabolomics is very sensitive, and the chemical reactions interconverting metabolites are very fast. Washing the cells very quickly at 4 °C is very important. Some metabolomics platforms also recommend flash freezing of adherent cells in culture dishes.
3. For mammalian cell culture, a minimum amount of at least ten million cells (of average size) per condition would be preferable for a reliable mass spectrometry analysis. Please note that the sizes of cells are different. Thus, for instance, the amount of small cells required for these experiments such as monocytes would be different to fibroblasts.
4. More than four independent biological replicates are required for robustness and statistical data analysis.
5. Wash cells well with ice-cold 1× PBS to remove serum proteins completely. This avoids the interference between medium and lysis buffer.
6. Protein lysates are sensitive to temperature and should be handled carefully at 4 °C.
7. Concentrated nitric acid is dangerous and should always be handled under a chemical hood with adequate safety precautions.
8. It is very important to clean the reaction vials properly with ultrapure 65% nitric acid and ultrapure water and dry the vials properly. This cannot be done with MilliQ water or similar grade water, but instead specific reagents that are trace metal free have to be employed.
9. Measurements have to be normalized to cell number or dry weight. It is crucial to measure this accurately. High precision scales should be used to determine the dry weight.
10. Samples should be fully digested in 65% nitric acid at 80 °C overnight. We recommend to use Teflon septa in the glass vials, as the acid can dissolve plastic screw caps.

11. The digested material in concentrated nitric acid has to be diluted to a final concentration of 0.475 N before analysis. Adding water to acid is dangerous due to possible projections. We recommend to let the nitric acid digested sample cool and then add the water slowly into each vial under a chemical hood facing away from the user to avoid any acid projections.
12. OsO_4 is very toxic. Handle with care!
13. Be careful when you remove the capsules with the hardened resin from the cover slips. They can break easily. You can put several embedding capsules onto a cover slip.
14. Make sure you prepare samples at the same time, section them to the same thickness and deposit them onto the silicon chip close to each other. This is crucial for comparative quantification.

References

1. Kooistra SM, Helin K (2012) Molecular mechanisms and potential functions of histone demethylases. Nat Rev Mol Cell Biol 13: 297–311
2. Dai Z, Ramesh V, Locasale JW (2020) The evolving metabolic landscape of chromatin biology and epigenetics. Nat Rev Genet 21: 737–753
3. Tran TQ, Lowman XH, Kong M (2017) Molecular pathways: metabolic control of histone methylation and gene expression in cancer. Clin Cancer Res 23:4004–4009
4. Teperino R, Schoonjans K, Auwerx J (2010) Histone methyl transferases and demethylases; can they link metabolism and transcription? Cell Metab 12:321–327
5. Agathocleous M, Meacham CE, Burgess RJ, Piskounova E, Zhao Z, Crane GM et al (2017) Ascorbate regulates haematopoietic stem cell function and leukaemogenesis. Nature 549:476–481
6. Müller S, Sindikubwabo F, Cañeque T, Lafon A, Versini A, Lombard B et al (2020) CD44 regulates epigenetic plasticity by mediating iron endocytosis. Nat Chem 12:929–938
7. Greer EL, Shi Y (2012) Histone methylation: a dynamic mark in health, disease and inheritance. Nat Rev Genet 13:343–357
8. Rice JC, Briggs SD, Ueberheide B, Barber CM, Shabanowitz J, Hunt DF et al (2003) Histone methyltransferases direct different degrees of methylation to define distinct chromatin domains. Mol Cell 12:1591–1598
9. Chakraborty AA, Laukka T, Myllykoski M, Ringel AE, Booker MA et al (2019) Histone demethylase KDM6A directly senses oxygen to control chromatin and cell fate. Science 363: 1217–1222
10. Janke R, Dodson AE, Rine J (2015) Metabolism and epigenetics. Annu Rev Cell Dev Biol 31:473–496
11. Husmann D, Gozani O (2019) Histone lysine methyltransferases in biology and disease. Nat Struct Mol Biol 26:880–889
12. Losman JA, Koivunen P, Kaelin WG Jr (2020) 2-Oxoglutarate-dependent dioxygenases in cancer. Nat Rev Cancer 20:710–726
13. Shi Y, Sawada J, Sui G, Affar EB, Whetstine JR et al (2003) Coordinated histone modifications mediated by a CtBP co-repressor complex. Nature 422:735–738
14. Shi Y, Lan F, Matson C, Mulligan P, Whetstine JR et al (2004) Histone demethylation mediated by the nuclear amine oxidase homolog LSD1. Cell 119:941–953
15. Chen H, Giri NC, Zhang R, Yamane K, Zhang Y et al (2010) Nickel ions inhibit histone demethylase JMJD1A and DNA repair enzyme ABH2 by replacing the ferrous iron in the catalytic centers. J Biol Chem 285:7374–7383
16. Müller S, Versini A, Sindikubwabo F, Belthier G, Niyomchon S et al (2018) Metformin reveals a mitochondrial copper addiction of mesenchymal cancer cells. PLoS One 13: e0206764

17. Niwa M, Hirayama T, Okuda K, Nagasawa H (2014) A new class of high-contrast Fe (II) selective fluorescent probes based on spirocyclized scaffolds for visualization of intracellular labile iron delivered by transferrin. Org Biomol Chem 12:6590–6597
18. Turcu AL, Versini A, Khene N, Gaillet C, Cañeque T et al (2020) DMT1 inhibitors kill cancer stem cells by blocking lysosomal iron translocation. Chemistry 26:7369–7373
19. Guerquin-Kern JL, Wu TD, Quintana C, Croisy A (2005) Progress in analytical imaging of the cell by dynamic secondary ion mass spectrometry (SIMS microscopy). Biochim Biophys Acta 1724:228–238
20. Nunez J, Renslow R, Cliff JB 3rd, Anderton CR (2017) NanoSIMS for biological applications: current practices and analyses. Biointerphases 13:03B301

Part IV

Structural Investigation of Histone Methyltransferases

Chapter 7

Determination of Histone Methyltransferase Structure by Crystallography

Jon R. Wilson

Abstract

As discussed in previous chapters, the methylation of specific arginine and lysine side chains is carried out by two families of histone methyltransferases, the Protein Arginine Methyltransferase (PRMT) family for arginine, and the SET domain family for lysine. The methylation of H3K79 by Dot1 is a notable outlier. In all cases, X-ray crystallography has been a powerful technique that has provided the framework for understanding the enzyme mechanism, kinetics, regulation and specificity of these enzymes and is now a platform for the design of compounds aimed to inhibit their activity either to further understand their function or in a therapeutic setting. Notably, in combination with the structures of the complementary recognition domains that recognize their products, these structures have provided an important insight into how integral the number of methyl groups added to the acceptor amine is to making histone methylation a key process in epigenetic regulation of gene transcription. Here the concepts applied to determine their structure by X-ray crystallography are outlined, with particular emphasis on lysine methylation by the SET domain.

Key words Structural biology, X-ray crystallography, Molecular replacement, PRMT, SET domain, Recombinant protein expression

1 Introduction

X-ray crystallography has been used extensively to determine the molecular details of the architecture and mechanism of both the histone arginine and lysine methyltransferase families [1, 2]. In addition to addressing aspects of the basic function of methyl transfer, these studies revealed important insights into concepts surrounding targeting, regulation, and how the specificity of the number of added methyl groups added to the side chain amine is achieved [3]. Given their pivotal role in many cancers, histone methyltransferases are an important target in drug discovery and high-resolution X-ray structures are an essential tool in the process of compound development [4].

Raphaël Margueron and Daniel Holoch (eds.), *Histone Methyltransferases: Methods and Protocols*,
Methods in Molecular Biology, vol. 2529, https://doi.org/10.1007/978-1-0716-2481-4_7,

Histone methyltransferases are often multi-domain proteins that in addition to the catalytic site contain other components, for example, histone reader domains [5, 6]. Although the details of higher-level folding and interactions are ultimately important for understanding the broader physiological role of histone methyltransferases in the chromatin context, successful crystallization often necessitates the reduction of a protein to a minimal construct. X-ray studies that have focused on a recombinant construct containing only the catalytic domain have revealed significant insights into mechanism and function [1, 7]. However, in the case of some proteins, such as EZH2 and MLL1, a larger protein assembly is required to stabilize the active conformation, containing a larger portion of the methyltransferase protein itself and associated partners in the cognate multi-protein complex [8–10]. This highlights the importance of using functional studies to understand the properties of the target protein when optimizing constructs for structure determination.

In higher organisms, both the PRMT and SET domain proteins have evolved into varied protein families, with individual members that have unique substrate and product specificities or variations in the accompanying chromatin recognition domains. The arginine PRMT family conserved catalytic domain consists of two subdomains, a N-terminal Rossman fold which contains the SAM binding site and a C-terminal β-barrel (Fig. 1a) [1]. When designing constructs for structural biology, it is important to consider that the active protein depends on the formation of a homodimer; the two resulting active sites remain physically independent, even though only one is active [6]. The SET domain family contains a larger number of members and consequently displays greater diversity [2, 5]. However, the core catalytic domain is structurally well conserved and spans only around 130 amino acids (Fig. 1b). The size and composition of the regions flanking the SET domain show significant divergence as can the central substrate binding (SET-I) region [11]. These flanking regions may be integral for stabilizing the active conformation of the domain or may have a regulatory function, for example, with Nsd1 [12], and it is important to consider this when designing the minimal construct. The canonical SET domain family can be expanded further by inclusion of the closely related PRDM proteins, some of which now have confirmed methyltransferase activity [5, 13]. The H3K79-specific Dot1 methyltransferase is a classical Rossmann-like (Class I) SAM-dependent methyltransferase fold [14, 15]. The design of constructs for X-ray studies is now greatly aided by the availability of the structures of a broad range of examples of these proteins in the Protein Data Bank (http://www.wwpdb.org) [16].

A key to many of the important mechanistic insights derived from X-ray studies on histone methyltransferases has depended on the inclusion of the cofactor S-adenosyl methionine (SAM) and the

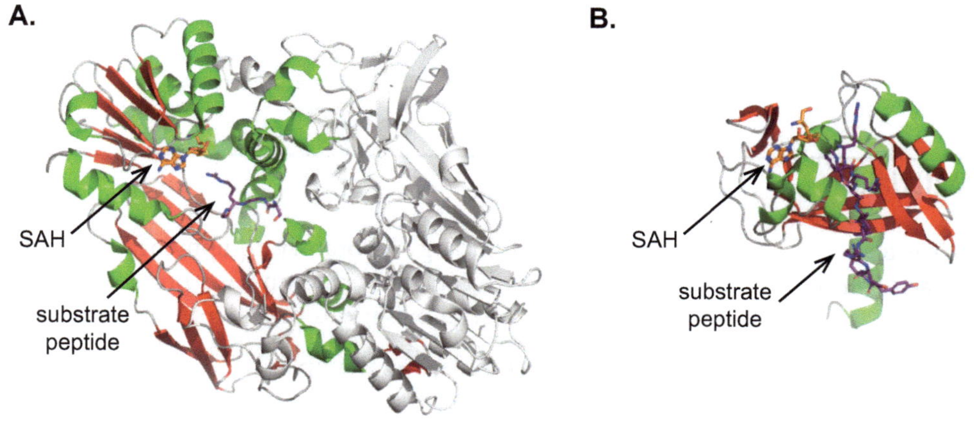

Fig. 1 Typical structures of protein methyltransferase catalytic domains. (**a**) PRMT7 from *Trypanosoma brucei* (PDB code: 4M38) in complex with SAH and substrate peptide. (**b**) Human PR-Set7 (PDB code 2BQZ) in complex with SAH and substrate peptide. Helices are shown in green, strands in red, SAH in orange, and peptide in purple

substrate, or their products or analogues, in the structure. To facilitate crystallization, rather than having a nucleosome in the ternary complex, it has been standard practice to reduce the substrate to a short peptide that spans the target residue on the histone. Due to stability issues, the cofactor is often substituted with its more stable reaction product S-adenosyl homocysteine (SAH). Significantly, although technically a product complex, inclusion of these components may stabilize the active conformation of the enzyme, especially in the area surrounding the active site. The presence of peptide substrate/product has revealed the basis of sequence recognition in the SET domain in structures where electron density of the peptide sequence flanking the target lysine is well defined and key interactions can be discerned. Intriguingly, to date, in X-ray structures of the PRMT family catalytic domains, it has been difficult to discern density for the residues flanking the target arginine side chain [6] with the ternary complex structure of PRMT7 from Trypanosoma brucei being a notable exception [17]. In pharmaceutical studies cofactor, substrate, or both are replaced with test compounds.

The strategy for crystallization, data collection, and structural determination for histone methyltransferases essentially follows standard protein X-ray crystallography methods (see for example this series, Volume 1607). Although early X-ray structure studies of these proteins required anomalous data collection for structure determination [18–21], there is now a sufficient range of structures available in the Protein Data Bank and sufficient conservation of the catalytic domain between family members that the Molecular Replacement method is likely to be viable. The method outlined below describes the expression, purification, crystallization, and structure determination of a generic SET domain construct. Further discussion of issues specific to histone methyltransferase enzymes is provided in Subheading 4.

2 Materials

2.1 Molecular Biology and Protein Purification

Standard molecular biology laboratory equipment for the cloning, expression, and purification of recombinant proteins is required. Materials can be sourced by various providers.

1. Expression plasmids, for example, pET, pGEX6P derived or equivalents for tagged expression in bacteria.
2. Competent *E. coli* strain for cloning, for example, DH5α.
3. Competent *E. coli* strain for protein expression, for example, BL21 (DE3).
4. Isopropyl-β-d-thiogalactopyranoside (IPTG) for the induction of expression if a T7 promoter is used.
5. Reagents and equipment for PCR cloning: DNA oligo synthesis, high fidelity DNA polymerase, thermocycler, microcentrifuge, DNA purification kit, agarose gels.
6. Equipment for protein purification: Shaking incubator, sonicator, a range of centrifuges—for cell pelleting a Beckman-Coulter J6MI with JS 4.2 rotor or equivalent; for cell lysis debris pelleting Beckman-Coulter Avanti J-26S XP with JA 25.50 rotor or equivalent; for protein concentration with centrifugal device Thermo Scientific Multifuge X3R or equivalent), chromatography equipment, for example, ÄKTA purification system (Cytiva Life Sciences), size exclusion column, for example (Superdex 75 (16/60) (Cytiva Life Sciences).
7. Consumables for protein purification: Protease inhibitor tablets, a nuclease, for example, Benzonase Nuclease (Merck), syringe filters (0.2 μm), affinity resin, Centrifugal concentrator (Millipore; molecular weight cut-off, 10 kDa).
8. Media: LB broth.
9. Purification buffers: Lysis buffer: 20 mM Hepes, 150 mM NaCl, 1 mM DTT, pH 8.0 (*see* **Note 1**); Gel filtration buffer: 20 mM Hepes, 150 mM NaCl, and 0.5 mM TCEP (pH 8.0).
10. Equipment and gels for protein electrophoresis and staining. For example, polyacrylamide (4–12%) bis-Tris NuPAGE system (Thermo Fisher Scientific) and Coomassie Brilliant Blue (Sigma).
11. Liquid nitrogen.

2.2 Protein Crystallography

Developments in the field mean that protein crystallization has become increasingly reliant on specialized equipment.

1. For crystallization screening a range of commercial screens from Qiagen, Hampton Research and Molecular Dimensions.
2. Crystallization plates, for example, Swissci MRC two-drop trays (Molecular Dimensions).
3. Mosquito or equivalent crystallization robot. For example, Mosquito LCP robot with a humidity chamber (TTP Labtech).
4. For crystal visualization, microscope with stereoscopic viewer. Or if available a crystal hotel system with an automated visualization system.
5. For crystal freezing MiTeGen lithographic loops, liquid nitrogen, liquid nitrogen transport dewar, liquid nitrogen storage dewar.

2.3 Data Collection and Structure Determination

Essential requirements are access to a focused X-ray radiation source and detector, and there is a requirement for a range of specialized software, although increasingly online servers are now available for general access.

1. If available, an in-house X-ray generator and detector are valuable for selecting the conditions producing the best diffracting crystals.
2. Access to a synchrotron radiation source (https://en.wikipedia.org/wiki/List_of_synchrotron_radiation_facilities) for the collection of a high-resolution dataset.
3. A range of software is required for indexing, processing, building, and refining the crystal structure. For indexing XDS [22]. For processing the CCP4 suite of programs [23]. For model building Coot Molecular Graphics [24]. Alternative software is available.

3 Methods

The strategy for structural determination of the histone methyltransferase catalytic domains effectively follows standard procedures for the production of recombinant proteins, crystallization, and structural determination. The protocol outlined below was developed for the crystallization of a recombinant SET domain construct that expresses well in *E. coli* and can be purified readily with an N-terminal affinity tag (*see* **Note 2**). The principles of this basic procedure have been effective for a number of targets.

3.1 Construct Design

1. Standard PCR cloning methods are used to generate a construct encompassing the catalytic domain of the target protein. Common expression plasmids such as the pET series for

inclusion of 6xHis tag or pGEX based vector for inclusion of GST tag as preferred with a T7 promoter to induce expression is standard. Ideally, a protease cleavage sequence, such as LEVLFQGP for rhinovirus 3C protease or ENLYFQG for TEV protease, is included in the linker between the tag and protein. This has the advantage of allowing the production of an untagged protein domain for the final structure solution (*see* **Note 3**).

3.2 Protein Production

1. Proteins are expressed in an *E. coli* expression strain, BL21 (DE3) or derivative, and grown in LB broth. An expression scale that has proved suitable to obtain sufficient protein for X-ray crystallization studies is six flasks of 800 mL culture.
2. After initial growth at 37 °C, expression in log-phase cultures is induced with 0.5 mM isopropyl-β-d-thiogalactopyranoside (IPTG). Cells are incubated overnight at 25 °C and harvested by centrifugation.
3. The cell pellet is resuspended with Lysis buffer supplemented with nuclease (*see* **Note 1**) at a ratio of 20 mL buffer per gram cell pellet.
4. The cell suspension is lysed by sonication and the debris pelleted by centrifugation.
5. The lysate is filtered using a 0.45-μm syringe filter (optional).
6. The lysate is then applied to the glutathione-Sepharose affinity resin (for GST tagged protein). Either batch or column methods are valid; however, the batch method is more adaptable for elution of the SET domain by direct cleavage on the resin to release the domain from the tag.
7. After several rounds of washing, the resin can be transferred to a column and connected to a purification system (e.g., an ÄKTA system or equivalent) and washed at a flow rate of 1 mL/min with buffer until the absorbance (280 nm) returns to baseline.
8. For cleavage the buffer flow is stopped, site-specific protease added directly to the resin, mixed, and incubated overnight. The column can be reconnected to a purification system to allow UV monitoring and collection of fractions or eluted in a fixed buffer volume.
9. The eluted protein is concentrated using centrifugal ultracentrifugation and applied to a size exclusion column, equilibrated in Gel Filtration buffer.
10. One of the most crucial factors in obtaining diffraction quality crystals is the quality of the protein preparation. An SDS-PAGE gel is run with a sample of an aliquot, 5 μL, from each step and stained with Coomassie to assess purity, with the expectation of

obtaining a single band at the final step. Typically the UV-visible spectrum obtained indicates that there is no co-purification of cofactor (SAM or SAH) (suggested techniques for more rigorous assessment of protein quality are discussed in **Note 4**).

11. The protein is concentrated by centrifugal ultrafiltration to a final concentration of ideally greater than 10 mg/mL, aliquoted, and then snap-frozen in liquid nitrogen, and stored at −80 °C (*see* **Note 5**).

3.3 Screening Crystallization Conditions to Obtain Well-Diffracting Crystals

Screening for conditions that promote crystal formation follows the standard vapor diffusion protocols and is normally performed in 96-well sitting drop format. These procedures are now greatly aided by automation. Initial screening with commercial screens is followed by optimization of hit conditions using iterative screening methods.

1. A co-crystallization strategy is employed here and has proved successful for determination of ternary complex structures in a number of targets. A benefit over the alternative soaking method is that the cofactors have a role in stabilizing the catalytically important conformation, which may help promote crystal formation. Prepare the cofactor and substrate peptide stocks as indicated by the known biology of the target domain. To generate a co-complex for crystallization either a substrate or product complex can be formed and is discussed further in **Note 6**. Mix the protein, cofactor, and substrate peptide in buffer at approximately 1:2:2 ratio to improve the likelihood of pushing the equilibrium toward the ternary complex in the final crystallization solution. The final concentration of the protein construct will depend on the target protein and should be determined empirically, but a concentration of 0.5 mM protein is a reasonable guide.
2. Drops are set up in a screening plate with a crystallization robot (Mosquito LCP robot). Typically, 0.1 μL protein mix plus 0.1 μL precipitant solution with a 50 μL reservoir. Trays are sealed and stored in a crystallization hotel at 18 °C and monitored for 12 h intervals for 4 days, then every 24 h.
3. Identify conditions that form crystals. Ideally, if an in-house generator/detector is available at this stage, crystals can be harvested and bathed in cryogenic solution based on the crystallization condition, flash-frozen in liquid nitrogen and tested to assess diffraction quality. This provides an additional criterion other than size and morphology to guide the direction of condition optimization.
4. Hit conditions are ranked, and the best conditions are subjected to further rounds of optimization of pH, salt

concentration, or precipitant concentration. Optimization can be conducted manually but more commonly will involve using a liquid handling system such as Formulator (Formulatrix). Several rounds of seeding may also be appropriate to iteratively improve crystal quality. Optimized crystals are harvested, bathed in cryogenic solution, flash-frozen, and stored under liquid nitrogen for data collection.

3.4 Data Collection and Structure Solution

1. Data collection. Crystals are shipped to a synchrotron X-ray source. It is assumed here that a single wavelength dataset will be collected for structure solution by the Molecular Replacement method. The reflections are indexed with XDS [22] and then scaled with programs from the CCP4i suite [23].
2. Molecular Replacement. An appropriate model for Molecular Replacement is selected, assuming a close match to a SET domain of known structure is available (*see* **Note 7**). The PHASER program within CCP4 or PHENIX suites is used to obtain a unique solution for the data [25].
3. Difference maps are used to rebuild and extend the initial model using the Coot molecular graphics package [24], and iterative cycles of building and refinement in REFMAC [26] are continued until satisfactory refinement statistics are obtained.

4 Notes

1. The most effective buffer conditions for a particular construct will vary. The selection of pH 8 reflects a consideration of the nuclear location of the histone methyltransferase and conditions that favor deprotonation of the target lysine for catalysis. A higher salt concentration (300–500 mM NaCl) has proved successful for the purification of some constructs, but this must be determined empirically.
2. Although the purification of many catalytic domains of histone methyltransferase proteins has proved amenable to straightforward tagged expression and purification in *E. coli*, this cannot be guaranteed for all targets. To overcome issues encountered in expression or purification, it is useful to explore variations in domain boundaries, different tagging systems, or *E. coli* strains. Alternative expression systems such as insect cells or 293F cells for mammalian expression may also offer effective alternatives. It may also be that a closely related homolog of the target protein from another species behaves better in the recombinant systems. If the biology so indicates then a strategy of multiprotein expression to reproduce the cognate physiological complex may be required.

3. For some proteins, it may be desirable to leave the tag in place as it may assist in maintaining the solubility of the minimal domain construct. In this case, standard elution procedures with glutathione in the case of GST tag or imidazole gradient in the case of a Histidine tag are used.
4. The minimal basis for accepting that the protein preparation is suitable for the next step in structure determination is single-band purity on an SDS-PAGE gel stained with Coomassie blue that corresponds to a symmetrical gel filtration elution peak. For samples that show signs of degradation, impurities, or co-purification with chaperone proteins, it is necessary to re-evaluate purification conditions and/or construct design and engage in an iterative process of optimization. For example, the inclusion of additional purification steps, typically DEAE or Q ion exchange chromatography, can be inserted following the affinity matrix elution step. Ideally an assessment of enzyme activity should be used to confirm that the selected domain boundaries are valid and that the purified construct represents an active catalytic domain (*see* Chapter 2). Further analysis steps that may be applied to confirm an optimal preparation and assess homogeneity include Multi-Angle Laser Light Scattering and Dynamic Light Scattering methods. The inclusion of glycerol at 5–10% is also a consideration at this stage and the requirement should be judged empirically against the naïve view that if it can be avoided, this may benefit crystallization. The best-behaved constructs can readily survive a single cycle of snap-freezing and thawing. For some constructs, glycerol is an essential additive, and others may require the addition of glycerol throughout purification process to maintain solubility.
5. Protein concentrations are determined by ultraviolet-visible absorbance spectroscopy using the calculated extinction coefficient at 280 nm derived from the tyrosine and tryptophan content using online tools. The concentration selected for protein stocks will depend on the construct, and some may tolerate a significantly higher concentration.
6. X-ray structures of the histone methyltransferase are most informative when a complex with cofactor and substrate is studied in order to visualize the key interactions in the active site and the basis for substrate recognition. A ternary substrate complex formed with SAM and an unmethylated (or partially methylated) substrate peptide can show the starting conformation of the reaction, but there is likely to be turnover during the crystallization incubation period, and this may result in a heterogeneous sample, which may compromise crystallization or reduce the resolution of the active site region. The product complex of SAH with methylated peptide has more stable components; however, the affinity of the SET domain for

SAH is significantly weaker than for SAM. Nevertheless, at the concentrations used in crystallization, the complex with SAH has proved sufficiently stable for structure determination. Both SAM and SAH stock solutions should be prepared fresh from solid. SAM is readily soluble in most buffers, while SAH is less soluble in some buffers and it may be necessary to prepare a stock solution in water. Stocks should be filtered with a 0.1-μm or smaller syringe filter after solubilizing. The concentration of the stock solutions can be estimated using UV-visible absorbance from the molar extinction coefficient at 260 nm (15.4 mM^{-1} cm^{-1} for SAM and 16 mM^{-1} cm^{-1} for SAH). When mixing cofactor and substrate to the SET domain, with some proteins, mild cloudiness is observed temporarily before rapidly returning to a clear solution. If precipitation is persistent, then it is necessary to explore conditions, such as pH and salt, to avoid this.

7. Both the PRMT and SET domain families are conserved in eukaryotes, but reflecting the increasing demands for gene regulation in higher-order species, there is a corresponding increase in the number of paralogs. There are now many structures in the Protein Data Bank representing a broad sample across these families. Selecting a suitable model for molecular replacement may accordingly be a straightforward matter of selecting a previous structure of the target protein or a close homolog from another species. If an obvious close homolog is not readily apparent and if Molecular Replacement has initially failed to find a solution, it may be useful to consider the anatomy of the catalytic domain to revise the search model [27]. Although small, the SET domain can be further divided into five subdomains. Two of these, the SET-N and SET-C, are strongly conserved β-structures. These are separated by the SET-I region. Although this region usually retains the canonical structure, it often diverges in sequence as its helical segment is the main region recognizing the sequence flanking the target substrate lysine. Although a subgroup of proteins has a conserved cysteine-rich zinc-binding domain, in others there may be considerable variability. The region flanking the N-terminus of the domain is highly variable. Trimming the model used for Molecular Replacement back to just the best conserved structural elements with a polyalanine sequence may be an effective strategy.

Acknowledgments

This work was supported by the Worldwide Cancer Fund (AM) and by the Francis Crick Institute which receives its core funding from Cancer Research UK (FC001078), the UK Medical Research Council (FC001078), and the Wellcome Trust (FC001078).

References

1. Tewary SK, Zheng YG, Ho MC (2019) Protein arginine methyltransferases: insights into the enzyme structure and mechanism at the atomic level. Cell Mol Life Sci 76:2917–2932
2. Dillon SC, Zhang X, Trievel RC, Cheng X (2005) The SET-domain protein superfamily: protein lysine methyltransferases. Genome Biol 6:227
3. Del Rizzo PA, Trievel RC (2011) Substrate and product specificities of SET domain methyltransferases. Epigenetics 6:1059–1067
4. Ferreira de Freitas R, Ivanochko D, Schapira M (2019) Methyltransferase inhibitors: competing with, or exploiting the bound cofactor. Molecules 24:4492
5. Herz HM, Garruss A, Shilatifard A (2013) SET for life: biochemical activities and biological functions of SET domain-containing proteins. Trends Biochem Sci 38:621–639
6. Morales Y, Caceres T, May K, Hevel JM (2016) Biochemistry and regulation of the protein arginine methyltransferases (PRMTs). Arch Biochem Biophys 590:138–152
7. Couture JF, Dirk LM, Brunzelle JS, Houtz RL, Trievel RC (2008) Structural origins for the product specificity of SET domain protein methyltransferases. Proc Natl Acad Sci U S A 105:20659–20664
8. Justin N, Zhang Y, Tarricone C, Martin SR, Chen S, Underwood E, De Marco V, Haire LF, Walker PA, Reinberg D, Wilson JR, Gamblin SJ (2016) Structural basis of oncogenic histone H3K27M inhibition of human polycomb repressive complex 2. Nat Commun 7:11316
9. Jiao L, Liu X (2015) Structural basis of histone H3K27 trimethylation by an active polycomb repressive complex 2. Science 350:aac4383
10. Li Y, Han J, Zhang Y, Cao F, Liu Z, Li S, Wu J, Hu C, Wang Y, Shuai J, Chen J, Cao L, Li D, Shi P, Tian C, Zhang J, Dou Y, Li G, Chen Y, Lei M (2016) Structural basis for activity regulation of MLL family methyltransferases. Nature 530:447–452
11. Xiao B, Wilson JR, Gamblin SJ (2003) SET domains and histone methylation. Curr Opin Struct Biol 13:699–705
12. Qiao Q, Li Y, Chen Z, Wang M, Reinberg D, Xu RM (2011) The structure of NSD1 reveals an autoregulatory mechanism underlying histone H3K36 methylation. J Biol Chem 286: 8361–8368
13. Vervoort M, Meulemeester D, Behague J, Kerner P (2016) Evolution of Prdm Genes in Animals: Insights from Comparative Genomics. Mol Biol Evol 33:679–696
14. Falnes PO, Jakobsson ME, Davydova E, Ho A, Malecki J (2016) Protein lysine methylation by seven-beta-strand methyltransferases. Biochem J 473:1995–2009
15. Sawada K, Yang Z, Horton JR, Collins RE, Zhang X, Cheng X (2004) Structure of the conserved core of the yeast Dot1p, a nucleosomal histone H3 lysine 79 methyltransferase. J Biol Chem 279:43296–43306
16. Berman H, Henrick K, Nakamura H (2003) Announcing the worldwide Protein Data Bank. Nat Struct Biol 10:980
17. Wang C, Zhu Y, Caceres TB, Liu L, Peng J, Wang J, Chen J, Chen X, Zhang Z, Zuo X, Gong Q, Teng M, Hevel JM, Wu J, Shi Y (2014) Structural determinants for the strict monomethylation activity by trypanosoma brucei protein arginine methyltransferase 7. Structure 22:756–768
18. Zhang X, Zhou L, Cheng X (2000) Crystal structure of the conserved core of protein arginine methyltransferase PRMT3. EMBO J 19: 3509–3519
19. Zhang X, Tamaru H, Khan SI, Horton JR, Keefe LJ, Selker EU, Cheng X (2002) Structure of the Neurospora SET domain protein DIM-5, a histone H3 lysine methyltransferase. Cell 111:117–127
20. Trievel RC, Flynn EM, Houtz RL, Hurley JH (2003) Mechanism of multiple lysine methylation by the SET domain enzyme Rubisco LSMT. Nat Struct Biol 10:545–552
21. Wilson JR, Jing C, Walker PA, Martin SR, Howell SA, Blackburn GM, Gamblin SJ, Xiao B (2002) Crystal structure and functional analysis of the histone methyltransferase SET7/9. Cell 111:105–115
22. Kabsch W (2010) XDS. Acta Crystallogr D Biol Crystallogr 66:125–132
23. Bailey S (1994) The Ccp4 suite—programs for protein crystallography. Acta Crystallographica Section D-Biological Crystallography 50:760–763
24. Emsley P, Cowtan K (2004) Coot: model-building tools for molecular graphics. Acta Crystallogr D Biol Crystallogr 60:2126–2132
25. McCoy AJ, Grosse-Kunstleve RW, Adams PD, Winn MD, Storoni LC, Read RJ (2007) Phaser crystallographic software. J Appl Crystallogr 40:658–674
26. Murshudov GN, Vagin AA, Dodson EJ (1997) Refinement of macromolecular structures by the maximum-likelihood method. Acta Crystallogr D Biol Crystallogr 53:240–255
27. Cheng X, Collins RE, Zhang X (2005) Structural and sequence motifs of protein (histone) methylation enzymes. Annu Rev Biophys Biomol Struct 34:267–294

Chapter 8

Determination of Histone Methyltransferase Structures in Complex with the Nucleosome by Cryogenic Electron Microscopy

Cathy J. Spangler and Robert K. McGinty

Abstract

Cryogenic electron microscopy (cryo-EM) has recently emerged as an optimal technique for the determination of histone methyltransferase-nucleosome complex structures. Histone methyltransferases are a group of enzymes that posttranslationally methylate histone lysine and arginine residues on the nucleosome, providing important epigenetic signals that regulate gene expression. Here we describe a protocol to solve the structure of histone lysine methyltransferase Dot1L bound to a chemically ubiquitylated nucleosome, including complex reconstitution, crosslinking, grid preparation, and data collection and analysis. Throughout, we discuss key steps requiring optimization to allow this protocol to serve as a starting point for the determination of new histone methyltransferase-nucleosome complex structures.

Key words Methyltransferase, Nucleosome, Histone, Cryogenic electron microscopy, Cryo-EM, Structural biology

1 Introduction

Histone methyltransferases are a class of epigenetic enzymes that catalyze methylation of histone lysine and arginine residues [1, 2]. The two major subclasses of histone methyltransferases are lysine methyltransferases, which mono-, di-, or trimethylate the terminal amino group in the side chain of lysine residues, and arginine methyltransferases, which methylate one or both of the terminal nitrogens in the guanidino group of arginine residues to generate mono-, asymmetric di-, or symmetric dimethyl arginine. Both subclasses rely on the cofactor *S*-adenosyl methionine (SAM) as a methyl donor and exhibit some specificity to both histone methylation position and degree of methylation.

Histone proteins are the basic building blocks of the nucleosome, the fundamental unit of chromatin that facilitates organization and regulation of eukaryotic genomes. Nucleosomes are made

Raphaël Margueron and Daniel Holoch (eds.), *Histone Methyltransferases: Methods and Protocols*, Methods in Molecular Biology, vol. 2529, https://doi.org/10.1007/978-1-0716-2481-4_8,

up of an octameric protein core that contains two copies of each of the four core histone proteins (H2A, H2B, H3, and H4), around which DNA wraps [3]. The nucleosome serves not only as a structural scaffold for genome compaction but also as a complex signaling hub that modulates the activity of numerous nuclear proteins. Posttranslational histone methylation functions as a molecular signal on the nucleosome to ensure proper genome function, and its misregulation is associated with diseases such as cancer [4–7]. Since histone methyltransferases function on nucleosome substrates, determining the precise molecular details of histone methyltransferase–nucleosome interactions is an essential step toward understanding their contributions to genome regulation in normal and disease states.

Due to the large size and flexibility of histone methyltransferase–nucleosome complexes, high-resolution structural details of these complexes were historically difficult to determine by X-ray crystallography. Instead, most early structural studies of histone methyltransferases and their substrates included the isolated catalytic domains of methyltransferases bound to peptides comprising short portions of their target histone sequences [8]. Although these studies provided valuable insight into histone methyltransferase function, they failed to characterize the interactions that must occur with nucleosomal histones and DNA in native chromatin. This is especially critical for methyltransferases, like Dot1L, that require nucleosome substrates to function and fail to bind to and methylate individual histones or histone peptides [9, 10]. Due to recent hardware and software advances that spurred the resolution revolution in cryogenic electron microscopy (cryo-EM), structural determination of entire histone methyltransferase-nucleosome complexes is now more attainable, and a handful of such complex structures have been determined for lysine methyltransferases [11–25] (Fig. 1).

Herein, we describe the sample preparation and cryo-EM data acquisition and analysis strategies that enabled us to solve the structure of the histone H3 lysine 79-specific methyltransferase Dot1L bound to an H2BK120 chemically ubiquitylated nucleosome [11]. Many general principles of sample preparation and data analysis are shared between our strategy and those leading to several other histone methyltransferase–nucleosome complex structures published to date. However, as each complex has its own unique challenges, optimization of steps along the experimental and computational pipelines is typically required. We annotate alternative approaches that have been successfully utilized for cryo-EM structure determination of other histone methyltransferase–nucleosome complexes throughout this protocol to highlight key steps that may require optimization for the determination of new structures.

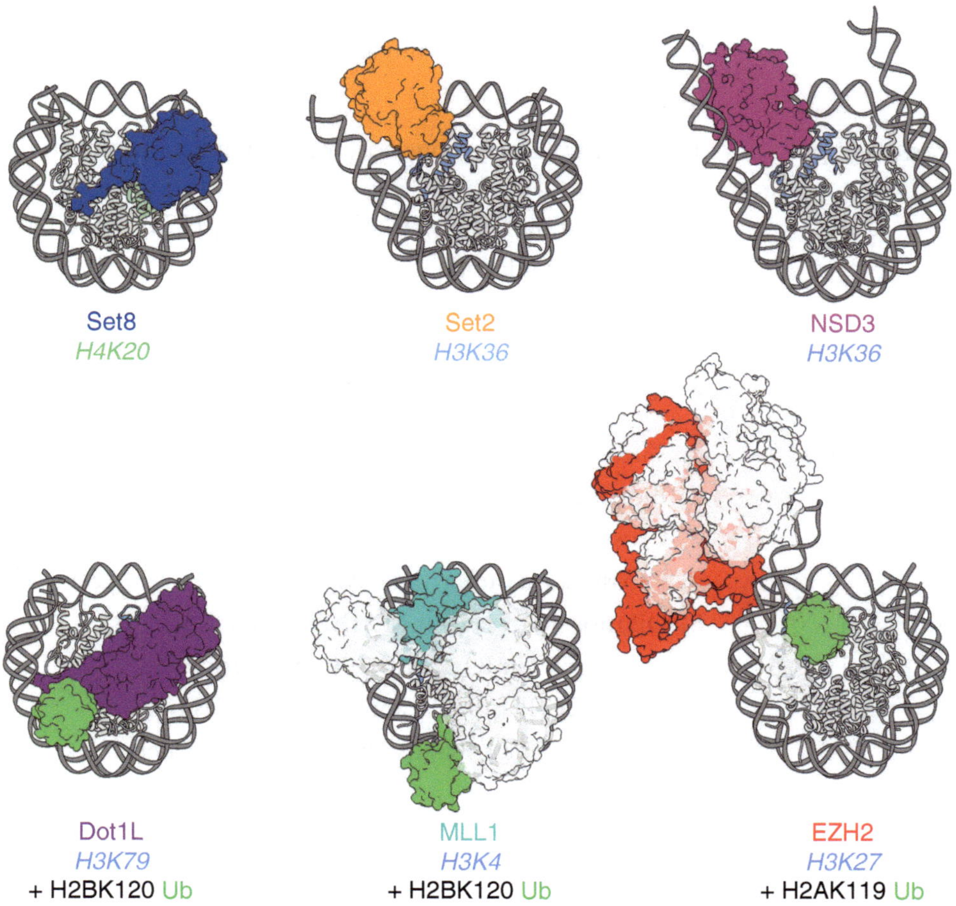

Fig. 1 Structures of histone methyltransferase–nucleosome complexes solved by cryo-EM. The following models are shown: Set8 (PDBID: 7D1Z), Set2 (PDBID: 6NZO), NSD3 (PDBID: 7CRP), Dot1L (PDBID: 6NN6), MLL1 (PDBID: 6KIU), EZH2 (PDBID: 6WKR)

2 Materials

2.1 Reconstitution

1. Superdex 200 10/300 increase column and AKTA Pure 25 M HPLC system.
2. Reconstitution buffer: 10 mM Tris–Cl, pH 7.6, 150 mM NaCl, 1 mM dithiothreitol (DTT), 0.1 mM phenylmethylsulfonylfluoride (PMSF).
3. 1 M Tris–Cl, pH 7.6.
4. 5 M NaCl.
5. 1 M DTT.
6. 100 mM PMSF (in 100% ethanol).
7. Purified methyltransferase enzyme.

8. Purified reconstituted nucleosome.
9. Vivaspin 500 centrifugal concentrator (30 kDa MWCO) (Vivascience).

2.2 Crosslinking

1. Methyltransferase–nucleosome complex concentrated to ≥1 mg/mL.
2. Crosslinking buffer: 20 mM HEPES, pH 7.5, 150 mM NaCl.
3. 25% glutaraldehyde.
4. Zeba spin desalting column (40 kDa MWCO), Micro (Thermo).
5. Quenching buffer: 0.5 M Tris–HCl, pH 7.5.

2.3 Grid Freezing

1. Vitrobot Mark IV (FEI).
2. Blotting paper.
3. Crosslinked methyltransferase-nucleosome complex.
4. Crosslinking buffer: 10 mM HEPES, pH 7.5, 150 mM NaCl.
5. Dilution buffer: 10 mM HEPES, pH 7.5.
6. Quantifoil Cu R1.2/1.3 300 mesh grids.
7. PDC-32G Plasma Cleaner (Harrick Plasma).
8. Traceable Total-Range Thermometer (Fisher).
9. Grid boxes.
10. Tweezers.
11. Glass slide.
12. Petri dish.
13. Nitrogen dewar.
14. Liquid nitrogen.
15. Liquid ethane.
16. Kimwipes.
17. Timer.

2.4 Grid Screening and Data Collection

1. Autogrid (C-clip) rings, C-clips, and autogrid assembly station (Thermo).
2. Electron microscope (FEI Titan Krios or Talos Arctica) equipped with suitable direct electron detector (Falcon III).
3. Microscope control software (e.g., EPU, SerialEM).
4. Live data processing pipeline (e.g., cryoSPARC live).

2.5 Data Analysis and Model Building

1. GPU Workstation (Single Particle Performance GPU Workstation with 4× NVIDIA GTX 1080Ti GPUs and 256GB ECC DDR4-2400 RAM) or access to a GPU-equipped cluster.

2. Cryo-EM data analysis software (RELION).
3. Model-building and visualization software (Phenix, Coot, Chimera, PyMOL).

3 Methods

3.1 Reconstitute, Purify, and Concentrate Methyltransferase–Nucleosome Complex

In order to reconstitute methyltransferase–nucleosome complexes in vitro, we combine nucleosomes with incremental additions of enzyme and purify the resulting complex by size exclusion chromatography. We have found this general approach to be successful with many protein–nucleosome complexes by helping to ensure successful complex formation and sample homogeneity before grid freezing. However, crosslinking the complex before purification or omission of complex purification entirely have also been successful strategies for some histone methyltransferase–nucleosome complexes [13, 18, 24]. Our procedure for the preparation of Dot1L-ubiquitylated nucleosome complex is detailed below.

1. In a microcentrifuge tube at room temperature, mix 2.5 molar equivalents of methyltransferase (*see* **Note 1**) with nucleosome (*see* **Note 2**) in Reconstitution buffer (*see* **Note 3**) at a total concentration of 6.25 mg/mL in 200 μL. Titrate methyltransferase into complex reconstitution in four equal additions, waiting for 5 min in between each addition to allow for complex equilibration (*see* **Note 4**).

1 M Tris–Cl, pH 7.5	5.0 μL
5 M NaCl	4.2 μL
1 M DTT	0.2 μL
100 mM PMSF (prepared fresh in 100% EtOH)	0.2 μL
H_2O	114.3 μL
H2BK120Ub nucleosome (93.5 μM; 20.9 mg/mL)	39.1 μL (3.7 nmol)
Dot1L methyltransferase (248.0 μM; 11.8 mg/mL)	36.9 μL (9.3 nmol) [in 4 additions of 9.2 μL, mixing after each addition]
Total volume	200 μL

2. Clear sample by centrifugation in microcentrifuge at 17,000 × *g* for 3 min at 4 °C.
3. Load sample onto Superdex 200 10/300 increase column equilibrated in Reconstitution buffer and analyze elution fractions on 18% SDS-PAGE gel to verify complex formation (Fig. 2). Determine fractions that have sufficiently bound, but

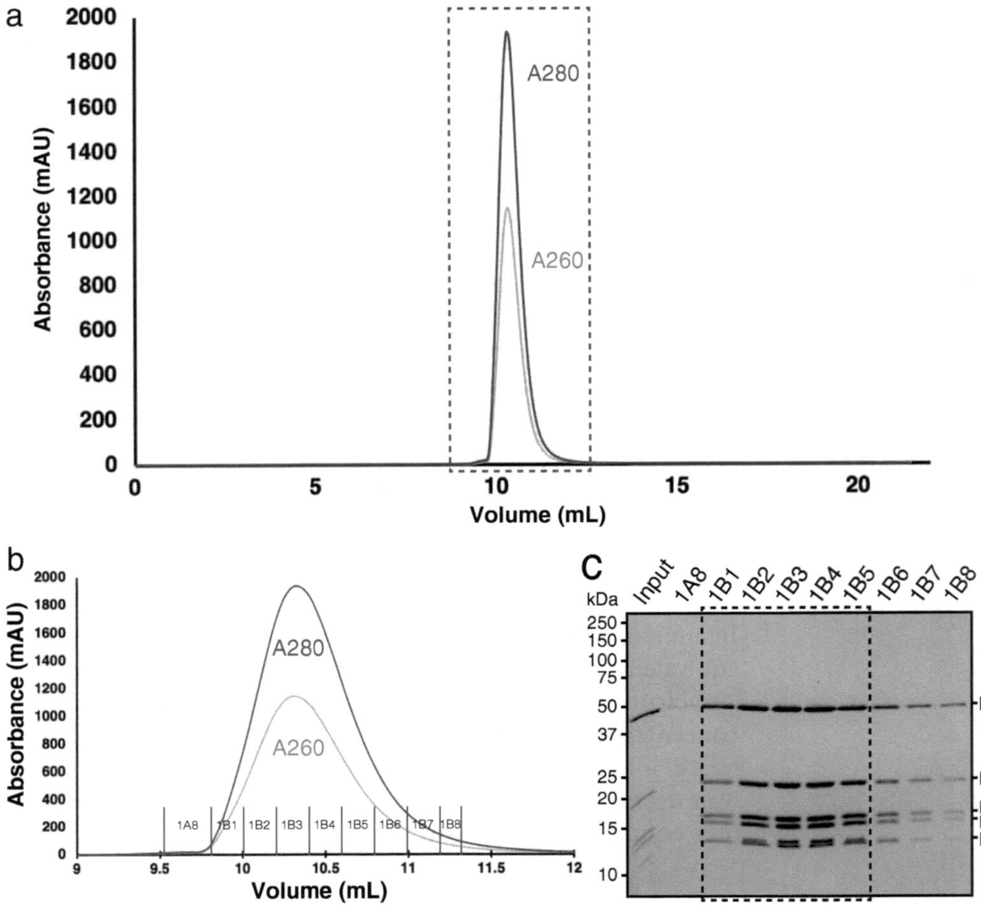

Fig. 2 Dot1L-H2BK120Ub nucleosome purification. (**a**) Chromatogram for size exclusion chromatography purification of Dot1L-H2BK120Ub nucleosome complex. (**b**) Zoom-in on region of chromatogram from (**a**) indicated by dotted grey box with annotated peak fractionation of fractions 1A8-1B8. (**c**) SDS-PAGE gel of peak elution fractions with pooled fractions indicated by dotted black box

not overbound, methyltransferase. We often exclude the first and/or last fraction of a peak if the peak has an asymmetric shoulder(s).

4. Pool fractions containing complex and concentrate to at least 1 mg/mL (we typically aim for ~10 mg/mL) using Vivaspin 500 (30K MWCO) at 9600 × *g* at 4 °C. Time required for concentration varies by complex identity, sample volume, and target concentration.

5. Quantitate complex. We quantitate nucleosome complexes using A260 measurements of two replicate 1:20 sample dilutions in 0.2 M NaOH and using an empirically derived extinction coefficient of 40 μg/mL DNA = 1 absorbance unit.

3.2 Crosslink Methyltransferase Complex

While it is typically ideal to visualize methyltransferase–nucleosome complexes in a native state, in our experience, these complexes frequently lack sufficient stability during cryo-EM sample preparation to allow structure determination without crosslinking. As an early stage in the optimization of cryo-EM samples, we prepare both native (i.e., uncrosslinked) and crosslinked complex samples. Below we describe our procedure for glutaraldehyde crosslinking of the Dot1L-ubiquitylated nucleosome complex (*see* **Note 5**).

1. Use Zeba Spin Desalting Column to buffer exchange sample into Crosslinking buffer (3 μL of ~1 mg/mL sample is required per cryo-EM grid) using manufacturer's protocol.
2. Prepare fresh dilution of 2% glutaraldehyde in Crosslinking buffer from 25% stock by adding 4 μL of 25% glutaraldehyde stock to 46 μL Crosslinking buffer.
3. Add 0.6 μL of 2% glutaraldehyde to a 12 μL sample to give a final concentration of 0.1% glutaraldehyde. Incubate the crosslinking reaction at room temperature for 5 min. Volume of crosslinking sample can be scaled as needed.
4. Quench reaction by the addition of 0.48 μL of Quenching buffer to a final concentration of 20 mM Tris–Cl followed by incubation of the sample at room temperature for an additional 15 min.
5. Use a Zeba Spin Desalting Column to buffer exchange sample into Crosslinking buffer (or another desired buffer) prior to grid freezing.
6. Requantitate sample as described in **step 5** in Subheading 3.1 for accurate post-crosslinking concentration. Typically, the sample is now at ~60–80% of the pre-crosslinking concentration. Requantitation at this stage is necessary for reproducibility of particle density on cryo-EM grids.
7. Although it is the best practice to proceed to grid preparation as quickly as possible after complex reconstitution, we have observed that the Dot1L–ubiquitylated nucleosome complex is stable overnight at 4 °C.

3.3 Prepare Cryo-EM Grids

Cryo-EM grid preparation techniques are often laboratory and device dependent, and numerous resources exist to aid in general grid preparation techniques [26, 27]. Here we describe our use of the Vitrobot Mark IV (FEI) grid freezing device to prepare cryo-EM grids with the Dot1L–ubiquitylated nucleosome complex. As there are many environment-dependent factors contributing to grid quality, it is recommended that users new to grid freezing ask frequent device users about the ideal settings on their local grid preparation instruments.

1. Install blotting paper, fill humidifier with 60 mL of water, turn on Vitrobot, and set freezing chamber to 4 °C and 100% humidity. About 30 min is required to fully cool and humidify chamber.
2. While waiting for Vitrobot to cool, assemble ethane cup with grid box holder and spider in styrofoam chamber, label and loosen grid boxes and place in grid box holder, and fill dewar with liquid nitrogen. Gently remove grids from storage box with tweezers and place onto glass slide inside of a Petri dish. Although we typically freeze in batches of eight grids, you may freeze up to 16 grids at a time with the Vitrobot (*see* **Note 6**).
3. Once Vitrobot is equilibrated, cool ethane chamber by filling ethane cup and surrounding styrofoam chamber with liquid nitrogen. Wait for at least 5 min for ethane cup to fully cool.
4. While ethane chamber is cooling, plasma clean grids. We used a PDC-32G basic plasma cleaner and cleaned grids for 60 s at maximum strength (*see* **Note 7**). Aim to be consistent with how long grids sit in between plasma cleaning and freezing. We routinely use all grids within 20–30 min of plasma cleaning.
5. Condense ethane in bronze ethane cup (*see* **Note 8**). Wipe off ethane-condensing tip with Kimwipe immediately before plunging into cup to minimize ice contamination. Be sure to fill ethane cup high enough for grids to be completely submerged upon plunging but avoid spilling ethane into surrounding liquid nitrogen. Once ethane is condensed, wait for about 2 min and then monitor temperature every 30 s. Remove spider with tweezers once ethane has reached −183 °C.
6. Set desired Vitrobot settings and freeze grids. We recommend diluting samples to their final sample concentration and buffer composition immediately before blotting to minimize any deleterious effects of the dilution and/or buffer change that may increase over time (e.g., aggregation or instability). For the Dot1L–ubiquitylated nucleosome complex, the sample was diluted with a mix of Crosslinking and Dilution buffers to a final concentration of 0.93 mg/mL at 50 mM NaCl. Optimal Vitrobot settings for our Dot1L–nucleosome complex sample are below, along with ranges used for similar structures (*see* **Note 9**).
   (a) Sample concentration: 0.93 mg/mL (0.5–1 mg/mL).
   (b) Sample volume: 3 μL per grid (2–3 μL).
   (c) Blot time: 4 s (3–4 s).
   (d) Wait time: 0 s.
   (e) Blot force: −15 (−15 to 0).
7. Store grids in grid box in liquid nitrogen storage dewar until use.

3.4 Grid Screening and Data Collection

As ice quality has significant impacts on data quality, one of the main goals of grid screening is to determine the ice thickness and resulting resolution limit of data collected on each grid. Efficiently utilized screening time can also answer important questions regarding complex stability and quality. For these reasons, we recommend using cryoSPARC live (or an equivalent live processing platform) during grid screening.

1. Clip grids with C-ring and C-clip in grid transfer station. This step is required for autoloader use on FEI Titan Krios, Talos Arctica, and equivalent microscopes. Assemble clipped grids in autoloader and load into microscope.
2. Screen grids to assess quality of sample, including particle density and orientational diversity, and ice thickness (Fig. 3).

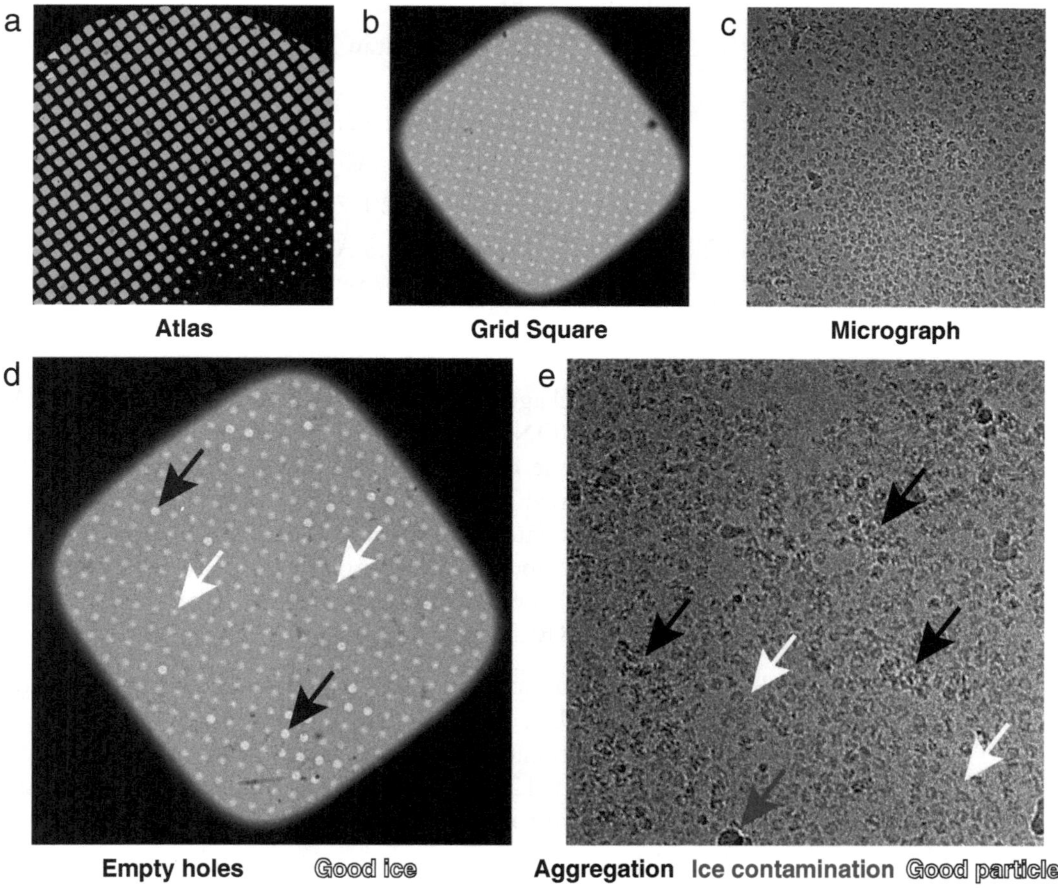

Fig. 3 Grid screening for Dot1L-ubiquitylated nucleosome complex. (**a**) Atlas of grid exemplifying typical Vitrobot ice gradient. All subsequent panels are images taken from this grid. (**b**) Grid square with desired ice thickness for Dot1L–ubiquitylated nucleosome complex. (**c**) Micrograph showing optimal particle density and distribution. (**d**) Grid square with examples of good ice and empty holes. (**e**) Micrograph with examples of particle aggregation and ice contamination, in addition to good particles

We favor the use of live data processing pipelines, such as cryoSPARC live, to quickly identify if the complex is stable on the grid and to calculate the approximate resolution limits of data during grid screening based on contrast transfer function (CTF) estimation. Use information gained from screening to iteratively improve both sample preparation and grid freezing parameters.

3. After optimizing conditions, we typically freeze and screen up to eight identically prepared grids and advance the best grids to data collection.
4. Collect dataset. The following are data collection instrumentation and parameters used for our Dot1L–nucleosome complex dataset and relevant alternatives and ranges used for other published histone methyltransferase–nucleosome complex structures.
   (a) Microscope: FEI Titan Krios (FEI Talos Arctica).
   (b) Detector: Falcon III (Gatan K2 or K3 or Thermo Falcon II or IV).
   (c) Number of micrographs: 1238 (1000–15,000).
   (d) Total exposure: 42 e/Å^2 (40–100 e/Å^2).
   (e) Number of frames: 30 (20–70).
   (f) Pixel size: 1.08 Å (0.4–1.3 Å).
   (g) Magnification: 75,000× (28,000–130,000×).
   (h) Defocus range: −1.25 to −2.75 μm (−1 to −3 μm).

3.5 Data Analysis

Many software packages can be used for single particle cryo-EM data analysis (RELION, cryoSPARC, etc.). For the purposes of this protocol, an example data analysis workflow in RELION will be discussed for our previously published 1238-micrograph dataset on the Dot1L–ubiquitylated nucleosome complex [11] (Fig. 4). This workflow has been updated from the structure's original publication, with the data being re-analyzed in the most recent major release version of RELION (3.1) [28].

1. Import, motion correct, and perform CTF estimation on micrographs. Use subset selection job to save micrographs with a CTF-estimated resolution less than 5 Å. In this example dataset, 1182 of 1238 micrographs pass this resolution threshold and proceed to particle picking.
2. Autopick micrographs using RELION's Laplacian-of-Gaussian blob detection option with the following settings. Diameters can be adjusted based on predicted size of complex being analyzed.

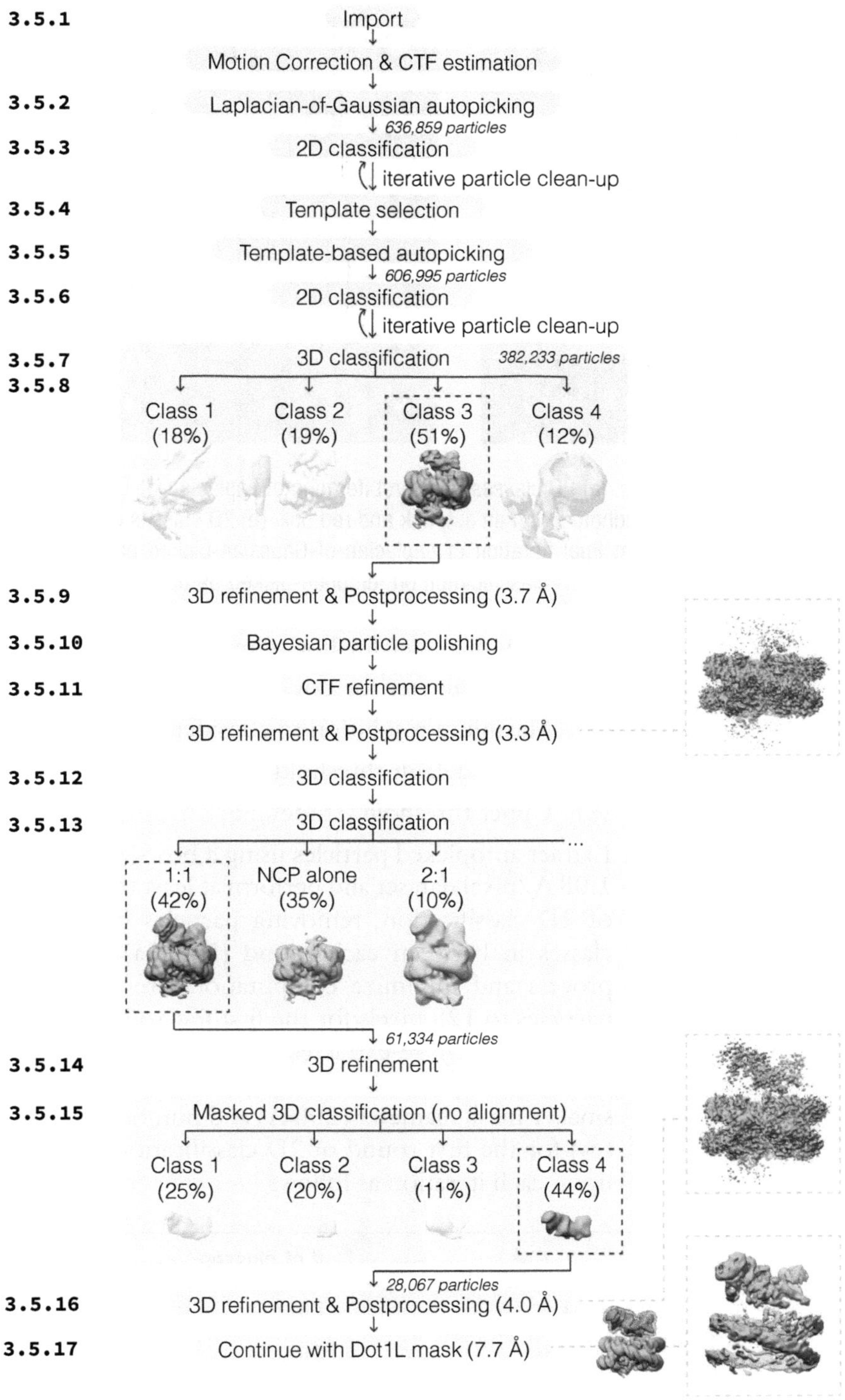

Fig. 4 Example data analysis workflow for Dot1L–ubiquitylated nucleosome complex dataset

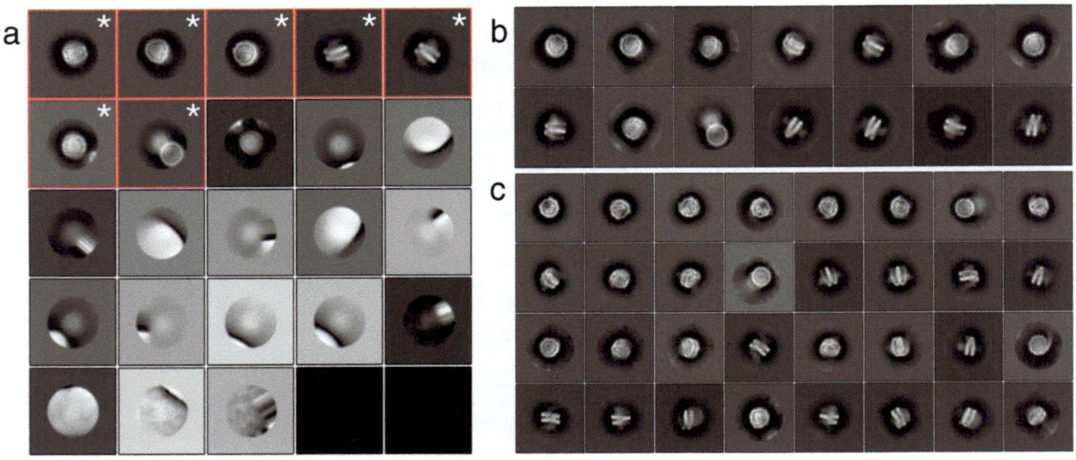

Fig. 5 2D classification examples. (**a**) 2D classes from first iteration of Laplacian-of-Gaussian-picked particle clean-up. Selected classes are denoted with an asterisk and red box. (**b**) 2D classes chosen as templates for template-based autopicking from final iteration of Laplacian-of-Gaussian-picked particle classification. (**c**) Selected classes from final round of 2D classification on template-picked particles

(a) Min. diameter for LoG filter (Å): 100.

(b) Max. diameter for LoG filter (Å): 180.

(c) Maximum resolution to consider (Å): 20.

(d) Adjust default threshold (stddev): 1.

(e) Upper threshold (stddev): 5.

3. Extract autopicked particles using a box size of 330 pixels for a 1.08 Å/pixel dataset and perform at least three iterative rounds of 2D classification, removing particles binned into inferior classes in between each round (Fig. 5a). To speed up this process and minimize computational resources, we rescaled particles to 128 pixels for the first few rounds of 2D classification and then re-extracted all remaining particles without rescaling only for the final iteration. Additionally, we use a smaller mask diameter (200 Å) and number of classes (25 classes) for the first round of 2D classification and increase these upon each iteration as follows.

	# of classes	Mask diameter (Å)
Round 1	25	200
Round 2	35	215
Round 3	45	230
Round 4 (final round)	95	245

4. Select the best classes from the final round of 2D classification to use as templates for template-based autopicking (Fig. 5b).

We typically choose 10–25 classes. Ensure that chosen templates are equally representative of different particle views so as not to skew template-based autopicking toward certain particle orientations. Sometimes, nonspecific DNA binding is evident in 2D classes. Nonspecific DNA binding is suggested by nondescript visible density near the nucleosomal DNA in a class with nucleosome density that is not well aligned rotationally with respect to nucleosome orientation (see 2D class third from the left in the bottom row of Fig. 5b).

5. Autopick micrographs with chosen templates using template-based autopicking in RELION with the following settings.
   (a) Picking threshold: 0.1.
   (b) Minimum inter-particle distance (Å): 140.
   (c) Maximum stddev noise: 1.1.
6. Extract particles and perform at least four iterative rounds of 2D classification on new autopicked particles as described above. Generally, we are less stringent with removing 2D classes at this stage, as we want to keep as many good particles in our analysis as possible and removing a slightly inferior class will likely remove many good particles along with the bad ones. Remaining bad particles will get filtered out of the analysis in subsequent 3D classification (Fig. 5c).
7. Run initial round of 3D classification into four classes, using a 50 Å low-pass filtered nucleosome map as a reference (PDBID: 6ESF). Initial reference maps can be made from PDB files in Chimera using the molmap command. Adjust box size and pixel size of the initial reference model to 330 Å and 1.08 Å, respectively, to match the dataset using relion_image_handler. Use this initial reference for all future 3D classification and refinement jobs.
8. View 3D classes in Chimera and keep particles in classes that show well-resolved Dot1L-bound nucleosomes (Class 3).
9. Run 3D refinement and postprocessing of the selected particles. This brings our reconstruction to 3.7 Å.
10. Run Bayesian polishing on particles and repeat 3D refinement and postprocessing. This increased our reconstruction resolution to 3.5 Å.
11. Run CTF refinement on particles. Beamtilt, trefoil, and tetrafoil corrections were run first, followed by anisotropic magnification correction and then CTF parameter fitting (defocus was fit per-particle and astigmatism was fit per-micrograph). Repeat 3D refinement and postprocessing. This increased our reconstruction resolution to 3.3 Å. At this point, the nucleosome in our map is well resolved. However, the density for both Dot1L and ubiquitin is less well resolved, likely due to the presence of

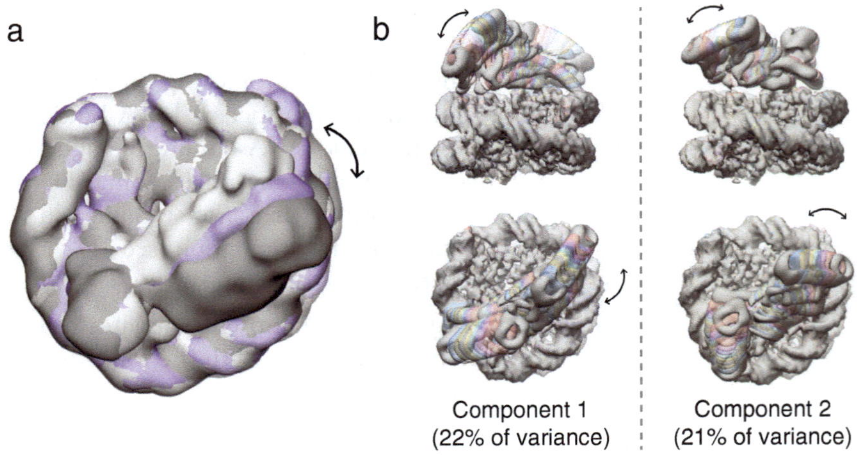

Fig. 6 Conformational heterogeneity exhibited by Dot1L. Mobility in Dot1L can be visualized by (**a**) 3D classification and (**b**) multibody analysis in RELION. 3D classes shown in (**a**) are from original data analysis in RELION 2.1 presented in Anderson et al. *Cell Rep.* 2019. Multibody analysis results shown in (**b**) are from updated analysis in RELION 3.1. Percentages of data variance explained by each of the first two principal components are noted

heterogeneity in this section of the complex (Fig. 6). To improve this region of the map, we carried out several additional rounds of 3D classification.

12. Run another round of 3D classification into four classes and use a subset selection job to select the best resulting class.
13. Run subsequent round of 3D classification into 10 classes. This classification step enabled the visualization of particles in the dataset that were unbound, bound 1:1, and bound 2:1. Use a subset selection job to select the particles in the 1:1 bound class.
14. Run 3D refinement of the selected particles.
15. Prepare a mask that covers the non-nucleosome regions of the map, corresponding to Dot1L and ubiquitin. We typically prepare masks using the Volume Eraser tool in Chimera, and then low-pass filter and smooth this mask using the Mask Create job in RELION. Run masked 3D classification without alignment of the remaining particles with the Dot1L/ubiquitin mask. Use a subset selection job to select particles with the best density corresponding to this masked region (Class 4).
16. Run 3D refinement and postprocessing of the selected particles. This results in a final 4.0 Å map that is slightly lower resolution overall than the map from **step 11** but has better density for the Dot1L/ubiquitin regions of the structure.
17. Prepare a mask that covers the non-nucleosome regions of the new map from **step 16** (as described in **step 15**). Continue 3D refinement with this mask and postprocess the resulting map.

This produces a map with lower overall resolution and much reduced nucleosome density but improved secondary structural features in regions of the map corresponding to Dot1L and ubiquitin (*see* **Note 10**).

3.6 Model Building and Validation

The extent of model building and validation will vary greatly depending on the final map resolution and availability of complex component structures. Our approach for model building into the 3–4 Å Dot1L–ubiquitylated nucleosome map using the previously determined crystal structures of Dot1L (PDBID: 1NW3), ubiquitin (PDBID: 1UBQ), and the nucleosome (PDBID: 3LZ0) will be described.

1. Open up final map and pdb files of individual components in Chimera. Manually place pdb models into approximate docking position in map. Run fitmap command in Chimera to roughly dock pdb models into map. Save docked models as combined pdb file.
2. This Chimera-docked structure was refined using iterative rounds of Phenix real-space refine and manual model building in Coot, including adjusting for sequence differences between our complex and the reference structures. Reference model restraints were applied to Dot1L and ubiquitin regions of the model due to the lower resolution of the map in these regions.
3. Model validation and map-model correlation was performed using MolProbity integrated into Phenix.

4 Notes

1. Expression and purification of histone methyltransferase will vary substantially depending on the protein of interest. Generally, one should aim to acquire >0.5 mg of methyltransferase for reconstitution and structural determination in a nucleosome complex. Additionally, the portion of methyltransferase used for complex reconstitution and structure determination should be carefully considered. Although full-length methyltransferase is theoretically ideal, it may be difficult to obtain in sufficient quantities and unnecessary regions can reduce complex stability. Often, using the minimal portion of methyltransferase required for nucleosome binding and activity is preferred, as flexible regions that do not directly interact with the nucleosome will be averaged out and poorly visualized by cryo-EM. In the case of Dot1L, the minimal construct for this purpose was residues 1–416, which includes the structured catalytic domain and some adjacent basic residues necessary for nucleosome binding [29].

2. As protocols for nucleosome reconstitution have been extensively and thoroughly reported elsewhere [30], we will not go into details of nucleosome preparation here. However, some thought should be put into choosing the correct nucleosome to use for structural studies with the specific histone methyltransferase of interest. In some cases, alterations to wild-type canonical histone sequences, linker DNA, or histone post-translational modifications may need to be incorporated into the nucleosomes used for complex reconstitution in order to stabilize the complex for successful high-resolution structure determination. These alterations will depend on the specific complex and are often one of the main elements requiring optimization before successful complex reconstitution and structure determination. For example, mutation of the target lysine residue to either methionine or norleucine has been found to stabilize methyltransferase binding in several cases [13, 14, 17, 18, 24, 25]. Similarly, linker DNA helped stabilize methyltransferase binding in the case of histone lysine methyltransferases NSD2/3 and EZH2 (within the PRC2 complex), which required a specific linker DNA length in the context of a dinucleosome for optimal binding [12, 14]. In addition, nucleosome complex structures of histone lysine methyltransferases Dot1L/p, MLL1/3, and Set1 included ubiquitylation of histone H2B at lysine 120, which upregulates methyltransferase activity through direct contacts with the methyltransferases observed in these structures [11, 17–25]. Instructions detailing the chemical synthesis of modified histones are outside the scope of this protocol and can be found elsewhere [31–33].
3. Many buffers may be used for complex reconstitution, but it should be noted that nucleosome binding is often dominated by electrostatic interactions and complexes are therefore extremely and variably salt sensitive. Therefore, it is often best to initially attempt reconstitution at physiological nuclear pH and ionic strength (pH 7–8; ~150 mM NaCl) [34, 35]. If unsuccessful, one should attempt reconstituting at decreasing salt concentrations to find an ionic strength at which stable complex formation is observed, as transient and low affinity electrostatic interactions can be rendered more stable for structural analysis at lower salt concentrations. Additionally, if one plans to crosslink with a fixation agent that is reactive to amines such as glutaraldehyde, buffer constituents with primary amines, for example, Tris–Cl, should be avoided so that a buffer exchange step is not necessary before crosslinking. Finally, one should consider inclusion of histone methyltransferase cofactors (such as *S*-adenosyl methionine or *S*-adenosyl homocysteine) or cofactor analogs (5′-(diaminobutyric acid)-*N*-iodoethyl-5′-deoxyadenosine ammonium hydrochloride or

sinefungin) in the reconstitution buffer, which may help to stabilize the methyltransferase in either an active or an inactive bound state. Although we did not include cofactor in our Dot1L-ubiquitylated nucleosome reconstitution, the majority of histone H3 in our complex sample was monomethylated at H3K79, as evidenced by LCMS analysis. Likely, endogenous cofactor was bound to Dot1L in our sample, and many Dot1L molecules had performed one round of methylation on the nucleosomes.

4. A range of methyltransferase:nucleosome ratios have been used for successful complex reconstitutions, ranging from 2:1 to 10: 1. Choice of reconstitution ratio depends on the expected complex stoichiometry, affinity, concerns over nonspecific binding, and negative consequences of oversaturated complexes on sample behavior in subsequent steps. Since we have observed evidence of nonspecific DNA binding in the case of Dot1L, we favored a reconstitution ratio in slight excess of the expected complex stoichiometry (2:1 Dot1L:NCP). Additionally, take note of solution opacity during methyltransferase titrations. In some cases, we have observed the formation of white cloudy precipitate, immediately after methyltransferase addition, that clears upon mixing. Oversaturation can lead to precipitation that does not clear upon mixing and results in lower complex yields.
5. In some cases, crosslinking may not be required for complex stability. However, for the majority of histone methyltransferase–nucleosome complex structures crosslinking was required. The most commonly used fixation agent in these structures is glutaraldehyde, which has been successfully employed by simple incubation at room temperature followed by quenching (as described above) or through use in the Gra-Fix crosslinking method [36]. In a few cases, protein stabilizers (e.g., trehalose) or detergents (e.g., NP40) have helped stabilize complexes and optimize particle orientations on the grid, respectively [12, 15]. Although we often observe some particle aggregation on grids as a result of crosslinking (Fig. 3e), the added stability crosslinking provided for non-aggregated particles is ultimately beneficial and often necessary for structure determination.
6. Although Quantifoil Cu R1.2/1.3 300 mesh grids are the most commonly used grids in histone methyltransferase–nucleosome complex structures published to date, various other grid types (Cu or Au, R1.2/1.3 or R2/2, 200–400 mesh, and streptavidin coatings) have also been used successfully.
7. We now prefer plasma cleaning our grids with a Tergeo EM Plasma Cleaner (PIE Scientific) using the following settings—

Mode: Remote; RF Power (W): 15; Pulse ratio (N/255): 64; Gas Flow Rate (sccm): 2.5, 7.5 for Gas 2 (Oxygen) and Gas 3 (Argon), respectively; Time: 1 min; Purging: none.

8. For the Dot1L-ubiquitylated nucleosome complex, we froze grids using 100% ethane. However, we now favor using a 60–40 mixture of propane–ethane, as this mixture does not freeze as quickly at ideal grid-freezing temperatures (approximately −193 to −183 °C).
9. Freezing conditions will likely need to be optimized for different samples, grid types, and freezing environments. These parameters are meant to serve as a starting point for optimization and should be iteratively adjusted based on the results of grid screening. For example, if ice is too thick, consider increasing the blot force (and vice versa for thin ice). In general, we have not found small changes in blot time or volume to be as useful as changes in blot force. If particles are too densely packed on the grid, decrease the sample concentration, but be aware that small changes in concentration may have large effects on particle density, as these two properties do not necessarily scale linearly.
10. Many data analysis strategies exist to improve heterogeneous regions of EM maps [37], with masked refinement being one option that worked well for this dataset. We also carried out multibody refinement, which produced similar results for the regions of the map corresponding to Dot1L and ubiquitin, while also pushing the resolution of the nucleosome and indicating potential paths along which Dot1L and ubiquitin are mobile (Fig. 6b). However, both of these techniques significantly decreased the resolution at the interface between Dot1L and the nucleosome and were still limited in their abilities to improve the regions of the map corresponding to Dot1L and ubiquitin. Often, masked refinement and multibody analysis are not optimally successful when used for smaller methyltransferase–nucleosome complexes, as the masked methyltransferase region can be too small and noisy to efficiently guide image alignment without the nucleosome. These analysis strategies typically recommend having at least 150–200 kDa components inside of a masked region that gets independently refined [37]. Alternative approaches that we find to be more successful are variations of 3D classification, such as classification without alignment, classification with masked heterogeneous regions, and/or classification restrained to local angular searches. However, such extensive 3D classification often leads to a significant reduction of particles and may result in lower resolution simply due to a limited particle number. As a result, it may be advantageous to collect large datasets on complexes that are exhibiting troublesome heterogeneity. Ultimately, sample preparation strategies that

minimize complex flexibility should be pursued if heterogeneity cannot be addressed computationally.

Acknowledgments

This work was supported by the NIH awards from the NIGMS R35GM133498 to R.K.M. and NCI F99CA253730 to C.J.S.

References

1. Hyun K, Jeon J, Park K, Kim J (2017) Writing, erasing and reading histone lysine methylations. Exp Mol Med 49:1–22. https://doi.org/10.1038/emm.2017.11
2. Di Lorenzo A, Bedford MT (2011) Histone arginine methylation. FEBS Lett 585:2024–2031. https://doi.org/10.1016/j.febslet.2010.11.010
3. Luger K, Mäder AW, Richmond RK et al (1997) Crystal structure of the nucleosome core particle at 2.8 Å resolution. Nature 389: 251–260. https://doi.org/10.1038/38444
4. Albert M, Helin K (2010) Histone methyltransferases in cancer. Semin Cell Dev Biol 21: 209–220. https://doi.org/10.1016/j.semcdb.2009.10.007
5. Yi X, Jiang XJ, Li XY, Jiang DS (2015) Histone methyltransferases: novel targets for tumor and developmental defects. Am J Transl Res 7: 2159–2175
6. Song Y, Wu F, Wu J (2016) Targeting histone methylation for cancer therapy: enzymes, inhibitors, biological activity and perspectives. J Hematol Oncol 9:1–21. https://doi.org/10.1186/s13045-016-0279-9
7. Morera L, Lübbert M, Jung M (2016) Targeting histone methyltransferases and demethylases in clinical trials for cancer therapy. Clin Epigenetics 8:1–16. https://doi.org/10.1186/s13148-016-0223-4
8. Cheng X, Collins RE, Zhang X (2005) Structural and sequence motifs of protein (histone) methylation enzymes. Annu Rev Biophys Biomol Struct 34:267–294. https://doi.org/10.1146/annurev.biophys.34.040204.144452
9. Lacoste N, Utley RT, Hunter JM et al (2002) Disruptor of telomeric silencing-1 is a chromatin-specific histone H3 methyltransferase. J Biol Chem 277:30421–30424. https://doi.org/10.1074/jbc.C200366200
10. Feng Q, Wang H, Ng HH et al (2002) Methylation of H3-lysine 79 is mediated by a new family of HMTases without a SET domain. Curr Biol 12:1052–1058. https://doi.org/10.1016/S0960-9822(02)00901-6
11. Anderson CJ, Baird MR, Hsu A et al (2019) Structural basis for recognition of Ubiquitylated nucleosome by Dot1L methyltransferase. Cell Rep 26:1681–1690. https://doi.org/10.1016/j.celrep.2019.01.058
12. Poepsel S, Kasinath V, Nogales E (2018) Cryo-EM structures of PRC2 simultaneously engaged with two functionally distinct nucleosomes. Nat Struct Mol Biol 25:154–162. https://doi.org/10.1038/s41594-018-0023-y
13. Bilokapic S, Halic M (2019) Nucleosome and ubiquitin position Set2 to methylate H3K36. Nat Commun 10:1–9. https://doi.org/10.1038/s41467-019-11726-4
14. Li W, Tian W, Yuan G et al (2020) Molecular basis of nucleosomal H3K36 methylation by NSD methyltransferases. Nature 590:498–503. https://doi.org/10.1038/s41586-020-03069-8
15. Kasinath V, Beck C, Sauer P et al (2021) JARID2 and AEBP2 regulate PRC2 in the presence of H2AK119ub1 and other histone modifications. Science 371:eabc3393. https://doi.org/10.1126/science.abc3393
16. Ho C-H, Takizawa Y, Kobayashi W et al (2021) Structural basis of nucleosomal histone H4 lysine 20 methylation by SET8 methyltransferase. Life Sci Alliance 4(1–9):e202000919. https://doi.org/10.26508/lsa.202000919
17. Valencia-Sánchez MI, De Ioannes P, Wang M et al (2021) Regulation of the Dot1 histone H3K79 methyltransferase by histone H4K16 acetylation. Science 371:eabc6663. https://doi.org/10.1126/science.abc6663
18. Worden EJ, Hoffmann NA, Hicks CW, Wolberger C (2019) Mechanism of cross-talk between H2B ubiquitination and H3 methylation by Dot1L. Cell 176:1490–1501. https://doi.org/10.1016/j.cell.2019.02.002
19. Valencia-Sánchez MI, De Ioannes P, Wang M et al (2019) Structural basis of Dot1L

stimulation by histone H2B lysine 120 ubiquitination. Mol Cell 74:1010–1019. https://doi.org/10.1016/j.molcel.2019.03.029
20. Jang S, Kang C, Yang HS et al (2019) Structural basis of recognition and destabilization of histone H2B ubiquitinated nucleosome by DOT1L histone H3 Lys79 methyltransferase. Genes Dev 33:620–625. https://doi.org/10.1101/508663
21. Yao T, Jing W, Hu Z et al (2019) Structural basis of the crosstalk between histone H2B monoubiquitination and H3 lysine 79 methylation on nucleosome. Cell Res 29:330–333. https://doi.org/10.1038/s41422-019-0146-7
22. Xue H, Yao T, Cao M et al (2019) Structural basis of nucleosome recognition and modification by MLL methyltransferases. Nature 573: 445–449. https://doi.org/10.1038/s41586-019-1528-1
23. Park SH, Ayoub A, Lee YT et al (2019) Cryo-EM structure of the human MLL1 core complex bound to the nucleosome. Nat Commun 10:1–13. https://doi.org/10.1038/s41467-019-13550-2
24. Worden EJ, Zhang X, Wolberger C (2020) Structural basis for COMPASS recognition of an H2B-ubiquitinated nucleosome. Elife 9:1–23. https://doi.org/10.1101/841262
25. Hsu PL, Shi H, Leonen C et al (2019) Structural basis of H2B ubiquitination-dependent H3K4 methylation by COMPASS. Mol Cell 76:712–723. https://doi.org/10.1016/j.molcel.2019.10.013
26. Grassucci RA, Taylor DJ, Frank J (2007) Preparation of macromolecular complexes for cryo-electron microscopy. Nat Protoc 2:3239–3246. https://doi.org/10.1038/nprot.2007.452
27. Passmore LA, Russo CJ (2016) Specimen preparation for high-resolution Cryo-EM. Methods Enzymol 579:51–86. https://doi.org/10.1016/bs.mie.2016.04.011
28. Zivanov J, Nakane T, Scheres SHW (2020) Estimation of high-order aberrations and anisotropic magnification from cryo-EM data sets in RELION-3.1. IUCrJ 7:253–267. https://doi.org/10.1107/S2052252520000081
29. Min J, Feng Q, Li Z et al (2003) Structure of the Catalytic Domain of Human DOT1L , a Non-SET Domain Nucleosomal Histone Methyltransferase. Cell 112:711–723
30. Lugar K, Rechsteiner TJ, Richmond TJ (1999) Preparation of nucleosome Core particle from recombinant histones. Methods Enzymol 304: 3–19
31. Muir TW (2003) Semisynthesis of proteins by expressed protein ligation. Annu Rev Biochem 72:249–289. https://doi.org/10.1146/annurev.biochem.72.121801.161900
32. Nadal S, Raj R, Mohammed S, Davis BG (2018) Synthetic post-translational modification of histones. Curr Opin Chem Biol 45: 35–47. https://doi.org/10.1016/j.cbpa.2018.02.004
33. McGinty RK, Chatterjee C, Muir TW (2009) Semisynthesis of Ubiquitylated proteins. Methods Enzymol 462:225–243. https://doi.org/10.1016/S0076-6879(09)62011-5
34. Hooper G, Dick DAT (1976) Nonuniform distribution of sodium in the rat hepatocyte. J Gen Physiol 67:469–474. https://doi.org/10.1085/jgp.67.4.469
35. Moore RD, Morrill GA (1976) A possible mechanism for concentrating sodium and potassium in the cell nucleus. Biophys J 16: 527–533. https://doi.org/10.1016/S0006-3495(76)85707-4
36. Kastner B, Fischer N, Golas MM et al (2008) GraFix: sample preparation for single-particle electron cryomicroscopy. Nat Methods 5:53–55. https://doi.org/10.1038/nmeth1139
37. Scheres SHW (2016) Processing of structurally heterogeneous Cryo-EM data in RELION. Methods Enzymol 579:125–157. https://doi.org/10.1016/bs.mie.2016.04.012

Part V

Analysis of Histone Methylation in Cells and Genomes

Chapter 9

Development and Validation of Antibodies Targeting Site-Specific Histone Methylation

Lara Zorro Shahidian and Sylvain Daujat

Abstract

The development of specific anti-modification antibodies as research tools has revolutionized the way histone methylation is studied. Lack of stringent quality controls, however, led to the development of nonspecific antibodies, compromising their use. In this chapter, we provide a series of protocols that collectively will help those studying histone methylation to develop and thoroughly validate high-end sequence-specific and methylation-dependent antibodies.

Key words Rabbit polyclonal antibodies, Site-specific antibody development, Antibody validation, Affinity purification, Histone methylation

1 Introduction

The study of the role of histone methyltransferases (HMTs) is intimately associated with the investigation of the histone methylations they catalyze. Histone methylation occurs at various sites within all four core histones, and it plays a crucial role in all chromatin-dependent processes [1, 2]. However, several factors hinder the study of histone methylation via classical genetic approaches in higher eukaryotes, including the high organizational complexity of the histone genes [3, 4], the existence of multiple posttranslational modifications [5, 6]—many of which have no known writer enzyme—and the existence of multiple methyltransferases capable of methylating the same residue (s) [7, 8]. Consequently, alternative approaches have been developed, among which is the development of specific and sensitive antibodies capable of recognizing distinct states of methylation at specific sites in histones [9]. To truly fulfill their role, it is of vital importance that such antibodies are thoroughly tested and validated, as any off-target recognition could lead to erroneous conclusions and would potentially render their use counter-productive.

Raphaël Margueron and Daniel Holoch (eds.), *Histone Methyltransferases: Methods and Protocols*,
Methods in Molecular Biology, vol. 2529, https://doi.org/10.1007/978-1-0716-2481-4_9,

There are two main factors that hamper the development of high-end antibodies targeting histone methylations. The first one regards the existence of multiple lysine and arginine methylation states [10, 11]. All developed antibodies targeting a specific methylation must differentiate lysine mono- (Kme1), di- (Kme2), and tri-methylation (Kme3), as well as arginine N^G-monomethyl- (Rme1), asymmetric N^G,N^G-dimethyl- (Rme2a), and symmetric N^G,N'^G-dimethylarginine (Rme2s). The second factor is specific to histone H3, where the amino acid sequence -ARKS- is repeated in the histone tail. In fact, two of the best-characterized methylation sites, lysines 9 and 27, are located in this sequence motif [12]. Additionally, arginines 8 and 26, also located within this motif, can also be methylated [13]. This sequence repetition can lead to major cross-reactions between the antibodies targeting these modifications, and special attention must be taken when raising antibodies targeting methylations in these regions.

With the explosion in the number of posttranslational modifications (PTMs) identified within histones in recent years [5], it became clear that developing high-affinity antibodies against specific histone methylations was essential for laboratories in this area of research. This chapter will focus on methods that can be used when developing antibodies that specifically recognize histone methylations. It consists of a thorough, yet practical, protocol on how to raise rabbit polyclonal antibodies targeting histone lysine- or arginine-methylations (*see* **Note 1**). Furthermore, we have included a sequence of characterization steps, using in vitro and in vivo approaches, aimed to meticulously validate the sensitivity and specificity of such antibodies. Despite the very complex nature of histone modifications, and particularly histone methylation, we hope these detailed protocols and notes will provide a solid basis for the generation of specific antibodies against methylated histones.

2 Materials

2.1 Antibody Preparation and Preliminary Specificity Tests

1. Degassed ultrapure MilliQ water.
2. Peptides of interest from commercial companies (such as Biosyntan), including the immunizing peptide and as many additional control peptides as desired (*see* **Note 2**).
3. Imject™ maleimide activated mcKLH (Thermo Fisher Scientific, USA).
4. Rabbit sera as source of unpurified antibodies, generated by a specialized company (such as Biogenes) after rabbit immunization with custom peptide.
5. 0.5 M Ethylenediaminetetraacetic acid (EDTA).
6. 1.5- and 15-mL tubes.

7. Benchtop centrifuge.
8. 0.1 μm Protran® nitrocellulose Western blotting membrane (Cytiva, Austria).
9. Tris-buffered saline (TBS): 150 mM NaCl, 50 mM Tris–HCl pH 8.0.
10. TBST: TBS containing 0.1–0.25% TWEEN-20.
11. Blocking solution: TBST containing 4% (w/v) bovine serum albumin fraction V (BSA) as blocking reagent.
12. Rocking stand.
13. Plastic boxes of appropriate size for all the incubations.
14. Secondary HRP-conjugated anti-rabbit antibody.
15. Chemiluminescent reagents for Western blotting.

2.2 Affinity Purification

2.2.1 Peptide Coupling to Beads

1. SulfoLink™ resin/beads (Thermo Fisher Scientific, USA).
2. 1.5- and 2-mL tubes.
3. Benchtop centrifuge.
4. Coupling buffer: 50 mM Tris, 5 mM EDTA pH 8.5.
5. Immunizing peptide and control peptides (*see* Subheading 2.1).
6. Rotating mixer or rolling wheel.
7. Quenching buffer: 50 mM L-cysteine-HCl in coupling buffer.
8. Wash solution: 1 M NaCl.
9. TBS: 150 mM NaCl, 50 mM Tris–HCl pH 8.0.
10. 0.5 M EDTA.
11. Sodium azide.
12. Storage buffer: degassed TBS containing 5 mM EDTA and 0.05% sodium azide.
13. 0.1 μm Protran® nitrocellulose Western blotting membrane (Cytiva, Austria).
14. Ponceau S staining solution: 0.1% Ponceau (w/v) in 5% acetic acid solution.

2.2.2 Affinity Purification Schemes

1. Clarified rabbit sera as source of unpurified antibodies (*see* Subheading 3.1.3, Serum Preparation).
2. Immunizing peptide and control peptides coupled to SulfoLink™ resin/beads (*see* Subheading 3.2.1).
3. 1.5- and 15-mL tubes.
4. Rotating mixer or wheel.
5. Benchtop centrifuge.
6. TBS: 150 mM NaCl, 50 mM Tris–HCl pH 8.0.
7. 0.5 M EDTA.

8. Wash buffer: TBS supplemented with 5 mM EDTA.
9. Elution buffer: 0.1 M glycine pH 2.5.
10. 1 M Tris–HCl pH 9.5.
11. pH strips.
12. Immunizing peptide and control peptides (*see* Subheading 2.1).
13. TBST: TBS containing 0.25% TWEEN-20.
14. Blotting blocking solution: TBST containing 4% BSA.
15. Plastic boxes of appropriate size for all the incubations.
16. Secondary HRP-conjugated anti-rabbit antibody.
17. Chemiluminescent reagents for Western blotting.
18. TBS containing 5 mM EDTA.
19. Slide-A-Lyzer Dialysis Cassettes 20K MWCO, 0.5–12 mL (Thermo Fisher Scientific, USA).
20. Phosphate Buffer Saline (PBS): 137 mM NaCl, 2.7 mM KCl, 10 mM Na_2HPO_4, 1.8 mM KH_4PO_4.
21. BCA assay (for protein concentration measurement).
22. Ultrafiltration 30 kDa MWCO device (Vivaspin, Sartorius, Germany).
23. 1.5-mL screw-cap tubes.
24. Sodium azide.

2.2.3 Validation of Antibody Specificity

Native Versus Recombinant Histone Analysis by Protein Immunoblot

1. Mammalian cells.
2. 1.5- and 15-mL tubes.
3. PBS: 137 mM NaCl, 2.7 mM KCl, 10 mM Na_2HPO_4.
4. Triton X-100.
5. 4-(2-Aminoethyl)benzenesulfonyl fluoride hydrochloride (AEBSF).
6. cOmplete protease inhibitor cocktail (Roche, Switzerland).
7. Benchtop centrifuge.
8. 0.2 N HCl.
9. Rotating mixer or wheel.
10. BCA assay.
11. 1 M Tris–HCl pH 9.5.
12. Protein migration and Western blotting equipment for running and transferring the gel onto a membrane.
13. Reagents cited in Subheading 2.2.2 for immunoblot detection.
14. Laemmli SDS loading sample buffer 4×: 250 mM Tris–HCl pH 6.8, 20% β-mercaptoethanol, 8% SDS, 40% glycerol, 0.08% bromophenol blue.

15. Coomassie staining solution: 0.1% (w/v) Coomassie R-250, 40% (v/v) ethanol, 10% (v/v) acetic acid.
16. Coomassie destaining solution: 40% (v/v) ethanol, 10% (v/v) acetic acid.
17. BL21 *E. coli* strain.
18. pET-histone expression vector.
19. 1 M isopropyl β-D-1-thiogalactopyranoside (IPTG).
20. Water bath.
21. Histone wash buffer: 50 mM Tris–HCl pH 7.5, 100 mM NaCl, 1 mM EDTA pH 8.0, 5 mM β-mercaptoethanol.
22. Branson sonicator or equivalent.
23. Centrifuge compatible with SS34 rotor or equivalent.
24. 15-mL tubes compatible with SS34 rotor.
25. Histone wash buffer containing 1% Triton X-100.
26. Unfolding buffer: 7 M guanidium hydrochloride, 20 mM Tris–HCl pH 7.5, 10 mM DTT.
27. Ultrapure MilliQ water.
28. Ponceau S staining solution: 0.1% Ponceau (w/v) in 5% acetic acid solution.
29. Purified primary antibody to be tested.
30. 0.45 μm nitrocellulose membrane.
31. 1× Transfer buffer: 0.025 M Tris–HCl pH 8.3, 0.192 M glycine, 20% ethanol.
32. TBS: 150 mM NaCl, 50 mM Tris–HCl pH 8.0.
33. TBST: TBS containing 0.25% TWEEN-20.
34. Blotting blocking solution: TBST containing 4% BSA.

Protein Immunoblot and Peptide Competition

1. Purified native histones, prepared as described in Subheading "Native Versus Recombinant Histone Analysis by Protein Immunoblot".
2. Equipment and reagents cited in Subheadings 2.2.2 and "Native Versus Recombinant Histone Analysis by Protein Immunoblot" for protein migration and transfer, as well as for immunoblot detection.
3. Competing peptides used in solution as free competitors for the antibody recognition of the antigen immobilized on the nitrocellulose membrane. These competing peptides are usually the immunizing peptide and as many control peptides as desired (*see* **Note 2**).
4. Round-bottom 5-mL tubes.

2.3 Verification of the General Antibody Specificity and Activity In Vitro and In Vivo

2.3.1 HMT Assays Using Nuclear Extracts and Recombinant K-Mutant Histones as Substrates

1. Recombinant wild-type (WT) and specific point mutant core histones expressed in and purified from bacteria according to the protocol described in Subheading "Native Versus Recombinant Histone Analysis by Protein Immunoblot".
2. Reconstituted WT and mutant octamers using recombinant WT and mutant core histones mentioned in the above step according to the protocol described in Subheading "Native Versus Recombinant Histone Analysis by Protein Immunoblot" (*see* **Note 3**).
3. Nuclear extract (NE) from HeLa cells (*see* **Note 4**).
4. HMT buffer I 1×: 50 mM Tris–HCl pH 8.5, 20 mM KCl, 10 mM $MgCl_2$, 0.25 mM sucrose.
5. HMT buffer II 1×: 50 mM Tris–HCl pH 8.5, 5 mM $MgCl_2$, 4 mM DTT freshly added.
6. HMT buffer III 1×: 50 mM Tris–HCl pH 8.0, 100 mM NaCl, 1 mM DTT freshly added.
7. Tritiated adenosyl-L-methionine, *S*-[methyl-^{3}H] ([^{3}H]-SAM), 0.55 mCi/mL, 10 Ci/mmol (Perkin Elmer, USA).
8. Thermomixer.
9. Equipment and reagents cited in Subheadings 2.2.2 and "Native Versus Recombinant Histone Analysis by Protein Immunoblot" for protein migration and transfer, as well as for immunoblot detection.
10. Exposure cassette, BiomaxMS films and BioMax intensifying screen (VWR, USA), or equipment to detect radioactive signals, such as Typhoon Phosphorimager (Cytiva, Austria) or equivalent.
11. *S*-Adenosyl-L-methionine (SAM), 32 mM (New England Biolabs, USA).
12. Purified antibody to be tested.

2.3.2 Expression of Tagged K/R-Mutant Histones in Mammalian Cells

1. Mammalian cells to be transfected (mouse NIH3T3 or human HEK293T cells).
2. Cell culture petri dishes and reagents for cell culture.
3. Flag/HA-tagged WT or mutated histone expression vector.
4. Transfection reagents (e.g., Lipofectamine 2000).
5. Acid-extracted histones from transfected cells (*see* Subheading "Native Versus Recombinant Histone Analysis by Protein Immunoblot" for protocol).
6. Equipment and reagents cited previously for protein migration and transfer, as well as for immunoblot detection (*see* Subheadings 2.2.2 and "Native Versus Recombinant Histone Analysis by Protein Immunoblot").

7. M2 beads for Flag-tagged histones immunoprecipitation (Sigma, USA).
8. Purified antibody to be tested.
9. Anti-core histone antibody (i.e., anti-H3).

2.3.3 Native Antigen Recognition by Immunofluorescence

1. Mammalian cells.
2. Sterile coverslips, diameter between 12 and 18 mm.
3. Cell culture petri dishes and reagents for cell culture.
4. 0.1% gelatin in PBS.
5. Fixing solution: 4% paraformaldehyde (PFA) in PBS containing 2% sucrose.
6. PBS: 137 mM NaCl, 2.7 mM KCl, 10 mM Na_2HPO_4, 1.8 mM KH_4PO_4.
7. Quenching solution: 50 mM ammonium chloride in PBS.
8. Triton X-100.
9. IF blocking solution: 4% BSA, 0.2% TWEEN-20 in PBS.
10. Purified antibody to be tested.
11. Humid chamber.
12. Secondary anti-rabbit antibody, donkey, 488 nm-conjugated (Jackson ImmunoResearch, UK).
13. Secondary anti-mouse antibody, donkey, 546 nm-conjugated (Jackson ImmunoResearch, UK).
14. Ultrapure MilliQ water.
15. Vectashield mounting medium containing DAPI (Vector Laboratories, USA).
16. Nail polish.
17. Fluorescence microscope.

2.3.4 Native Antigen Recognition by Nucleosome Immunoprecipitation (IP)

1. Mammalian cells.
2. PBS: 137 mM NaCl, 2.7 mM KCl, 10 mM Na_2HPO_4, 1.8 mM KH_4PO_4.
3. Benchtop centrifuge with refrigeration.
4. Hypotonic buffer: 20 mM HEPES pH 7–7.5, 5 mM KCl, 1.5 mM $MgCl_2$.
5. Complete protease inhibitor cocktail (Roche, Switzerland).
6. Dounce homogenizer with corresponding "tight" pestle.
7. Crystal violet.
8. Isolation buffer: 10 mM Tris–HCl pH 7.5, 1.5 mM $MgCl_2$, 1 mM $CaCl_2$, 0.25 M sucrose.
9. 5 M NaCl.

10. Spectrophotometer for OD_{260} measurements.
11. Centrifuge compatible with SS34 rotor or equivalent.
12. 100 U/μL micrococcal nuclease (MNase, Thermo Fisher Scientific, USA).
13. 1 M $CaCl_2$.
14. EDTA.
15. Solubilization solution: 10 mM Tris–HCl pH 6.8, 5 mM EDTA, 650 mM NaCl.
16. 5% and 40% sucrose solutions: 10 mM Tris–HCl pH 6.8, 5 mM EDTA, 650 mM NaCl containing 5% or 40% (w/v) sucrose.
17. Sucrose gradient maker to make the 5–40% sucrose gradient.
18. 14 mL ultracentrifugation tubes compatible with the SW40Ti rotor.
19. Ultracentrifuge with SW40Ti swinging-bucket rotor or equivalent.
20. Equipment and reagents cited previously for protein migration and transfer, as well as for immunoblot detection (*see* Subheadings 2.2.2 and "Native Versus Recombinant Histone Analysis by Protein Immunoblot").
21. Coomassie R-250 powder.
22. 100% ethanol.
23. 100% acetic acid.
24. Coomassie staining solution: 0.1% (w/v) Coomassie R-250, 40% (v/v) ethanol, 10% (v/v) acetic acid.
25. Coomassie destaining solution: 40% (v/v) ethanol, 10% (v/v) acetic acid.
26. 20% SDS.
27. Agarose powder.
28. Equipment and reagents for agarose gel electrophoresis and staining.
29. 1× TBE: 89 mM Tris, pH 8.0, 89 mM boric acid, 2 mM EDTA.
30. Ethidium bromide.
31. IP dilution buffer: 50 mM Tris–HCl pH 8, 150 mM NaCl.
32. 10 kDa MWCO dialysis membrane (the 3.5 kDa MWCO dialysis membrane also works).
33. 30 kDa MWCO ultrafiltration device (Vivaspin, Sartorius, Germany).
34. Rolling Wheel.
35. 1.5-mL low-binding tubes.
36. Purified antibody to be tested.

37. NP-40.
38. Single-stranded (s/s) salmon sperm DNA.
39. Bovine serum albumin fraction V (BSA).
40. Competitor peptides which can be either the immunizing peptide or any other control peptides as described in Subheading "Protein Immunoblot and Peptide Competition".
41. BSA and salmon sperm DNA blocked A/G beads.

3 Methods

3.1 Peptide Design, Antibody Preparation, and Preliminary Specificity Tests

3.1.1 Peptide Design and Aqueous Solubilization

The successful generation of antibodies targeting specific histone methylations is greatly dependent on the antigen's design. Below, we include a few guidelines that should be taken into consideration when designing immunizing peptides. There are several available commercial companies, such as Biosyntan (Berlin, Germany), that synthesize custom peptides.

1. Peptides intended for rabbit immunization should be 10–15 amino acids long.
2. Where possible, center the amino acid to be methylated.
3. To facilitate peptide coupling to proteins or beads, and therefore affinity purification assays, add a cysteine to the C-terminus of the peptide and include a 1–2 glycine linker between the cysteine (*see* **Note 5**) and the remaining peptide sequence (*see* Subheadings 3.1.2 and 3.2.1).
4. Dissolve the peptide in degassed ultrapure MilliQ water, or appropriate vehicle, at 5 mg/mL for subsequent experiments (*see* **Notes 6** and 7).

3.1.2 Peptide Conjugation to Carrier Protein and Rabbit Immunization

1. To couple the immunizing peptide to keyhole limpet hemacyanin (KLH; *see* **Note 8**), we suggest the use of Imject™ maleimide-activated mcKLH. The manufacturer's instructions should be followed.
2. The antigen is ready after it has been coupled to KLH. There are several companies, such as Biogenes (Berlin, Germany) that immunize rabbits with custom peptides to generate polyclonal antibodies (*see* **Note 9**).
3. Different immunization schedules can be used, and often each company employs its own. The following immunization schedule generally results in good production of antibodies against modified, particularly methylated, histones:
   (a) Day 1: first injection.
   (b) Day 21: first boost.
   (c) Day 42: second boost.

(d) Day 49: third boost.
(e) Day 56: fourth boost.
(f) Day 63: first bleeding (yields ~15–20 mL of serum).
(g) Day 70: second bleeding.
(h) Day 77: fifth boost.
(i) Day 84: bleeding.
(j) Day 91: final bleeding (yields ~50–60 mL of serum).

3.1.3 Bleed Handling and Initial Serum Characterization in Peptide Immuno-Dot Blot

As described in Subheading 3.1.2, during the immunization process, intermediate and final bleeds are collected and used to determine if a rabbit's serum contains antibodies specific toward the immunizing peptide. Upon receipt, the bleeds must be stored appropriately (*see* **Note 10**). Peptide immuno-dot blots, described below, are a quick and easy approach to check the specificity of the polyclonal antibody population (*see* **Notes 11** and **12**).

Serum Preparation

1. Transfer an appropriate volume of serum to a new tube (usually 1–5 mL).
2. Add EDTA to a final concentration of 5 mM, and mix carefully.
3. Clarify by centrifugation at 10,000 × *g* for 15 min at 4 °C.
4. Transfer the supernatant to a new tube. Discard any precipitate.

Membrane Preparation by Peptide Spotting (Dot Blot)

5. Make serial dilutions of the peptides you would like to test, including the peptide carrying the methylation of interest and different control peptides (*see* **Note 2**).
6. Spot 2 μL of each peptide dilution on a 0.1 μm nitrocellulose membrane (*see* **Note 13**). Ideally spots should be separated by 1 cm.
7. Air-dry the membrane for a minimum of 30 min at room temperature (RT).

Peptide Immuno-Dot Blot

8. Block the membrane with blocking solution on a rocking stand for 1 h at RT.
9. Incubate the membrane with the diluted serum (usually ranging from 1:500 to 1:2500) in blocking solution on a rocking stand overnight (O/N) at 4 °C.
10. Wash the membrane three times in TBST for 5 min each at RT.
11. Incubate the membrane with the secondary anti-rabbit antibody diluted according to the manufacturer's instructions in blocking solution, for 1 h at RT.

12. Wash the membrane as described in **step 10**.
13. Carefully transfer the membrane to an appropriate support and proceed with chemiluminescence detection according to the manufacturer's instructions.

The results of the initial characterization can help decide which sera are promising—considering the ratio between specific and nonspecific signals—and design the best affinity purification scheme, one that will allow eliminating any existing cross-reactivities (*see* Figs. 1 and 2). We also recommend checking the different bleeds and not only the final one, as occasionally intermediate bleeds can be better sources for the purification of antibodies (*see* Fig. 3a).

3.2 Affinity Purification

Before starting, equilibrate all reagents at RT.

3.2.1 Coupling of Peptides to Beads

Peptide Coupling to Resin

1. Carefully resuspend the SulfoLink resin/beads slurry by gently stirring and swirling the bottle.
2. Using a wide-bore tip, transfer an appropriate volume of slurry to a 2-mL centrifuge tube to obtain 1 mL of resin bed (*see* **Note 14**).
3. Centrifuge the beads for 1 min at 850 × g in a benchtop centrifuge.
4. Without disturbing the beads, carefully remove the supernatant.
5. Wash beads with 1 mL of coupling buffer. Spin, aspirate, and repeat wash three times.
6. Dilute the peptide to be coupled in coupling buffer at a final concentration of 1 mg/mL.
7. After the final wash, resuspend the beads in 1 mL of diluted peptide solution (1 mg/mL). Keep 10 μL of peptide solution for testing the coupling efficiency.
8. For coupling: mix for 15 min on a slowly rotating mixer at RT. Then place the tube in a standing rack for an additional 30 min.
9. Centrifuge the sample as before (**step 3**).
10. Transfer the supernatant to a new tube. The supernatant will be used to check coupling efficiency.
11. Wash beads with 1 mL of coupling buffer. Spin, aspirate, and repeat wash three times.

Blocking of Nonspecific Binding, Washing, and Storage

12. After the last wash, resuspend the beads in 500 μL of quenching buffer (*see* **Note 15**).

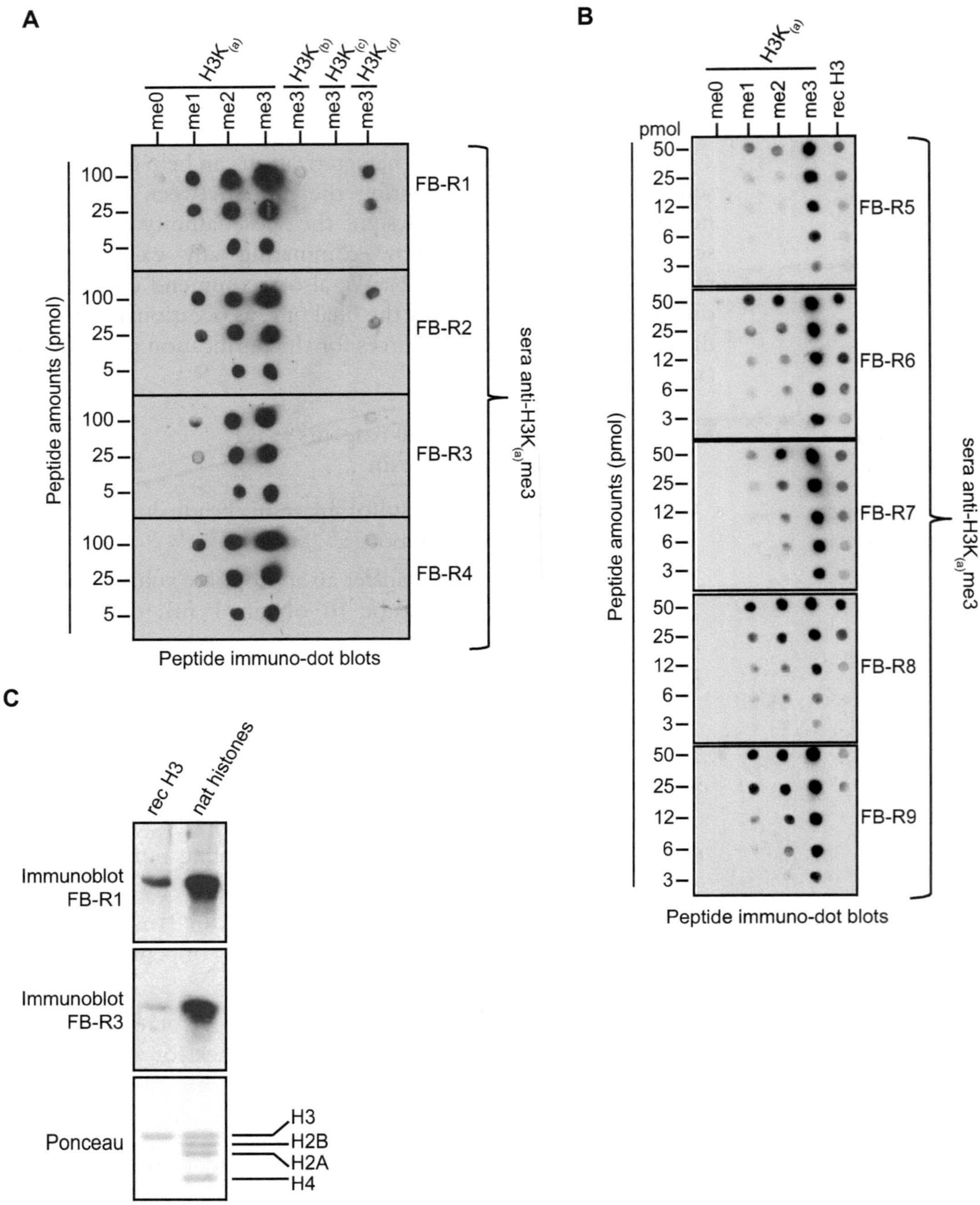

Fig. 1 Initial characterization of the reactivity of different anti-sera raised against a tri-methylated lysine in histone H3 ($H3K_{(a)}me3$). (**a**) Peptide immuno-dot blot assay showing the reactivity of final bleeds from different rabbits (FB-R1–4, 1:2000 dilution) immunized with a $H3K_{(a)}me3$ peptide. Serial dilutions of indicated linear H3 peptide amounts were spotted on the membrane, including the unmodified (me0), mono-methylated (me1), di-methylated (me2), or tri-methylated (me3) for $K_{(a)}$ and tri-methylated peptides at different residues ($K_{(b)}$, $K_{(c)}$, and $K_{(d)}$). Although all four rabbits provided final bleeds showing good reactivity toward the $H3K_{(a)}me3$ peptide, they also presented cross-reactivities toward $H3K_{(a)}me2$ and $H3K_{(a)}me1$. (**b**) Peptide immuno-dot blot assay showing the reactivity of final bleeds from different rabbits (FB-R5-9, 1:1000 dilution) immunized with a $H3K_{(a)}me3$ peptide. Serial dilutions of indicated linear H3 peptides were spotted on the membrane. The five

13. Mix for 15 min on a rotating mixer at RT. Then place the tube in a standing rack for an additional 30 min.
14. Centrifuge the sample as before.
15. Wash the beads six times with 1 mL of wash solution.
16. After the final wash, resuspend the beads in degassed storage buffer (1:1 (v/v)).

    The peptides are now covalently coupled to the beads via the Cys sulfhydryl groups at their C-termini, and they can be used for affinity purifications. Store the bead slurry at 4 °C.

Testing the Coupling Efficiency Through Ponceau Staining (*See* Note 16)

1. Spot 2 μL of the initial peptide solution (from **step 6**) and the supernatant after coupling (from **step 9**) on a 0.1 μm nitrocellulose membrane.
2. Let the membrane air-dry for 30 min.
3. Quickly ponceau stain the membrane for 10–15 s.

If coupling was successful, the intensity of the stain will be much lower or absent with the supernatant sample (*see* **Note 17**).

3.2.2 Affinity Purification Schemes and Preliminary Assessment of Purified Antibody Specificity

Rabbit polyclonal antibodies have the advantage that they contain a mixture of several antibody populations with slightly different specificities. However, it must also be considered that these populations can also contain different cross-reactivities. Affinity purification protocols based on peptide affinity can be used to separate unwanted antibody populations (*see* **Note 18**).

While the steps below aim to provide a general affinity purification scheme for obtaining antibodies against specific methylation states, we recommend adapting it according to the nature of the sera, using the examples provided in Figs. 2 and 3 (*see* **Note 19**) as a guide.

Purification Toward Specific Antibodies

1. If several rabbits were immunized, before starting the purification, it can be useful to do a pre-selection to determine which sera to focus on. We recommend an initial test comparing the reactivity of the clarified sera toward recombinant and native

Fig. 1 (continued) final bleeds show good reactivity toward the H3K$_{(a)}$me3 peptide, with R5 and R7 already appearing to be the most specific and promising. (**c**) Initial protein immunoblot assays comparing the final bleeds' recognition (1:2000 dilution) of recombinant bacterially produced H3 (rec H3; left lane) and native mammalian histones (nat histones, right lane). Positions of the different core histones are indicated. Ponceau S staining shows similar loading of both recombinant and native H3 (bottom panel). Altogether, these initial characterizations allow a first sorting among the different rabbit sera and rather qualify R3, R4, R5, and R7 for affinity purifications (*see* Fig. 2)

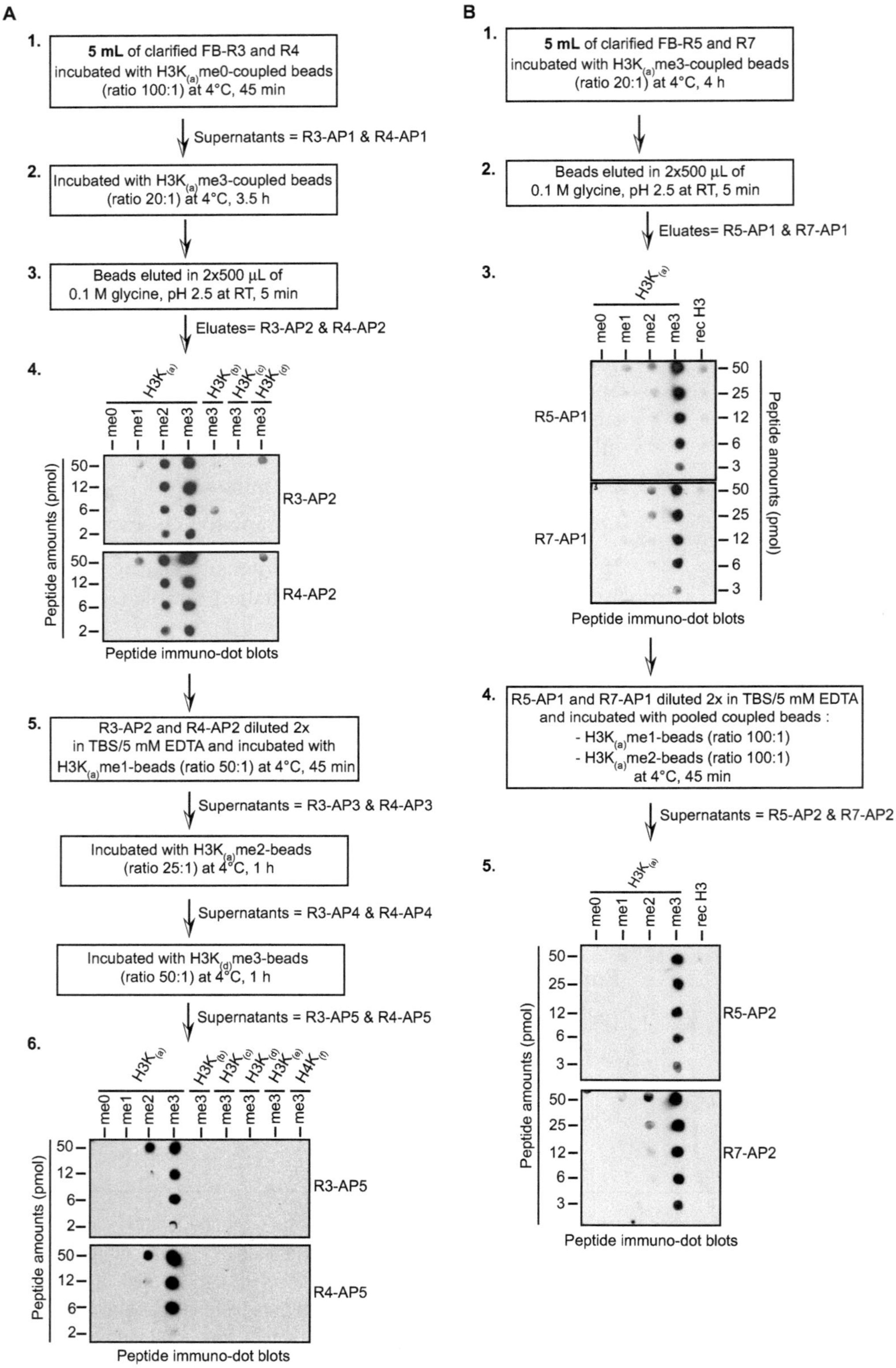

Fig. 2 Affinity purification scheme and characterization of antibodies raised against H3K(a)me3. (**a**) Affinity purification steps taken to purify anti-H3K(a)me3 antibodies (*see* Fig. 1a) from rabbits 3 and 4's final bleeds.

histones (*see* Subheading 3.2.3 for details on how to prepare these samples). A preferential recognition of native histones at this stage will help selecting the sera to be purified (*see* Fig. 1c and **Note 20**).

2. Add the clarified serum (*see* Subheading 3.1.3, **steps 1–4**) to the immobilized immunizing peptide using a serum:beads ratio of 20:1 (v/v; *see* **Notes 21** and **22**). To 5 mL of clarified serum add 250 μL of immunizing peptide beads (Subheading 3.2.1).
3. Incubate the sample for 1–4 h at 4 °C on a rotating mixer (*see* **Note 23**).
4. Centrifuge the sample for 1 min at 850 × *g*.
5. Transfer the supernatant to a new tube (*see* **Note 24**).
6. Wash beads with 5 mL of wash buffer for 5 min at RT on a rotating mixer. Repeat wash a total of three times.
7. After the last wash, resuspend the beads in 500 μL of elution buffer.
8. Incubate the sample on a rotating wheel for 5 min. In the meantime, prepare two 1.5-mL tubes with 17.5 μL of 1 M Tris–HCl, pH 9.5 (*see* **Notes 25** and **26**) and place them on ice.
9. Precisely 5 min after starting the incubation, centrifuge the samples for 1 min at 850 × *g*.
10. Carefully remove the supernatant (the eluate) without disturbing the beads and transfer to one of the prepared ice-cold tubes containing 1 M Tris–HCl pH 9.5. Carefully mix. This represents the first eluate of affinity purification 1, AP1.
11. Repeat **steps 7–10** one more time. In the meantime, check the pH of the neutralized purified antibodies using a pH strip and adjust the pH to 7.5–8.0 if necessary.

Fig. 2 (continued) The initial indicated steps (boxes A1–3, *see* Subheading 3.2.2) generated the first (R3- and R4-AP1) and the second affinity purifications (R3- and R4-AP2) that were then checked by performing a peptide immuno-dot blot as shown in A4 (1:10,000 dilution). From the results, it was deemed necessary to further purify the antibodies to minimize cross-reaction to $H3K_{(a)}me1$, $H3K_{(a)}me2$, and $H3K_{(d)}me3$. The scheme used is described in box A5. The final affinity-purified antibodies (R3-AP5 and R4-AP5) were tested by peptide immuno-dot blots as shown in A6 (1:5000 dilution). The results indicate that, although not complete, a significant reduction in the cross-reactivity of the antibody preparations toward $H3K_{(a)}me2$ was achieved. Indeed, at lower and mid-range concentrations, R3-AP5 and R4-AP5 are highly specific for the $H3K_{(a)}me3$ antigen. (**b**) Affinity purification steps taken to purify anti-$H3K_{(a)}me3$ antibodies (*see* Fig. 1b) from rabbit 5 and 7's final bleeds. The initial indicated steps (boxes B1–2, *see* Subheading 3.2.2) generated the first affinity purification (R5- and R7-AP1) that was checked by peptide immuno-dot blot, as shown in B3 (1:5000 dilution). The results guided the subsequent affinity purification steps, described in box B4, and then the final affinity-purified antibodies (R5-AP2 and R7-AP2) were tested by peptide immuno-dot blots, as shown in B5 (1:5000 dilution). Pooling of $H3K_{(a)}me1$- and $H3K_{(a)}me2$-coupled beads facilitated efficient depletion of the cross-reactivities of R5-AP1, without loss of specific reactivity

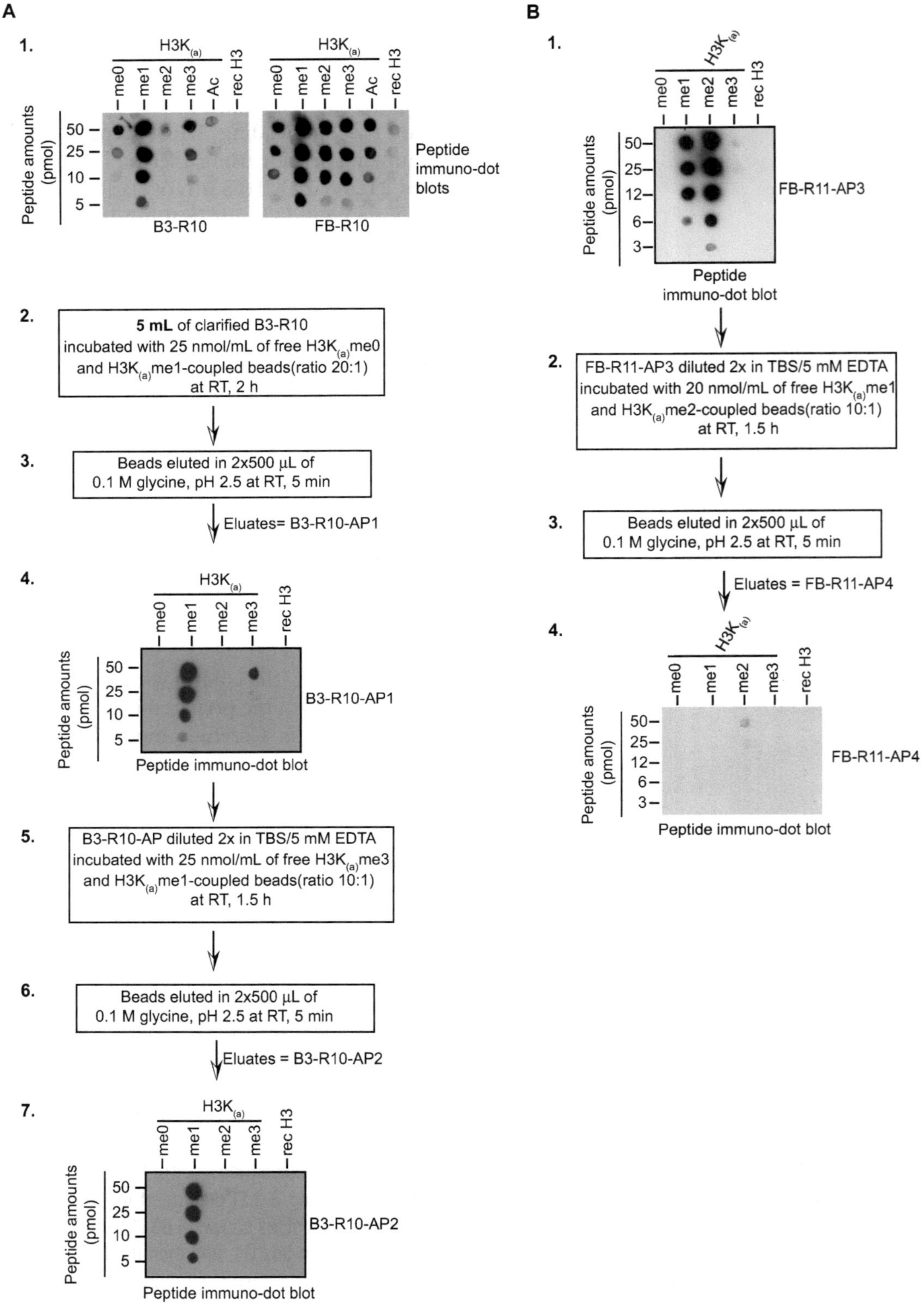

Fig. 3 Affinity purification scheme and characterization of antibodies raised against mono-methylated $K_{(a)}$ or di-methylated $K_{(a)}$ in histone H3 ($H3K_{(a)}me1$ or $H3K_{(a)}me2$). (**a**) Characterization and affinity purification scheme of the anti-sera from rabbit 10, raised against a mono-methylated $H3K_{(a)}$ peptide. In A1, a peptide

Specificity Check

12. Check the specificity of the two purified antibody eluates by peptide immuno-dot blot (*see* Subheading 3.1.3).
13. Usually both the eluates present comparable specificity and sensitivity to the antigen of interest, so they can be pooled together (1 mL total), otherwise they should be kept separate.
14. Depending on the peptide immuno-dot blot results, you can immediately proceed to **step 21** or continue with the purification.

Purification from Nonspecific Antibodies

If the peptide immuno-dot blot reveals nonspecific recognition of one or more control peptides, including an unmethylated version of the immunizing peptide, you may choose to specifically deplete the previously purified antibody from the nonspecific populations by sequential affinity purifications (*see* Figs. 2 and 3).

15. Before proceeding with further affinity purifications, dilute the purified serum (AP1 in step 10 above) twice in TBS containing 5 mM EDTA.
16. Add 20–50 μL of beads coupled to the peptide carrying the cross-reactivity you would like to eliminate to the 2 mL of purified antibodies AP1 (following a serum:beads ratio comprised between 40:1 and 100:1).
17. Incubate for 30–60 min on a rotating mixer at 4 °C.

Fig. 3 (continued) immuno-dot blot assay was used to compare the specificities of the third (B3-R10) and final (FB-R10) bleeds from rabbit 10 (1:1000 dilution). The different peptides tested, together with the various amounts spotted onto the nitrocellulose membrane, are indicated. Interestingly, the third bleed demonstrated robust recognition of $H3K_{(a)}me1$, with a much higher specificity than the final bleed. This is a great example highlighting that a final bleed is not always the best and that purification steps should be performed with intermediate bleeds. The first affinity purification of B3-R10 (B3-R10-AP1), performed according to the scheme described in A2–3, was checked by peptide immuno-dot blot as shown in A4 (1:5000 dilution). The amounts of the indicated linear H3 peptides and recombinant H3 spotted on the membrane are shown. B3-R10-AP1 was then completely depleted of any $H3K_{(a)}me3$ cross-reacting antibodies by following the scheme described in boxes A5–6. The peptide immunoblot assay shown in A7 (1:5000 dilution) revealed the production of a very specific antibody against $H3K_{(a)}me1$. (**b**) Affinity purification scheme of a failed anti-$H3K_{(a)}me2$ antibody purification. The immuno-dot blot shown in B1 shows the reactivity of a third affinity purification of rabbit 11's final bleed (FB-R11-AP3, 1:1000 dilution). FB-R11-AP3 showed a high sensitivity toward the immunizing peptide $H3K_{(a)}me2$ accompanied by a considerable cross-reactivity toward $H3K_{(a)}me1$. An attempt to remove antibody populations recognizing $H3K_{(a)}me1$ was carried out following the scheme described in boxes B2–3. The resulting FB-R11-AP4 preparation was then tested by peptide immuno-dot blot (1:1000 dilution), which revealed the loss of antibodies specific for $H3K_{(a)}me2$, as well as those cross-reacting with $H3K_{(a)}me1$. These results suggest that there are no specific antibodies targeting only $H3K_{(a)}me2$ and that the anti-serum contained one antibody population that efficiently recognizes both mono- and di-methylated $H3K_{(a)}$

18. Centrifuge the sample for 1 min at 850 × *g*.
19. Transfer the supernatant (AP2) to a new tube. AP2 can be further affinity-purified if other cross-reactivities were previously identified by peptide immuno-dot blot (Figs. 2a and 3a).
20. Analyze the final AP by peptide immuno-dot blot taking **Note 2** into consideration for peptide choice (Fig. 2a-6 and b-5; *see* **Notes 27** and **28** for alternative affinity purifications).

Antibody Concentration and Storage

21. Dialyze the antibody suspension overnight against 1 L of PBS at 4 °C using a Slide-A-Lyser dialysis cassette, 20K MWCO or equivalent.
22. Transfer the dialyzed antibody suspension to a new screw-cap tube.
23. Measure antibody concentration by BCA. Dialyzed antibody can be concentrated if necessary in an ultrafiltration 30 kDa MWCO device for instance.
24. Add sodium azide to 0.05% before storing the purified antibodies (*see* **Note 29**).

3.2.3 Validation of Antibody Specificity

Having carried out a first assessment of the specificity and sensitivity of the antibody, a thorough characterization is now necessary. The use of short peptides is optimal for the generation of antibodies aimed at detecting specific histone PTMs, and they are useful for the initial screening of their specificity. However, due to their simple low complexity sequence and structure, recognition of peptides does not necessarily accurately reflect the specificity and sensitivity of the antibody toward full-length histones.

Here, we suggest a series of protein immunoblot-based assays that together aim to address the specificity of the affinity-purified antibodies. The suggested assays grow in complexity and gradually determine whether the antibody can be reliably used for the precise identification of the methylation under study.

Native Versus Recombinant Histone Analysis by Protein Immunoblot

To exclude the possibility that an anti-methylated histone antibody recognizes an unmodified histone, an assay comparing its affinity toward cellular histone (bearing a full repertoire of modifications) and unmodified recombinant histone must be carried out. For this, we suggest testing the antibody with two types of substrates: (A) native histones extracted from mammalian cells, where the modification is present, and (B) recombinant histones expressed in and extracted from *E. coli* bacteria [14], which largely lack methylation.

Native Histone Preparation by Acid Extraction from Mammalian Cells

1. Harvest cells and wash them twice with ice-cold PBS.
2. Resuspend cells in PBS containing 0.5% Triton X-100 (v/v), 2 mM AEBSF, and cOmplete protease inhibitors at a cell density of 10^7 cells/mL.
3. Lyse cells on ice for 10 min with gentle agitation from time to time.
4. Centrifuge at 400 × *g* for 10 min at 4 °C and discard the supernatant.
5. Resuspend the pellet in 0.2 N HCl at a cell density of 4×10^7 cells/mL.
6. Acid-extract the histones (3–4 h to O/N) at 4 °C on a rotating wheel.
7. Centrifuge samples at 10,000 × *g* for 5–10 min at 4 °C.
8. Discard pellet and determine protein concentration in the supernatant (containing the histones) by a BCA assay.
9. Neutralize histone extracts by adding Tris–HCl pH 9.5 at an approximate final concentration of 200 mM.
10. Check extraction of histones by resolving a few micrograms using a 17.5% polyacrylamide SDS gel.
11. To gauge the purity of extracted histones, stain the gel with enough Coomassie solution to cover it. Incubate for 30–60 min at RT with gentle shaking.
12. Decant Coomassie solution and add enough destaining solution to fully cover the gel. Incubate at RT with gentle shaking until the protein bands are revealed and the background is clear. The gel can be stored in water.
13. Store aliquots at −80 °C and avoid too many freeze/thaw cycles.

Recombinant Histone Preparation from Bacterial Inclusion Bodies

Canonical recombinant human histones are expressed and purified from *E. coli* essentially as described by Luger et al. [14].

14. Transform BL21 bacteria with the pET-histone expression plasmid following standard procedures.
15. Induce histone expression in transformed BL21 bacteria by adding IPTG to a final concentration of 0.5 mM and incubate for 3–3.5 h at 37 °C.

16. After pelleting bacteria at 4000 × *g* for 10 min, resuspend pellet in 25 mL histone wash buffer supplemented with cOmplete protease inhibitors and freeze at −80 °C for at least 30 min.
17. Rapidly thaw at 37 °C in a water bath and immediately place the samples on ice once the suspension has finished thawing. Unless specified otherwise, the next steps should be performed on ice.
18. Sonicate the cell suspension. If using a Branson sonicator or equivalent, sonicate three times for 1 min, 0.5 s ON 0.5 s OFF, 100% amplitude.
19. Centrifuge lysate for 15 min, 4 °C at 17,000 × *g* (e.g., in a SS34 rotor) to pellet the inclusion bodies.
20. Wash the inclusion bodies twice in 20 mL histone wash buffer containing 1% Triton X-100 and twice in 20 mL wash buffer without Triton X-100.
21. Resuspend the inclusion bodies in 10–20 mL of unfolding buffer and denature histones by rotating at RT for 1 h.
22. Centrifuge at 17,000 × *g* for 15 min at RT and transfer the supernatant, containing the histones, to a new tube. Repeat centrifugation if the supernatant is not clear.
23. For migration in SDS-PAGE followed by immunoblot, recombinant histones do not require further purification: dialyze O/N against MilliQ water at 4 °C.
24. Centrifuge to discard any pelleted debris and measure protein concentration.
25. Check recombinant histone integrity by migration through a 17.5% SDS polyacrylamide gel.

Using Native and Recombinant Histones to Test Antibody Specificity

26. Resolve equal amounts (Fig. 1c) of recombinant and native histones, prepared as described above, in a 17.5% polyacrylamide SDS-PAGE gel.
27. Western blot the samples to a nitrocellulose membrane and Ponceau stain the membrane to check transfer.
28. Incubate the membrane in blotting blocking solution for 1 h at RT.
29. Incubate with the diluted antibody to be tested in blotting blocking solution O/N at 4 °C on a rocking platform.
30. Wash the membrane three times, each for 5 min, in TBST at RT.

31. Incubate the membrane with the secondary HRP-conjugated anti-rabbit antibody diluted according to the manufacturer's instructions in blotting blocking solution, for 1 h at RT.
32. Repeat **step 30**.
33. Proceed with the chemiluminescence detection according to the manufacturer's instructions. If the antibody specifically recognizes a methylated site within the histone, then it should only recognize the histone purified from cells and not the bacterially synthesized unmodified protein (Fig. 1c, 4a and b).

Protein Immunoblot and Peptide Competition

The previous step assessed whether an antibody recognizes modified histones versus unmodified ones; however, it did not explore the specificity toward a particular methylation site. For that purpose, we suggest performing a protein immunoblot in the presence of a competing peptide.

1. Proceed as you would for a normal protein Western blot and load repeatedly in different lanes of the same 17.5% SDS polyacrylamide gel equal amounts of native histones, prepared as previously described.
2. Run the gel and transfer the samples to a nitrocellulose membrane. Ponceau stain the membrane to check transfer efficiency.
3. Cut the lanes into strips and incubate them in blotting blocking solution for 1 h at RT on a rocking platform.
4. Incubate each strip individually in a round-bottom 5-mL tube with the diluted purified antibody in blotting blocking buffer together with 100 nM of each modified peptide to be tested O/N at 4 °C on a rocking platform (*see* **Note 30**).
5. Keeping the strips separate, wash the strips three times for 5 min each in TBST at RT.
6. Incubate the strips with the secondary HRP-conjugated anti-rabbit antibody diluted according to the manufacturer's instructions in blocking solution, for 1 h at RT.
7. Repeat **step 5**.
8. Carefully transfer the membrane strips to an appropriate support and proceed with chemiluminescence detection according to the manufacturer's instructions (Fig. 4b and c).

3.3 Verification of the General Antibody Specificity and Activity In Vitro and In Vivo

3.3.1 HMT Assays Using Nuclear Extracts and Recombinant K-Mutant Histones as Substrates

The peptide competition assays previously described are great assays to assess antibody specificity. However, often the amount of antigen used in these assays is very high compared to physiological conditions, resulting in the potential masking of minor cross-reactions. In vitro HMT assays, using full-length WT and mutated histones (on the residue of interest) are an excellent alternative. Although these assays are not equivalent to in vivo conditions, they can assess the antibody's specificity against WT and mutant histones (i.e., of the modified site) methylated in vitro by a full complement

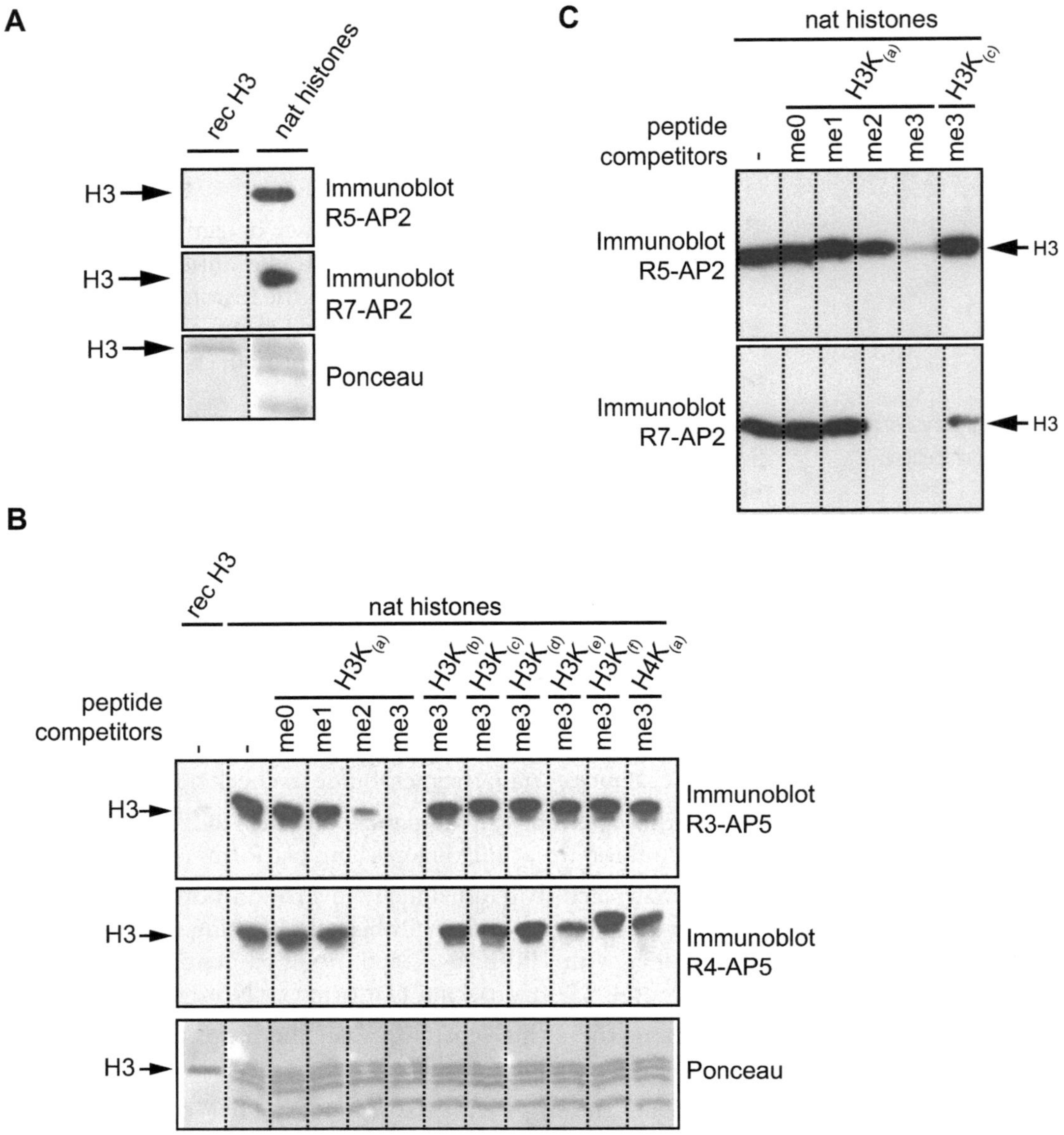

Fig. 4 Validation of the specificity of H3K$_{(a)}$me3 affinity-purified antibodies. (**a**) The reactivities of R5- and R7-AP2 (*see* Fig. 2b) were further tested on recombinant H3 and native histones extracted from HeLa cells. Both the antibodies show strong and specific recognition of native histone H3 at 1:2000 dilution in immunoblot assays, and they do not detect an equal amount of rec H3. Ponceau S staining is shown as loading control. (**b**) Peptide competition assays followed by immunoblots (*see* Subheading 3.2.3) were used to test the specificity of R3- and R4-AP5 (*see* Fig. 2a). Equal amounts of recombinant H3 and acid-extracted histones from HeLa cells were loaded as indicated, separated by SDS–PAGE and transferred to nitrocellulose membranes. Each lane of the membrane was then cut (black dotted lines) into a strip, and they were individually incubated with the indicated AP5 preparation (1:4000 dilution) that had been preincubated with 100 nM of the indicated competitor peptide. Ponceau S staining of a membrane with the same batch of acid-extracted histones is shown as loading reference (bottom panel). Note that the H3K$_{(a)}$me3 competitor peptide completely abrogates the signal, confirming the specificity of both AP5 antibodies for H3K$_{(a)}$me3. R3-AP5 appears to be the most specific out of the two antibody preparations, although it does possess mild cross-reactivity to H3K(a)me2. (**c**) Peptide competition assays followed by immunoblots to test the specificity of R5- and R7-AP2 (*see* Fig. 2b). The methodology used was the same as that described for panel (**b**), R5- and R7-AP2 (1:2000 dilution) were preincubated with 100 nM of the indicated competitor peptide. Note that both R5- and R7-AP2 antibodies are specific for H3K$_{(a)}$me3. However, the R7-AP2 antibody cross-reacts with H3K$_{(a)}$me2 and, to a lesser extent, H3K$_{(c)}$me3

of nuclear HMTs. If the antibody is truly specific against a single modification site, then it will not recognize the mutated histone, even though the mutated protein is widely modified at other sites. We recommend mutating the residue of interest to an unmethylatable amino acid, such as alanine or glycine, using a classical mutagenesis strategy.

Histone Methyltransferase Assay

1. Prepare the recombinant WT and mutant core histones as described in Subheading "Native Versus Recombinant Histone Analysis by Protein Immunoblot" in order to reconstitute and purify WT and mutant octamers (*see* **Note 3**).
2. Grow sufficient HeLa S3 cells, or any other cells where the enzymatic activity of interest is present, and prepare NE (*see* **Note 4**).
3. For a 25 μL HMT reaction, dilute 1 μg of purified recombinant human octamers (containing either WT or mutated histones) and 10–20 μg of HeLa NE in 1x methylation buffer (*see* **Note 31**) supplemented with 2.5–5 μM of ^{3}[H]-SAM (usually 1–2 μL). Incubate for 1 h at 30 °C.
4. Add SDS loading sample buffer, to a final concentration of 1× to stop HMT reactions. Boil samples for 5 min.
5. Run and separate proteins in a 17.5% SDS polyacrylamide gel.
6. Blot proteins to a 0.45 μm nitrocellulose membrane and expose membrane to BiomaxMS films using a BioMax intensifying screen at −80 °C or alternatively use a phosphorimager apparatus for radioactive signal detection. Different exposure periods should be performed to assess HMT activities on histones (Fig. 5a).
7. To avoid over-manipulating radioactivity, once the HMT assay has been proven to work, we recommend carrying out cold HMT assays to test the specificity of the antibody. Proceed as described in **steps 4–6** using a final concentration of 5 μM cold SAM as methyl donor and the appropriate controls.
8. After SDS-PAGE separation and protein transfer, the nitrocellulose membrane is subjected to immunoblot detection of histone methylation by the purified antibody under evaluation as described above (Fig. 5b).

3.3.2 Expression of Tagged K/R-Mutant Histones in Mammalian Cells to Assess Antibody Specificity

The generation of mutant mammalian cell lines expressing Flag/HA-tagged K/A,G,R-mutant histones might be the closest approach to physiological conditions. As with the in vitro HMT assay previously described, this specific setup can test for potential cross-reactions within the same histone in which other residues can be methylated, but in this case in vivo. The Flag/HA tag allows us to distinguish between the mutated exogenous histones and the WT endogenous ones, as the tagged version migrates more slowly

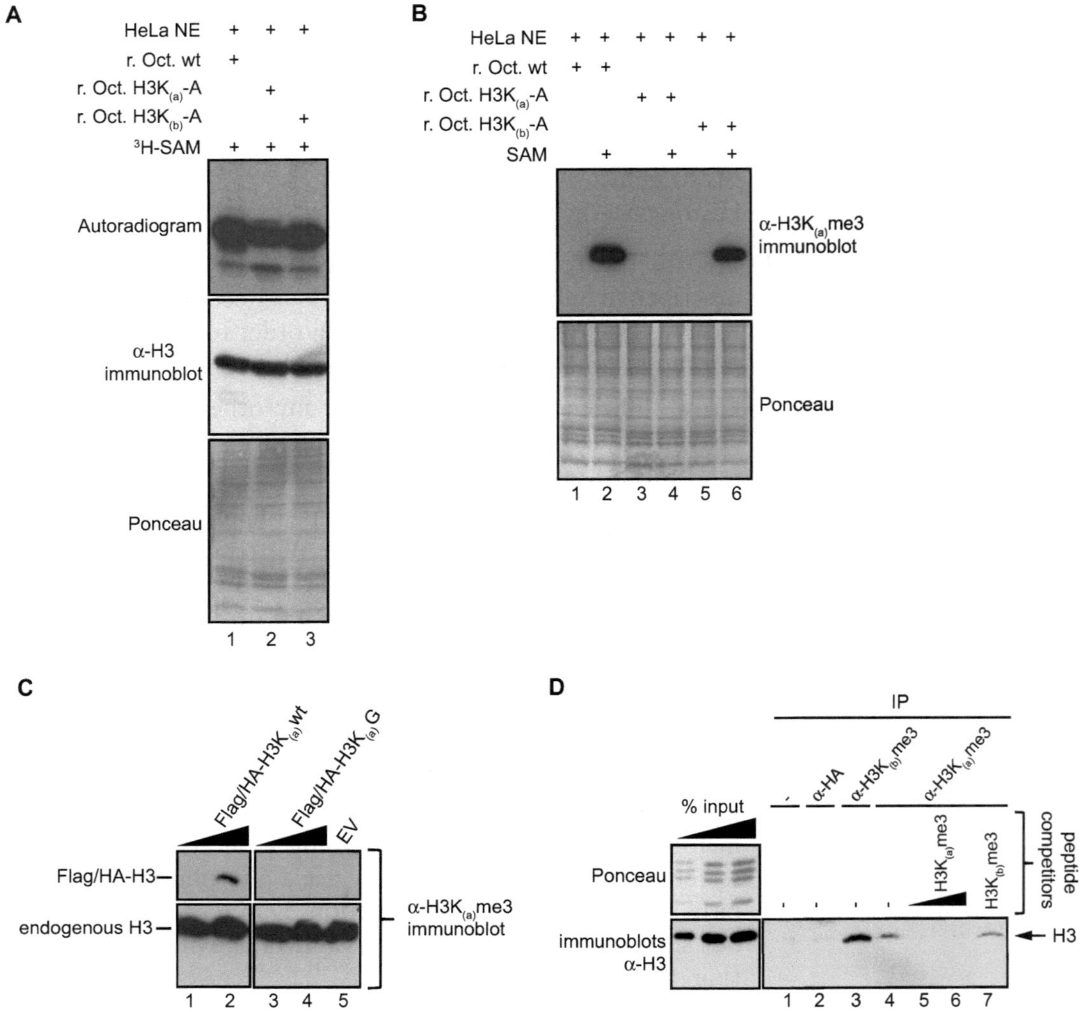

Fig. 5 Possible approaches to verify the specificity and activity of affinity-purified antibody against H3K$_{(a)}$me3. (**a**) Radioactive in vitro HMT assay with WT recombinant histone octamers (r. Oct wt) or mutant recombinant histone octamers in which H3K$_{(a)}$ is substituted to alanine (r. Oct H3K$_{(a)}$A) or H3K$_{(b)}$ is substituted to alanine (r. Oct H3K$_{(b)}$A). HeLa NE was used as a source of methyltransferases. Assay products were resolved on a 17.5% SDS-PAGE gel, blotted to nitrocellulose membrane and assessed by autoradiography. The signals detected on the autoradiogram demonstrate radioactive methyl group incorporation into histones and testify to the presence of efficient HMT activities in the HeLa NE. A similar immunoblot probed with an antibody against general H3 (middle panel) or stained with Ponceau S (lower panel) demonstrate equal loading. (**b**) A non-radioactive in vitro HMT assay with the octamers described in (**a**) and the enzymatically active HeLa NE. The samples were resolved and blotted as described in (**a**). The immunoblot analysis using the affinity-purified antibody against H3K$_{(a)}$me3 (1:2000 dilution) shows the presence of this tri-methylation in r. Oct wt (lane 2) and r. Oct. H3K$_{(b)}$A (lane 6) when the assay was performed in the presence of non-radioactive SAM as methyl donor. Note the absence of detected signal in the mutated r. Oct H3K$_{(a)}$A (lane 4) and in the absence of SAM (lanes 1, 3, and 5). (**c**) Histones were acid-extracted from NIH3T3 transfected cells expressing Flag-HA-tagged H3K$_{(a)}$wt, H3K$_{(a)}$G mutant or an empty vector (EV). The samples were analyzed by protein immunoblots using the affinity-purified anti-H3K$_{(a)}$me3 antibody (1:2000 dilution). As expected, the purified antibody detected endogenous tri-methylated H3K$_{(a)}$ in all samples, but only detected it in the exogenously expressed Flag-HA-tagged WT H3 (lane 2). (**d**) Native HeLa nucleosomes (40 μg) were immunoprecipitated by the purified anti-H3K$_{(a)}$me3 (4 μg) or different control antibodies, as indicated (anti-HA and H3K$_{(b)}$me3), in

through a SDS-PAGE gel. Moreover, the Flag/HA tag enables the immunoprecipitation of tagged histones, which can then be probed by the purified antibody to detect the presence of the methylation of interest as a function of whether the corresponding residue is WT or mutant.

1. Transfect mammalian cells with vectors expressing Flag/HA-tagged WT or mutated histones (*see* **Note 32**).
2. After 48–72 h, harvest the cells and acid-extract the histones as described in Subheading "Native Versus Recombinant Histone Analysis by Protein Immunoblot".
3. Use the extracted samples on a protein immunoblot.
4. Following the immunoblotting procedure described in Subheading "Native Versus Recombinant Histone Analysis by Protein Immunoblot", **steps 26–32**, tagged or endogenous histones can be detected using the antibody under evaluation by immunoblot (Fig. 5c).
5. Acid-extracted histones from transfected cells can then be immunoprecipitated by an anti-tag antibody (such as M2 beads), migrated through a 17.5% SDS polyacrylamide gel and transferred to a nitrocellulose membrane.
6. Detect immunoprecipitated histones by the antibody under evaluation by immunoblot using a general anti-histone antibody (anti-H3 for instance). An H3 signal detected only when the Flag-HA-tagged H3K(a)wt is immunoprecipitated will indicate an appropriate specificity of the antibody under evaluation.

3.3.3 Native Antigen Recognition by Immunofluorescence

Immunofluorescence (IF) not only enables the visualization of the natural distribution of a particular histone methylation, but can also be very informative regarding the specificity of a given antibody. Mouse cells are particularly useful for this purpose, as DAPI staining readily distinguishes between euchromatin and heterochromatin. Antibody staining of particular nuclear regions will constitute a first clue to the potential function of the methylation of interest. Peptide competitions can be carried out as a proof of specificity of the purified antibodies for their native nondenatured epitopes in histones.

Fig. 5 (continued) presence or not of the indicated competitor peptides (200 and 400 nM for $H3K_{(a)}me3$ and 200 nM for $H3K_{(b)}me3$). The proteins immunoprecipitated by the different antibodies, together with increasing amounts of input material (1, 2, and 5%), were resolved by SDS-PAGE and blotted on a nitrocellulose membrane, before immunoblotting was performed using an anti-H3 antibody (bottom panels). This shows that the purified antibody preparation specifically recognizes and immunoprecipitates tri-methylated $H3K_{(a)}$-containing nucleosomes, since the H3 signal is competed away only by the $H3K_{(a)}me3$ peptide, and not the $H3K_{(b)}me3$ peptide

1. Install up to 12 sterile coverslips in a 100 mm diameter dish.
2. Coat the coverslips with 0.1% gelatin by incubation at RT for a minimum of 30 min.
3. Grow mammalian cells on coverslips to approximately 50% confluence.
4. Discard medium and incubate cells in fixing solution (*see* **Note 33**) for 15 min at RT.
5. Quench cross-linking reaction with 50 mM ammonium chloride in PBS.
6. Wash coverslips three times with PBS (*see* **Note 34**).
7. Permeabilize cells with 0.6% Triton X-100 in PBS for 15 min at RT.
8. After washing the cells, incubate coverslips in IF blocking solution for 1 h at RT.
9. Add a 100 μL drop of diluted purified antibody in blocking solution and incubate the coverslips upside down on it in a humid chamber O/N at 4 °C (*see* **Notes 35** and **36**).
10. Wash coverslips three times for 5 min each with blocking solution at RT.
11. Carefully place a 100 μL drop of fluorophore-conjugated secondary antibody (Alexa Fluor 488- or Alexa Fluor 546-conjugated) diluted 1:250 in blocking solution. Incubate coverslips upside down on the drop for 1 h at RT, in a dark and humid chamber.
12. Wash coverslips three times for 5 min each with blocking solution at RT.
13. Quickly rinse coverslips once each with PBS and with MilliQ water to eliminate any salt crystals.
14. Coverslips can now be mounted onto Vectashield and sealed with nail polish for storage at 4 °C in the dark, prior to conventional analysis by fluorescent microscopy.

3.3.4 Native Antigen Recognition by Nucleosome Immunoprecipitation

The generation of specific antibodies recognizing particular histone methylations allows the determination of their genomic distribution with a high degree of precision using techniques such as chromatin immunoprecipitation and all its recently developed variants. Before embarking on such experiments, alternative approaches can be carried out to ensure that the purified antibody can be used with confidence. This entails the preparation of clean cell nuclei for the isolation of nucleosomes that can be used in IP experiments with the purified antibody under evaluation.

The following protocol was inspired and adapted from the protocol described in [15]. All steps should be done on ice unless indicated otherwise, and all centrifugations should be carried out at 4 °C.

1. Harvest 1–5 × 10^8 cells. Wash the cell pellet with PBS and centrifuge at 750 × *g* for 5 min.
2. Resuspend cell pellet in 5 mL of hypotonic buffer containing cOmplete protease inhibitors.
3. After centrifugation at 750 *g* for 5 min, repeat **step 2** with half the volume of hypotonic buffer and incubate on ice for 10 min.
4. Transfer to a Dounce homogenizer and disrupt cell membrane to free nuclei with approximately 12 strokes of a "tight" pestle. Incubate on ice for 30 min. The efficiency of nuclei preparation can be checked with crystal violet staining.
5. After centrifugation at 750 × *g* for 10 min, resuspend the nuclei in 1 mL of isolation buffer containing 100 mM NaCl and incubate on ice for 10 min (*see* **Note 37**).
6. Centrifuge at 750 × *g* for 5 min and resuspend the nuclei in 1 mL of isolation buffer supplemented with 250 mM NaCl. Incubate on ice for 10 min.
7. Centrifuge nuclei at 4000 × *g* for 10 min and resuspend nuclei in 1 mL of isolation buffer.
8. Measure the nucleosomal DNA concentration at OD_{260nm} by diluting nuclei 500 times in 1 mL isolation buffer.
9. Add 50 units of micrococcal nuclease per mg of nucleosomal DNA and 3 mM $CaCl_2$ final concentration to the 1 mL suspension of nuclei.
10. Incubate for 30 min at 37 °C with agitation from time to time.
11. Add 5 mM EDTA to stop reaction and centrifuge at 10,000 × *g* for 10 min.
12. Resuspend pellet in 1 mL of solubilization buffer and store on ice for 30 min.
13. Centrifuge the solubilized pellet at 16,500 × *g* for 10 min and store supernatant containing released digested nucleosomes on ice.
14. Prepare a 5–40% sucrose gradient with a gradient maker in 14 mL ultracentrifugation tubes for the SW40Ti swinging-bucket rotor.
15. Carefully load the supernatant containing the nucleosomes on the 5–40% sucrose gradient.
16. Centrifuge at 200,000 × *g* for 16 h at 4 °C.
17. Collect 600 μL fractions and migrate 5–20 μL of each fraction in a 17.5% SDS polyacrylamide gel. Following **steps 11–12** described in Subheading "Native Versus Recombinant Histone Analysis by Protein Immunoblot", Coomassie-stain the gel to check for the presence of core histones in the fractions (*see* **Note 38**).

18. In parallel, load 5–10 μL of each fraction, previously supplemented with SDS to a final concentration of 0.5%, on a 1% agarose gel and migrate it in 0.5× TBE. Stain the gel after migration by incubating it for 30 min at RT with gentle shaking in 0.5× TBE solution containing 0.5 μg/mL ethidium bromide to check for the presence and size of nucleosomal DNA (*see* **Note 39**).
19. All fractions containing mono- or di-nucleosomes should be pooled and dialyzed (using 10 kDa MWCO dialysis membrane) O/N against 2 L of dilution buffer at 4 °C.
20. Dialyzed nucleosomes can be concentrated if necessary, in an ultrafiltration 30 kDa MWCO device to ideally reach a nucleosomal DNA concentration of 200–400 ng/μL. DNA concentration should be measured at 260 nm.
21. Use nucleosomes containing the equivalent of 20–40 μg of DNA for each IP and incubate O/N on the rolling wheel with 1–5 μg of purified antibody (which can be empirically determined) in a final volume of 500 μL of IP dilution buffer supplemented with 0.1% NP-40, 100 μg of BSA and 25 μg of salmon sperm DNA, in 1.5 mL low-binding tubes (*see* **Note 40**).
22. After O/N incubation, add 20–40 μL of salmon sperm DNA and BSA-blocked protein-A/G beads and incubate on the rolling wheel for 2 h.
23. Centrifuge beads at 400 × *g* and wash them three times in IP dilution buffer supplemented with 0.1% NP-40 and 250 mM NaCl.
24. After centrifugation, wash beads once with IP dilution buffer supplemented with 0.1% NP-40 and 375 mM NaCl (*see* **Note 41**).
25. After the final wash, centrifuge the sample at 400 × *g*.
26. Elute immunoprecipitated complexes from the beads by boiling them in 25 μL of Laemmli SDS loading sample buffer at 95 °C for 5 min.

The samples can now be used on protein immunoblots as previously described (Fig. 5d). The quality of the results will indicate whether the purified antibody is adequate for chromatin immunoprecipitation and related methodologies.

3.4 Concluding Remarks

Generating highly specific antibodies capable of distinguishing different histone methylation states is not an easy task, but one of extreme importance when studying HMTs. Although rabbit polyclonal antibodies are composed of a mixture of antibodies that can lead to nonspecific antigen recognition, they benefit from being highly active in a wide range of applications, while at the same time representing an affordable alternative to rabbit monoclonal antibodies. Independently of the origin of the antibodies used (self-made versus commercial) high vigilance is required when validating their specificity. The golden control is often considered testing the antibody of interest in a WT versus mutant background. In the case of canonical histones that would mean creating cell lines that do not express WT core histone, but only a mutant version specifically mutated on the methylatable residue under study. Unfortunately, due to the high copy number of genes and the very complex organization of histone gene loci in higher eukaryotes this is very challenging, if not impossible. While it is possible to create mutant cells where the HMT establishing the methylation of interest has been silenced, many mammalian HMTs are redundant, and the same residue can be methylated by several HMTs [6, 8]. Additionally, recent improvements in mass spectrometry techniques have resulted in the identification of new methylation sites on histones for which no HMT (or only a subset of thereof) has yet been identified. In these cases, classical enzyme silencing approaches would not be possible or informative. Together, these issues highlight the constant need to improve existing approaches and generate new alternatives for the generation and validation of antibodies targeting specific histone methylations.

4 Notes

1. Compared to monoclonal antibodies, rabbit polyclonal antibodies are easier and cheaper to generate. They also benefit from the advantage that several epitopes in the same antigen can be recognized, often resulting in higher sensitivity and affinity. However, due to their nature, they are more prone to cross-reactions and suffer from batch-to-batch variability. Therefore, when possible, we recommend the development of rabbit monoclonal antibodies, which not only tend to present fewer off-target recognitions, but also provide an unlimited source of specific antibodies.
2. Depending on the antibody being tested, different peptide dilution ranges are necessary, typically ranging from 0.2 to 50 pmol/μL. At the very least, equimolar amounts of a peptide carrying the methylation under study together with its unmethylated counterpart should be spotted. Ideally, other methylated states of the same peptide will also be tested. As many additional peptides as desired, carrying other lysines/arginines methylated to the same or different levels can also be used. The availability and cost of peptides are significant limitations when choosing the best strategy. Therefore, as a compromise, we usually spot histone peptides containing residues with the same methylation state as the one under study to make sure the purified antibody does not recognize the methyl moiety alone.

3. We recommend carrying out HMT assays on reconstituted octamers as they have the advantage of containing all core histones, allowing the detection of possible cross-reactions with other histones. A detailed protocol for octamer reconstitution and purification has been described by Luger et al. [14]. We understand, however, that reconstituting and purifying recombinant octamers require substantial effort and access to state-of-the-art equipment and may not be within the reach of all laboratories. Instead, the HMT assays can be performed on an equimolar mix of all four core histones.
4. HeLa S3 cells are generally a good source of HMT activity. They can be grown in suspension giving the opportunity to easily get a high number of cells, up to 10^{10} cells. This allows the preparation of large quantities of NE, which can be prepared following a classical protocol that can be found in [16], usually providing protein concentrations of 5–10 mg/mL.
5. If the antigen is derived from the very C-terminus of a protein, it is better to add the cysteine to the peptide's N-terminus to maintain a "natural" C-terminus end.
6. The peptide's solubility depends on its nature (acidic, basic, or hydrophobic). We recommend testing a small amount of the peptide for solubility in water, and if the peptide does not dissolve properly, the following tricks can be used: add a small amount of diluted aqueous acetic acid (for basic peptides), diluted ammonium hydroxide (for acidic peptides), or DMSO (for hydrophobic peptides), in a stepwise manner, until the peptide has dissolved and supplement with water until the desired concentration is reached.
7. Lyophilized peptides can be stored at −80 °C, protected from light, for several years. Once dissolved, the peptide's shelf-life decreases significantly, and the following recommendations should be taken into account: (A) cysteines in peptides are particularly susceptible to oxidation when exposed to air, which can prevent the efficient coupling to matrices; therefore, dissolving the lyophilized peptide in degassed water—or appropriate solvent—helps reduce oxidation; (B) whenever possible, solubilize only the necessary amount of peptide, using small plastic devices to weigh the powder; (C) aliquoting dissolved peptides helps reduce the number of freeze–thaw cycles resulting in better peptide conservation; (D) if degassed ultrapure water was used to dissolve the peptide, it can be re-lyophilized before storage.
8. Peptides are usually too small to trigger a sufficient immune response when inoculated alone. It is, therefore, recommended to conjugate them to larger immunogenic proteins, such as KLH, in order to obtain the production of specific antibodies.

KLH has a high molecular mass and can trigger a stronger immune response when compared to other protein carriers.

9. Often these companies inject KLH-coupled custom peptides in an adjuvant (such as complete Freud's adjuvant) to enhance the immune response.
10. To extend the shelf life of the bleed samples, upon receipt they must be supplemented with thimerosal or sodium azide, to a final concentration of 0.02–0.05% (w/v). While keeping a working aliquot at 4 °C, make aliquots of the bleeds and store them at −20 °C for optimal shelf life.
11. Alternatively, new high-throughput technologies, such as peptide arrays [17, 18], can be used to identify potential cross-reactions. Compared to traditional peptide immuno-dot blots, these approaches have the advantage that a large number of peptides can be screened at once, potentially guiding the choice of the affinity purification scheme. This advantage can also facilitate the testing of the antibody against peptides carrying different modifications in neighboring amino acids, which could prevent the proper recognition of the antigen of interest.
12. Enzyme-linked immunosorbent assays (ELISA) are another viable alternative to peptide immuno-dot blots. ELISA has the potential to be a semi-high-throughput assay; however, we tend to prefer immuno-dot blots because they are much easier to implement and just as meaningful for initial detections of possible cross-reactivities.
13. PVDF membranes can also be used; however, we usually prefer 0.1 μm nitrocellulose membranes, as they are easier to handle and work equally well.
14. If you only have regular tips, you can cut off the tip end using scissors to obtain a wide-bore tip.
15. Prepare the quenching buffer freshly before use.
16. Some peptides do not stain well with Ponceau. Alternatively, coupling efficiencies can be determined through other methods based on the absorbance of peptides. A reliable way to measure the concentration of free peptides in solution is with Ellman's reagent [19], also known as DNTB (5,5-dithio-bis-(2-nitrobenzoic acid)). Ellman's reagent reacts with free sulfhydryl groups, in other words with reduced cysteine residues that are present in the peptides and produces a yellow dye. This allows the relative quantification of the amount of free cysteine residues before and after coupling. For each sample to be quantified in a 96-well microtiter plate's well, 5 μL of Ellman's reagent (50 mM sodium acetate, 2 mM DTNB), 10 μL of 1 M Tris pH 8.0, and 75 μL of water should be mixed with 10 μL of peptide fraction. Coupling efficiencies can also be assessed by comparing the color of the input to that of the supernatant

after coupling to beads and to the no peptide control. More precisely, the absorption of the dye at 412 nm can be measured, and OD ratios of the different fractions give the percentage of peptide that has been coupled to the beads.

17. If you face low coupling efficiency, this might be due to the oxidation of the peptide and the formation of disulfide bonds. Immediately before coupling, reduce peptides by adding Tris (2-carboxyethyl) phosphine (TCEP) to a final concentration of 25 mM in coupling buffer. This will not interfere with subsequent iodoacetyl coupling to the SulfoLink resin.

18. While we provide a protocol for affinity batch purifications, similar efficiencies can be achieved using empty spin or gravity flow columns (Bio-Rad, USA) to which coupled peptides are added. These columns can be handled as tubes and incubated on a rotating mixer or wheel.

19. We often start by selecting antibody populations that react with the immunizing methylated peptide, followed by selecting out those populations that react with different methylation states on the same residue or at different residues, depending on the results obtained from peptide immune-dot blots. One should keep in mind that too many serial affinity purifications could result in the reduction of the antibody titer or avidity without any significant gain in specificity. This could be explained by the deterioration of the antibody's properties over the many affinity purification cycles or, simply, by the presence of a single truly cross-reacting antibody population. If a sensitivity–specificity balance cannot be found, it is advisable to immunize new rabbits or to change the approach used to generate specific antibodies (e.g., through the generation of monoclonal antibodies).

20. Although the preferential recognition of native histones, at this stage, is generally a good indicator, in the case of rare or inducible methylations its absence should not necessarily lead to serum disqualification.

21. The combination of affinity purification steps may be changed and depends on the exact characteristics of the serum/antibodies. We recommend testing different conditions in preliminary assays. The serum can be first purified over the immunizing peptide and then over specific cross-reacting peptides or vice versa (Figs. 2a, b and 3a).

22. The optimal ratio serum:beads for the affinity purification in batch may vary between 10:1 and 20:1 for the immunizing specific peptide and between 25:1 and 100:1 for unspecific peptide.

23. The incubation of the serum with the peptide slurry can be done at RT or 4 °C. In some cases, it is better to perform the incubation at 4 °C, thereby increasing the time needed for specific recognition of antigen by the antibody and thus reducing nonspecific binding.
24. Often, the serum is not fully depleted of specific antibodies after a single purification and several rounds of purification can be done from the same serum aliquot before it is fully depleted. We recommend testing the serum after each purification in order to assess whether further antibodies can be purified from it.
25. The quick neutralization of the elution buffer's low pH is very important, as it can permanently denature the eluted antibodies. Moreover, ice-cold Tris–HCl is also critical, as the neutralization reaction is exothermic and could destroy the antibody. Therefore, it is crucial that **steps 11–13** are performed as quickly as possible and on ice. Usually, adding Tris–HCl pH 9.5 to a final concentration of 35 mM promotes efficient neutralization, but nevertheless, this should be checked, with pH strips for example.
26. Most antibodies can be eluted in low-pH glycine-containing buffers, but some antibodies can only be eluted by high-pH conditions. In this case, buffers containing 100 mM triethanolamine at pH 11–12 can be used and neutralized afterwards. In some rare cases, very-high-affinity antibodies cannot be eluted from the antigen. In these cases, we recommend trying the use of crude sera.
27. If the serum shows cross-reactivity to several particular peptides, you might be interested in specifically depleting the nonspecific antibodies by pooling the different beads, individually coupled to each cross-reacting peptides (*see* Fig. 2b-4).
28. Sometimes the purification against nonspecific antibodies is not efficient. Alternatively, the cross-reacting peptide(s) (at a final concentration of 20–25 μM) can be added uncoupled to the serum while it is being affinity-purified over the immunizing specific peptide. Figure 3a shows an example of a successful affinity purification that followed this alternative protocol, while Fig. 3b shows an unsuccessful attempt.
29. Some antibodies are more stable when stored at 4 °C, whereas others prefer −20 °C (in 50% glycerol). Both the conditions should be tested, and if no alteration of the antibody reactivity is detected after storage, we recommend storing at −20 °C.
30. For peptide competition, the amount of competing peptide needs to be determined empirically in each case. We suggest starting with 100 nM of competing peptide. After numbering the strips for identification in the following steps, we recommend pre-incubating the purified antibody with each

competitor peptide to be tested for 30 min at 4 °C. For the number of peptides to be used, the same considerations as described in **Note 2** should be applied. A control with no competing peptide should be included. If the antibody is specific toward the methylation under study, only in the presence of a peptide carrying the specific methylation of interest should the signal be lost.

31. We recommend testing different HMT buffers (*see* Materials for examples) to allow different HMT enzymes to function optimally. The ^{3}[H]SAM used in HMT assays is 0.55 mCi/mL, 10 Ci/mmol (corresponding to a SAM concentration of 55 μM). This should be taken into consideration when calculating the appropriate SAM concentration used in nonradioactive assays. Since SAM is not stable at RT, we recommend keeping the stock solution in small aliquots at −20 °C, in order to decrease the number of freeze–thaw cycles. For this same reason, SAM can be added twice to HMT reactions, at the beginning and halfway through the assay.
32. WT and mutant histones can be cloned into a mammalian expressing vector under the control of the EF1α promoter and fused to a tag comprising one Flag and two HA. This allows proper levels of histone expression. NIH3T3 mouse cells or human HEK293T cells can be transfected for 48–72 h, allowing the exogenous expression of the different tagged histones.
33. We recommend fixing the cells in PFA containing 2% sucrose to better preserve the cell nuclear structure and integrity.
34. Cells can be stored at 4 °C in PBS for a few days.
35. The dilution of the purified antibody depends on its reactivity determined during the procedure described in Subheading 3.2.3. Usually, a 10 times lower dilution than the one used in immunoblot detection constitutes a good starting point.
36. If peptide competitions are to be carried out, before starting **step 9** add 100–500 nM of the histone competitor peptide (s) to the diluted antibody. Then proceed as described in **step 9**.
37. Extraction of nuclei with isolation buffer containing increasing concentrations of salt, such as NaCl, allows the removal of most loosely bound proteins from chromatin and yields cleaner nucleosomes.
38. Core histones should be enriched mostly in fractions 8–12.
39. The addition of SDS to a final concentration of 0.5% allows the denaturation of the nucleosome structure and frees the DNA for proper migration. Stain the gel with ethidium bromide only after migration to avoid interference with SDS. The

nucleosomal DNA is protected from MNase digestion and is released as mono-, di-, tri-nucleosomes, and so on; mononucleosome are ~150 bp, di-nucleosomes ~300 bp, etc.

40. Peptide competitions can be carried out during the IP: add 100–500 nM of the histone competitor peptide(s) to the diluted antibody and proceed as described in **step 21**.

41. Washing conditions can be adjusted depending on the background corresponding to unspecific binding of beads or irrelevant antibody control detected by immunoblot. Extra steps with dilution buffer containing 500 mM NaCl and 250 mM LiCl can be carried out if necessary.

Acknowledgments

We would like to thank Robert Schneider for his support and people from his laboratory for helpful discussions. We would like to thank Andrew Bannister for critical reading an earlier version of the manuscript. This work was supported by the Agence Nationale de la Recherche (CoreAc).

References

1. Martin C, Zhang Y (2005) The diverse functions of histone lysine methylation. Nat Rev Mol Cell Biol 6:838–849. https://doi.org/10.1038/nrm1761
2. Jambhekar A, Dhal A, Shi Y (2019) Roles and regulation of histone methylation in animal development. Nat Rev Mol Cell Biol 20:625–641. https://doi.org/10.1038/s41580-019-0151-1
3. Sierra F, Lichtler A, Marashi F et al (1982) Organization of human histone genes. Proc Natl Acad Sci U S A 79:1795–1799. https://doi.org/10.1073/pnas.79.6.1795
4. Marzluff WF, Gongidi P, Woods KR et al (2002) The human and mouse replication-dependent histone genes. Genomics 80:487–498. https://doi.org/10.1006/geno.2002.6850
5. Chan JC, Maze I (2020) Nothing is yet set in (hi)stone: novel post-translational modifications regulating chromatin function. Trends Biochem Sci 45:829–844. https://doi.org/10.1016/j.tibs.2020.05.009
6. Kouzarides T (2007) Chromatin modifications and their function. Cell 128:693–705. https://doi.org/10.1016/j.cell.2007.02.005
7. Yang W, Ernst P (2017) Distinct functions of H3K4 methyltransferases in normal and malignant hematopoiesis. Curr Opin Hematol 24:322–328. https://doi.org/10.1097/MOH.0000000000000346
8. Black JC, Van Rechem C, Whetstine JR (2012) Histone lysine methylation dynamics: establishment, regulation and biological impact. Mol Cell 48:1–31. https://doi.org/10.1016/j.molcel.2012.11.006
9. Perez-Burgos L, Peters AHFM, Opravil S et al (2003) Generation and characterization of methyl-lysine histone antibodies. Methods Enzymol 76:234–254
10. Rice JC, Allis CD (2001) Histone methylation versus histone acetylation: new insights into epigenetic regulation. Curr Opin Cell Biol 13:263–273. https://doi.org/10.1016/s0955-0674(00)00208-8
11. Zhang Y, Reinberg D (2001) Transcription regulation by histone methylation: interplay between different covalent modifications of the core histone tails. Genes Dev 15:2343–2360. https://doi.org/10.1101/gad.927301
12. Rice JC, Briggs SD, Ueberheide B et al (2003) Histone methyltransferases direct different degrees of methylation to define distinct chromatin domains. Mol Cell 12:1591–1598. https://doi.org/10.1016/S1097-2765(03)00479-9
13. Di Lorenzo A, Bedford MT (2011) Histone arginine methylation. FEBS Lett 585:2024–

2031. https://doi.org/10.1016/j.febslet.2010.11.010

14. Luger K, Rechsteiner TJ, Richmond TJ (1999) Expression and purification of recombinant histones and nucleosome reconstitution. Methods Mol Biol 119:1–16
15. O'Neil LP, Turner BM (2003) Immunoprecipitation of native chromatin: NChIP. Methods 31:76–82. https://doi.org/10.1016/s1046-2023(03)00090-2
16. Dignam JD, Martin PL, Shastry BS, Roeder RG (1983) Eukaryotic gene transcription with purified components. Methods Enzymol 101: 582–598. https://doi.org/10.1101/pdb.prot5330
17. Kudithipudi S, Kusevic D, Weirich S, Jeltsch A (2014) Specificity analysis of protein lysine methyltransferases using SPOT peptide arrays. J Vis Exp 1–8:e52203. https://doi.org/10.3791/52203
18. Mauser R, Jeltsch A (2019) Application of modified histone peptide arrays in chromatin research. Arch Biochem Biophys 661:31–38. https://doi.org/10.1016/j.abb.2018.10.019
19. Ellman GL (1959) Tissue sulfhydryl groups. Arch Biochem Biophys 82:70–77. https://doi.org/10.1016/0003-9861(59)90090-6

Chapter 10

Genetic, Genomic, and Imaging Approaches to Dissect the Role of Polycomb Group Epigenetic Regulators in Mice

Nayuta Yakushiji-Kaminatsui, Takashi Kondo, Yasuhide Ohinata, Junichiro Takano, and Haruhiko Koseki

Abstract

Among the most important histone methyltransferases for metazoan development are EZH1/2 and their homologs, which methylate histone H3 lysine 27 and act as part of a highly conserved set of chromatin regulators called Polycomb Group (PcG) proteins. Reaching a precise understanding of the roles of PcG proteins in the orchestration of differentiation and the maintenance of cell identity requires a variety of genetic and molecular approaches. Here, we present a full suite of methods for the study of PcG proteins in early murine development, including mutant strain generation, embryonic stem cell derivation, epigenomic profiling, and immunofluorescence and *in situ* hybridization.

Key words Polycomb group proteins, Embryonic stem cells, Embryonic tissues, CUT&Tag, Immuno-DNA FISH

1 Introduction

Polycomb group (PcG) factors were first identified in the fruit fly *Drosophila melanogaster* as regulators of anterior-posterior specification, lineage restriction, and position effect variegation and subsequently were found to be conserved among many other metazoan species [1, 2]. It is widely accepted that in all of these diverse species, PcG-class epigenetic regulators contribute to downregulation of development- or differentiation-related genes and, thereby, to spatiotemporally restrict the expression of key transcriptional regulators and signaling molecules to facilitate proper execution of developmental programs. Particularly, as PcG factors were genetically identified as maintenance repressors for homeotic selector gene clusters in *Drosophila*, they have long been believed to contribute to maintenance of a downregulated status of target genes [3]. However, recent studies in mammals revealed that PcG factors are also involved in mediating the

Raphaël Margueron and Daniel Holoch (eds.), *Histone Methyltransferases: Methods and Protocols*,
Methods in Molecular Biology, vol. 2529, https://doi.org/10.1007/978-1-0716-2481-4_10,

transition of transcriptional status of developmental regulators, which accompanies cellular differentiation and/or compartmentalization of fetal tissues [4–6]. We, therefore, expect a dual role for PcG factors, not only to maintain cellular phenotypes, but also to promote their differentiation during development. Consistent with this model, PcG factors have been reported to play critical roles for both maintenance of stem/progenitor cells and their differentiation to maintain the hematopoietic system and barrier organs, such as skin and gut [7–9].

The molecular mechanisms underlying PcG-mediated gene silencing have been addressed by combining biochemical and genetic approaches to study embryonic stem (ES) cells harboring mutant alleles of respective PcG factors. PcG proteins form at least two distinct multimeric protein complexes, Polycomb repressive complexes-1 (PRC1) and -2 (PRC2). PRC1 is known to catalyze mono-ubiquitination of histone H2A at lysine 119 (H2AK119ub1) via RING1A/B, while PRC2 associates with histone methyltransferase activities on histone H3 lysine 27 (H3K27me) via EZH1/2 [10–13]. These catalytic activities have been shown to be directed toward CpG islands (CGIs) associated with PRC target genes and contribute to downregulation of CGI-associated gene promoters. However, consistent with the above-mentioned functional diversity of PcG factors, recent studies have revealed further structural complexity for both PRC1 and PRC2 in mammals. RING1A/B, the catalytic components of PRC1, were shown to heterodimerize with at least six different PCGF proteins (PCGF1-6) and, accordingly, contribute to form six PRC1 sub-complexes [14, 15]. Similarly, PRC2 exists as at least two distinct complexes, namely PRC2.1 and PRC2.2, defined by their specific accessory proteins [16]. Both PRC1 and PRC2 are shown to recognize CGIs via PCGF1-PRC1 and PRC2.2, respectively, and mediate H2AK119ub1 and H3K27me3 to facilitate their mutual interaction [17, 18]. H2AK119ub1 facilitates recruitment of PRC2.2, whereas H3K27me3 is bound by PRC1 incorporating PCGF2 or -4 (PCGF2/4-PRC1, also termed canonical PRC1) through recognition of H3K27me3 by chromobox (CBX) proteins to mediate target gene condensation. Different from PCGF1-PRC1, PRC1 incorporating PCGF3 or -5 (PCGF3/5-PRC1) and PCGF6-PRC1 mediate H2AK119ub1, but their catalytic activities are not directed toward CGIs. Notably, however, both PCGF3/5- and PCGF6-PRC1 have been shown to synergize with PCGF1-PRC1 to accomplish full deposition of H2AK119ub1 at CGIs [19, 20]. Remarkably, a similar synergy is also seen between PRC2.1 and PRC2.2 to coordinate H3K27me3 at CGIs. Such structural diversity of PcG factors and functional redundancy and interplay among them are expected to confer both robustness and reversibility to PcG-mediated gene silencing that may play a critical role to facilitate development.

There can be no doubt that murine ES cells have provided a pivotal experimental platform to explore molecular mechanisms underpinning PcG-mediated gene silencing in the last 20 years. ES cells allowed us to easily converge the expertise of genetics, biochemistry, genomics, imaging, and other technologies and integrate data sets from different layers to gain comprehensive views. However, as ES cells only represent epiblasts that transiently appear in the blastocyst stage of pre-implantation development, the genetic circuitry revealed in ES cells may not represent a complete picture of PcG-mediated regulation [19]. Indeed, although depletion of canonical PRC1 in developmental contexts is well-known to lead to inappropriate differentiation and/or development, its removal in ES cells elicits only a very minor changes in gene expression. We thus expect that PcG-mediated gene regulation may harbor further functional complexity in developing tissues. This would contribute to optimize gene expression profiles in respective cell lineages to execute developmental programs appropriately by adapting to a huge diversity of developmental and environmental stimuli. Although still recognizing the value of ES cells, we find it increasingly important to visit the "real" developmental processes to explore the action modes of PcG factors during cellular differentiation upon receipt of developmental and environmental cues.

Clearly, it is not as easy as in ES cells to assess the impacts of PcG-mediated regulation in a quantitative manner during developmental processes, mainly because of the structural complexity of the respective fetal tissues and their gradually transitioning nature. As PcG factors are, more or less, generally expressed during development, we need to disrupt PcG activity in a spatiotemporally restricted manner to explore the role of PcG factors in respective developmental processes. Practically, we need to generate conditional mutant lines for relevant PcG factors and isolate tissues of interest, which would then be subjected to a series of experimental manipulations. To this end, we would need to be equipped with expertise in mouse genetics, tissue isolation, derivation of cell lines, genomics, imaging, and other techniques. In this chapter, we would like to describe four methods that are routinely used by our team to explore how PcG factors function in mice, that is, how to generate mutant mice and ES cells [21], chromatin profiling using Cleavage Under Targets & Tagmentation (CUT&Tag) with mouse embryonic tissues [22–24] and immuno-DNA FISH on histological sections [25].

2 Materials

2.1 Generation of PcG Mutant Mice

1. Tamoxifen (Sigma-Aldrich, T5648)(*see* **Note 1**).
2. Corn oil.
3. 1.5-mL tube.
4. 1-mL syringe.
5. 18-gauge needle.
6. Tube rotator.

2.2 Derivation of Embryonic Stem Cells from Blastocysts

1. Male mice (8–20 weeks old).
2. Female mice (around 6–16 weeks old are preferred).
3. NDiff227 medium: 1:1 mixture of DMEM/F12 (Thermo Fisher Scientific, 11330-032) and Neurobasal (Thermo Fisher Scientific, 10888-022), 0.5× N-2 Supplement (Thermo Fisher Scientific, 17502-048), 0.5× B-27 Supplement (Thermo Fisher Scientific, 12587-010), 1% GlutaMAX Supplement (Thermo Fisher Scientific, 35050-061), 0.05% BSA (Sigma-Aldrich, A3156), 0.1 mM 2-mercaptoethanol (Thermo Fisher Scientific, 31350-010) and 1% Penicillin–Streptomycin (Thermo Fisher Scientific, 15140-122).
4. ES medium: NDiff227 supplemented with 1000 U/mL LIF (Merck, ESG1107), 3 μM CHIR99021 (Stemgent, 04-0004), and 1 μM PD0325901 (Stemgent, 04-0006).
5. Tyrode's Solution, Acidic (Sigma-Aldrich, T1788).
6. PBS.
7. Cell freezing medium (CELLBANKER 2, ZENOAQ RESOURCE).
8. TrypLE Select Enzyme (1×) (Thermo Fisher Scientific, 12563029).
9. Gelatin from porcine skin (Sigma-Aldrich, G2500).
10. Human plasma fibronectin (Sigma-Aldrich, FC010).
11. Sterile petri dish.
12. 24-well plate.
13. 48-well plate.
14. 15-mL tube.
15. 21-gauge needle.
16. 10-mL syringe.
17. Scissors.
18. Tweezer.
19. Dissection microscope.

20. CO_2 incubator.
21. Centrifuge.
22. Primers for genotyping: *Zfy*-antisense, 5′-CTA TGA AAT CCT TTG CTG CAC ATG T-3′; *Zfy*-sense, 5′-GTA AAG CTT ACA TAA TCA CAT GGA-3′.
23. Thermal cycler.

2.3 Epigenomic Profiling of Mouse Embryonic Tissues by Using Quantitative CUT&Tag

2.3.1 Cell Dissociation and Cell Freezing

1. Mouse embryos.
2. 293T cells.
3. 10% FBS/PBS.
4. Accumax (Innovative Cell Technologies, AM105).
5. Accutase (Innovative Cell Technologies, AT104).
6. 0.1% BSA/PBS.
7. 1.5 mL DNA LoBind Tube.
8. ThermoMixer, Eppendorf.
9. Cell strainer.
10. 10% DMSO/FBS or cell freezing medium (CELLBANKER 1 plus, ZENOAQ RESOURCE).
11. Cell freezing container.
12. Swinging bucket rotor.
13. Centrifuge.

2.3.2 Preparation of pAG-Tn5 or Tn5 Adapter Complex

1. Homemade pAG-Tn5 (*see* **Note 2**).
2. Homemade Tn5 (Addgene plasmid # 60240; http://n2t.net/addgene:60240; RRID:Addgene_60240, [26]).
3. Adapter oligos (Integrated DNA Technologies): Mosaic end-reverse (ME-R), 5′-[phos] CTGTCTCTTATACACATCT-3′; Mosaic end- adapter A (ME-A), 5′-TCGTCGGCAGCGTCAGATGTGTATAAGAGACAG-3′; Mosaic end- adapter B (ME-B), 5′-GTCTCGTGGGCTCGGAGATGTGTATAAGAGACAG-3′.
4. Thermal cycler.
5. Tube rotator.

2.3.3 Spike-in CUT&Tag

1. Binding Buffer: 20 mM HEPES-KOH pH 7.5, 10 mM KCl, 1 mM $CaCl_2$, 1 mM $MnCl_2$. Store at 4 °C for 6 months.
2. Concanavalin A-coated magnetic beads (Bangs Laboratories, BP531).
3. 0.2% Formaldehyde/PBS.
4. 1.25 M Glycine.

5. Wash Buffer: 20 mM HEPES-KOH pH 7.5, 150 mM NaCl, 0.5 mM spermidine, 1× protease inhibitor cocktail. Store at 4 °C for up to 1 week.
6. Antibody Buffer: Mix 10 μL 5% digitonin (*see* **Note 3**), 4 μL 0.5 M EDTA pH 8.0, 13.4 μL 7.5% BSA with 1 mL Wash Buffer. Store at 4 °C for up to 2 days.
7. Dig-wash Buffer: Mix 10 μL 5% digitonin with 1 mL Wash Buffer. Store at 4 °C for up to 2 days.
8. pAG-Tn5 adapter complex (preparation using materials in Subheading 2.3.2 and procedure in Subheading 3.3.3).
9. Dig-300 Buffer: 20 mM HEPES-KOH pH 7.5, 300 mM NaCl, 0.5 mM spermidine, 0.01% digitonin, 1× protease inhibitor cocktail. Store at 4 °C for up to 2 days.
10. Tagmentation Buffer: Mix 10 μL 1 M $MgCl_2$ with 1 mL Dig-300 Buffer.
11. Primary antibody.
12. Secondary antibody.
13. Proteinase K (10 mg/mL).
14. Phenol/Chloroform/Isoamyl alcohol (25:24:1).
15. Glycogen (20 mg/mL).
16. 0.5 M EDTA pH 8.0.
17. 10% SDS.
18. TE: 10 mM Tris–HCl pH 8.0 + 1 mM EDTA.
19. 10 mM Tris–HCl pH 8.0.
20. 100% Ethanol.
21. DNase/RNase-free distilled water.
22. RNase A (10 mg/mL).
23. 1.5 mL DNA LoBind Tube.
24. LoBind PCR plate 96-well skirted.
25. Flat 8 or 12 cap strips.
26. Phase-Lock Gel, Light 2 mL (Quanta Biosciences, 2302820).
27. Swinging bucket rotor.
28. Centrifuge.
29. Magnetic tube/plate separator.
30. Tube/plate rotator or Nutator.
31. ThermoMixer, Eppendorf.
32. Thermal cycler.

2.3.4 Spike-in Control Genomic DNA Library Preparation

1. Lysis Buffer: 10 mM KCl, 20 mM Tris–HCl pH 8.0, 10 mM $(NH_4)_2SO_4$, 1 mM EDTA pH 8.0, 0.1% Triton X-100.
2. 4× Tagmentation Buffer: 40 mM Tris–HCl pH 8.0, 40 mM $MgCl_2$.
3. 2× Tagmentation Buffer: Mix 50 μL Tagmentation Buffer with 50 μL 100% dimethylformamide. Prepare very fresh just before use.
4. Homemade Tn5 adapter complex (preparation using materials in Subheading 2.3.2 and procedure in Subheading 3.3.3).
5. Phenol/Chloroform/Isoamyl alcohol (25:24:1).
6. Glycogen (20 mg/mL).
7. 100% ethanol.
8. 0.2% SDS.

2.3.5 Library Amplification for Spike-in CUT&Tag and Spike-in Control Genomic DNA

1. Custom Barcodes Adapter 1 (index i5) and Custom Barcodes Adapter 2 (index i7) [27].
2. Q5 Hot Start High-Fidelity 2× Master Mix (New England BioLabs, M0494).
3. SPRIselect (Beckman Coulter, B23319).
4. 10 mM Tris–HCl pH 8.0.
5. Library Quantification Kit (Takara Bio, 638324).
6. 1.5-mL tube.
7. PCR tube strips.
8. Qubit Fluorometer, Thermo Fisher Scientific.
9. TapeStation, Agilent Technologies.
10. Magnetic tube/plate separator.
11. Thermal cycler.
12. Real-Time PCR thermal cycler.

2.4 Immuno-fluorescence and Fluorescence In Situ Hybridization (Immuno-FISH)

1. Nick Translation kit (Sigma Aldrich, 10976776001): Control DNA pBR322, 0.4 mM dATP, 0.4 mM dCTP, 0.4 mM dGTP, 0.4 mM dTTP, 10× Nick Translation Buffer, Enzyme Mixture.
2. dCTP-Cy3 (Sigma-Aldrich, GEPA53021).
3. dCTP-Cy5 (Sigma-Aldrich, GEPA55021).
4. dUTP-Alexa488 (Thermo Fisher Scientific, C11397).
5. Cy3-mixture: 2 μL of 10× Nick Translation Buffer, 2 μL of dATP, 2 μL of dGTP, 2 μL of dTTP, 0.4 μL of dCTP, 1 μL of dCTP-Cy3, 4 μL of Enzyme Mixture, 0.5 μg of DNA (*see* Subheading 3.4.2 **Probe Preparation**) and add water to 20 μL.

6. Cy5-mixture: 2 μL of 10× Nick Translation Buffer, 2 μL of dATP, 2 μL of dGTP, 2 μL of dTTP, 0.4 μL of dCTP, 1 μL of dCTP-Cy5, 4 μL of Enzyme Mixture, 0.5 μg of DNA (*see* Subheading 3.4.2 **Probe Preparation**) and add water to 20 μL.
7. Alexa488-mixture: 2 μL of 10× Nick Translation Buffer, 2 μL of dATP, 2 μL of dGTP, 0.4 μL of dTTP, 2 μL of dCTP, 1 μL of dUTP-Alexa488, 4 μL of Enzyme Mixture, 0.5 μg of DNA (*see* Subheading 3.4.2 **Probe Preparation**) and add water to 20 μL.
8. Salmon sperm DNA (10 mg/mL).
9. Yeast tRNA (5 mg/mL).
10. Mouse Cot-1 DNA (1 mg/mL) (Thermo Fisher Scientific, 18440016).
11. 4 M Ammonium Acetate.
12. 100% EtOH.
13. 70% EtOH.
14. Formamide (biochemical grade).
15. 1.5-mL tube.
16. DNase/RNase-free distilled water.
17. 20× saline sodium citrate (SSC) pH 4.5.
18. BSA (20 mg/mL) (Sigma-Aldrich, 10711454001).
19. 50% dextran sulfate (Sigma-Aldrich, D8906): Dissolve in DNase/RNase-free distilled water and store at 4 °C for 1 week.
20. Hybridization buffer: Mix 5 μL of 20× SSC, 5 μL of BSA, 10 μL of 50% dextran sulfate with 5 μL of DNase/RNase-free distilled water and store at 4 °C for 1 week.
21. Probe mixture: Mix 1.5 μL of Cy3-labeled probe, 1.5 μL of Cy5-labeled probe (or formamide if you do not have second probe), 1.5 μL of Alexa488-labeled probe (or formamide) with 1.5 μL of Cot-1 DNA.
22. Tween 20.
23. Parafilm.
24. Mounting medium.
25. 5% Normal Horse Serum (NHS)/PBS.
26. Primary antibody in 0.1% BSA/PBS.
27. Secondary antibody in 0.1% BSA/PBS.
28. Xylene.
29. 4% PFA/PBS.
30. HistoVT One (nacalai tesque, 06380-05).

31. DAPI (Sigma-Aldrich, D9542): Dissolve in water to be 10 mg/mL.
32. Glass slides.
33. Glass cover slips.
34. Coplin staining jar.
35. Slide staining rack.
36. Staining dish.
37. Heat block.
38. Incubator.
39. Low temperature incubator.
40. Water bath.
41. Microscope.

3 Methods

3.1 Generation of PcG Mutant Mice

3.1.1 How to Search for Mutant Mice from the Resource

The International Mouse Strain Resource (IMSR) (http://www.findmice.org/index), to which around 30 repositories over the world are contributing, is an online database of available mouse and ES cell lines. First, we can easily inquire from the website whether a floxed mouse of your PcG gene of interest exists and, if it does, we can check on the details. As shown in Table 1, for example, several PcG flox strains and Cre recombinase-expressing strains are available from The Jackson Laboratory (https://www.jax.org) and RIKEN BRC (https://mus.brc.riken.jp/en/).

3.1.2 Breeding Condition for Generation of Conditional Knockout Embryos/Mice

Although the Cre-loxp-mediated conditional knockout system is a powerful genetic tool, we still need to pay attention to the potential of undesired deletion of the target genes, particularly due to cryptic expression of Cre recombinase in the germline, as well as to the deletion efficacy in the experimental setup. As a detailed description for each Cre deleter line is provided on the websites, we recommend checking these issues before you send purchase orders. These instructions are also needed to plan breeding schemes. In our laboratory, in general, we prepare homozygous flox females and homozygous or heterozygous males harboring the Cre transgene as mating pairs and cross them to generate both conditional knockout embryos and control littermates at the same time. This mating scheme is particularly important when using drug-inducible *Cre* deleter alleles (such as B6.129-$Gt(ROSA)26Sor^{tm1(cre/ERT2)Tyj}$/J) to deplete genes of interest in the fetus by treating pregnant females with Tamoxifen, as we can avoid deletion of target genes in the mothers upon Tamoxifen treatment. We use the same mating scheme to derive conditional mutant ES cells from blastocysts (*see* Subheading 3.2).

Table 1
Available PcG flox and Cre recombinase-expressing mutant strains

PcG fl/fl mouse		
Strain name	**Stock no**	**Target gene**
B6;129S1-*Ezh2*^{tm2Sho}/J	022616	*Ezh2*
B6;129S1-*Eed*^{tm1Sho}/J	022727	*Eed*
B6;129-*Jarid2*^{tm1Yskl}/J	031141	*Jarid2*
STOCK *Bmi1*$^{tm1.1Sjm}$/J	028974	*Bmi1*
FVB.Cg-*Kdm2b*$^{tm1.1Bes}$/J	029345	*Kdm2b*

Cre mouse		
Strain name	**Stock no**	**Specificity**
C57BL/6-Tg(Zp3-cre)93Knw/J	003651	Female germ cells
STOCK Tg(Stra8-icre)1Reb/J	008208	Postnatal, premeiotic, male germ cells
129-*Alpl*$^{tm1(cre)Nagy}$/J	008569	Embryonic primordial germ cells
B6.Cg-Tg(Prrx1-cre)1Cjt/J	005584	Developing limb bud mesenchyme and a part of craniofacial mesenchyme
B6.Cg-*E2f1*$^{Tg(Wnt1-cre)2Sor}$/J	022501	Developing neural crest and midbrain
B6.129S2(Cg)-*Msx1*$^{tm2.1cre/ERT2)Bero}$/J	027850	Developing palate, teeth, and nails
STOCK Tg(KRT14-cre)1Amc/J	004782	Ectoderm and its derivatives such as the skin and dental epithelium
B6.Cg-Tg(Lck-cre)548Jxm/J	003802	Thymocyte
B6.Cg-Tg(Cd4-cre)1Cwi/BfluJ	022071	CD4-expressing T cells
B6.129-*Gt(ROSA)26Sor*$^{tm1(cre/ERT2)Tyj}$/J	008463	Tamoxifen-inducible Cre activity in whole body
B6.Cg-Tg(CAG-cre/Esr1*)5Amc/J	004682	Tamoxifen-inducible Cre activity in whole body

3.1.3 Tamoxifen Treatment for Obtaining Mutant Embryos

1. Dissolve 15 mg of Tamoxifen in 1 mL corn oil in a foil-wrapped 1.5-mL tube by rotating at 4 °C overnight. Store at 4 °C for up to 1 month.
2. 100 μL of Tamoxifen (15 mg/mL) is administered to the pregnant mother via intraperitoneal injection. Repeat this injection every 24 or 48 h as needed.

3.2 Derivation of Embryonic Stem Cells from Blastocysts

3.2.1 Mating

1. One male should be kept in each cage for mating. In the evening, check the estrus status of the females used for mating. Select females in estrus with large vulvae and keep them with one male and one or two females per cage.
2. Next early morning, check the female mice. If they have mated, the vagina will be filled with the copulatory plug. Noon of the plug confirmation day is defined as 0.5-day post coitum (d.p.c).

3.2.2 Collecting Blastocysts from the Uterus

1. Dissect the female at 3.5 d.p.c. and remove the uterus using scissors and tweezers that have been sterilized in an autoclave or with alcohol. The uterus can be separated into the left and right sides or it can be kept connected. Excess fatty tissues should be removed.
2. Place the uterus in a sterile petri dish. Using a 21-gauge needle and a 10-mL syringe, reflux from the ovarian side of the uterus using NDiff227 medium. Under a dissection microscope, collect blastocysts by using a micropipette and transfer them to a small drop (100–200 μL of NDiff227 medium) in a new dish.

3.2.3 Removal of the Zona Pellucida

1. Prepare four small droplets of about 100–200 μL each in the lid of a culture dish in the following order: first, NDiff227, second and third, Tyrode's Solution, fourth, NDiff227.
2. Transfer the recovered embryos to the first drop, then immediately move to the second and third drops. During these transfers, carryover of the previous solution should be minimized as much as possible.
3. The embryos can be transferred to the fourth drop when the zona pellucida is sufficiently dissolved. Embryos become sticky once the zona pellucida is removed. To prevent embryos from sticking to each other or to the lid, we usually keep the medium and Tyrode's Solution cooler than room temperature and also move the embryos by gentle pipetting.

3.2.4 Culture of Blastocysts

1. Coat 48-well plates with 0.2% gelatin/PBS. Add 250 μL/well of the coating solution and let it stand for 15 min at room temperature. Aspirate the coating solution and add 250 μL/well of ES medium.
2. Transfer one blastocyst to each well and culture them in 5% CO_2 in a humidified incubator at 37 °C. After about a week, an outgrowth will be formed from each blastocyst.

3.2.5 Passaging of Blastocyst Outgrowth

1. Aspirate the medium gently and add 100 μL/well of TrypLE select, then incubate at 37 °C for 10 min. As serum-free medium is used, it is not necessary to wash the well with PBS.

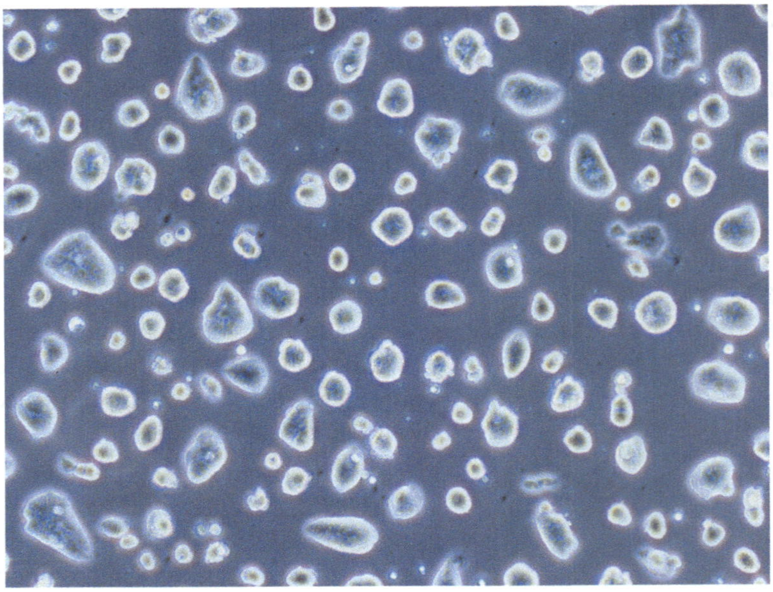

Fig. 1 Colony morphology of ES cells immediately after their establishment

2. Dissociate the cells by pipetting and suspend in 2 mL of NDiff227 medium. Centrifuge at 500 × *g* for 3 min at room temperature and then aspirate the supernatant. This washing process by centrifugation is essential to remove the cell dissociation enzymes. Without this step, the plating efficiency of the dissociated cells may be considerably reduced.
3. Suspend the cells in 500 μL of ES medium and transfer them on to a new gelatin- or fibronectin-coated 24-well plate. At this stage, establishment of ES cells can be confirmed by visual inspection of colony growth (Fig. 1, *see* **Note 4**).
4. Repeat passages up to the required scale and determine genotypes before cryopreservation of respective cell lines. At this stage, we determine genotypes and sex of established cell lines (*see* Subheading 3.2.6). We prefer to use male ES cells as female ES cells tend to suffer from genomic instability, particularly of the X chromosome, and genome-wide DNA hypomethylation after a number of passages.

3.2.6 Genotyping for Identification of Male ES Cells

The *PCR* temperature cycling *conditions* are as follows: initial denaturation at 94 °C for 3 min; 30 cycles of denaturation at 94 °C for 45 s, annealing at 60 °C for 45 s, and elongation at 72 °C for 45 s; additional elongation at 72 °C for 3 min. The PCR product size is about 600 bp.

3.3 Epigenomic Profiling of Mouse Embryonic Tissues by Using Quantitative CUT&Tag

3.3.1 Cell Dissociation from Mouse Embryonic Tissues and Cell Freezing

1. Dissect each tissue into 300 μL 10% FBS/PBS in a 1.5-mL tube (*see* **Note 5**).
2. Centrifuge at 400 × *g* for 1 min at room temperature in the swinging bucket rotor.
3. Remove 10% FBS/PBS and add 300 μL Accumax, then place the tube on the ThermoMixer at 37 °C at 1200 rpm for 5–15 min. Shake the tube vigorously every 5 min (*see* **Note 6**).
4. Add 300 μL of 10% FBS/PBS into the 1.5-mL tube and gently suspend the cells. Pass cells through a cell strainer into a new LoBind tube.
5. Centrifuge at 400 × *g* for 3 min at room temperature in the swinging bucket rotor.
6. Remove the supernatant and resuspend cells in 1 mL 0.1% BSA/PBS. Centrifuge at 400 × *g* for 3 min at room temperature in the swinging bucket rotor.
7. Remove the supernatant and resuspend cells in 300–500 μL 10% DMSO/FBS or cell freezing medium. Place tubes in the freezing container and store at −80 °C until you have enough cells for your experiment.

3.3.2 Harvest 293T Cells and Cryopreservation

1. Culture 293T cells in 5% CO_2 in a humidified incubator at 37 °C until they reach around 80% confluency.
2. Remove the media and wash with PBS, then add 500 μL of Accutase. Wait for 5–10 min at 37 °C.
3. Pipet gently to resuspend the cells while separating them from one another. Add 5 mL of PBS and take everything into a 15-mL tube. Add another 5 mL of PBS into the dish and take all into the 15-mL tube.
4. Centrifuge at 400 × *g* for 3 min at room temperature with the swing bucket rotor.
5. Remove the supernatant and resuspend cells in 5 mL PBS, then count number of cells. Take cells as much as needed when you use fresh 293T cells for spike-in CUT&Tag experiment.
6. When you want to freeze 293T cells, centrifuge cells at 400 × *g* for 3 min at room temperature with the swing bucket rotor. Remove the supernatant and resuspend cells in 300–500 μL 10% DMSO/FBS or cell freezing medium. Place tubes in the freezing container and store at −80 °C.

3.3.3 Prepare pAG-Tn5 or Tn5 Adapter Complex

Dissolve the lyophilized adapter oligos in annealing buffer (50 mM NaCl, 40 mM Tris–HCl pH 8.0) to be 100 μM. Anneal ME-A and ME-B oligonucleotides with ME-R oligonucleotides by using a thermal cycler with following program: 95 °C for 5 min, to 65 °C 0.1 °C/s, 65 °C for 5 min, to 4 °C 0.1 °C/s. Mix 16 μL of 100 μM equimolar mixture of pre-annealed ME-A and ME-B oligonucleotides with 100 μL of ~0.2 mg/mL homemade pAG-Tn5 fusion protein or Tn5 protein. Incubate the mixture on a rotator for 1 h at room temperature and then store at −20 °C.

3.3.4 Prepare Precoated 96-Well Plate and Concanavalin A-Coated Magnetic Beads

1. Add 100 μL of 1% BSA/PBS into each well of the plate and close the cap strip. Rotate the plate while Concanavalin A-coated magnetic beads are prepared.
2. Add 10 μL of Concanavalin A-coated magnetic beads per reaction into a 1.5-mL tube and place it in the magnetic separator.
3. Remove the supernatant and suspend the beads in 1 mL of Binding Buffer. Place in the magnetic separator and then remove the Binding Buffer. Add 1 mL of Binding Buffer again and repeat this wash step twice.
4. Remove the Binding Buffer and add 20 μL of Binding Buffer per reaction and then mix completely by pipetting.
5. Remove the 1%BSA/PBS from 96-well plate carefully and then add 20 μL of Concanavalin A-coated magnetic beads into each well. Keep on ice until cells are ready.

3.3.5 Defrost Cells Derived from Embryonic Tissues and 293T Cells

1. Place the tubes at 37 °C until it is half melted, add 0.1% BSA/PBS and then centrifuge at 400 × *g* for 3 min at room temperature in the swinging bucket rotor.
2. Remove the supernatant carefully and resuspend cells in 0.1% BSA/PBS. If the exact number of cells from embryonic tissues is known, add 1/10 the number of 293T cells as spike-in controls. If it is unknown, add roughly 1/10 the number of 293T cells and DNA-seq should be performed to measure the ratio between mouse cells and 293T cells (*see* Subheading 3.3.15).
3. Mix sample completely with 293T cells and centrifuge at 400 × *g* for 3 min at room temperature in the swinging bucket rotor. Remove the supernatant and resuspend the cells in 500 μL PBS.

3.3.6 Light Fixation

1. Add 500 μL 0.2% formaldehyde/PBS to the 1.5-mL tube to a final concentration of 0.1%, mix very well and then immediately place tube on a rotation wheel at room temperature for 2 min.
2. Add 100 μL 1.25 M glycine immediately to quench the reaction and centrifuge at 400 × *g* for 3 min at room temperature in a swinging bucket rotor. Remove the supernatant carefully and wash cells once with PBS.

3.3.7 Bind Cells to Concanavalin A-Coated Magnetic Beads

1. Centrifuge at 400 × *g* for 3 min at room temperature in the swinging bucket rotor. Prepare 100 μL × (sample number + 1) of Wash Buffer. Remove the supernatant and resuspend cells in100 μL of Wash Buffer.

2. Add 100 μL of cells to each well and then place on a rotator or nutator at room temperature for 10–20 min. Keep remaining cells in the tube at −20 °C for DNA-seq if exact cell number is not known (continue to Subheading 3.3.15).

3.3.8 Bind Primary Antibody

After rotation, place the plate in a magnetic stand to clear and remove the supernatant and then add 100 μL of Primary Antibody Buffer containing 1–10 μL primary antibody into each sample (*see* **Notes** 7 and **8**). Place on a rotator or nutator at 4 °C overnight.

3.3.9 Bind Secondary Antibody

1. Place the plate in a magnetic stand to clear and remove the supernatant and then wash with Dig-wash Buffer once.
2. Place the plate in a magnetic stand and remove the supernatant and then add 100 μL of Dig-wash Buffer containing 1 μL secondary antibody into each sample. Place on a rotator or nutator at room temperature for 30–60 min.
3. Place the plate in a magnetic stand and remove the buffer and then wash with Dig-wash Buffer. Close the cap strip perfectly and invert the plate 10 times and centrifuge at 300 × *g* briefly in the swing rotor. Place the plate in a magnetic stand. Repeat this wash step twice.

3.3.10 Bind the Protein AG-Tn5 Adapter Complex

1. Remove the Dig-wash Buffer and add 100 μL of Dig-300 Buffer containing 0.8 μL of pAG-Tn5 into each well and then place on a rotator or nutator at room temperature for 1 h.
2. Place the plate in a magnetic stand to clear and remove the supernatant and then add 100 μL of Dig-300 Buffer. Place the plate on a rotator or nutator for 3 min, then place it back to a magnetic stand to remove the supernatant. Repeat this wash at least three times.
3. Add 100 μL of Dig-300 buffer and place all samples into new wells before tagmentation to completely remove the unbound pAG-Tn5 adapter complex.

3.3.11 Tagmentation and De-Crosslinking

1. Remove the Dig-300 Buffer and add 100 μL of Tagmentation Buffer and then incubate at 37 °C on ThermoMixer at 600 rpm for 30–60 min (*see* **Note 9**).
2. Add 3.3 μL of 0.5 M EDTA and 1 μL of 10% SDS to each well immediately. Incubate at 65 °C on a ThermoMixer at 600 rpm overnight.

3.3.12 DNA Extraction

1. Add 2 μL of Proteinase K to each well and incubate at 37 °C on a ThermoMixer at 600 rpm for 30 min.
2. Transfer the DNA into the Phase-Lock gel 2 mL tube and add 200 μL of 0.1× TE and 300 μL of Phenol/Chloroform/Isoamyl alcohol, then mix completely by vortexing. Centrifuge at 16,000 × *g* for 10 min.

3. Transfer the aqueous layer to a 1.5 mL LoBind tube containing 750 μL of ice-cold 100% EtOH and 1 μL of glycogen and then centrifuge at 16,000 × *g* for 20 min at 4 °C.
4. Remove the EtOH and wash with 300 μL of 100% EtOH, then centrifuge at 16,000 × *g* for 10 min at 4 °C. After centrifugation, carefully remove the EtOH and dry the pellet at room temperature.
5. Dissolve the pellet in 22 μL of 0.1× TE containing 25 μg/mL RNase A and incubate at 37 °C on a ThermoMixer at 600 rpm for 10 min.

3.3.13 Library DNA Amplification

1. Mix 21 μL of DNA with 25 μL of Q5 Hot Start High-Fidelity 2× Master Mix and 4 μL of premix of Custom Barcodes Adapter 1 (index i5) and Custom Barcodes Adapter 2 (index i7) (10 μM) in each well of the PCR strip.
2. Mix completely and spin down, then immediately transfer the tube strip to a thermal cycler with a heated lid: 72 °C for 5 min (gap filling), 98 °C for 30 s, followed by at least 15 cycles with 98 °C for 10 s (denaturation) and 63 °C for 15 s (annealing and extension), then 72 °C for 1 min (*see* **Note 10**).
3. Add 0.9 volume (45 μL) of homogenous SPRIselect beads to each 50 μL of DNA sample, mix by pipetting at least 10 times and then incubate at room temperature for 3 min.
4. Place the PCR strip in the magnetic separator to collect the beads, wait for 5 min, then discard the clear supernatant.
5. Apply 180 μL of fresh 85% EtOH to the beads, wait for 30 s, then remove the EtOH. Repeat this wash twice. After removal of the EtOH, spin down briefly and put the PCR strip in the magnetic separator and then completely remove the remaining EtOH.
6. Let the beads air dry for 3 min, add 20 μL of 10 mM Tris–HCl to each well, then mix well by vortexing and briefly spin down. Incubate for 3–5 min at room temperature.
7. Place the PCR strip in the magnetic separator until the solution is clear and then transfer the cleared supernatant to a new tube. Measure the quantity of library DNA by Qubit and check the fragment size by using TapeStation according to the respective manufacturer's instructions.

3.3.14 Library DNA Quantification for Sequencing

We regularly measure an accurate concentration of DNA library by using the library quantification kit which allows the specific quantification of DNA that are bound to adapters, and then make a library pool for sequencing. We normally prepare 2 nM library with unique dual indexes and perform 50 bp paired-end sequencing with NovaSeq6000.

If you know the cell ratio between the mouse sample and the 293T cells, the following steps are optional. However, when you do not know the exact cell number of your sample derived from dissecting embryonic tissues, these steps are needed to measure the ratio between mouse sample and the 293T cells.

3.3.15 Genomic DNA Extraction for Spike-in Control Genomic DNA Library Preparation

1. Add 200 μL of Lysis Buffer and 3 μL of Proteinase K to the tube obtained in Subheading 3.3.7 and incubate at 55 °C overnight.
2. Add the contents of the tube into a Phase-Lock gel 2 mL tube, add 300 μL of Phenol/Chloroform/Isoamyl alcohol, then mix completely by vortexing. Centrifuge at 16,000 × *g* for 10 min.
3. Transfer the aqueous layer to a 1.5-mL LoBind tube containing 1 μL of glycogen and 750 μL of 100% EtOH. Mix completely and centrifuge at 16,000 × *g* for 20 min at 4 °C.
4. Remove the EtOH and wash with 300 μL of 100% EtOH and centrifuge at 16,000 × *g* for 10 min at 4 °C. Remove the EtOH carefully and dry the pellet, then dissolve in 30 μL of pure water.
5. Measure the DNA quantity by using Qubit and dilute the gDNA to 0.5 ng/μL.

3.3.16 Tagmentation

1. Carefully mix 3 μL of gDNA with 5 μL of 2× Tagmentation Buffer and 2 μL of the Tn5 adapter complex (*see* **Note 11**) in a PCR strip tube.
2. Immediately transfer the PCR tube strip to a thermal cycler with a heated lid and start the program: 55 °C for 5 min followed by 10 °C.
3. When the temperature reaches 10 °C, immediately add 2.5 μL of 0.2% SDS into each well to stop the reaction. Vortex and incubate at room temperature for 5 min, then keep the tube on ice.

3.3.17 Spike-in Control Genomic DNA Library Amplification

1. Mix 12.5 μL of DNA with 25 μL of Q5 Hot Start High-Fidelity 2× Master Mix, 8.5 μL of pure water and 4 μL of premix of Custom Barcodes Adapter 1 (index i5) and Custom Barcodes Adapter 2 (index i7) (10 μM) in each well of a PCR strip.
2. Mix completely and spin down, then immediately transfer the tube strip to a thermal cycler with a heated lid: 72 °C for 5 min (gap filling), 98 °C for 30 s, followed by 17–20 cycles with 98 °C for 10 s (denaturation) and 63 °C for 15 s (annealing and extension), then 72 °C for 1 min.

3. Add 0.9 volume (45 μL) of homogenous SPRIselect beads to each 50 μL of DNA sample, mix by pipetting at least 10 times, then incubate at room temperature for 3 min. Place the PCR strip in the magnetic separator to collect beads, wait for 5 min, then discard the clear supernatant.
4. Apply 180 μL of fresh 85% EtOH to the beads, wait for 30 s, then remove the EtOH. Repeat this wash twice. After removal of the EtOH, spin down briefly and put the PCR strip in the magnetic separator and then carefully remove the remaining EtOH.
5. Let the beads air dry for 3 min, add 20 μL of 10 mM Tris–HCl to each well, then mix well by vortexing and briefly spin down. Incubate for 3–5 min at room temperature.
6. Place the PCR strip in the magnetic separator until the solution is clear and then remove the cleared supernatant to a new tube. Measure the quantity of library DNA by Qubit and check the fragment size by using TapeStation according to the respective manufacturer's instructions. For library quantification, follow Subheading 3.3.14.

3.4 Immunofluorescence and Fluorescence In Situ Hybridization (Immuno-FISH)

3.4.1 Probe Design

Fosmid clones containing ~30 kb of genomic sequence can be used as DNA probes. However, fosmids contain repetitive sequences that give background and lower the signal-to-noise ratio for detection. In order to reduce the background, we use mouse genomic DNA to generate PCR products consisting of ~7 kb genomic sequence designed from about 10–15 kb of target sequence. Repetitive sequences are eliminated from the 15 kb target sequences by RepeatMasker (http://www.repeatmasker.org). From these repeat-masked sequences, DNA sequences without repetitive sequences are designed as probes. About 20 sequences (average size about 300 bp) will be produced by PCR; mix them in equimolar amounts for one probe/target sequence.

3.4.2 Probe Preparation

Here we used three colors, i.e., Cy3, Cy5, and Alexa488 for FISH probes. The side chains of nucleotide compounds greatly influence the incorporation ratio of labeled compounds into probes. Therefore, you need to be careful if you change the labeled compounds.

1. Prepare Cy3- or Cy5- or Alexa488-mixtures in 1.5-mL tubes, respectively, then incubate at 15 °C for 3–4 h.
2. Place the tube at 65 °C for 10 min for inactivation and then put on ice.
3. Add 1.5 μL of salmon sperm DNA, 3 μL of yeast tRNA, 5 μL of mouse Cot-1 DNA, 3 μL of 4 M ammonium acetate and 80 μL of 100% EtOH and mix well.
4. Keep the tube at −80 °C for 20 min and then centrifuge at 20380 × *g* for 20 min at 4 °C.

5. Remove the supernatant and wash the pellet once with 500 μL of 70% EtOH.
6. Centrifuge at 20380 × *g* for 10 min at 4 °C.
7. Aspirate the supernatant carefully and ensure that all liquid is completely removed (*see* **Note 12**).
8. Suspend a pellet in 20 μL of formamide and store at −80 °C before use.

3.4.3 Section Preparation

Paraffin sections need to be deparaffinized before the following processes: xylene for 5 min three times, 100% EtOH for 5 min three times and then wash with running water for 20 min.

Frozen sections need to be fixed before the following processes: 4% PFA/PBS for 15 min and wash with running water for 20 min.

3.4.4 Section Treatment for Antigen Retrieval and DNA Hybridization

This process is required for both antigen retrieval and DNA hybridization.

1. Place slides in a slide staining rack and dip it in a staining dish filled with HistoVT One and incubate in a 100 °C water bath for 20 min, then immediately place a staining rack in a staining dish containing water. Wash the slides with running water for 20 min (*see* **Note 13**).

   Proceed to the next step if you are performing FISH or continue to Subheading 3.4.5 if you are performing immuno-FISH.
2. Place slides in a glass jar containing cold 70% EtOH, incubate for 5 min, then air dry the slides. Continue to Subheading 3.4.6.

3.4.5 Immunohistochemistry

For immuno-FISH, we usually run immunohistochemistry before the FISH process.

1. Place glass slides in a Coplin staining jar containing 5% NHS/PBS for 20 min at room temperature for blocking.
2. Add an aliquot of primary antibody in 0.1%BSA/PBS and incubate at 4 °C overnight.
3. Wash glass slides with PBS and add an aliquot of secondary antibody in 0.1% BSA/PBS, then incubate at room temperature for 30 min.
4. Wash glass slides with PBS and fix with 4% PFA/PBS for 15 min at room temperature, then wash with PBS.
5. Dip slides in cold 70% EtOH for 5 min, proceed to 100% EtOH for 5 s, then air dry.

   Continue to Subheading 3.4.6.

3.4.6 Hybridization

1. Place a tube containing 6 μL of Probe mixture at 80 °C for 10 min and then keep it on ice. Add an equal volume (6 μL) of hybridization buffer and mix well.
2. Put an aliquot of the hybridization cocktail prepared in **step 1** onto the mouse section and seal with parafilm. The volume of the aliquot depends on the size of the section and normally we use 2.5 μL of hybridization cocktail per section of an E9.5 mouse and a piece of parafilm cut into 5 mm squares.
3. Hybridize at 37 °C for more than 48 h (usually about 72 h).
4. Remove the parafilm carefully and wash the slides with 50% formamide/2× SSC at 42 °C for 5 min twice, then wash with 0.1% Tween20 2×SSC at 42 °C for 7 min twice.
5. Wash with PBS and perform DAPI staining (5 μg/mL), then wash with PBS.
6. Dip in 70% EtOH for 5 min and proceed to 100% EtOH for 5 min, then air dry and seal with mounting medium.

4 Notes

1. When we perform tamoxifen treatment with ES cells, we add 4-hydroxytamoxifen (Sigma-Aldrich, H7904) dissolved in ethanol to culture medium to a final concentration of 800 nM.
2. Protein AG-Tn5 was constructed by using both pAG-MNase (Addgene plasmid # 123461; http://n2t.net/addgene:123461; RRID:Addgene_123461, [28]) and pA-Tn5 (Addgene plasmid # 124601; http://n2t.net/addgene:124601; RRID: Addgene_124601, [22]) plasmid DNAs.
3. Before use, if some precipitation is observed, heat at 70 °C until it is completely melted.
4. Using this protocol, ES cell derivation efficacy is about 95% with the C57Bl/6 inbred strain.
5. Keep each remaining embryonic tissue or yolk sac in another tube for genotyping when you use mutant embryos.
6. Accumax used in this protocol is a mixture of protease, collagenase and DNase, and it works perfectly with collecting tissues at E9.0 to E11.5. However, when you plan to use embryos at a very early embryonic stage, you may need to find the appropriate dissociation reagents instead of Accumax. Using other reagents such as Trypsin or TypLE (Thermo Fisher Scientific) would be another option.
7. When you use a 1.5-mL LoBind tube for the assay, increase the reaction volume to 300 μL, allowing the reaction solution to mix completely during rotation.

8. When histone mark antibodies such as H3K27me3 (Cell Signaling Technologies, 9733) and antibodies for PcG factors such as RING1B (Cell Signaling Technologies, 5694) are used, we normally use 1:50 to 1:100 dilution.
9. If the 1 h reaction for tagmentation seems to reach saturation, you may reduce the reaction time as an option. To determine the reaction time for tagmentation, we perform the reaction at several time duration, such as 30 min, 40 min, 50 min, and 1 h, with samples using IgG and histone mark antibody (for example, H3K27me3), and then we do library amplification and measure the quantity of library by Qubit. We select the reaction time in which the difference in the amount of DNA between IgG and histone mark antibody sample is around 10–100 times.
10. The number of cycles depends on the initial cell number, antibodies, and whether the cells are fixed or not. If the cell number per assay is below 1×10^4, the annealing and extension time can be extended to 20 s as an option.
11. The quantity depends on the lot of homemade Tn5. To obtain the amount of DNA library required for sequencing, modify the quantity of Tn5 adapter complex for the reaction or the number of PCR cycle properly. At this point, we do not care about over-tagmentation since we just want to know the ratio between mouse and 293T genomes for spike-in normalization. If you use Illumina Tagment DNA Enzyme and Buffer, follow the manufacturer's protocol.
12. If the pellet is overdried, it will not dissolve in formamide.
13. Using a microwave may result in insufficient heat treatment, thus we strongly recommend using a water bath. After antigen retrieval, wash the glass slides with water very quickly, otherwise the following experiment will not work well due to the remaining detergent on the section.

Acknowledgments

We would like to thank Dr. Hiroki Sugishita for discussion and advice on CUT&Tag modifications. This work was supported by the Japan Agency for Medical Research and Development (AMED-CREST) (13417643 to H.K.), Grant-in-Aid for Scientific Research on Innovative Areas (JP19H05745 to H.K.), JSPS KAKENHI (JP19K22695, JP21K09784 to N.Y.-K.), and Takeda Science Foundation (to N.Y.-K.).

References

1. Lewis EB (1978) A gene complex controlling segmentation in Drosophila. Nature 276: 565–570
2. Pirrotta V, Rastelli L (1994) White gene expression, repressive chromatin domains and homeotic gene regulation in Drosophila. Bioessays 16:549–556
3. Boyer LA, Plath K, Zeitlinger J et al (2006) Polycomb complexes repress developmental regulators in murine embryonic stem cells. Nature 441:349–353
4. Almeida M, Pintacuda G, Masui O et al (2017) PCGF3/5-PRC1 initiates Polycomb recruitment in X chromosome inactivation. Science 356:1081–1084
5. Yakushiji-Kaminatsui N, Kondo T, Hironaka K-I et al (2018) Variant PRC1 competes with retinoic acid-related signals to repress Meis2 in the mouse distal forelimb bud. Development 145:dev166348
6. Morey L, Santanach A, Blanco E et al (2015) Polycomb regulates mesoderm cell fate-specification in embryonic stem cells through activation and repression mechanisms. Cell Stem Cell 17:300–315
7. Iwama A, Oguro H, Negishi M et al (2004) Enhanced self-renewal of hematopoietic stem cells mediated by the polycomb gene product Bmi-1. Immunity 21:843–851
8. Ezhkova E, Pasolli HA, Parker JS et al (2009) Ezh2 orchestrates gene expression for the stepwise differentiation of tissue-specific stem cells. Cell 136:1122–1135
9. Chiacchiera F, Rossi A, Jammula S et al (2016) Polycomb complex PRC1 preserves intestinal stem cell identity by sustaining Wnt/β-catenin transcriptional activity. Cell Stem Cell 18: 91–103
10. Cao R, Wang L, Wang H et al (2002) Role of histone H3 lysine 27 methylation in Polycomb-group silencing. Science 298:1039–1043
11. Fischle W, Wang Y, Jacobs SA et al (2003) Molecular basis for the discrimination of repressive methyl-lysine marks in histone H3 by Polycomb and HP1 chromodomains. Genes Dev 17:1870–1881
12. Wang H, Wang L, Erdjument-Bromage H et al (2004) Role of histone H2A ubiquitination in Polycomb silencing. Nature 431:873–878
13. de Napoles M, Mermoud JE, Wakao R et al (2004) Polycomb group proteins Ring1A/B link ubiquitylation of histone H2A to heritable gene silencing and X inactivation. Dev Cell 7: 663–676
14. Gao Z, Zhang J, Bonasio R et al (2012) PCGF homologs, CBX proteins, and RYBP define functionally distinct PRC1 family complexes. Mol Cell 45:344–356
15. Tavares L, Dimitrova E, Oxley D et al (2012) RYBP-PRC1 complexes mediate H2A ubiquitylation at polycomb target sites independently of PRC2 and H3K27me3. Cell 148:664–678
16. Healy E, Mucha M, Glancy E et al (2019) PRC2.1 and PRC2.2 synergize to coordinate H3K27 Trimethylation. Mol Cell 76: 437–452.e6
17. Højfeldt JW, Hedehus L, Laugesen A et al (2019) Non-core subunits of the PRC2 complex are collectively required for its target-site specificity. Mol Cell 76:423–436.e3
18. Blackledge NP, Farcas AM, Kondo T et al (2014) Variant PRC1 complex-dependent H2A ubiquitylation drives PRC2 recruitment and polycomb domain formation. Cell 157: 1445–1459
19. Fursova NA, Blackledge NP, Nakayama M et al (2019) Synergy between variant PRC1 complexes defines Polycomb-mediated gene repression. Mol Cell 74:1020–1036.e8
20. Endoh M, Endo TA, Shinga J et al (2017) PCGF6-PRC1 suppresses premature differentiation of mouse embryonic stem cells by regulating germ cell-related genes. Elife 6:e21064
21. Ying Q-L, Wray J, Nichols J et al (2008) The ground state of embryonic stem cell self-renewal. Nature 453:519–523
22. Kaya-Okur HS, Wu SJ, Codomo CA et al (2019) CUT&Tag for efficient epigenomic profiling of small samples and single cells. Nat Commun 10:1930
23. Kaya-Okur HS, Janssens DH, Henikoff JG et al (2020) Efficient low-cost chromatin profiling with CUT&Tag. Nat Protoc 15:3264–3283
24. Eto H, Kishi Y, Yakushiji-Kaminatsui N et al (2020) The Polycomb group protein Ring1 regulates dorsoventral patterning of the mouse telencephalon. Nat Commun 11:5709
25. Kondo T, Isono K, Kondo K et al (2014) Polycomb potentiates meis2 activation in midbrain by mediating interaction of the promoter with a tissue-specific enhancer. Dev Cell 28:94–101
26. Picelli S, Björklund AK, Reinius B et al (2014) Tn5 transposase and tagmentation procedures for massively scaled sequencing projects. Genome Res 24:2033–2040
27. Buenrostro JD, Wu B, Litzenburger UM et al (2015) Single-cell chromatin accessibility reveals principles of regulatory variation. Nature 523:486–490
28. Meers MP, Bryson TD, Henikoff JG et al (2019) Improved CUT&RUN chromatin profiling tools. Elife 8:e46314

Chapter 11

Profiling Histone Methylation in Low Numbers of Cells

Julie Brind'Amour and Matthew C. Lorincz

Abstract

Chromatin immunoprecipitation (ChIP) enables the study of DNA–protein interactions. When coupled with high-throughput sequencing (ChIP-seq), this method allows the generation of genome-wide profiles of the distribution of specific proteins in a given cellular context. Typical ChIP-seq experiments require millions of cells as input material and thus are not ideal to study many in vivo cell populations. Here, we describe an ultra-low-input native ChIP-seq method, ULI-NChIP-seq, to profile histone modification patterns in as low as 150 cells.

Key words Chromatin, Histone modifications, Epigenetics, Low-input, Embryo, Oocyte, Methylation, Chromatin immunoprecipitation

1 Introduction

Chromatin immunoprecipitation [1, 2] followed by sequencing (ChIP-seq) has been widely used in recent years to study DNA–protein interactions at the genome-wide level. When the method was initially developed [3–5], generating quality sequencing libraries required relatively large sample sizes, on the order of ten million cells per assay. The need to generate histone modification profiles in cell populations of low abundance has pushed the development of new adaptations to the classic ChIP-seq protocols. Methods based on fragmentation of formaldehyde-cross-linked DNA–protein complexes [6–9], or enzymatic fragmentation of native chromatin [10–13], and even targeted fragmentation using enzyme-coupled antibodies [14, 15] have reduced the required input size for ChIP-seq of histone modifications to as few as hundreds of cells. While cross-linked ChIP methods offer a higher sample stability and may be better suited for histone modifications with lower stability (i.e., phosphorylation), ChIP methods based on enzymatic fragmentation of native chromatin generally yield more complex libraries, with the added advantage of enabling the mapping of nucleosome positioning in parallel.

Raphaël Margueron and Daniel Holoch (eds.), *Histone Methyltransferases: Methods and Protocols*,
Methods in Molecular Biology, vol. 2529, https://doi.org/10.1007/978-1-0716-2481-4_11,

Here, we present an updated description for Ultra-Low-Input Native ChIP-seq (ULI-NChIP-seq) [11], a protocol suitable to generate genome-wide maps of histone modifications from a range of 100–100,000 cells per assay. Included and detailed are procedures for sample isolation and storage of various types of cells, sample pooling and chromatin preparation, ChIP and elution as well as library preparation and finally pooling for sequencing.

2 Materials

All solutions should be prepared with ultrapure water (deionized water purified to a sensitivity of 18.2 MΩ-cm at 25 °C) and analytical grade reagents. Prepare and store reagents at room temperature unless otherwise indicated.

2.1 Sample Isolation and Storage

1. Nuclear Isolation Buffer (Sigma-Aldrich NUC-101) (*see* **Note 1**).
2. EDTA-free protease inhibitor cocktail (PIC, Roche). Prepare a 25× stock solution by diluting one tablet in 2 ml of water. Store at 4 °C for up to 1 week (*see* **Note 2**).
3. 200 mM Phenylmethylsulfonyl fluoride (PMSF) stock. Weigh 0.174 g of PMSF powder and dilute in 5 ml of ethanol. Store at 4 °C for up to 1 week (*see* **Note 3**).
4. Complete Nuclear isolation buffer: Nuclear isolation buffer, 1× PIC, 1 mM PMSF. Add 16 μl 25× PIC stock and 2 μl of 200 mM PMSF stock to 382 μl of Nuclear Isolation Buffer. Prepare fresh and keep on ice.
5. [For oocytes and early embryos isolation, Subheading 3.1.2] M2 media. Store at 4 °C in sterile aliquots.
6. Acid Tyrode's Solution. Store at −20 °C in sterile aliquots.
7. Phosphate-buffered saline (PBS): 137 mM NaCl, 2.7 mM KCl, 10 mM Na_2HPO_4, and 1.8 mM KH_2PO_4. Stir 800 mg NaCl, 20 mg KCl, 144 mg Na_2HPO_4 and KH_2PO_4 into 80 ml of ultrapure water using a magnetic stirrer. Once all reagents are in solution, add ultrapure water until the final solution volume reaches 100 ml.
8. Nuclease-free low adherence 1.5-ml sample tubes.
9. Liquid nitrogen.

2.2 Chromatin Preparation and Antibody–Bead Complex Preparation

1. 10× Nuclear lysis buffer: 10% sodium deoxycholate, 10% Triton X-100, in water. Store at room temperature in a container protected from light.
2. [Optional] SNAP-ChIP K-MetStat Panel (*see* **Note 4**). Store in 2 μl aliquots at −20 °C.

3. 1.25× Small Input Micrococcal Nuclease (MNAse) Master Mix: 75 μl 10× MNAse digestion buffer (NEB), 75 μl PEG 6000 50% solution, 5.6 μl 100 mM Dithiothreitol, 444.4 μl ddH_2O, 1 μl MNAse enzyme (NEB). Prepare fresh and keep on ice (*see* **Note 5**).
4. 500 mM EDTA stock solution: Stir 93.05 g disodium ethylenediaminetetraacetate•2H_2O into 400 ml of water using a magnetic stirrer. Add 8–10 NaOH pellets slowly to adjust the pH to 8.0. The EDTA will go into solution as the pH nears 8.0. Dilute the solution to 500 ml with water.
5. MNAse Stop Solution: 100 mM EDTA. Dilute 1 ml of 500 mM EDTA stock solution with 4 ml of water.
6. [Optional] Trichostatin A (TSA) stock solution: add 500 μl of DMSO to a 1 mg vial of Trichostatin A powder and mix vigorously. Store in aliquots at −20 °C (*see* **Note 6**).
7. Magnetic rack suitable for 1.5-ml assay tubes.
8. Magnetic rack suitable for PCR tubes.
9. Protein A and G magnetic beads (*see* **Note** 7). Store at 4 °C.
10. 1 M Tris–HCl pH 8.0 stock. Stir 60.55 g of Tris base into 400 ml of water and slowly add drops of concentrated HCl until the pH nears 8.5, then switch to adding drops of 1 M HCl until reaching pH 8.0. Dilute the solution to 500 ml with water.
11. 2 M NaCl stock. Weigh 58.44 NaCl and add water to 500 ml. Mix until completely dissolved.
12. 10% sodium dodecyl sulfate (SDS) solution: Weigh 1 g of SDS powder, add water to 10 ml. Mix to dissolve the powder (*see* **Note 8**).
13. Native ChIP buffer: 20 mM Tris–HCl pH 8.0, 2 mM EDTA, 150 mM NaCl, 0.1% Triton X-100. Mix 1 ml 1 M Tris–HCl pH 8.0, 3.75 ml 0.5 M EDTA solution, 50 μl Triton X-100, and 45.2 ml water (*see* **Note 9**).
14. Complete Native ChIP buffer: Native ChIP buffer, 1× EDTA-free protease inhibitor cocktail, 1 mM PMSF. Add 200 μl 25× PIC stock (Subheading 2.1, **item 2**) and 25 μl 200 mM PMSF stock to 4.75 ml of Native ChIP Buffer. Store at 4 °C and use within 48 h.
15. [Optional] Immersion sonicator.
16. Vortex.
17. Magnetic rack suitable for 1.5-ml assay tubes.
18. Magnetic rack suitable for PCR tubes.

2.3 Chromatin Immunoprecipitation

1. Low Salt Wash Buffer: 20 mM Tris–HCl pH 8.0, 2 mM EDTA, 150 mM NaCl, 1% Triton X-100, 0.1% SDS. Add 1 ml of 1 M Tris–HCl pH 8.0, 200 μl 0.5 M EDTA, 3.75 ml 2 M NaCl, 50 μl Triton X-100 and 500 μl of 10% SDS solution to 44.5 ml of ultrapure water and mix well (*see* **Note 8** *on dissolving Triton X-100*).
2. PCR strip tubes with a volume capacity of 200 μl (assay tubes).
3. High Salt Wash Buffer: 150 mM Tris–HCl pH 8.0, 2 mM EDTA, 500 mM NaCl, 1% Triton X-100, 0.1% SDS. Add 1 ml of 1 M Tris–HCl Ph 8.0, 200 μl 0.5 M EDTA, 12.5 ml 2 M NaCl, 50 μl Triton X-100, and 500 μl of 10% SDS solution to 35.75 ml of ultrapure water and mix well (*see* **Note 8** *on dissolving Triton X-100*).
4. ChIP elution buffer: 100 mM $NaHCO_3$, 1% SDS. Prepare fresh (*see* **Note 10**).
5. Elution Buffer: 10 mM Tris–HCl pH 8.0 (*see* **Note 11**).
6. Magnetic rack suitable for 1.5-ml assay tubes.
7. Magnetic rack suitable for PCR tubes.

2.4 DNA Purification

1. Solid Phase Reversible Immobilization (SPRI) beads. Store at 4 °C.
2. Ethanol wash solution: 70% ethanol. Prepare fresh.
3. Elution Buffer: 10 mM Tris–HCl pH 8.0 (*see* **Note 11**).

2.5 Library Construction

1. NEBNext Ultra II Library Prep Kit for Illumina (*see* **Note 12**).
2. End Prep Master Mix: For each sample: 3.5 μl 10× NEBNext End Repair Reaction Buffer and 1.5 μl NEBNext End Prep Enzyme Mix. Prepare fresh as a master mix for (*n* samples + 10%) and keep on ice until use.
3. Elution buffer: 10 mM Tris–HCl pH 8.0.
4. Adaptor Working Stock: dilute NEBNext Adaptor for Illumina 1:15 in Elution Buffer. Prepare fresh and keep on ice.
5. Adaptor Ligation Master Mix (*see* **Note 13**). For each sample: 0.5 μl NEBNext Ligation Enhancer; 15 μl NEBNext Ultra II Ligation Master Mix; 1.25 μl Adaptor Working Stock. Prepare fresh as a master mix for (*n* samples + 10%) and keep on ice until use.
6. USER enzyme.
7. Solid Phase Reversible Immobilization (SPRI) beads. Store at 4 °C.
8. Ethanol wash solution: 70% ethanol. Prepare fresh.
9. NEBNext Multiplex Oligos for Illumina (Universal PCR Primer/i5 Primer, Index Primers/i7 Primers).

10. Phusion High-Fidelity DNA polymerase (*see* **Note 14**).
11. 5× Phusion HF Buffer.
12. 10 mM dinucleotide mix (dNTPs).
13. Dimethyl sulfoxide (DMSO).
14. PCR Master Mix. For each sample: 5 μl 5× HF Phusion Reaction Buffer; 1 μl 10 mM dNTPs; 0.75 μl DMSO; 2.5 μl Universal PCR Primer/i5 Primer; 0.25 μl Phusion High Sensitivity DNA polymerase. Prepare fresh as a master mix for (*n* samples + 10%) and keep on ice until use.
15. Thermal cycler.
16. Agilent TapeStation system with Agilent High Sensitivity D1000 Screentape and reagents (*see* **Note 15**).

2.6 Library Pooling and Size Selection

1. Solid Phase Reversible Immobilization (SPRI) beads. Store at 4 °C.
2. Ethanol wash solution: 70% ethanol. Prepare fresh.
3. Elution buffer: 10 mM Tris–HCl pH 8.0.
4. Invitrogen E-gel EX precast agarose gel.
5. Invitrogen E-gel apparatus.
6. DNA Zymoclean Gel DNA Recovery Kit (*see* **Note 16**).
7. Agilent TapeStation system with Agilent High Sensitivity D1000 Screentape and reagents (*see* **Note 15**).

3 Methods

3.1 Sample Isolation and Storage (Select Appropriate Section for Tissue)

ULI-NChIP-seq is a dilution-based method which aims to reduce as much as possible sample loss due to repeated washing and resuspension steps. The method can be used for a wide variety of samples difficult to obtain in large quantities. We provide here three sample isolation procedures (Subheading 3.1.1, 3.1.2, or 3.1.3) that should cover the preparation and storage of nuclei for a range of sample types (*see* **Note 17**).

3.1.1 FACS-Sorted Cells

1. Sort cells by flow cytometry directly into a 1.5-ml sample tube containing 20 μl freshly prepared Complete Nuclear Isolation Buffer (*see* **Notes 18** and **19**).
2. Estimate total volume in the tube to calculate the proportion of FACS sheath buffer to nuclear isolation buffer. If the proportion of sheath buffer is 1:4 or higher, add additional Complete Nuclear Isolation Buffer to reach this ratio.
3. Spin down sample at 13,000 × *g* for 30 s.
4. Remove liquid in excess of 20 μl (*see* **Note 20**).

5. Flash freeze the sample by submerging the isolation tube in a small volume of liquid nitrogen.
6. Store the sample tube(s) at −80 °C for a few hours to a few years (<5 years), until ready for chromatin isolation.

3.1.2 Oocytes and Early Embryos

1. Using a mouth or hand pipette to transfer isolated oocytes or pre-implantation embryos, wash samples by transferring through three to four drops of M2 media, then through two drops of PBS (*see* **Note 21**).
2. To remove the Zona Pellucida, transfer the samples through three drops of Tyrode's acid, spending approximately 5 s in each drop.
3. Once the Zona Pellucida is removed, rapidly transfer oocytes or embryos back to a fresh drop of M2 media (*see* **Note 22**).
4. Wash embryos through three drops of M2 media.
5. Transfer embryos into the bottom of a 1.5-ml sample isolation tube, taking care to transfer as little liquid volume as possible (<5 μl).
6. Spin the sample down on a tabletop mini-centrifuge for 30 s to ensure it is at the bottom of the sample tube.
7. Top the isolated samples to 20 μl with Complete Nuclear Isolation Buffer.
8. Flash freeze the sample by submerging the isolation tube in a small volume of liquid nitrogen.
9. Store the sample tube(s) at −80 °C for a few hours to a few years (<5 years), until ready for chromatin isolation.

3.1.3 Dissected Tissue or Postimplantation Embryo (<5000 Cell-Equivalents)

1. Once the tissue sample is dissected out, transfer to a 1.5-ml sample isolation tube.
2. Wash sample with PBS and pulse-spin down at 13,000 × *g* for 30 s.
3. Add 20 μl of Complete Nuclear Isolation Buffer.
4. Flash freeze the sample by submerging the isolation tube in a small volume of liquid nitrogen.
5. Store the sample tube(s) at −80 °C for a few hours to a few years (<5 years), until ready for chromatin preparation.

3.2 Chromatin Preparation

Chromatin preparation is one of the most critical steps of a native ChIP protocol. Insufficient chromatin fragmentation can lead to low yields of immunoprecipitated DNA and increased levels of background signal. An overview of this step is presented in Fig. 1.

3.2.1 Pooling of Samples Stored in Multiple Tubes

Skip to Subheading 3.2.2 if your sample is contained in a single tube.

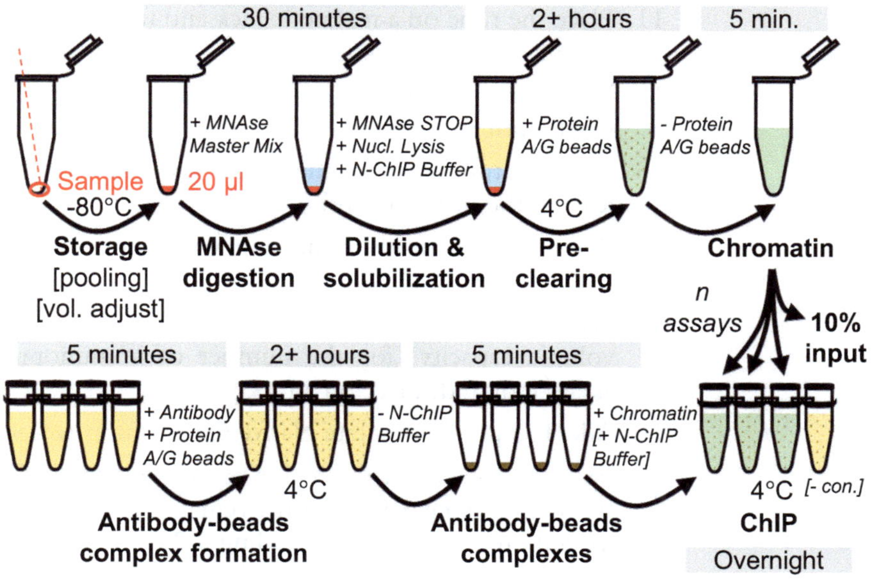

Fig. 1 Quick guide to chromatin preparation and immunoprecipitation setup. Time estimate and guide to chromatin preparation (top row) and antibody–bead complexes formation (bottom row)

1. Thaw samples on ice.
2. Spin samples at 13,000 × *g* for 30 s.
3. Line up samples so as to have the sample tube with the most cells last.
4. For each tube, resuspend nuclei in the Complete Nuclear Isolation Buffer already in the tube by gently pipetting up and down, avoiding introducing bubbles into the sample. Transfer the sample to the last tube.
5. Once all the samples are pooled in the last tube, take a single 50 μl fresh Complete Nuclear Isolation Buffer aliquot and rinse each sample tube one after the other. Place this buffer in the last tube containing all the pooled samples.
6. Mix the samples well by pipetting up and down gently.
7. Spin the pooled samples for 30 s at 13,000 × *g* to pellet the nuclei. The pellet may not be visible.
8. Take out the liquid volume except for the last 20 μl (*see* **Note 20**).
9. Resuspend nuclei in the remaining 20 μl of nuclear isolation buffer.

3.2.2 Preparation of Antibody–Bead Complexes

10. Prepare a 1:1 solution of protein A: protein G magnetic beads. Calculate the amount of protein A: Protein G Dynabeads for your assay: (Number of individual immunoprecipitation assays + 10%) × 10 μl. *If starting material exceeds 25,000 cell-equivalent, see* **Note 23**.

11. Place the tube on a magnetic rack and take out all the liquid, leaving the beads behind.
12. Resuspend the beads in the same volume of Complete Native ChIP buffer.
13. Place the tubes on a magnetic rack and take out all the liquid, leaving the beads behind.
14. Resuspend the beads in the same volume of Complete Native ChIP buffer.
15. Separate enough tubes from a clean strip of PCR tubes (200 μl volume capacity) for the number of immunoprecipitation assays required. Place on a rack.
16. Add 100 μl of Complete Native ChIP Buffer to each tube.
17. For each assay tube, add 5 μl of protein A: protein G magnetic beads from **step 14**. Keep the remaining magnetic beads on ice until required (*see* Subheading 3.2.3, **step 32**). *If starting material exceeds 25,000 cell-equivalent*, add 10 μl of protein A: protein G magnetic beads.

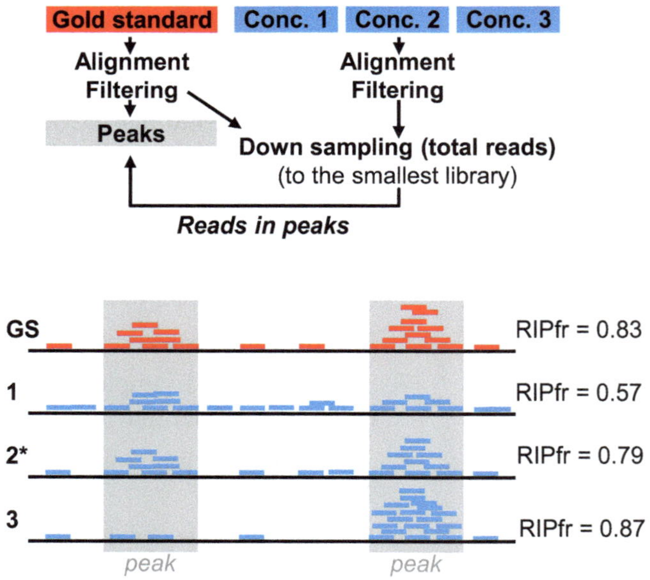

Fig. 2 Antibody titration procedure. *See also* **Note 24**. Steps to calculate the Reads In Peaks fraction (RIPfr) in a cell line where the histone modification has previously been profiled (Gold Standard –GS–). The ideal antibody concentration (2*) has the maximal RIPfr value while maintaining the ability to detect all peaks. An antibody concentration that is too high will lead to a low RIPfr (1) and an antibody concentration too low (3) will generate libraries with a high RIPfr but that is not sensitive enough to detect all peaks

18. For each assay tube, add an appropriate amount of previously titrated antibody. *See* **Note 24** and Fig. 2 for a description of the antibody titration procedure. We recommend starting with 0.25 ng antibody per assay. *If using > 10,000 cell-equivalent per assay, we recommend starting with 0.5 ng antibody per assay.*
19. Rotate at 4 °C for at least 2 h and up to 10 h.

3.2.3 Micrococcal Nuclease (MNAse) Digestion

20. Place samples on a rack at room temperature.
21. Prepare a fresh batch of 1.25× Small Input MNAse Master Mix. *If starting material exceeds 25,000 cell-equivalent, see* **Note 25**.
22. [Optional] Add SNAP-ChIP K-MetStat Panel to each sample. For inputs <2000 cells, use 0.25 μl per 1000 cell-equivalents. For inputs 2000–20,000 cells, use 1 μl. For inputs 20,000–100,000 cells, use 2 μl.
23. Add 80 μl of 1.25× Small Input MNAse Master Mix to the sample tube.
24. Mix gently by pipetting up and down, avoiding introducing bubbles into the sample.
25. Incubate at room temperature (21–25 °C) for 7.5 min (*see* **Note 26**).
26. Arrest the enzymatic reaction by adding 10 μl of MNAse Stop Solution.
27. Mix well by pipetting up and down and place the sample on ice.
28. Add 12 μl of 10× Nuclear Lysis Buffer to the digested chromatin. Incubate on ice for 15 min.
29. If you do not have access to an immersion sonicator: vortex the sample 2× 20 s at the highest setting of a benchtop vortex. If you have access to an immersion sonicator, skip this step.
30. Add 378 μl of Complete Native ChIP Buffer to your sample.
31. If you have access to an immersion sonicator, pulse-sonicate your sample 3 × 10 s at low setting.
32. For each immunoprecipitation assay to be done with the chromatin preparation, add 5 μl of the remaining protein A: protein G magnetic beads (**step 17**).
33. Rotate at 4 °C for at least 2 h and up to 10 h.

3.3 Chromatin Immunoprecipitation

3.3.1 Immunoprecipitation

1. Calculate the volume of chromatin that will be used for each immunoprecipitation assay: total volume of chromatin preparation/(number of immunoprecipitation assays + 15%). *For 3 assays: 500 μl chromatin preparation/3.45 = 144.9 μl chromatin per assay.*

2. Retrieve the antibody:bead complexes (*see* Subheading 3.2.2, **step 19**).
3. Briefly spin the tubes containing the antibody:bead complexes to remove any liquid from the caps.
4. Place the tubes containing the antibody:bead complexes on a magnetic rack and remove the liquid, leaving the beads behind.
5. Briefly spin the tube containing the digested chromatin (Subheading 3.2.3, **step 32**) on a tabletop mini-centrifuge and place on a magnetic rack. Transfer the pre-cleared chromatin to a fresh 1.5-ml sample tube.
6. Add the volume of chromatin calculated in **step 1** to each assay tube containing the antibody:bead complexes. The volume of chromatin should be between 100 μl and 175 μl. Complete Native ChIP buffer can be used here to top up if chromatin is diluted in an insufficient volume.
7. Cap the assay tubes, making sure there are no leaks, and rotate at 4 °C overnight.
8. Transfer 10% of the chromatin volume used for each assay to a separate 1.5-ml sample tube, as an input control. *For three assays, a single 14.5 μl aliquot of chromatin will be taken aside.* Keep at 4 °C overnight.

3.3.2 Washes and Elution

9. Briefly spin the assay tubes to remove any liquid from the caps.
10. Place assay tubes on magnetic rack.
11. Discard liquid portion, taking care not to disturb the magnetic beads aggregated toward the magnet.
12. Transfer assay tubes onto a non-magnetic rack.
13. Gently resuspend magnetic beads in 100 μl Complete Native ChIP Buffer and transfer to a new 1.5-ml assay tube.
14. Using 100 μl Complete Native ChIP Buffer, rinse the assay tube to collect all remaining magnetic beads and transfer to the same sample tube.
15. Repeat for each sample.
16. Leave the sample tubes at room temperature for 2 min.
17. Transfer the 1.5-ml sample tubes onto a magnetic rack and allow the magnetic beads to aggregate toward the magnet.
18. Take out the liquid portion and remove the sample tubes from the magnetic rack.
19. Resuspend the magnetic beads in 200 μl Low Salt Wash Buffer.
20. Repeat **steps 15–17** with a second wash in Low Salt Wash Buffer and two washes in High Salt Wash Buffer (*see* **Note 27**). [If doing ChIP on a non-histone protein (i.e., RNA polymerase II), *see* **Note 28**].

21. Once the last wash is removed, resuspend the magnetic beads in 25 μl of freshly prepared ChIP Elution Buffer.
22. Also add 25 μl of ChIP Elution Buffer to the input aliquots set aside in Subheading 3.3.1, **step 6**.
23. Transfer the ChIP and input sample tubes into a water bath at 67 °C.
24. Incubate for 75 min, taking care to vortex at medium intensity every 15–20 min to resuspend the magnetic beads.
25. Briefly spin the samples in a countertop micro-centrifuge to collect all the condensation at the bottom of the sample tube.
26. Place the sample tubes on a magnetic rack and allow the beads to aggregate toward the magnet.
27. Carefully retrieve the ChIP Elution Buffer and transfer to a new 1.5-ml sample tube.
28. Resuspend the magnetic beads into 50 μl Elution Buffer (not ChIP Elution Buffer).
29. Place the sample tubes on a magnetic rack and combine this Elution Buffer with the ChIP Elution Buffer transferred in **step 25** for each sample.

3.4 DNA Purification and Elution

1. Resuspend SPRI bead slurry and leave at room temperature for at least 20 min.
2. Add 150 μl of SPRI beads to the eluted ChIP material. Resuspend by gently pipetting up and down, taking care not to create froth.
3. Incubate at room temperature for 5–10 min (*see* **Note 29**).
4. Place samples on a magnetic rack and allow 5 min for the beads to aggregate toward the magnet.
5. Remove the liquid portion completely.
6. Transfer the tubes containing the magnetic beads to a nonmagnetic rack.
7. Re-suspend the beads in 200 μl of freshly prepared Ethanol Wash solution (*see* **Note 30**).
8. Transfer the tubes back onto a magnetic rack.
9. Remove the ethanol wash completely.
10. Repeat **steps 7–9** with a second ethanol wash.
11. After the second ethanol wash removed, close the tube caps and briefly spin the sample tubes on a countertop micro-centrifuge to bring any remaining liquid down.
12. Using a fine pipette tip, remove the last of the Ethanol Wash.

13. Leave the sample tubes caps open on a nonmagnetic rack and allow the magnetic beads to dry at room temperature for 5–10 min (*see* **Note 31**).
14. Gently place 25 μl of Elution Buffer onto the magnetic bead pellet.
15. Allow the beads to re-hydrate for 5 min, then resuspend the beads gently using a pipette. Leave on a nonmagnetic rack for an additional 2 min.
16. Place the sample tubes on a magnetic rack and allow the beads to aggregate toward the magnet.
17. Using a pipette with a fine tip, carefully transfer the DNA eluate into a clean PCR strip tube, leaving the magnetic beads behind (*see* **Note 32**).
18. At this point, the samples can be stored at 4 °C for up to 48 h, or at −20 °C for a few weeks to a few months.

3.5 Low Input DNA Library Construction

3.5.1 End Prep and Adaptor Ligation

1. Add 5 μl of End-Prep Master Mix to 25 μl of eluted ChIP DNA (*see* Subheading 3.4, **step 17**).
2. Place the sample tubes into a thermal cycler and run the following program: 30 min at 20 °C, 30 min at 65 °C, hold at 4 °C.
3. Once complete, add 1.25 μl of 1 μM Adaptor Working Stock and 15.5 μl of Adaptor Ligation Master Mix.
4. Place the sample tubes into a thermal cycler and run the following program: 30 min at 20 °C, hold at 4 °C.
5. Add 3 μl USER enzyme to each sample.
6. Place the sample tubes into a thermal cycler and run the following program: 15 min at 37 °C, hold at 4 °C.

3.5.2 DNA Purification and Size Enrichment

1. Carefully transfer the ligation product to a new 1.5-ml sample tube.
2. Add 49.75 μl of SPRI bead slurry (*see* **Note 33**). Resuspend by gently pipetting up and down, taking care not to create froth.
3. Incubate at room temperature for 5–10 min.
4. Place samples on a magnetic rack and allow 5 min for the beads to aggregate toward the magnet.
5. Remove the liquid portion completely.
6. Rinse the beads with 200 μl freshly prepared Ethanol Wash solution.
7. Remove the ethanol wash completely.
8. Repeat **steps 6** and 7 with a second ethanol wash.

9. After the second ethanol wash, close the tube caps and briefly spin the sample tubes on a countertop microcentrifuge to bring any remaining liquid down.
10. Place the sample tubes on a magnetic rack. Using a fine pipette tip, remove the last of the Ethanol Wash.
11. Leave the sample tubes caps open on a nonmagnetic rack and allow the magnetic beads to dry at room temperature for 5–10 min.
12. Gently place 12 μl of Elution Buffer onto the magnetic bead pellet.
13. Allow the beads to re-hydrate for 5 min, then resuspend the beads gently using a pipette. Leave on a nonmagnetic rack for an additional 2 min.
14. Place the sample tubes on a magnetic rack and allow the beads to aggregate toward the magnet.
15. Using a pipette with a fine tip, carefully transfer the DNA eluate into a clean PCR strip tube, leaving the magnetic beads behind. At this point, the samples can be stored at −20 °C until ready for PCR amplification.

3.5.3 PCR Amplification of Low-Input Libraries

1. Add 2.5 μl of Index Primer/i7 Primer to each sample (*see* Subheading 3.5.2, **step 15**). Make sure not to repeat the same index for samples that will be pooled together for sequencing.
2. Add 9.5 μl of PCR Master Mix to each sample and mix well by pipetting up and down five to ten times, carefully avoiding generating froth.
3. Briefly spin samples on a countertop micro-centrifuge.
4. Place the sample tubes in a thermal cycler and run the following program: 4 min at 98 °C [30 s at 98 °C, 30 s at 65 °C, 30 s at 72 °C], repeat 8–21 cycles (*see* **Note 34**), 5 min at 72 °C, hold at 4 °C.
5. In a post-PCR work area, transfer the amplified libraries to clean 1.5-ml sample tubes.
6. Carefully add 20 μl of SPRI bead slurry. Resuspend by gently pipetting up and down, taking care not to create froth.
7. Incubate at room temperature for 5–10 min.
8. Place samples on a magnetic rack and allow 5 min for the beads to aggregate toward the magnet.
9. Remove the liquid portion completely.
10. Rinse the beads with 100 μl freshly prepared Ethanol Wash solution.
11. Remove the ethanol wash completely.

12. Repeat **steps 6** and **7** with a second ethanol wash.
13. After the second ethanol wash removed, close the tube caps and briefly spin the sample tubes on a countertop microcentrifuge to bring any remaining liquid down.
14. Place the sample tubes on a magnetic rack. Using a fine pipette tip, remove the last of the Ethanol Wash.
15. Leave the sample tube caps open on a nonmagnetic rack and allow the magnetic beads to dry at room temperature for 5–10 min.
16. Gently place 15–20 μl of Elution Buffer onto the magnetic bead pellet.
17. Allow the beads to re-hydrate for 5 min, then resuspend the beads gently using a pipette. Leave on a nonmagnetic rack for an additional 2 min.
18. Place the sample tubes on a magnetic rack and allow the beads to aggregate toward the magnet.
19. Using a pipette with a fine tip, carefully transfer the DNA eluate into a clean PCR strip tube, leaving the magnetic beads behind. At this point, the constructed libraries can be quality-controlled and stored at −20 °C until ready to be pooled and sequenced.

3.5.4 Quantitation and Quality Control of Constructed Libraries

1. Run 1 μl of each sample on an Agilent High Sensitivity D1000 Screentape (Agilent TapeStation) or on a High Sensitivity DNA kit (Agilent Bioanalyzer).
2. Using the Agilent Software, measure the molarity of each library, including the nucleosomal + adaptor and carefully avoid including adaptor dimers in the measure (*see* Fig. 3; **Note 35**).
3. To control for proper library fragmentation, calculate the Mono-nucleosomal Fraction (MFr) of the input libraries: $\text{MFr} = \frac{\text{mono-nucleosomes (nM)}}{\text{nucleosomal fraction (nM)}}$. It is expected that ChIP of certain histone modifications, in particular but not exclusive to heterochromatin, will enrich for larger fragments than the input prepared from the same sample.
4. If input or ChIP sequencing libraries are not as expected, please refer to the troubleshooting guide to find a probable cause and solution (Table 1).

3.6 Pooling and Size Selection of Multiplexed Low-Input Libraries for Sequencing

The representation of each library within a sequencing pool should be determined according to the desired number of reads per library, the sequencing platform used, the availability of each constructed library, and the minimal volume and library pool concentration requirements of the sequencing facility. Whenever possible, aim

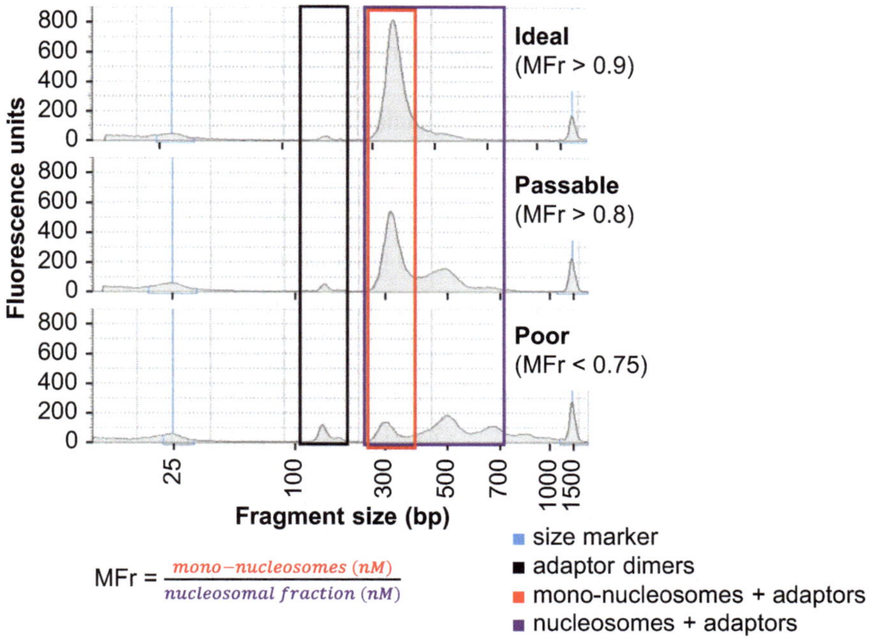

Fig. 3 Evaluating chromatin fragmentation on constructed input libraries. Chromatin prepared from a few hundred to a few thousand cells does not yield sufficient material to visualize fragmentation prior to library construction. Therefore, an aliquot of the constructed input libraries is used to estimate fragmentation efficiency. To generate ChIP-seq libraries with a representative coverage, particularly of heterochromatic regions, we recommend using starting chromatin with a minimum mono-nucleosomal fraction (MFr) > 0.9, although good results can be obtained with starting material >0.8 MFr. It is normal and expected that some libraries constructed from ChIP material (especially heterochromatin) will have a lower MFr than the input chromatin from which they are prepared. *MFr* mono-nucleosomal fraction/nucleosomal fraction. Blue: size markers (Agilent TapeStation); Black: adaptor dimers (~130 bp); Purple: nucleosomal fraction + adaptors; Red: mono-nucleosomal fraction + adaptors

for five to ten million sequenced fragments for histone modifications with a narrow distribution (i.e., H3K4me3 and H3K27ac), 20–30 million fragments for modifications with a broad distribution (i.e., H3K9me3 and H3K27me3) and 60+ million fragments for histone modifications with a domain-like distribution (i.e., H3K9me2 and H3K36me2).

1. Calculate the volume of each library to include into the sequencing pool using the formula provided below. *Note that you should account for sample loss during size selection and DNA purification*

$$\text{volume} = \frac{\text{desired reads for library}}{\text{total sequenced reads in pool}} * \frac{\text{desired pool concentration} * \text{vol.}}{\text{library concentration}}$$

Table 1
Troubleshooting constructed ULI-NChIP and input libraries

Problem	Solution
1. Improper chromatin fragmentation	
1.1 Underdigestion	
1.1.1 MNAse has lost its potency.	*1.1.1* Fresh batch of MNase, use benchtop cooler.
1.1.2 Insufficient Ca+ in MNAse digestion buffer.	*1.1.2* Aliquot MNAse digestion buffer to avoid freeze–thaw cycles. Add enzyme to Master Mix immediately before proceeding to digestion. Keep on ice while distributing.
1.1.3 MNAse has no access to the nucleus.	*1.1.3* Nuclear Isolation Buffer should be >0.75 sample volume. Add 0.1 vol. of [10% Triton; 10% Deoxycholate] solution prior to adding MNase Master Mix.
1.1.4 Final EDTA concentration too high in sample.	*1.1.4* If EDTA present in sample isolation buffer, spin down sample, remove all but 5 μl sample isolation buffer, add 45 μl Complete Nuclear Isolation Buffer and remove volume in excess of 20 μl.
1.1.5 Insufficient digestion time.	*1.1.5* Increase digestion time.
1.1.6 Digestion temperature too low.	*1.1.6* Proceed to digestion at 37 °C.
1.2 Overdigestion	
1.2.1 MNAse concentration is too high for cell type.	*1.2.1* Reduce MNAse concentration by half.
1.2.2 Digestion time is too long for cell type.	*1.2.2* Reduce digestion time or temperature.
1.3 Combination of under-and overdigestion	
1.3.1 Sample is improperly mixed.	*1.3.1* Swirl the sample while pipetting up and down 15–20 times when adding the MNAse Master Mix. Vortex lightly.
1.3.2 MNAse has gone through too many freeze–thaw cycles.	*1.3.2* Fresh batch of MNase (test each new batch).
2. Unexpectedly low yield of constructed library	
2.1 Adapter dimers present in the constructed library	
2.1.1 Loss of raw ChIP material.	*2.1.1* Make sure the Ethanol Wash solution is freshly prepared with the appropriate ethanol:water ratio. Ensure the Elution Buffer has the appropriate concentration of Tris–HCl.
2.1.2 Excess SDS and/or salt in eluted ChIP material.	*2.1.2* Make sure the SPRI beads are resuspended in the Ethanol washes. Make sure to remove as much of the ethanol wash from the beads prior to drying.

(continued)

Table 1 (continued)

Problem	Solution
2.1.3 Mark only that covers a small portion of the genome.	*2.1.3* If a mark covers <2% of the genome, 2–3 additional PCR cycles could be required during library construction.
2.2 No adapter dimers present in the constructed library	
2.2.1 Low-quality Illumina adapters.	*2.2.1* Use new batch of adapters.
2.2.2 Inefficient adapter ligation step.	*2.2.2* Make sure all components of the adapter ligation Master Mix are present. Use fresh batch of enzymes.
2.2.3 Amplification of constructed libraries failed.	*2.2.3* Use fresh PCR buffer and a fresh dilution of dNTPs.
3. High background (sequencing)	
3.1 Antibody concentration is too high.	*3.1* Titrate down antibody concentration (see Note 23).
3.2 Non-specific signal is insufficiently washed.	*3.2* Increase the volumes of the washes. Make sure all of the wash is removed from the magnetic beads.
3.3 Nature of the histone mark.	*3.3* Histone marks (i.e., H3K9me2) with a broader distribution have a lower signal-to-noise ratio and should be sequenced more deeply.
4. Low ChIP signal	
4.1 Antibody concentration is too low.	*4.1* Titrate up the antibody concentration (see Note 23).
4.2 Poor antibody quality.	*4.2* Only use ChIP-grade antibodies. Aliquot antibodies to avoid repeated freeze–thaw cycles. Only use fresh antibody dilutions.
4.3 Antigen is very rare.	*4.3* Increase the input size for ChIP on this antigen.
4.4 Post-immunoprecipitation washes are too harsh for the ChIP target (non-histone protein).	*4.4* Replace the two high salt washes by two additional low salt washes.

2. Pool all libraries to be sequenced in a 1.5-ml sample tube according to the volume calculated in **step 1**.
3. Run the libraries on a 2% agarose gel to separate the nucleosomal + adaptor fraction from the adaptor dimers fraction. *The E-Gel electrophoresis system (using E-gel EX 2% precast agarose gels) from ThermoFisher Scientific is a good alternative to a traditional tank electrophoresis system in order to reduce the possibility of contaminating a laboratory space with PCR products* (*see* **Note 36**).
4. Cut the 200–650 bp fraction of the library pool.

5. Proceed to DNA purification using a column-based agarose gel DNA cleanup kit. Elute in 15–20 μl.
6. Control for appropriate size selection by running 1 μl of the sample on an Agilent Screentape. *If some adaptor dimers remain in the library pool, repeat* ***steps 3–6****.*
7. Quantify exact library molarity using a qPCR-based method.

4 Notes

1. This reagent comes as a kit for nuclear extraction. Only the nuclear isolation buffer is used in this protocol.
2. If another vendor is used for this protocol, it is important to select an EDTA-free protease inhibitor cocktail, as EDTA would interfere with MNAse digestion in later steps. Once prepared, 25× PIC stock can be aliquoted and stored at −20 °C until use.
3. Once prepared, PMSF stock can be aliquoted and stored at −20 °C until use.

   PMSF may come out of solution at −20 °C but can easily be re-dissolved by bringing the aliquot to room temperature and mixing vigorously.
4. ChIP-seq methodologies evaluate the relative enrichment of a specific protein or histone modification in a genome-wide manner. SNAP-ChIP spike-ins allow quantitative measurement of histone enrichment between different samples. A SNAP-ChIP K-AcylStat Panel is also available if immunoprecipitating acetylated histone modifications (*see* also **Note 6** about including histone deacetylase inhibitors in the assay).
5. This protocol has been optimized using the MNAse enzyme and buffers sold by New England Biolabs (NEB). It is important to note that, if selecting a different vendor, some companies may use different units to measure enzymatic activity.
6. If an acetylated histone is the target, we recommend adding this HDAC inhibitor to the fragmented chromatin and to the Complete Native ChIP buffer.
7. Although protein A or protein G magnetic beads respectively have a higher affinity for certain antibody isotypes, including both A and G beads can simultaneously reduce nonspecific binding and allow the immunoprecipitation of different antibody isotypes in parallel assays.
8. Wear a mask when weighing SDS. Gently transfer the powder to a glass bottle and rinse the weighing boat with water. Placing the bottle in a 37 °C water bath can help the powder getting into solution. The resulting 10% stock solution should be

completely clear. You should discard if the SDS precipitates over time or if the solution has a yellow tinge. Once the SDS has come out of suspension, re-dissolving it by warming the solution is not recommended, as it will lead to a poor recovery yield in subsequent steps.

9. Make sure to completely transfer the Triton X-100 by pipetting up and down several times until the pipette tip is clear. Allow to the Triton X-100 to dissolve for at least 2 h before using the solution. This is also true for the preparation of the Low Salt Wash and High Salt Wash solutions.
10. Prepare the ChIP elution buffer fresh and keep at room temperature. The buffer should not be placed on ice, as the reagents in solution will precipitate at 4 °C.
11. The pH of the elution buffer is important to maximize the recovery of the material. Alternatively, Elution Buffer from Qiagen can be used here.
12. Although we have had good results with the NEB Next Ultra II library prep kit, other library construction protocols suited to picogram levels of fragmented double stranded DNA can also be used here. The original ULI-NChIP-seq protocol was published with a custom library construction protocol, but the availability of suitable commercial library construction kits has eliminated the need for this more tedious procedure.
13. The Adaptor Ligation Master Mix is very viscous. Make sure to pipette slowly and accurately all components, and mix well by pipetting up and down 10–15 times.
14. Another high-fidelity DNA polymerase previously validated for library construction can be used here instead.
15. Alternatively, the Agilent Bioanalyzer with the High Sensitivity DNA kit can be used.
16. Another column-based gel recovery kit can be used here. Note that the size of the column filter may impact the recovery of very small volumes.
17. The sample isolation step aims to prepare nuclei which can be later utilized for MNAse chromatin fragmentation. Unlike a standard nuclear isolation, the cytoplasmic material will not be completely removed from the suspension. This is because pelleting nuclei and removing the entirety of the liquid portion will result in loss of a significant portion of the sample. If working with pelleted, frozen samples, adding 20 μl of Complete Nuclear Isolation Buffer to the pellet, re-suspending and incubating cells for 5 min at room temperature will result in a nuclear preparation appropriate for MNAse digestion.
18. The nozzle size utilized on the flow cytometer will impact the volume of sheath buffer that will be transferred along the

sample. Ideally, one should use as small a nozzle as possible to reduce this volume as much as possible.

19. It is important to properly adjust the stream of the flow cytometer toward the bottom center of the Eppendorf tube to maximize cell recovery.

20. To estimate sample volume, prepare calibration tubes with known volume of water: 10, 20, and 30 μl. Hold aside your sample tube as you remove excess liquid. This step does not need to be excessively precise, as the subsequent MNAse digestion step will work well with a range of ±5 μl.

21. Make sure there are no somatic cells attached to the samples, in particular, oocytes. If so, a treatment with hyaluronidase prior to the M2 is required.

22. At this point, the oocyte/embryos become very sticky. It is preferable to rinse the glass pipette with M2 media prior to collecting the samples from the last drop of Tyrode's acid to prevent the samples adhering to the glass.

23. If sample for chromatin preparation exceeds 25,000, double the amount of protein A/G magnetic beads are utilized: (Number of individual immunoprecipitation assays + 10%) × 20 μl.

24. Antibody titration procedure (*see* also Fig. 2). Ideally, for titrating a new antibody, you will work with a cell line for which the histone modification has already been profiled ("gold standard") and for which peaks have already been identified.

    (1) Call peaks in "gold standard" library.

    (2) Perform ULI-NChIP in replicates, using three distinct antibody concentrations (i.e., 0.15 ng, 0.25 ng, and 0.5 ng antibody). *The same chromatin preparation should be divided for all nine assays and the input.*

    (3) Sequence at lower resolution (i.e., one million reads for a histone modification with a narrow distribution such as H3K4me3 or five million reads for a modification with broad distribution such as H3K9me3).

    (4) Align data and filter out duplicate and unmapped reads.

    (5) Down sample all libraries to be compared to obtain equal numbers of reads, including the "gold standard" library.

    (6) Identify all peaks detected in the "gold standard" library at the evaluated sequencing depth.

    (7) Calculate the Reads In Peaks fraction (RIPfr):

$$\text{RIPfr} = \frac{\text{reads overlapping peaks (6)}}{\text{total reads in library (5)}}$$

    (8) The ideal antibody concentration will have the highest RIPfr value, provided that all (or most) peaks are detected. If

not all peaks are detected, a higher antibody concentration is recommended. If there is a low RIPfr value, a lower antibody concentration is recommended.

25. If sample exceeds 25,000 cell-equivalents, it is preferable to proceed to the MNAse digestion in the absence of PEG 6000. 2× PEG-free MNAse Master Mix: 75 μl 10× MNAse digestion buffer (NEB), 5.6 μl 100 mM Dithiothreitol, 294.4 μl ddH_2O, and 1 μl MNAse enzyme (NEB). Follow these alternate steps for MNAse digestion.

    (22) Add 20 μl of 2× PEG-free MNAse Master Mix to the sample tube.

    (23) Mix gently by pipetting up and down, avoiding to froth the sample.

    (24) Incubate at 37 °C for 7.5 min.

    (25) Arrest the enzymatic reaction by adding 4 μl of MNAse Stop Solution.

    (26) Mix well by pipetting up and down and place the sample on ice.

    (27) Add 4.4 μl of 10× Nuclear lysis buffer to the digested chromatin. Incubate on ice for 15 min.

    Proceed with **step 28** as described in Subheading 3.2.3.

26. For each new batch of enzyme, proper digestion using these conditions must be ensured. Digestion of chromatin using PEG as a crowding agent is more sensitive to poor MNAse digestion efficiency. Therefore, it is insufficient to benchmark a new batch with large amounts of cells, and it is preferable to use this chromatin preparation protocol and verify digestion efficiency on constructed libraries (*see* also Fig. 3).

27. To further reduce nonspecific signal, it is recommended to transfer beads to a new 1.5-ml sample tube after the first High Salt wash. Remove the first High Salt Wash solution, re-suspend beads in high Salt Wash Buffer and transfer to a fresh tube. When removing the last wash, close the tubes and briefly spin beads on a tabletop micro centrifuge. Place the tubes back on the magnetic rack and remove the last drops of wash buffer.

28. High salt concentrations are too stringent for the majority of DNA-associated non-histone proteins. It is possible to perform ULI-NChIP-seq targeting non histone proteins that are bound to DNA (i.e., RNA polymerase II). To do so, simply replace the two washes in High Salt Buffer by two additional washes in Low Salt Buffer.

29. Although a 5-min incubation is usually sufficient, there is a small improvement in recovered yield when incubating for 10 min.

30. It is very important to resuspend the beads properly in the ethanol wash at this step and for the second wash. Residual SDS and salt left on the DNA sample will negatively impact end prep and library construction (as well as any qPCR step).
31. Make sure the beads are completely dried before adding the Elution Buffer. Residual ethanol will negatively impact enzymatic activity during library end prep. The beads are dry when the pellet has no shine, and begins to show small cracks.
32. It is sometimes difficult to separate the eluted sample from the SPRI beads. In such case, maximizing sample recovery is preferable over transferring an eluate completely free of beads. Residual SPRI beads will not negatively impact library end prep.
33. This step aims to remove library preparation reagents from the reaction, as well as to enrich for fragments larger than 200 bp to reduce the fraction of dimerized adaptors prior to PCR amplification. In order to more stringently enrich for the desired fragments (nucleosomes + adaptors), the volume of SPRI beads slurry can be reduced to 39.8 μl (0.8 volume): it is critical to accurately pipette the beads slurry, as using a lower volume will lead to a loss of a portion of the mononucleosomes + adaptors fraction (*see* also Fig. 3).
34. The number of required PCR cycles will depend on the ChIP target and on the input size. For a broad mark such as H3K9me3, we recommend 8–10 cycles for 10,000 cell-equivalent per assay, 10–12 cycles for 1000 cell-equivalent per assay, and 18 cycles for 100–200 cell-equivalent per assay. For a mark that covers a smaller fraction of the genome such as H3K4me3, we recommend 10–12 cycles for 10,000 cell-equivalent per assay, 12–14 cycles for 1000 cell-equivalent per assay, and 18–21 cycles for 100–200 cell-equivalent per assay.
35. In order to reduce library loss and over-amplification artifacts, adaptor dimers will only be removed after pooling all of the libraries which will be sequenced together. Although there is a size enrichment following the Adaptor Ligation step, the small input size of ULI-ChIP means that there will often still be some adaptor dimers left. Further reducing adaptor concentration during the ligation step to avoid this will usually lead to a reduced ligation efficiency.
36. The E-gel EX pre-cast agarose gels are run on the E-gel system without the need for a buffer tank. Pooled libraries are loaded in one to three consecutive wells (20 μl per well), and 20 μl of Elution Buffer is loaded in all the empty wells. The gel casts can be cracked open at the seams using the spatula tool provided with the E-gel electrophoresis system.

References

1. Gilmour DS, Lis JT (1984) Detecting protein-DNA interactions in vivo: distribution of RNA polymerase on specific bacterial genes. Proc Natl Acad Sci U S A 81(14):4275–4279. https://doi.org/10.1073/pnas.81.14.4275
2. Solomon MJ, Varshavsky A (1985) Formaldehyde-mediated DNA-protein crosslinking: a probe for in vivo chromatin structures. Proc Natl Acad Sci U S A 82(19): 6470–6474. https://doi.org/10.1073/pnas.82.19.6470
3. Mikkelsen TS, Ku M, Jaffe DB et al (2007) Genome-wide maps of chromatin state in pluripotent and lineage-committed cells. Nature 448(7153):553–560. nature06008 [pii]
4. Barski A, Cuddapah S, Cui K et al (2007) High-resolution profiling of histone methylations in the human genome. Cell 129(4): 823–837. S0092-8674(07)00600-9 [pii]
5. Albert I, Mavrich TN, Tomsho LP et al (2007) Translational and rotational settings of H2A.Z nucleosomes across the Saccharomyces cerevisiae genome. Nature 446(7135):572–576. nature05632 [pii]
6. Sachs M, Onodera C, Blaschke K et al (2013) Bivalent chromatin marks developmental regulatory genes in the mouse embryonic germline in vivo. Cell Rep 3(6):1777–1784. https://doi.org/10.1016/j.celrep.2013.04.032
7. Dahl JA, Jung I, Aanes H et al (2016) Broad histone H3K4me3 domains in mouse oocytes modulate maternal-to-zygotic transition. Nature 537(7621):548–552. https://doi.org/10.1038/nature19360
8. Lara-Astiaso D, Weiner A, Lorenzo-Vivas E et al (2014) Immunogenetics. Chromatin state dynamics during blood formation. Science 345(6199):943–949. https://doi.org/10.1126/science.1256271
9. Jakobsen JS, Bagger FO, Hasemann MS et al (2015) Amplification of pico-scale DNA mediated by bacterial carrier DNA for small-cell-number transcription factor ChIP-seq. BMC Genomics 16:46-014-1195-4. https://doi.org/10.1186/s12864-014-1195-4
10. Gilfillan GD, Hughes T, Sheng Y et al (2012) Limitations and possibilities of low cell number ChIP-seq. BMC Genomics 13:645-2164-13-645. https://doi.org/10.1186/1471-2164-13-645
11. Brind'Amour J, Liu S, Hudson M et al (2015) An ultra-low-input native ChIP-seq protocol for genome-wide profiling of rare cell populations. Nat Commun 6:6033. https://doi.org/10.1038/ncomms7033
12. Zhang B, Zheng H, Huang B et al (2016) Allelic reprogramming of the histone modification H3K4me3 in early mammalian development. Nature 537(7621):553–557. https://doi.org/10.1038/nature19361
13. van Galen P, Viny AD, Ram O et al (2016) A multiplexed system for quantitative comparisons of chromatin landscapes. Mol Cell 61(1): 170–180. https://doi.org/10.1016/j.molcel.2015.11.003
14. Skene PJ, Henikoff S (2017) An efficient targeted nuclease strategy for high-resolution mapping of DNA binding sites. elife 6. https://doi.org/10.7554/eLife.21856
15. Kaya-Okur HS, Wu SJ, Codomo CA et al (2019) CUT & Tag for efficient epigenomic profiling of small samples and single cells. Nat Commun 10(1):1930-019-09982-5. https://doi.org/10.1038/s41467-019-09982-5

Chapter 12

Automated CUT & RUN Using the KingFisher Duo Prime

Setareh Aflaki, Raphaël Margueron, and Daniel Holoch

Abstract

Elucidating the biological function of histone methyltransferases requires knowledge of the genomic sites at which they act. CUT&RUN represents a valuable alternative to chromatin immunoprecipitation for the mapping of histone methylation patterns, generally producing results of equivalent quality while requiring less sequencing depth, less starting material and less effort. Automated CUT&RUN procedures have been developed to further facilitate chromatin profiling. Here we describe our automated CUT&RUN protocol using the Thermo Fisher KingFisher Duo Prime system.

Key words Automated CUT&RUN, Chromatin profiling, Histone methylation, Easy CUT&RUN, KingFisher

1 Introduction

In order to understand the biological roles of histone methyltransferases, it is vital to determine the precise genomic loci to which their catalytic activities are deployed in relevant cell types, developmental contexts, disease states, and experimental conditions. Chromatin immunoprecipitation coupled to massively parallel DNA sequencing (ChIP-seq) has proved instrumental in mapping histone methylation marks for the last 15 years [1]. Nevertheless, ChIP-seq can exhibit certain limitations, including artifacts linked to differential availability of specific genomic regions following the crosslinking and chromatin fragmentation procedure [2–4], as well as the requirement for substantial starting material and time at the bench. While powerful techniques have been developed to adapt ChIP-seq to low numbers of cells (Chapter 11 of this volume) [5] and to calibrate ChIP-seq results with high quantitative precision and specificity [6], an attractive alternative approach termed "cleavage under targets and release using nuclease" (CUT&RUN) has also emerged and spread rapidly through the field in recent years [3, 7]. In contrast to ChIP, CUT&RUN (and its forerunner Chromatin Immunocleavage or ChIC [8]) does not require total

Raphaël Margueron and Daniel Holoch (eds.), *Histone Methyltransferases: Methods and Protocols*,
Methods in Molecular Biology, vol. 2529, https://doi.org/10.1007/978-1-0716-2481-4_12,

fragmentation of the genome but instead involves selective micrococcal-nuclease-mediated cleavage and release of fragments of interest identified by a specific antibody after cell permeabilization [3]. Since undigested chromatin not carrying the feature of interest remains insoluble, the assay has a remarkably low background, and as a result, good-quality results can be obtained using a lower sequencing depth and smaller quantity of biological starting material than for ChIP-seq [3]. The procedure is also generally less labor-intensive than ChIP and can be coupled to downstream construction of next-generation sequencing libraries with relative ease, particularly using the Tn5-based variant CUT&Tag for concurrent adapter ligation [9].

While generally very convenient for analyzing histone methylation patterns, CUT&RUN employs magnetic beads coupled to Concanavalin A, a plant lectin known to promote cell agglutination, particularly after treatment with trypsin [10]. The resulting clumping of the beads, also proposed to stem from cell lysis [7], complicates sample handling during the washing steps and makes automation of the procedure desirable. A robust and very high-throughput automated CUT&RUN method using a Beckman Coulter Biomek FX liquid handling robot was published by the original developers of CUT&RUN [11]. Here we propose an alternative automated CUT&RUN protocol based on the Thermo Scientific KingFisher Duo Prime system that is reliable and straightforward to implement and that we use to map histone methylation patterns in our laboratory.

2 Materials

Prepare the Wash Buffer before removing cells from culture and keep at room temperature until use. Concurrently, prepare 5% digitonin and keep on ice (*see* **Notes 1** and **2**).

2.1 Cells

Equipment, supplies, media, and reagents required for cell culture are assumed.

1. Freshly collected cells of interest for chromatin profiling. (The protocol was optimized using mouse embryonic stem cell line E14 (ES E14), 10^6 cells per sample; *see* **Note 3**.)
2. Freshly collected heterologous cells to be used as spike-in for normalization. (The protocol was optimized using *Drosophila* S2 cells, 10^5 cells per sample.)
3. Cell counting device (e.g., Beckman Coulter Vi-CELL XR or hemocytometer).

2.2 Buffers

1. Wash Buffer: 20 mM HEPES-NaOH pH 7.5, 150 mM NaCl and 0.5 mM spermidine, supplemented just before use with 1 μg/mL leupeptin, 10 μg/mL aprotinin, 1 μM pepstatin A, and 1 mM phenylmethanesulfonyl fluoride.
2. Binding Buffer: 20 mM HEPES-KOH pH 7.9, 10 mM KCl, 1 mM $CaCl_2$, and 1 mM $MnCl_2$ (*see* **Note 4**).
3. 5% digitonin (EMDMilipore, cat. no. 300410) in water (*see* **Notes 1** and **2**; use warm water to dissolve and then keep on ice).
4. Dig Wash Buffer: Wash Buffer supplemented with 0.1% digitonin using 5% digitonin solution (prepare just before use; *see* **Notes 5** and **6**).
5. Antibody Buffer: Dig Wash buffer supplemented with 2 mM EDTA (prepare just before use).
6. Stop Solution: 340 mM NaCl, 20 mM EDTA, 4 mM EGTA, 0.02% digitonin (use 5% digitonin stock solution above), 0.5 μg/mL Rnase A.

2.3 Reagents

1. Concanavalin A beads (BioMag, cat. no. 86057-10). 10 μL beads are used for each condition.
2. pA-MNase: Protein A-Micrococcal Nuclease fusion protein (*see* **Note 7**). The final concentration for the digestion step is 700 ng/mL.

2.4 Antibodies

Relevant antibodies are to be supplied by the user according to the biological question. For the development of the protocol the following antibodies were used:

1. Anti-H3K27me3 (Cell Signaling Technology, cat. no. 9733, clone C36B11).
2. Anti-H3K36me3 (Cell Signaling Technology, cat. no. 9763, polyclonal).
3. Anti-H3K4me3 (Cell Signaling Technology, cat. no. 9727, polyclonal).
4. IgG (rabbit immunoglobulin G, 1 mg/mL homemade stock solution).

2.5 Equipment and Supplies

1. Magnetic stand for 1.5-mL tubes.
2. KingFisher Duo Prime Purification System (Thermo Scientific, cat. no. 5400110) and accompanying BindIt Software.
3. KingFisher deep-well 96 plates (Thermo Scientific, cat. no. 95040450; *see* **Note 8**).
4. KingFisher Duo 12-tip comb for 96duo plate (Thermo Scientific, cat. no. 97003500).
5. NucleoSpin Gel and PCR clean-up kit (Macherey-Nagel, cat. no. 740609.50).

3 Methods

Carry out all procedures at room temperature unless otherwise specified (*see* **Note 9**). Before removing cells from culture (Subheading 3.2), make sure that the Wash Buffer is ready and that the KingFisher instrument is operational with the program to be used for CUT&RUN properly loaded (Subheading 3.1). In contrast to manual CUT&RUN in which buffers can each be prepared immediately before use, automated CUT&RUN requires several buffers to be prepared at the same time and loaded into the KingFisher in advance.

3.1 Setting Up the Automated KingFisher Protocol

Using the BindIt software supplied with the instrument, program the KingFisher Duo Prime to perform the steps described in Table 1 (Fig. 1).

3.2 Cell Preparation

1. Recover cells of interest from culture vessel using a suitable method (e.g., for ES E14 cells grown in N2B27 + 2i + LIF medium, Accutase treatment for 2 min).
2. Collect cells by centrifugation for 3 min at 600 × *g* at room temperature (*see* **Note 10**).
3. Resuspend the cells in appropriate medium and count.
4. Distribute cells into individual tubes as appropriate according to the experimental design. (The protocol was optimized using 10^6 ES E14 cells per condition.)
5. Concurrently, harvest and count cells for spike-in normalization. (The protocol was optimized using 10^5 *Drosophila* S2 cells per condition; *see* **Note 11**.)
6. Collect both the aliquoted cells of interest and the spike-in cells by centrifugation for 3 min at 600 × *g* at room temperature.
7. Wash each cell pellet with 1 mL Wash Buffer and collect by centrifugation for 3 min at 600 × *g* at room temperature.
8. Resuspend each aliquot of cells of interest in 900 μL Wash Buffer and the spike-in cells in the appropriate volume of Wash Buffer such that 100 μL will contain the number intended to be added to each sample (e.g., resuspend at 10^6/mL if 10^5 cells are to be added to each sample).
9. Transfer 100 μL spike-in cells to each suspension of cells of interest, resuspending the spike-in cells thoroughly before withdrawing each 100 μL aliquot (*see* **Note 12**).

3.3 Concanavalin A Bead Preparation

1. Fully resuspend the Concanavalin-A-coated beads in their closed plastic bottle by gentle tapping; then transfer the total volume of slurry needed (10 μL per experimental condition) to a 1.5-mL tube (*see* **Note 13**).

Table 1
Standard automated CUT&RUN program for KingFisher DuoPrime instrument

Step	Description
1. Pick up the Duo 12-tip comb from row B of the deep-well 96 plate.	
2. Binding of the cells to Concanavalin A beads in row C (*see* Fig. 1).	• Beginning of step. – Precollect: No. – Release beads: No. • Mixing/heating. – Mix time: 00:00:05, Speed: medium. – Mix time: 00:04:55, Speed: slow. – Loop count: 2. – Heating during mixing: No. • End of step. – Postmix: No. – Collect count: 4. – Collect time (s): 5. – Post-temperature: No.
3. Antibody binding in row D (*see* **Note 21**).	• Beginning of step. – Precollect: No. – Release beads: Yes. • Mixing/heating. – Mix time: 00:00:05, Speed: medium. – Mix time: 00:04:55, Speed: slow. – Loop count: 12. – Heating during mixing: No. • End of step. – Postmix No. – Collect beads: No. – Post-temperature: No.
4. First pause, for manual addition of Dig Wash Buffer to rows F, G and H, and pA-MNase to row E.	• Dispense 1000 μL Reagent: – Name: Dig Wash Buffer, Volume (μL): 1000. • Dispense 50 μL Reagent: – Name: pA-MNase, Volume (μL): 50.
5. Collection of Concanavalin A beads from row D.	• Beginning of step. – Precollect: Yes. – Release beads: No. • Mixing/heating. – Mix time: 0:00:05, Speed: Medium. – Heating during mixing: No. • End of step. – Postmix: No. – Collect count: 4. – Collect time (s): 5. – Post-temperature: No.
6. Post-antibody wash in row F.	• Beginning of step. – Precollect: No. – Release beads: Yes.

(continued)

Table 1
(continued)

Step	Description
	• Mixing/heating. – Mix time: 0:02:00, Speed: Slow. – Heating during mixing: No. • End of step. – Postmix: No. – Collect count: 4. – Collect time (s): 5. – Post-temperature: No.
7. pA-MNase binding in row E.	• Beginning of step. – Precollect: No. – Release beads: Yes. • Mixing/heating. – Mix time: 00:00:05, Speed: Medium. – Mix time: 00:04:55, Speed: Slow. – Loop count: 2. – Heating during mixing: No. • End of step. – Postmix: No. – Collect count: 4. – Collect time (s): 5. – Post-temperature: No.
8. First post-pA-MNase wash in row G.	• Beginning of step. – Precollect: No. – Release beads: Yes. • Mixing/heating. – Mix time: 0:01:00, Speed: Slow. – Heating during mixing: No. • End of step. – Postmix: No. – Collect count: 4. – Collect time (s): 5. – Post-temperature: No.
9. Second post-pA-MNase wash in row H.	• Beginning of step. – Precollect: No. – Release beads: Yes. • Mixing/heating. – Mix time: 0:01:00, Speed: Slow. – Heating during mixing: No. • End of step. – Postmix: No. – Collect beads: No. – Post-temperature: No.
10. Second pause, for manual addition of cold Dig Wash Buffer containing $CaCl_2$ to row A (*see* **Note 22**).	• Dispense 102 μL Reagents: – Name: Dig Wash Buffer +2 mM $CaCl_2$, Volume (μL): 102.

(continued)

Table 1
(continued)

Step	Description
11. Cooling in row A.	• Beginning of step. – Precollect: No. – Release beads: No. • Mixing/heating. – Mix time: 0:03:00, Speed: Paused. – Tip position when paused: Outside well/tube. – Heating during mixing: Yes. – Block temperature: 4 °C. • End of step. – Postmix: No. – Collect beads: No. – Post-temperature: 4 °C.
12. Collection of Concanavalin A beads from row H.	• Beginning of step. – Precollect: Yes. – Release beads: No. • Mixing/heating. – Mix time: 0:00:05, Speed: Medium. – Heating during mixing: No. • End of step. – Postmix: No. – Collect count: 4. – Collect time (s): 5. – Post-temperature: No.
13. MNase digestion in row A.	• Beginning of step. – Precollect: No. – Release beads: Yes. • Mixing/heating. – Mix time: 00:09:57, Speed: Slow. – Mix time: 00:00:03, Speed: Medium. – Loop count: 3. – Heating during mixing: Yes. – Block temperature: 4 °C. • End of step. – Postmix: No. – Collect beads: No. – Post-temperature: 37 °C.
14. Third pause, for manual addition of Stop Buffer to row A.	• Dispense 100 μL Reagents: – Name: Stop Solution, Volume: 100 μL.
15. Chromatin recovery in row A.	• Beginning of step. – Precollect: No. – Release beads: No. • Mixing/heating. – Mix time: 00:00:05, Speed: Medium. – Mix time: 00:09:55, Speed: Slow. – Heating during mixing: Yes. – Block temperature: 37 °C.

(continued)

Table 1
(continued)

Step	Description
	• End of step. – Postmix: No. – Collect count: 5. – Collect time (s): 5. – Post-temperature: No.
16. Leave the Duo 12-tip comb in row B.	

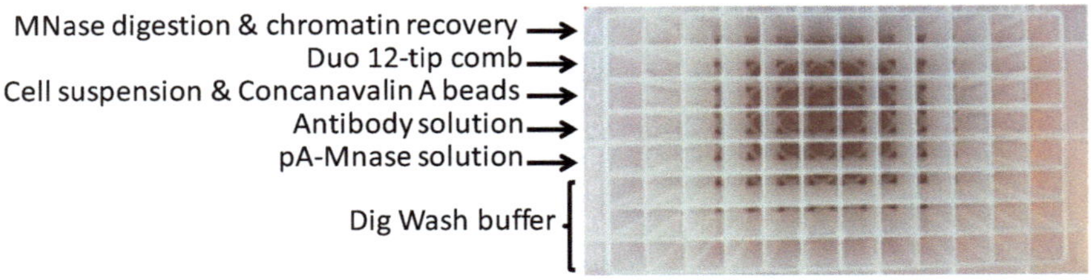

Fig. 1 KingFisher deep-well 96 plate. It contains eight rows, each of which is used for a distinct step of the CUT&RUN procedure. The capacity for each automated CUT&RUN is therefore 12 samples

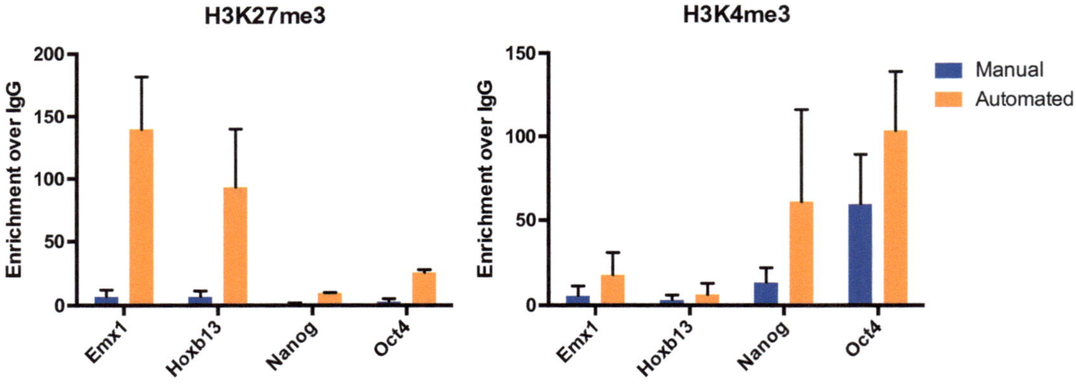

Fig. 2 Automated CUT&RUN can be more sensitive than manual CUT&RUN. CUT&RUN was performed on one million mouse embryonic stem cells using IgG and antibodies against the indicated methylated histones (*see* Subheading 2). Enrichment over IgG at the indicated loci was measured by quantitative PCR

2. Place on magnetic stand until solution clears (30 s to 2 min).
3. Remove liquid and wash beads by gentle pipetting using 1 mL Binding Buffer.
4. Repeat **step 2**.
5. Remove liquid and suspend the beads in a total volume of Binding Buffer representing 10 μL per experimental condition (i.e., the same volume as the slurry withdrawn in **step 1**).

3.4 Cell Binding to Concanavalin A Beads and Antibody Binding

1. Add 10 μL Concanavalin A beads prepared in Subheading 3.3 to each cell suspension prepared in Subheading 3.2.
2. Mix well by gentle pipetting and transfer each suspension to a well in row C of the KingFisher deep-well 96 plate.
3. Prepare Antibody Buffer (*see* **Note 14**).
4. Prepare each antibody solution by diluting the stock solution 1: 100 in Antibody Buffer, 50 μL per relevant experimental condition (*see* **Note 15**).
5. Transfer 50 μL of the appropriate antibody solution to row D of the KingFisher deep-well 96 plate (*see* **Note 16**).
6. Place the KingFisher Duo 12-tip comb in row B, load into KingFisher Duo Prime and start the program defined in Subheading 3.1 (*see* Table 1). The instrument will pause when the antibody binding step is complete, after about 1 h 10 min.

3.5 pA-MNase Binding

1. During the first pause in the KingFisher program (Table 1, **step 4**), prepare:
   (a) Dig Wash buffer (3 mL plus excess per experimental condition).
   (b) Dig Wash buffer containing pA-MNase at 700 ng/mL final concentration (50 μL plus excess per experimental condition).
2. Add 1 mL Dig Wash Buffer from **step 1a** to rows F, G, and H of the KingFisher deep-well 96 plate in the wells corresponding to each experimental condition (*see* **Note 17**).
3. Add 50 μL pA-MNase solution from **step 1b** to row E in the wells corresponding to each experimental condition.
4. Instruct the KingFisher Duo Prime to resume the program. The instrument will pause when the pA-MNase binding and subsequent wash steps are complete, after about 15 min.

3.6 MNase Digestion

1. During the second pause in the KingFisher program, transfer Dig Wash Buffer, 100 μL per sample (plus excess of one sample), into a tube and add 100 mM $CaCl_2$, 2 μL per sample (plus excess of one sample). Chill in a small ice-water bath.
2. Add 102 μL of the resulting ice-cold mixture to row A of the KingFisher deep-well 96 plate in the wells corresponding to each experimental condition (*see* **Note 18**).
3. Instruct the KingFisher Duo Prime to resume the program. The instrument will pause when the digestion step is complete, after about 30 min.

3.7 Digestion Stop and Chromatin Recovery

1. During the third pause in the KingFisher program, add 100 μL Stop Solution to each sample in row A of the KingFisher deep-well 96 plate.
2. Instruct the KingFisher Duo Prime to resume the program. The program will conclude when the digested chromatin has been released and the bead-cell complexes have been removed by the magnet, after about 10 min (*see* **Note 19**).

3.8 DNA Extraction and Downstream Analysis

1. Extract DNA from the samples in row A of the KingFisher deep-well 96 plate using NucleoSpin Gel and PCR clean-up kit according to the manufacturer's instructions. (Alternatively, use another preferred extraction method such as phenol-chloroform.)
2. Perform downstream analysis on purified DNA (e.g., next-generation sequencing library construction or qPCR; *see* **Note 20**) (Fig. 2).

4 Notes

1. Digitonin is toxic, and care should be taken especially when weighing out the powder. A digitonin stock may be prepared in dimethylsulfoxide (DMSO), but be aware that DMSO can absorb through the skin.
2. 5% digitonin stock (in water) stored at 4 °C can be used for up to a week, but it will need reheating as the digitonin slowly precipitates.
3. Different numbers of cells were also tested during protocol development, and as few as 10^5 give results that are comparable between manual and automated CUT&RUN for H3K4me3, H3K27me3, and H3K36me3.
4. Binding Buffer can be kept at 4 °C for several weeks.
5. When preparing Dig Wash Buffer, add digitonin last to prevent precipitation of digitonin in the buffer.
6. The effectiveness of digitonin varies between batches, so testing cell permeability using Trypan blue is recommended to determine the concentration to use for a given cell type.
7. Provided in 50% glycerol by the authors upon request. Store it at −20 °C.
8. Up to 12 samples (1 per column of wells) can be processed on one plate. If certain columns of wells are left unused, they can be covered with a plastic film after the run and the plate can be saved for use in a subsequent experiment.
9. All steps prior to the addition of the antibody are performed at room temperature to minimize stress on the cells. Because it is

crucial to minimize DNA breakage throughout the protocol, we recommend that cavitation during resuspension and vigorous vortexing be avoided.

10. This centrifugation step is important because some time may elapse between cell detachment and completion of counting. Resuspending cells in fresh medium before proceeding with counting minimizes cell stress.
11. Spike-in cells are used to normalize results for factors or chromatin features whose conservation allows dual reactivity of the antibodies, so adding a precisely quantified number of cells to each experimental condition is critical for data interpretation. Alternatively, residual *Escherichia coli* DNA from the purification of the recombinant pA-MNase may be used to control for variations in sample handling that occur after addition of the fusion protein [12].
12. Hereafter, it is no longer necessary to work under sterile conditions.
13. It is not necessary to take an excess of beads. For example, for five samples, withdraw 50 μL bead slurry.
14. The presence of EDTA during antibody treatment is necessary to chelate excess divalent cations from the Binding Buffer used to activate the Concanavalin A and thereby avoid prematurely initiating chromatin cleavage after addition of pA-MNase.
15. The final concentration in the general protocol is 1:100 or the manufacturer's recommended concentration for immunofluorescence [3]. For certain antibodies, we obtain equivalent results using dilutions as low as 1:1000; the minimal concentration required for reliable results must be determined for each antibody.
16. Add the relevant antibody solution to the well in row D corresponding to the sample of interest in row C.
17. Three washes are required: one after the antibody binding step and prior to pA-MNase binding, and two after the pA-MNase binding step, to minimize carryover of antibody and pA-MNase that could result in higher background cleavage during digestion step.
18. In the manual protocol [3, 7], cells are chilled to 0 °C and digestion begins with Ca^{2+} addition. In automated CUT&RUN, we prepare Dig Wash Buffer containing Ca^{2+} and cool it to 0 °C before transferring it to the deep-well plate at 4 °C. The KingFisher Duo Prime cannot cool below 4 °C, but we nevertheless found during protocol development that digestion at this temperature produces results similar to those obtained using manual CUT&RUN.

19. At the end of this step, beads are collected by the 12-tip comb, which is then returned to row B. So in row A we have the cleaved DNA fragments of interest in solution without the beads.
20. Keep the purified DNA samples frozen at −20 °C for long-term storage, but do not freeze samples prior to the DNA extraction step in Subheading 3.8.
21. During the first pause prepare pA-MNase solution for row E and Dig Wash Buffer for rows F, G, and H. The most significant advantage of the KingFisher protocol is that the cell-bead complexes do not need to be washed manually.
22. In the KingFisher Duo instrument, the temperature can only be changed in row A, since only this row is in contact with a heat block. This is why digestion (at 4 °C) and release (at 37 °C) of chromatin must be programmed to occur in this row. An important difference between manual and automated CUT&RUN is the temperature of the digestion step. In the automated method, the temperature is 4 °C (the lowest supported by the instrument) instead of 0 °C as traditionally performed in the manual protocol [3, 7].

Acknowledgments

The authors thank E. Guillard of Thermo Fisher Scientific for advice and troubleshooting in adapting the KingFisher Duo Prime to perform CUT&RUN. Work in the Margueron laboratory was supported by the FRM (Fondation pour la Recherche Médicale), the ARC (Fondation pour la Recherche sur le Cancer), the ANR (AMetHist), and the Labex DEEP. D.H. was supported by a postdoctoral fellowship from the FRM (SPF20150934266).

References

1. Barski A, Cuddapah S, Cui K et al (2007) High-resolution profiling of histone methylations in the human genome. Cell 129:823–837
2. Park D, Lee Y, Bhupindersingh G, Iyer VR (2013) Widespread misinterpretable ChIP-seq bias in yeast. PLoS One 8:e83506
3. Skene PJ, Henikoff S (2017) An efficient targeted nuclease strategy for high-resolution mapping of DNA binding sites. elife 6:576
4. Teytelman L, Thurtle DM, Rine J, van Oudenaarden A (2013) Highly expressed loci are vulnerable to misleading ChIP localization of multiple unrelated proteins. Proc Natl Acad Sci U S A 110:18602–18607
5. Brind'Amour J, Liu S, Hudson M et al (2015) An ultra-low-input native ChIP-seq protocol for genome-wide profiling of rare cell populations. Nat Commun 6:6033
6. Grzybowski AT, Shah RN, Richter WF, Ruthenburg AJ (2019) Native internally calibrated chromatin immunoprecipitation for quantitative studies of histone post-translational modifications. Nat Protoc 14: 3275–3302
7. Skene PJ, Henikoff JG, Henikoff S (2018) Targeted in situ genome-wide profiling with high efficiency for low cell numbers. Nat Protoc 13:1006–1019

8. Schmid M, Durussel T, Laemmli UK (2004) ChIC and ChEC; genomic mapping of chromatin proteins. Mol Cell 16:147–157
9. Kaya-Okur HS, Wu SJ, Codomo CA et al (2019) CUT&Tag for efficient epigenomic profiling of small samples and single cells. Nat Commun 10:1930–1910
10. Inbar M, Sachs L (1969) Interaction of the carbohydrate-binding protein concanavalin A with normal and transformed cells. Proc Natl Acad Sci U S A 63:1418–1425
11. Janssens DH, Wu SJ, Sarthy JF et al (2018) Automated in situ chromatin profiling efficiently resolves cell types and gene regulatory programs. Epigenetics Chromatin 11:74–14
12. Meers MP, Bryson TD, Henikoff JG, Henikoff S (2019) Improved CUT&RUN chromatin profiling tools. elife 8:e46314

Chapter 13

Bioinformatics Methods for ChIP-seq Histone Analysis

Nicolas Servant

Abstract

The field of genomics and genome-wide analysis has exploded since around 2008 with the development of high-throughput omics approaches, largely driven by the emergence of the next-generation sequencing technologies. Among the different biological applications supported by recent sequencing technologies, ChIP-seq (Chromatin ImmunoPrecipitation followed by Sequencing) is one of the most powerful techniques which has dramatically changed our view of the epigenetics landscape of cells.

In this chapter, I will present and discuss the main steps of bioinformatic and biostatistical analysis of ChIP-seq data (Fig. 1). While this technique has been widely used to study transcription factor binding sites, I will focus here on the analysis of histone modifications.

Key words Bioinformatics, ChIP-seq, Quality controls, Differential binding, Peak calling

1 Experimental Design

Before presenting in detail the different steps of data processing and downstream analysis, it is always good to take time to discuss the key points of the experimental design, which must be defined before starting any ChIP-seq sequencing project (Fig. 1). To cite a famous quote by Sir Ronal Fisher: "*(...)* To *consult the statistician after an experiment is finished is often merely to ask him to conduct a post-mortem examination. He can perhaps say what the experiment died of (...)* [1]". This rule still holds true for sequencing data analysis, and until now, the experimental design remains a fundamental step (if not the most important one) of any project.

In the framework of sequencing data applications, one of the first questions which usually arise is about the sequencing protocol and the number of reads (or sequencing depth) which must be generated per sample. For ChIP-seq data and histone modifications, the sequencing depth mainly depends on the histone mark which is targeted and of the signal-to-noise ratio of the antibody. Therefore, it is important to distinguish the profiles of active histone marks, which usually lead to sharp and local enrichment

Raphaël Margueron and Daniel Holoch (eds.), *Histone Methyltransferases: Methods and Protocols*,
Methods in Molecular Biology, vol. 2529, https://doi.org/10.1007/978-1-0716-2481-4_13,

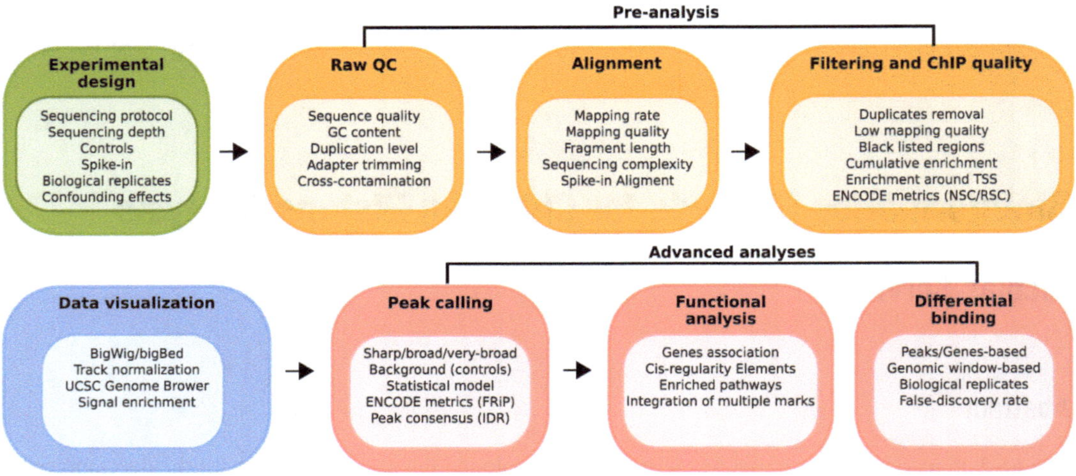

Fig. 1 ChIP-seq analysis workflow for histone mark analysis. Overview of a typical histone mark analysis. Four major steps are listed, including experimental design, pre-analysis, data visualization, and advanced analysis. Pre-analysis includes the quality control of raw sequencing data, the alignment of reads to the reference (and spike-in) genome(s), as well as the filtering of aligned data and the assessment of ChIP quality. Advanced analyses include the detection of enriched genomic regions (or peak calling), the functional annotation of these regions for biological interpretation, and the statistical comparison of binding affinity across different biological conditions. All these steps are further detailed in the different sections of this chapter

around transcription start site (TSS) or cis-regulatory elements, from repressive histone marks which occupy much broader domains, with a diffuse signal spanning several kilo- or megabases. In general, it is good to keep in mind that the number of called peaks increases with the sequencing depth [2]. For human samples and active histone marks, 20–40 million sequenced fragments are commonly recommended, while more fragments (50–70 million) are advised for repressive marks as the signal is spread along much larger genomic regions. A fragment can be sequenced in two different ways using either single-end (only one extremity of the fragment is read) or paired-end sequencing (both fragment ends are read). The single-end sequencing strategy is still frequently used with a read size between 50 and 100 nucleotides. This is typically enough to uniquely align a read on the mappable regions of the reference genome. However, as the sequencing cost is still decreasing, paired-end sequencing tends to be the standard for almost all applications. Regarding ChIP-seq analysis, paired-end sequencing has at least two advantages: (1) a better estimation of the fragment size which is an important parameter of the downstream analysis and can thus directly be inferred as the distance between the two paired reads, and (2) a better alignment on the reference genome, especially around repeated regions, which are known to be part of large heterochromatin domains.

Another important question to address before starting the experiments is the design of controls. As DNA fragmentation by sonication is not a truly random process, controls are mandatory to estimate the nonspecific (i.e., the background) signal across the genome. Two types of controls are commonly used in ChIP-seq studies. The “Input” control sample is made of DNA which has been cross-linked and sonicated but not immunoprecipitated. The “IgG” control sample uses an antibody that will not target any nuclear proteins. Although both types of controls should represent the nonspecific signal, “Input” controls are more widely used than “IgG.” The design of the controls also raises the question of their number and sequencing depth. When several histone marks are studied at the same time, from the same cell line or sample, a single “Input” control for all ChIP experiments is enough. Generating several controls only makes sense in a scenario in which the nonspecific signal could vary from one experimental condition to another. For instance, it can be worth generating controls for both wild-type and mutant conditions. Then, as the control represents the background signal, with no enrichment, a higher sequencing depth is usually required to have uniform coverage along the genome. In practice, the recommendation is to use at least the same number of reads for the controls as for the ChIP experiments. Another strategy which is frequently used is to generate several controls and to merge them for downstream analysis, thus ensuring a high-coverage nonspecific control.

In addition, ChIP-seq data analysis remains challenging when the experiment significantly affects global levels of histone marks. To address this question, several studies have proposed to use spike-in strategies. The idea is to add to each reaction a spike-in antibody designed to pull-down a small fraction of chromatin from an exogenous genome (*Drosophila*, *Escherichia coli*, etc.) which is consistent across all samples [3–5]. Sequencing reads coming from the spike-in data can then be extracted and processed to infer a scaling factor for each sample (*see* Subheading 4). The choice of whether to add spike-in controls to the experiment is crucial and must be discussed at the early stages of the project.

The last important question related to the experimental design is the number of biological replicates to generate per condition which mainly depends on the biological questions covered by the project. In 2012, the ENCODE consortium recommended two biological replicates per experimental condition [6]. But this guideline was defined for binding site identification only as a binary parameter (bound or not bound). For any question related to quantitative comparisons (binding affinity) of groups of samples, more biological replicates are necessary to capture both the biological and the technical variance inherent to ChIP-seq experiments. The notion of variance is central in statistics to compare measurements between two or more groups of samples. The higher the number of replicates, the better the estimation of the variance and the more precise the estimation of the mean binding affinity.

Importantly, the independence of replicates is an essential requirement for most statistical methods. Increasing the number of replicates consistently increases the statistical power of the study which is its ability to detect significant differences between conditions. In practice, we all know that cost, time, as well as the availability of the biological material are important considerations. Nevertheless, working with too few biological replicates can compromise the ability to extract significant results from the data.

Finally, as for any sequencing projects, it is crucial to limit as much as possible any blocking and confounding effects in the design. In this regard, attention must be paid to use the same protocol for all samples and to limit as much as possible uncontrolled variations such as experiments on different machines, different operators, or multiple days of the week. In statistics, we talk about confounding effects when two factors cannot be separated from each other. A typical example would be to perform all wild-type experiments on day 1 and all treated experiments on day 2. In this case, the treatment and day effects would be confounded so that we would not be able to separate out which factor (day or treatment) is causing the observed changes.

2 Pre-analysis Steps

2.1 Quality Controls on Raw Data and Adapter Trimming

During the last decade, technology based on sequencing-by-synthesis has continued to evolve to ultimately reach an overall excellent quality, with an error rate below 0.1% per nucleotide [7]. However, it is also known that, even if well standardized, any sequencing experiment can suffer from technical and systematic bias that may arise during the library preparation and sequencing processes. These issues can include variation in sequencing quality score, base composition biases, fraction of duplicated sequences, or contamination by other species. Thus, it is always worth checking the overall quality of the raw sequencing data for each sample before any downstream analysis. Importantly, all samples must have nearly the same profile. A sample harboring some quality metrics that are very different from the others must be carefully checked and validated.

Raw sequencing data are available as FASTQ files, which are text-based files storing both the nucleotide sequence and its corresponding quality scores. The quality values Q are encoded as *Phred* scores which are related to the base-calling error probability P as follows (Eq. 1):

$$Q = -10 \log_{10} P \tag{1}$$

As an example, a *Phred* quality score of 20 is associated with a probability of incorrect base call of 1 for 100 bp. In other words, it means that the base call accuracy is 99%, and that every 100 bp, a sequencing read is likely to contain an error. Among all tools developed for sequencing quality controls, the *FastQC* software is the most widely used by the community [8]. *FastQC* performs several quality controls including the sequence quality distribution at each nucleotide position. As the read length for a ChIP-seq experiment is commonly between 50 and 100 bp, the quality should remain above 20 (or even 30) at all positions. For longer reads, if a drop in the sequencing quality is observed after a certain read length, it might be useful to trim the 3′ end of the reads to restrict the analysis to high-quality bases. In the same way, the nucleotide content is expected to be the same along the length of the reads with no strong bias. Finally, *FastQC* also reports metrics on the duplication level, which is an important point of ChIP-seq data analysis (*see* Subheading 2.3).

In addition to the standard raw quality controls presented above, it is usually recommended to trim the reads for potential adapter sequence in their 3′ end. The fraction of reads containing an adapter sequence depends on the DNA fragment length and on the sequencing length. Working with shorter fragments and longer reads will increase the probability of sequencing part of the adapter. Together with *FastQC*, tools like *TrimGalore* [9], *Trimmomatic* [10], or *fastp* [11] can easily be applied to automatically detect and trim part of adapter sequence at the 3′ end of the reads. Of note, recent mapping strategies relying on soft-clipping of reads are also able to manage these adapter sequences during the alignment to the reference genome.

Finally, the last interesting control to run before starting to process the data is potential cross-species contamination. At this stage, the idea is to quickly check that the sequencing reads match the expected reference genome and that no contamination from external genomes is observed. This cross-contamination check can be performed on a subset of reads using tools such as *FastqScreen* [12].

2.2 Alignment to a Reference Genome

Once the raw sequencing data are clean, the next step is to align them to a reference genome. The goal of this process (also called read mapping) is to figure out where, in the genome, a sequence is from. This step requires specifying a genome version to use as a reference. The closer the reference genome to the organism used for the experiment, the better the alignment is expect to be. The reference genome must be indexed before running the alignment itself. Indexing a reference genome has the same goal as indexing a book, it makes it possible to quickly find short sequences embedded within it. Several indexing strategies have been proposed over the last years, together with different read aligners. The Burrows-Wheeler transform is one of the most popular indexing methods, which is for instance used by popular read alignment software such as *Bowtie2* [13] or *BWA* [14].

ChIP-seq data alignment is not particularly complex, and software such as *BWA-mem* or *Bowtie2* is frequently used for this task. It is however important to know how alignment positions are reported. In a ChIP-seq experiment, a large majority of reads are expected to be uniquely aligned to the reference genome. This fraction usually depends on the read length and the complexity of the genome. Reporting only uniquely aligned reads ensures a high-quality alignment but excludes from the analysis all repeated regions of the genome. This could make sense for active histone marks which are expected to be enriched around enhancer or promoter regions but could be limited for repressive marks covering large heterochromatin regions. In the latter case, alignment software usually reports one random alignment over all possible positions. This strategy thus improves the coverage of repeated regions [15].

Once the alignment is performed, the results are available in a *BAM* file, which is a binary format, summarizing the reads' sequence and quality, as well as where and how they are aligned to the reference genome. Note that BAM files can be visualized or filtered with the *samtools* software [16] and require an index file (.*bai*) to be quickly browsed. Each read alignment is associated with a mapping quality score (*MAPQ*) which reports the probability that its position on the genome is correct. The mapping quality is encoded using the *Phred* scale as previously described. The higher the *MAPQ*, the better the alignment. This value can also be used to filter out reads aligned with low confidence. For instance, fixing a minimum *MAPQ* of 20 will remove all reads with more than 1% chance of being aligned incorrectly. The probability is calculated as follows:

$$P = 10^{(-\mathrm{MAPQ}/10)}$$

Importantly, reads aligned to multiple positions usually have a MAPQ close to zero. Thus, filtering on the MAPQ value may lead to the removal of multiple alignments from the BAM file.

Once the sequencing reads are aligned to the reference genome, a few additional controls are recommended to validate the quality of the experiment. The mapping statistics are the first interesting metrics to investigate. The fraction of reads aligned to the reference genome is expected to be high (>80%), and most reads are expected to be aligned with a good mapping quality (>20). In the context of paired-end sequencing data, most reads are expected to be aligned together with their mate in a concordant way (following expected distance and orientation between mates). The fraction of reads aligned with low mapping quality or having multiple hits in the genome can vary from one histone mark to another. For instance, a broad histone mark spreading over large domains enriched in repeated elements will necessarily present a higher fraction of multi-aligned reads. Low mapping statistics can usually be explained either by some remaining adapter sequences at the end of the reads or by some contamination issue. In the same way, for paired-end data, it can be worth checking the insert size distribution of aligned read pairs. This can be done with *Picard* tools and should match the fragment size distribution observed during the library preparation (for instance with a Bioanalyzer). Finally, it is also useful to check the complexity of the sequencing libraries. Estimating the complexity is equivalent to estimating the number of redundant reads for a given sequencing depth (Fig. 2a). The *Preseq* package can be used to estimate library complexity but also to predict what fraction of redundant reads can be expected from additional sequencing [17]. When the sequencing depth increases, the number of new unique molecules detected is expected to reach a plateau, meaning that all unique molecules of the library have been sequenced. This estimate is very useful to optimize the sequencing depth, to avoid low complexity samples, and to assess whether additional sequencing could be of interest.

2.3 Duplicate Removal

Because of the small amount of biological material used for the sequencing library preparation, most protocols are still based on PCR amplification, introducing the possibility that a unique original fragment will ultimately be sequenced several times. When the duplication level is excessive, it can severely impact the downstream analysis, thus leading to the question of duplicate removal. While there is a consensus on the fact that these technical duplicates are artefactual and must be removed from the analysis, there is also a risk that two identical sequences were real biological sequences coming from distinct original fragments. Thus, when identifying and removing duplicated reads, the main issue is that, with standard protocols, we are currently not able to distinguish technical from biological duplicates. While, in a large majority of studies, duplicates are removed during ChIP-seq data analysis, the question is still open, with no real consensus in the community.

There are two main reasons why technical duplicates should be removed. First, they can artificially increase the number of observations, leading to a higher rate of false positives during downstream analysis such as peak calling or differential binding, Second, if the effect of technical duplication is not the same for all fragments, it can also introduce a systematic bias in the data. For these reasons, it is usually recommended to remove the duplicates during ChIP-seq data analysis. This can be done with tools such as *Picard MarkDuplicates* [18], which will call as duplicates all but one of the reads starting exactly at the same position and which are aligned in the same way.

On the other hand, a few other studies advice to not filter duplicates [19]. They rely on the idea that duplicates should be treated in relation to the read density. When the read density

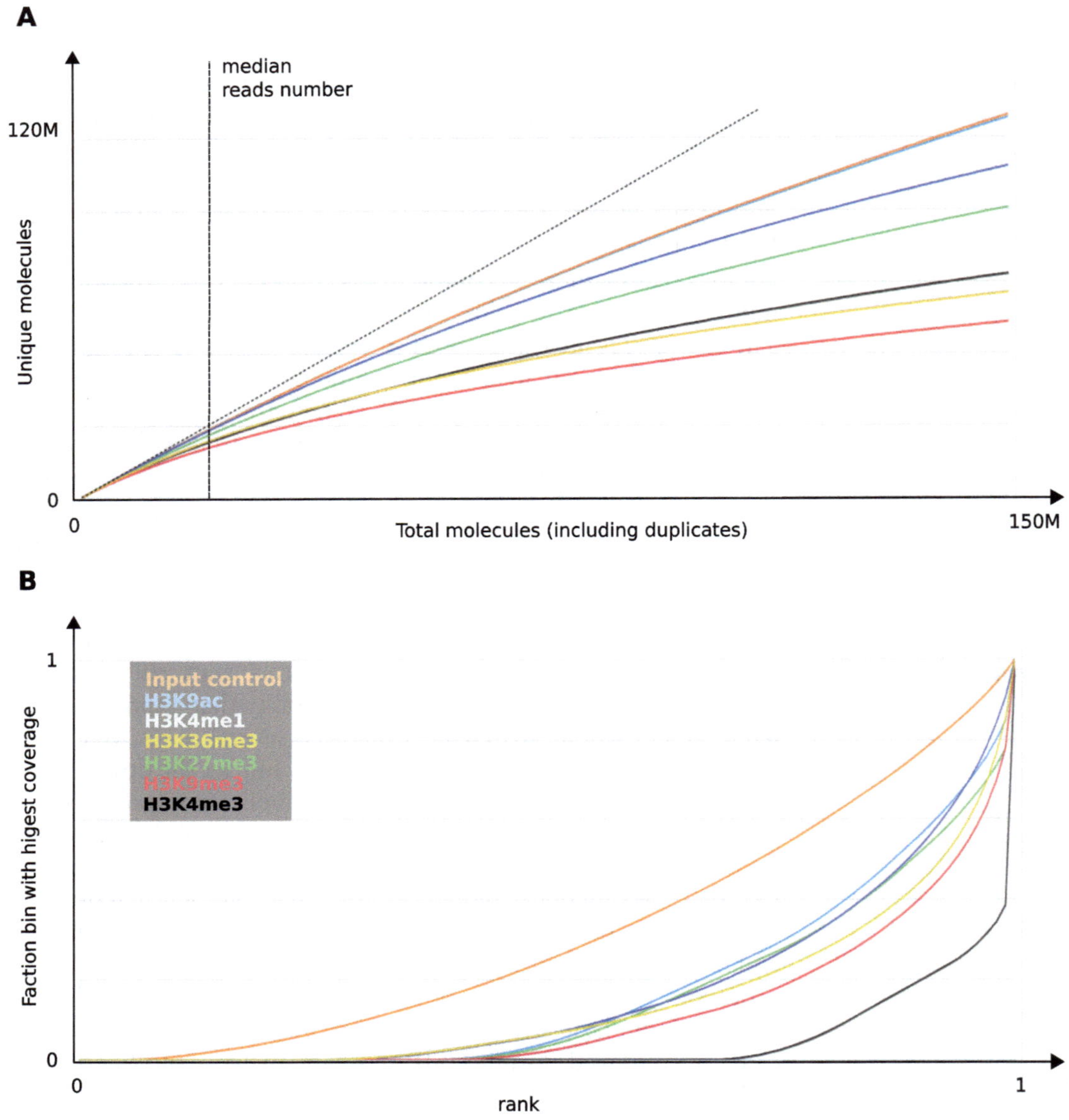

Fig. 2 Library complexity and cumulative enrichment. (**a**) The library complexity is presented as the number of unique molecules over the sequencing depth. The dashed line represents the actual median sequencing depth. The curve below this line thus represents the complexity extrapolation and is an informative metric to optimize the sequencing depth, to avoid low complexity samples, and to assess if additional sequencing could be of interest. (**b**) Cumulative enrichment plot generated with *deepTools - plotFingerprint* [Fidel et al. 2016], representing the fraction of reads with highest coverage (i.e., enrichment) over all genomic windows. As expected, the input control sample has a profile closer to a uniform distribution of reads along the genome, while the different ChIP-seq profiles show an enrichment of reads in a subset of genomic bins. Both profiles were generated using Human H1 cell line ChIP-seq data downloaded from the NCBI Roadmap Epigenomics project

increases, the probability of observing the same sequence several times increases, and so the number of biological duplicated sequences should also increase. In contrast, the level of technical duplicates should remain uniform regardless the read density. Thus, when duplicates are enriched in peaks, they are likely to represent true signal, and removing them can lead to an underestimation of the signal and impact the detection of changes across samples and conditions.

2.4 ChIP-seq Quality Metrics

The pre-analysis steps and quality controls presented above can, in theory, be applied to almost all sequencing applications. In contrast, the additional metrics presented below were designed to specifically evaluate the quality of ChIP-seq experiments.

2.4.1 Cumulative Enrichment

Cumulative enrichment curves (also called fingerprints) are an application of Lorenz curves and a standard way to visualize the enrichment of a ChIP-seq experiment (Fig. 2b). Their aim is to evaluate how well the enrichment signal of a sample can be differentiated from the background signal of the control. The curves represent the cumulative percent of total reads found in a given percent of all genomic bins. Under the assumption of random expectation and perfectly uniform distribution of reads along the genome, the cumulative enrichment curve should be a straight diagonal line. A close profile is expected from the input control samples for which we do not expect any strong enrichment. On the other hand, an increase of the cumulative sum of reads toward the highest fraction of genomic bins (i.e., a bottom-right elbow on the curve) indicates a strong ChIP enrichment. This is particularly true for active histone marks for which we could expect narrow and sharp enrichments. For broader profiles like repressive marks, the cumulative enrichment curves can be less clear, or at least closer to the input signal. This could also be informative if the type of signal to expect is not known from the experiments, giving an indication of the signal-to-background profiles. In practice, the Lorenz curves can be easily generated from tools like *deepTools—plotFingerprint* [20].

2.4.2 Enrichment Around Transcriptionally Active Regions

Active histone marks are known to be associated with transcriptionally active regions, both at proximal and at distal regions of TSS. A simple way to validate the quality of ChIP-seq experiments of active histone marks is therefore to check if the signal is well enriched at these loci. The main idea is then to calculate the average coverage of the ChIP-seq signal over a set of genomic regions such as known protein-coding genes. Depending on which histone mark is studied, the expected enrichment profile is not the same (Fig. 3). For instance, H3K4me3 or H4K9ac marks are expected to be enriched at the promoter level, while H3K4me1 or H4K36me3 is usually respectively enriched at TSS-distal regions or within the gene body.

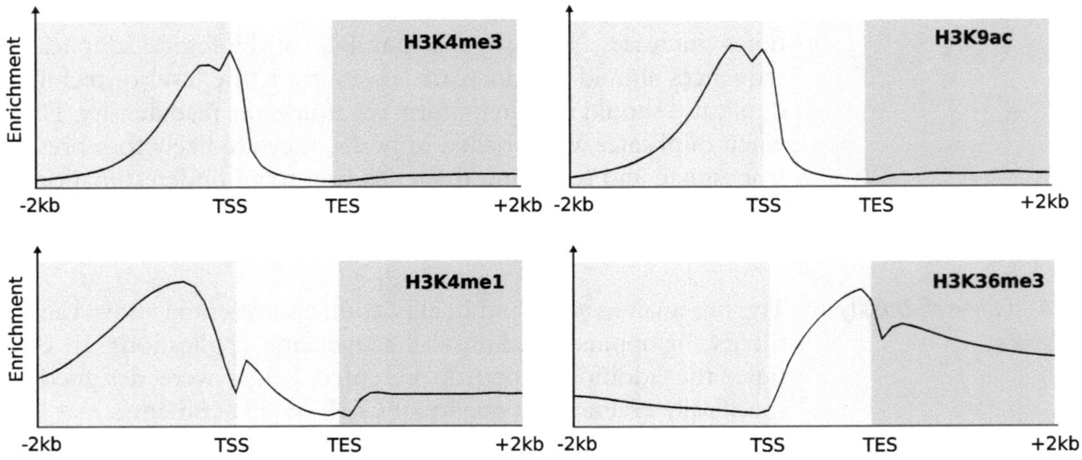

Fig. 3 ChIP-seq enrichment around TSS. ChIP-seq histone enrichment profiles around protein-coding genes from Human H1 cell line. Data was collected from the NCBI Roadmap Epigenomics project. The ChIP-seq enrichment profiles were generated using the *deepTools* suite [Fidel et al. 2016]. According to the histone marks, we can observe a signal enrichment at TSS level (H3K4me3, H3K9ac), within (H3K36me3), or upstream of (H3K4me1) the gene body

The *deeptools* suite, and more precisely the *computeMatrix, plotProfile, or plotHeatmap* commands, can compute the signal over genomic regions and plot them for visualization. Two different modes are available: the first one (reference-point) calculates the signal distribution of upstream/downstream regions relative to a single point, and the second (scale-regions) scales all features to the same length and compute the average distribution. Of note, these tools are also very useful during downstream analysis with a specific list of features/genes to explore condition-specific effect.

2.4.3 ENCODE Guidelines

More than 10 years ago, the ENCODE consortium published some guidelines and quality controls for ChIP-seq transcription factor and histone mark experiments which are still a reference in the field [6].

The strand shift profile (SSP) was proposed as a robust method to estimate the signal-to-noise ratio of an experiment without peak calling. It is based on the observation that, at an enriched locus, the read density is expected to accumulate on both forward and reverse strands around the binding-site center. The distance from the binding-site center mainly depends on the fragment size distribution. Thus, the ENCODE consortium proposed computing a strand cross-correlation profile relying on the calculation of Pearson correlations of read density between forward and reverse strands, while shifting one strand by k base pairs (Fig. 4). The cross-correlation values typically reach a maximum at the fragment length distance. Another peak is expected at the read length (also called a "phantom" peak). The absolute and relative height of the

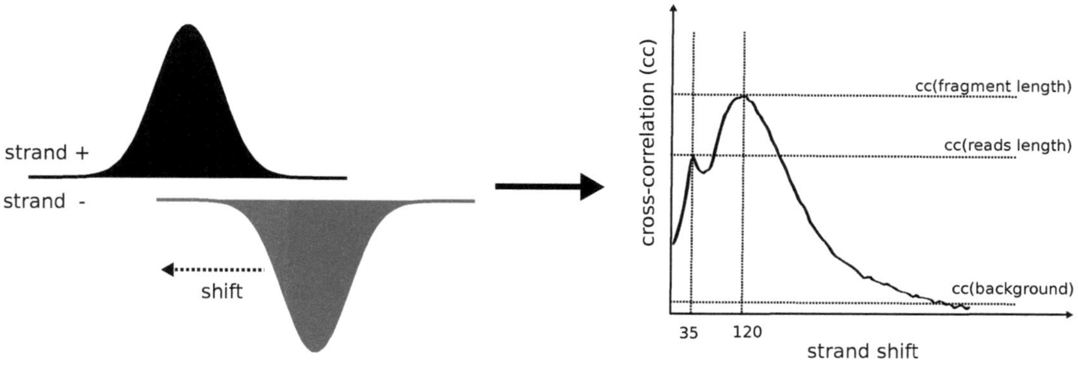

Fig. 4 Strand cross-correlation profile. The ENCODE consortium proposed to compute a strand cross-correlation profile relying on the calculation of Pearson correlations of read density between forward and reverse strands, while shifting one strand by *k* base pairs. The absolute and relative height of the read and fragment length peaks can be used to calculate the normalized strand cross-correlation coefficient (NSC) and the relative strand cross-correlation coefficient (RSC). The NSC is the ratio between the cross-correlation value at the fragment length and the background cross-correlation, defined as the minimum cross-correlation. The RSC is the ratio between the cross-correlation values at the fragment length and at the read length. Example of one sample (ENCFF262NPS) from ENCODE Human H9 cell line

two peaks can be used to derive two useful metrics: the normalized strand cross-correlation coefficient (NSC) and the relative strand cross-correlation coefficient (RSC). The NSC is the ratio between the cross-correlation value at the fragment length and the background cross-correlation, defined as the minimum cross-correlation. The RSC is the ratio between the cross-correlation values at the fragment length and at the read length. A good quality ChIP-seq experiment should be characterized by a peak at the fragment length which is much higher than the one at read length. This method has been proposed in the *PhantomPeak* package. Successful ChIP-seq experiments are expected to have an NSC > 1.05 and an RSC > 0.8, although these values can vary in some biological contexts or for very broad histone marks.

Interestingly, other SPP-based methods have recently been proposed [21]. For instance, Nakato et al. proposed a strand-shift measure based on the Jaccard index and implemented in the *SSP* software. After validation on more than a thousand public ChIP-seq datasets, they recommended an NSC threshold >5.0 (resp. 1.5) for sharp (resp. broad) histone marks, and an NSC threshold <2.0 for input samples.

3 Data Visualization

Data visualization is a mandatory and important step to intuitively assess the quality and to interpret the biological meaning of ChIP-seq data (Fig. 5). It offers a quick overview of the signal-to-noise

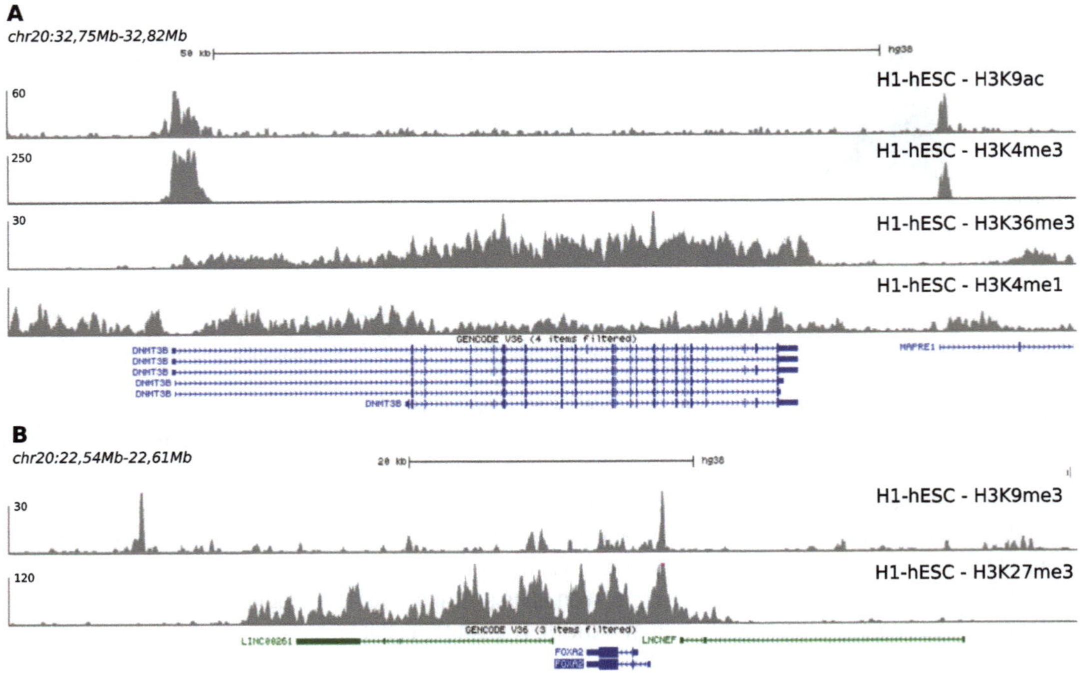

Fig. 5 ChIP-seq data visualization. Screenshots of Human H1 ChIP-seq profiles loaded into the UCSC genome browser. Each mark has a distinct enrichment profile. (**A**) The H4K9ac and H3K4me enrichments are associated with active transcription and located around TSS regions, while H3K36me3 is enriched along gene bodies. H3K4me1 presents here a more diffuse pattern. (**B**) Repressive marks are usually associated with broader domains of heterochromatin (H3K9me3) or repressed gene expression (H3K27me3)

ratio, of the enrichment landscape at genes of interest, or of any potential bias in blacklisted regions.

The main file format for visualizing ChIP-seq data is the BIGWIG format [22], a binary format converted from the WIG format, which displays continuous, genome-wide signal. BIGWIG files can be uploaded in various visualization tools or browsers such as the Integrative Genomic Browser (IGV, https://software.broadinstitute.org/software/igv/) or the UCSC genome Browser (https://genome.ucsc.edu/). Its main advantage is that only the parts of the file needed to display a particular region are transferred, thus allowing much faster performance than regular WIG files. However, when displayed on a genome browser, the BIGWIG files must be hosted on a web-accessible server (http, https, ftp), therefore requiring access to a computational infrastructure.

The BEDGRAPH format is related but not identical to the WIG format. The goal is the same, being able to represent continuous data with a value for each interval. The BEDGRAPH format is derived from the BED format and is a simple text file with the chromosome, start, end, and score of all intervals to display. The main differences between WIG and BEDGRAPH files relate to track visualization. WIG tracks use several compression levels which can reduce the precision of the input data, while BEDGRAPH tracks represent the exact original data. In practice, it is thus not always straightforward to choose one format or the other. In general, the BIGWIG format is used to represent very large dense continuous data at the genome-wide scale, such as read coverage, GC content, etc. For sparse data, with fewer data points, the BEDGRAPH format is preferred. In addition, it is also very powerful for displaying data with intervals of varying sizes, while the WIG format is better suited for genomic windows of constant size. Of note, the conversion from one format to another can easily be performed with the *ucscTools* [22].

Several bioinformatic tools can be used to generate these files. Among them, the *BEDtools genomecov* [23] or *deeptools bamCoverage* commands are very popular for generating genome-wide coverage files from aligned read data. Note that both tools can either extend the read length to an expected fragment size or manage paired-end data by setting the interval size to the maximum distance between two paired reads. When generating ChIP-seq visualization tracks, it is important to pay attention to signal normalization to put all samples in the same range and thus make them comparable. To do so, the simplest and most common approach is to normalize the total read number to one million using a scaling factor defined as 1,000,000/total read number. Of note, the scaling factor can also be calculated only from background regions, by first detecting and excluding enriched regions [24, 25]. Alternatively, several normalization options are directly available in *Deeptools*, such as the RPGC method (reads per genomic content) which normalizes the data to a 1× coverage. Although these methods can correct for different sequencing depths, they do not consider potential differences in the signal-to-noise ratio between experiments. Recent methods such as the *S3norm* package have thus been proposed to normalize both the ChIP backgrounds and signals across multiple epigenomic data [26].

4 The Value of Spike-in Data

Although commonly used, the normalization methods presented above are not appropriate when treatments or conditions affect the global level of signal. In this case, the use of spike-in data is recommended to detect and correct for changes in global levels of histone modifications [3, 5, 27]. The main principle is to add exogenous chromatin at a constant level in each sample before immunoprecipitation. The amount of exogenous chromatin detected in each sample can then be used to adjust to observed signal and directly compare the samples. While the spike-in strategy offers the possibility of detecting global changes in binding affinity, it also relies on strong assumptions. First, the ratio of spike-in to endogenous chromatin must be the same in each sample. Then, the spike-in chromatin is assumed to behave in the same way as endogenous chromatin during the entire protocol. This usually requires an accurate experimental procedure which is not always easy to establish.

The data processing for spike-in experiments is very close to that previously described. The main difficulty here is to be able to distinguish between the sequencing reads coming from the exogenous chromatin and those coming from the endogenous chromatin. To do so, the mapping strategy must be adapted. The first option is to align the data on a meta-genome built from the concatenation of both the reference and the exogenous genomes. In this case, by competition between the two genomes, the sequencing reads are expected to directly align with the highest mapping score to the genome they originate from. Attention should be paid to homologous regions between the two genomes, as corresponding reads might be classified as multiple hits and be assigned to one genome or the other in a random fashion. After mapping, reads aligned to the two genomes can easily be separated. In practice, this approach is a bit painful as it requires combining all possible spike and reference genomes and indexing them for read alignment. Alternatively, sequencing reads can be independently aligned to the reference and spike-in genomes. The two alignment files must then be processed to assign the reads to a given genome using their respective mapping scores and thus to remove the reads that align equally well on the two genomes. As before, the mapping will result in two alignment files, respectively, associated with the spike-in and reference genomes.

After alignment, a scaling factor can be calculated from the spike-in reads and directly applied to the ChIP signal for normalization. The simplest approach would be to calculate a scaling factor based on the total number of reads aligned to the spike-in genome. Another, more robust approach would be to calculate the spike-in read coverage per genomic bin and to use RNA-seq-based methods such as the DESeq2 [28] or edgeR [29] packages to infer size factors which can then be applied to normalize the ChIP signal.

5 Enrichment Calling

5.1 The Concept of Peak Calling

Once the data have been filtered and aligned to the reference genome, the first step of the downstream analysis is enrichment calling (also called peak calling). The principle is to use the read coverage to detect regions which are significantly enriched in the sample and are likely to represent binding sites or modified histones. Importantly, although some peak calling methods propose to estimate a random background model, it is strongly recommended to use a control for peak calling.

The task of peak calling remains challenging for several reasons. First, many enriched genomic loci can be simply explained by experimental noise or a greater chromatin accessibility. Thus, peak calling methods must be able to separate signal-to-noise in the read coverage and to assess a statistical significance for each candidate region by comparing the ChIP and the control signal. Another difficulty is that the shape of the ChIP-seq signal can be completely different from one type of experiment to another (Fig. 5), which means that dedicated methods are usually required for each type of peak. So far, we commonly distinguish three types of peak profiles:

- The sharp peaks located around specific short genomic regions, and usually obtained from ChIP of transcription factors, or active histone marks such as H3K27ac or H3K4me3.
- The broad (or very broad) domains which correspond to large chromatin regions spanning up to several kilobases. These wider signals are frequently observed for repressive histone marks spreading over large genomic regions such as H3K27me3 or H3K9me2 modifications.
- A mixture of sharp and broad regions which are frequently observed with RNA Polymerase II profiles.

During the last decade, many different bioinformatic tools have been developed to detect peaks in ChIP-seq profiles [30, 31]. Some of them were developed to detect sharp peaks, while others were developed to detect broad histone modifications domains, or even both types of peaks. In general, peak calling algorithms rely on several analysis steps up to the generation of a final list of significantly enriched genomic regions [32] (Fig. 6). As the aligned reads originate from the 3′ or 5′ ends of 150–300 bp long DNA fragments, reads are first shifted toward the center of the fragment to precisely determine the genomic bases involved in protein binding or histone modification. Then, enriched regions can be called by comparing the read coverage in each genomic window to a background model usually generated from the input control. Peaks can be scored in many ways using various statistical models based on Poisson, Negative Binomial, or Binomial distributions. Most algorithms calculate a p-value for each peak which is further transformed into a false discovery rate (FDR), representing the expected proportion of incorrectly identified sites among those that are found to be significant. Lastly, additional filtering of peaks can be performed to reduce the list of final peaks to the most reliable regions.

Choosing the best peak calling algorithm remains difficult as the answer mainly depends on the type of experiment. Among the most popular algorithms, the *MACS2* software [33] is widely used by the community. In addition to being efficient for sharp peak

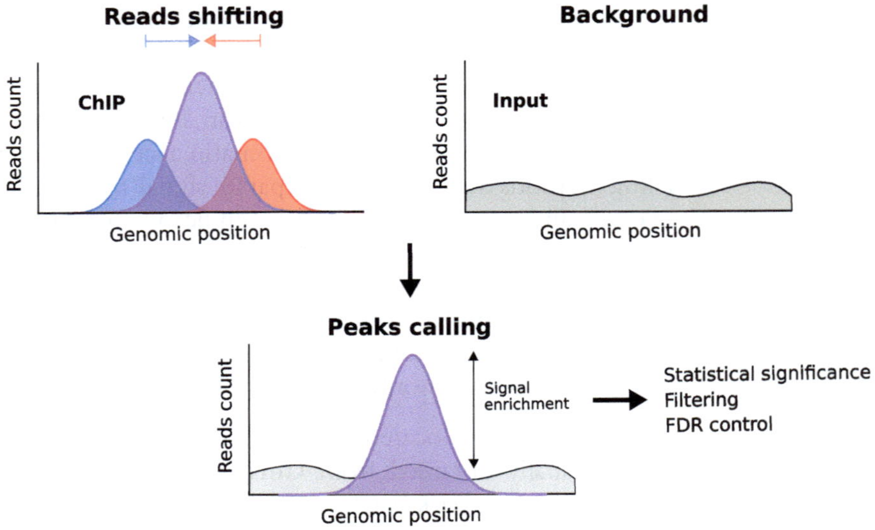

Fig. 6 Principle of peak calling. As the aligned reads originate from the 3′ or 5′ ends of DNA fragments, reads are first shifted toward the center of the fragment to precisely determine the genomic bases where the histone modification is located. Then, enriched regions can be called by comparing the read coverage in each genomic window to a background model generated from the input control. Peak significance can be assessed using various statistical models and corrected for multiple testing to control the false discovery rate

detection, *MACS2* remains easy to use and includes a mode dedicated to the detection of broader domains. Briefly, *MACS2* removes duplicates reads, and builds a read-shift model. It then slides the ChIP and input genomes using a Poisson distribution model to call significantly enriched regions and calculates an empirical FDR. The detection of broad histone modification peaks is more challenging. Among the available tools, *SICER* was developed to call large chromatin domains [34]. The method is based on the detection of clusters (or islands) that are unlikely to appear by chance. Windows and gap size can be adjusted according to the type of signal and chromatin modification. *SICER* uses a random background model to identify candidate clusters and the input sample to infer statistical significance, followed by multiple testing correction and FDR control.

Results of peak calling are available in BED-like formats. The BED format is the simplest format for displaying genomic intervals, characterized by a chromosome, start and end positions. Additional information, such as strand, name, or score can also be added. The narrowPeak and broadPeak formats have also been proposed by the ENCODE consortium and are very close to the BED format. In addition to the principal fields derived from the BED format (chromosome, start, end, name, score, strand), the narrowPeak and broadPeak formats also store for each peak the signal value, the *p*-value, *q*-value, and the peak summit. All these file formats are supported by browsers such as UCSC or IGV and can be visualized in parallel with the ChIP-seq density signal.

5.2 Peak Quality Controls

Peak calling results can be greatly affected by many parameters such as the statistical model used, the sequencing depth, or the quality of the input. Thus, after peak calling, it is always recommended to check whether the final list of peaks matches your biological expectations. Visualizing the ChIP-seq tracks together with the peaks' location can help validate whether the most striking enriched regions have been called and whether they fall in the expected genomic regions (TSS, enhancers, etc.).

The ENCODE consortium proposed to evaluate peak calling results using the fraction of reads in peaks (FRiP). The FRiP is calculated as a ratio of the number of reads falling in peak regions over the total number of reads [6] and is thus correlated with the number and intensity of the detected peaks. The ENCODE consortium first proposed to consider as good experiments those with a FRiP >1%. However, in practice, the FRiP value is highly variable from one transcription factor or histone mark to another. Thus, the FRiP is more useful for comparing results obtained with the same antibody across different samples for which the peak calling has been performed with the same tool and parameters.

The Irreproducible Discovery Rate (IDR) is another way to assess the quality and validity of peak calling results [35]. For a given pair of ChIP-seq experiments (i.e., biological replicates), the IDR compares the ranked list of peaks and assigns a score that reflects its reproducibility. The IDR aims at separating high confidence peaks from the background, and thus the input ranked list of peaks should not be pre-filtered. The main idea is that real peaks should have higher enrichments and should be consistent between the two replicates. On the other hand, peaks with low significance are likely to be noise and are expected to have low consistency. Then, the method fits a bivariate rank distribution to separate real peaks from noise. In addition, the method assigns an IDR score to each peak, which reflects the posterior probability that the peak belongs to the irreproducible group. The IDR score can then be used to set a threshold for consistent called peaks across replicates, exactly like an FDR value, to define a set of high-quality consensus peaks between two replicates.

Importantly, the IDR method is not yet applicable for broad peaks and was only tested for transcription factors and sharp histone marks by the ENCODE consortium. In addition, the IDR is highly affected by the worst replicate, meaning that if one replicate is of bad quality, many good peaks from the high-quality sample may be rejected.

5.3 Functional Annotation of Peaks

Once peaks have been detected from the ChIP-seq signal, it is usually interesting to explore the genomic annotations and functions associated with these regions. A common strategy would be to assign each peak to the nearest gene or TSS. Reporting details on the precise peak location (exon, intron, UTR, intergenic, etc.)

could also be informative. Once the peaks have been associated with a gene annotation, all standard functional enrichment analyses which are commonly used for gene expression analysis can also be applied here. Biological ontologies and pathways such as Gene Ontology (GO) or Reactome Pathway Database can be used to perform overrepresentation analysis associated with a list of genes. Many different tools and web interfaces have been proposed to perform such analysis. Among the most popular tools used for functional annotation of ChIP-seq data, we can cite the R packages *ChIPpeakAnno* [36] or *ChIPseeker* [37] and the *HOMER* [38] suite, or the *GREAT* [39] and *SCREEN* [40] web interfaces.

Conversely, ChIP-seq experiments can also be used to annotate new genomic regions. For instance, several histone marks have been reported to be associated with active enhancers (H3K27ac, H3K4me1), active promoters (H3K27ac, H3K4me3), or poised enhancers (H3K4me1). Likewise, repressive histone marks such as H3K9me3 or H3K27me3 are associated with heterochromatin and repressed promoters. Thus, there is a growing interest in combining these histone marks to better characterize the chromatin states of the genome and potentially detect novel classes of elements. In this context, the *chromHMM* [41] and *seqway* [42] tools have been used by the large consortia such as ENCODE to integrate and summarize their chromatin datasets into 15 chromatin states [43]. All results have been integrated in the UCSC genome browser and represent an interesting resource to compare and annotate ChIP-seq experiments. The *chromHMM* tool can also be directly applied to aligned ChIP-seq data or to a set of pre-computed lists of peaks. As output, it returns the state assignments for each genomic position. Of course, the number and nature of predicted states mainly depend on the type and diversity of the ChIP-seq histone marks provided in input.

6 Differential Analysis

Detecting differential signal between several conditions is a common question of sequencing data analysis. Differential analysis has been extensively studied in the last decade, mainly in the framework of RNA-seq experiments and gene expression. In the context of ChIP-seq data, the question is the same. Can we detect genomic regions with a significant differential binding affinity between conditions? As compared to RNA-seq, the differential analysis of ChIP-seq data presents a few additional particularities which make this question even more challenging.

6.1 Working on a Common Set of Prespecified Regions

Several experimental biases can occur during ChIP-seq experiments (cross-linking, immunoprecipitation, etc.) that will affect the level of noise from one experiment to another. Some methods recommend subtracting the normalized input signal before running the direct comparison of ChIP samples across conditions. Of course, this only makes sense if an input is available for all conditions, and if the background is expected to vary from a condition to another. However, this input subtraction has been further debated in other studies and has to be applied with caution as it could favor small count values and violate the assumptions of the statistical models used in differential analysis packages such as *DESeq2* or *edgeR* [44]. An alternative would be for instance to use the input control to first detect enriched regions in the background which can then be blacklisted during differential analysis.

Next, a set of genomic regions of interest must be defined across different replicates and conditions, in which the differential tests are to be applied (Fig. 7a). In the context of peak calling, a naïve approach would be to intersect or to take the union of all peaks detected on all replicates and conditions. More sophisticated approaches can be applied in some cases [45], to avoid the generation of large peak sets or to deal with large domains covered by repressive marks. Note that in some cases, the analysis can also be performed on a set of known genomic features, such as promoter regions for active histone marks, thus avoiding the issue of peak calling. Finally, it is important to keep in mind that many methods for differential binding analysis were developed for a particular type of data (active or repressive histone marks) and are therefore not efficient for all data types.

Recently, Steinhauser et al. performed an in-depth comparison of 14 different tools designed for differential ChIP-seq analysis using both real and simulated data [46]. Among the different tools tested by the authors, some of them were designed to work with or without replicates, on narrow or broad peaks, and are based on different statistical models. Surprisingly, comparison of these tools demonstrates a huge variability in the results. Among the different parameters tested, the presence of independent biological replicates appears to be one of the most important factors with a strong impact on the sensitivity and on the fraction of false positive calls. Thus, as it is now established for differential gene expression analysis, differential binding analysis of ChIP-seq data must rely on independent biological replicates. Although there is no standard for the number of replicates per condition, it is common to require at least three biological replicates to account for both the experimental and the biological variance.

Among the most popular and efficient methods, the *diffBind* R Bioconductor package performs differential ChIP-seq analysis between one or several groups [47, 48]. Briefly, the package includes functions to define a common peak set across the entire

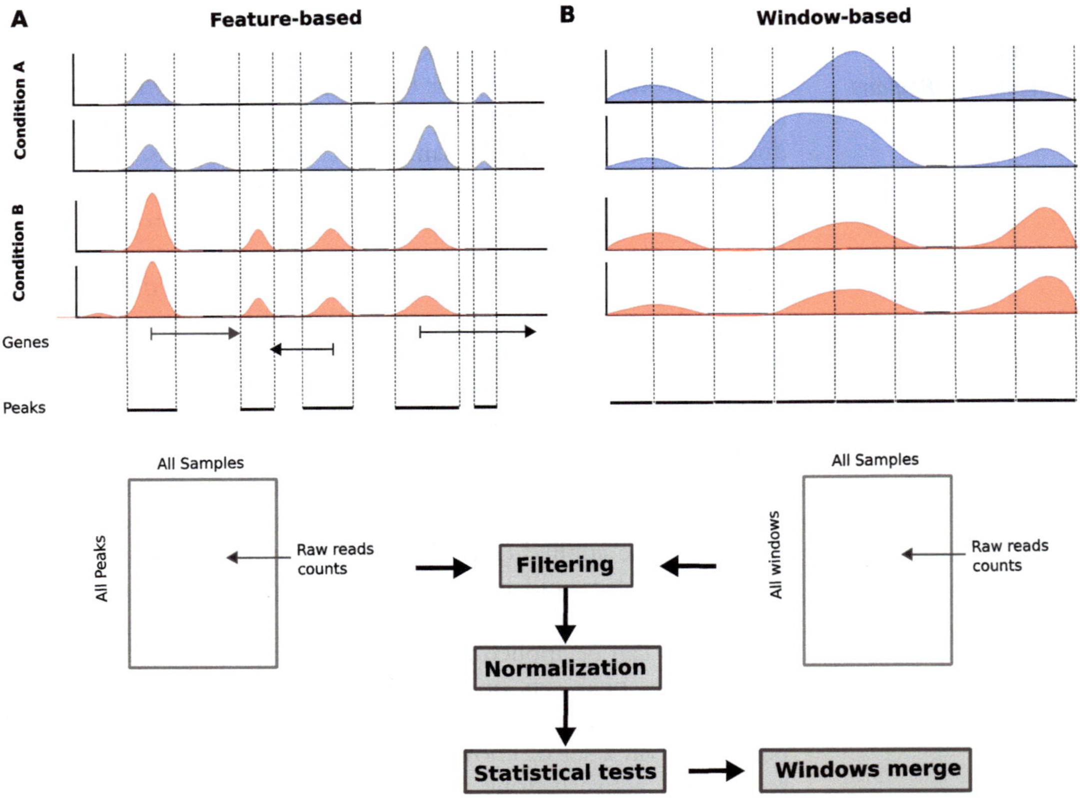

Fig. 7 Approaches for differential binding. (**a**) Feature-based approach which relies on the prior definition of regions of interest such as consensus peaks across samples or genes annotations (TSS or gene body). The binding affinity is then estimated by counting the number of reads per feature. Noninformative features are then filtered using the read counts or blacklisted regions. Finally, count-based normalization and statistical methods mainly derived from RNA-seq analysis can be applied to detect regions with differential binding (*see* the R *diffBind* package for example). (**b**) The genomic-window-based approach is particularly useful when peak calling is challenging. In this case, the genome is split into (overlapping) windows. Noninformative windows (i.e., background) can be filtered out using absolute threshold or global/local enrichment methods. The raw count data are then normalized and tested for differential binding. Significant windows can then be merged in a way that maintains FDR control at the level of the regions (*see* the R *csaw* package [Lun et al. 2016] for details)

dataset, to count the reads overlapping in the peak sets and to identify significantly differentially bound regions. Peaks overlapping blacklisted regions or showing enrichment in the input samples can be removed. Then, the statistical framework uses normalization and statistical methods first developed for counts data and that are the standard for RNA-seq differential analysis through the well-known *DESeq2* or *edgeR* packages. Finally, peaks with an FDR below a given threshold (for instance 5%) are called significantly differentially bound. Importantly, these methods assume that most features (peaks here) are not differential between conditions and are therefore not able to detect global chromatin changes.

6.2 Working on Genomic Windows

As previously discussed, most methods for differential binding analysis start with a set of consensus peaks to work on. Therefore, the quality of the peak calling analysis is crucial and can have a strong impact on the results. While several efficient methods have been proposed to detect sharp peaks from active histone marks, the detection of broad domains from repressive marks remains much more challenging, thus affecting the results of differential binding analysis.

To avoid these calling limitations, other approaches based on genomic windows are available. Here, the idea is to first split the genome into (overlapping) bins of fixed size using a sliding window approach (Fig. 7b). Each window is then tested for differential binding using the methods and packages mentioned earlier for count statistics. Among the different available tools, the *csaw* R Bioconductor package is one of the most flexible and powerful resources [49, 50]. A typical analysis framework with *csaw* could be summarized with the following steps. First, sequencing reads are counted into sliding genomic windows of fixed size. The distance between windows can be defined and blacklisted regions can be discarded from the analysis. Then, low-abundance windows must be filtered out to restrict the number of regions to test for differential binding. This filtering step can be done in many ways including proportion, global or local enrichment approaches, or using the input controls. The read counts in the selected windows are then normalized to remove any experimental bias using scaling factors inferred from RNA-seq-based methods (TMM—Trimmed Mean of M-values from the *edgeR* package) and assuming that most windows are not differentially bound. Alternatively, scaling factors can also be calculated from background regions using larger genomic bins or from exogenous spike-in chromatin. The choice of the normalization strategy remains the key step of this analysis and will have a strong impact on the results. Some understanding of the biological context is particularly useful to choose the appropriate method. Once normalized, windows are tested for differential binding using the *edgeR* framework, and p-values are corrected to control the FDR. Here, it is important to remember that commonly used multiple testing adjustment methods such as the popular Benjamini–Hochberg method [51] assume independence of statistical tests. If this assumption is already questionable for gene expression analysis, it becomes even more problematic when the statistical tests are performed on (overlapping) genomic windows. To address this statistical challenge, the *csaw* package offers different strategies to merge the window statistics into region-level FDR control. Interestingly, as the size of the genomic window is an important parameter of such approaches, results from different window sizes can also be combined to consolidate the final list of differentially bound regions.

7 Analysis of Cut&Run and Cut&Tag Data

The Cut&Run protocol (Cleavage Under Targets and Release Using Nuclease) was first introduced by Skene and Henikoff in 2017 [52]. Briefly, the main idea is to use a pA-MNase (i.e., a recombinant protein A—microccocal nuclease fusion) to guide the chromatin cleavage to antibodies targeting the protein of interest. Compared to standard ChIP-seq protocols, Cut&Run is easier to set up, with a lower cost, and usually requires fewer cells. In 2019, Kaya-Okur and colleagues extended the technique, proposing the Cut&Tag protocol (Cleavage Under Targets and Tagmentation). The advantage here is to use a Tn5 transposase to simultaneously cut the chromatin and insert the sequencing adapters, as is also done in chromatin accessibility assays (ATAC-seq). Thus, Cut&Tag is easier to set up and can even work with lower cell numbers, down to single cells [53].

Moving to bioinformatic analysis, most of the steps and quality controls presented above for ChIP-seq data analysis can also be applied for Cut&Run and Cut&Tag experiments, with two main differences. First, due to the cleavage strategies of these protocols, the DNA fragment length is usually smaller than for ChIP-seq protocols. In practice, the libraries are sequenced in paired-end but with smaller read lengths. Accordingly, the strategy to remove the potential adapter sequences must be checked or adapted if necessary. Then, because of smaller fragment sizes, it is common that the two mates from a read pair overlap each other. In some extreme cases, the mates can even overtake the beginning of the other mate (also called dovetailed reads). Thus, the alignment strategy may have to be adapted to be sure that such overlapping read pairs are considered for the alignment. Recent studies commonly used the *Bowtie2* software for read alignment with appropriate parameters ("*--no-mixed --dovetail*") [54].

The second main difference between Cut&Run/Cut&Tag and ChIP-seq experiments is the signal-to-noise ratio. In Cut&Run/Cut&Tag experiments, the background is expected to be much lower. This difference can affect the efficiency of traditional peak calling models. Although the *MACS2* peak caller can still be used, dedicated methods such as the *SEACR* software were specifically designed for low-background experiments and can lead to more accurate results [55]. In addition, spike-in normalization can almost always be used for Cut&Run experiments due to the residual Escherichia Coli DNA from the production of pA-MNase which is present in all samples [56]. Thus, adding exogeneous chromatin for spike-in normalization as for standard ChIP-seq protocols is not required in this case.

Finally, for the Cut&Tag protocol, an additional step can be performed to correct for the 9 bp duplication created by the Tn5 transposition. To do so, reads must be shifted from +4 bp and –5 bp for the positive and negative strands, respectively, to represent the center of the transposon binding event [57]. This can easily be achieved using the *Deeptools alignmentSieve* tool. Of note, this correction is expected to have a limited impact on histone mark analysis where single-base-pair resolution is generally not required. However, it can be extremely useful for transcription factor analysis and motif discovery such as footprinting analysis.

8 Conclusion

The goal of this chapter was to present the main steps of a standard ChIP-seq analysis workflow from the quality controls of raw sequencing reads to chromatin-state annotation and differential binding analysis, highlighting the key points associated with each step. Of course, the list of bioinformatic tools and methods mentioned here is far from exhaustive but should be a good starting point for any ChIP-seq analysis project (Table 1). Recently, the *nf-core* community has put forward a complete ChIP-seq pipeline including the main steps of a typical analysis process [58, 59]. This pipeline is built using the *Nextflow* workflow management system [60] and can run all analysis steps in a single command line. Importantly, all *nf-core* pipelines are designed to ensure portability and reproducibility, following the FAIR (findable, accessible, interoperable, and reusable) research principles. Hundreds of bioinformaticians and developers have joined the community, thus ensuring a very high quality of the available pipelines in terms of both reliability and methods/tools implemented. In the same way, the web-based platform Galaxy also makes it possible to run most of these analyses through a user-friendly interface [61]. Galaxy offers the possibility of building simple analysis workflows, applying them to multiple samples in a reproducible way, and sharing them with other users. In addition, the Galaxy community provides examples and tutorials including ChIP-seq data analysis.

The application of ChIP-seq and related methods is still an active field of research both in epigenetics and bioinformatics. Among the challenges for the future, the analysis of single-cell (sc) ChIP-seq data is a growing field of interest. Several scChIP-seq protocols have recently been established, requiring the development of dedicated computational methods. Finally, designing methods able to integrate ChIP-seq data with other sequencing assays such as DNA methylation, ATAC-seq, or 3D chromatin organization (Hi-C), remains a priority to combine the advantages of all these techniques and to better understand underlying gene regulatory mechanisms [62].

Table 1
Tool categories, descriptions, and examples. Summary of the difference steps of a typical ChIP-seq analysis for histone marks. The list of bioinformatic tools and methods mentioned here is far from exhaustive but should be a good starting point for any project

Tool categories	Description	Function and examples
Raw quality controls	Validation of sequencing quality	FastQC [8], FastqScreen [12]
Reads trimming	Remove adapter sequence in 3′ end	TrimGalore [9], Trimmomatic [10], fastp [11]
Alignments	Align reads on the reference genome and filter low quality alignments	Alignment: Bowtie2 [13], BWA [14] Filtering: samtools [16]
Duplicates removal	Remove duplicated sequences which are likely to be PCR artefacts	Picard [18]
ChIP-seq quality metrics	Compute quality metrics to validate the ChIP enrichment	Complexity curves: Preseq [17] Fingerprint/enrichment: deeptools [20] ENCODE metrics: phantompeakqualtools [6]
Data visualization	Visualize ChIP-seq profile along the genome	Visualization: IGV, UCSC genome browser Tracks manipulation: ucscTools [22], deeptools [20], bedTools [24]
Peak calling	Detect enriched ChIP-seq signal compared to a background control	Peaks calling: Macs [35], SICER [36], csaw [51] Peaks quality: IDR [37], FriP [6]
Peak annotation	Overlap peaks with known genomic annotations	ChIPpeakAnno [38], ChIPseeker [39], HOMER [40] GREAT [41], SCREEN [42]
ChIP-seq integration	Define chromatine states using multiple ChIP-seq profiles	chromHMM [43], seqway [44]
Differential binding	Detect differential binding affinity by comparing groups of samples	diffBind [49], csaw [51]

References

1. Fisher RA (1938) Indian statistical congress. 4: 17
2. Sims D, Sudbery I, Ilott NE et al (2014) Sequencing depth and coverage: key considerations in genomic analyses. Nat Rev Genet 15: 121–132
3. Orlando DA, Chen MW, Brown VE et al (2014) Quantitative ChIP-Seq normalization reveals global modulation of the epigenome. Cell Rep 9(3):1163–1170. https://doi.org/10.1016/j.celrep.2014.10.018
4. Egan B, Yuan CC, Craske ML et al (2016) An alternative approach to ChIP-Seq normalization enables detection of genome-wide changes in histone H3 lysine 27 trimethylation upon EZH2 inhibition. PLoS One 11(11): e0166438. https://doi.org/10.1371/journal.pone.0166438
5. Bonhoure N, Bounova G, Bernasconi D et al (2014) Quantifying ChIP-seq data: a spiking method providing an internal reference for sample-to-sample normalization. Genome Res

24(7):1157–1168. https://doi.org/10.1101/gr.168260.113

6. Landt SG, Marinov GK, Kundaje A et al (2012) ChIP-seq guidelines and practices of the ENCODE and modENCODE consortia. Genome Res 22(9):1813–1831. https://doi.org/10.1101/gr.136184.111
7. Fox EJ, Reid-Bayliss KS, Emond MJ et al (2014) Accuracy of next generation sequencing platforms. Next Gener Seq Appl 1: 1000106. https://doi.org/10.4172/jngsa.1000106
8. Andrews S (2010) FastQC: a quality control tool for high throughput sequence data. http://www.bioinformatics.babraham.ac.uk/projects/fastqc
9. Krueger F Trim Galore! http://www.bioinformatics.babraham.ac.uk/projects/trim_galore/
10. Bolger AM, Lohse M, Usadel B (2014) Trimmomatic: a flexible trimmer for illumina sequence data. Bioinformatics 30(15): 2114–2120. https://doi.org/10.1093/bioinformatics/btu170
11. Chen S, Zhou Y, Chen Y, Jia G (2018) fastp: an ultra-fast all-in-one FASTQ preprocessor. Bioinformatics 34(17):i884–i890. https://doi.org/10.1093/bioinformatics/bty560
12. Wingett SW, Andrews S (2018) FastQ Screen: a tool for multi-genome mapping and quality control. F1000 Res 7:1338. https://doi.org/10.12688/f1000research.15931.2
13. Langmead B, Salzberg S (2012) Fast gapped-read alignment with Bowtie 2. Nat Methods 9: 357–335
14. Li H (2013) Aligning sequence reads, clone sequences and assembly contigs with BWA-MEM. arXiv:1303.3997v2
15. Teissandier A, Servant N, Barillot E et al (2019) Tools and best practices for retrotransposon analysis using high-throughput sequencing data. Mob DNA 10:52. https://doi.org/10.1186/s13100-019-0192-1
16. Li H, Handsaker B, Wysoker A et al (2009) The Sequence Alignment/Map format and SAMtools. Bioinformatics 25(16): 2078–2079. https://doi.org/10.1093/bioinformatics/btp352
17. Daley T, Smith AD (2013) Predicting the molecular complexity of sequencing libraries. Nat Methods 10(4):325–327. https://doi.org/10.1038/nmeth.2375
18. Picard tools (2018) Broad Institute. http://broadinstitute.github.io/picard/
19. Tian S, Peng S, Kalmbach M et al (2019) Identification of factors associated with duplicate rate in ChIP-seq data. PLoS One 14(4): e0214723. https://doi.org/10.1371/journal.pone.0214723
20. Fidel R, Ryan D, Grüning B, Bhardwaj V, Kilpert F, Richter AS, Heyne S, Dündar F, Manke T (2016) deepTools2: a next generation web server for deep-sequencing data analysis. Nucleic Acids Res 44(W1):W160–W165. https://doi.org/10.1093/nar/gkw257
21. Nakato R, Sakata T (2021) Methods for ChIP-seq analysis: a practical workflow and advanced applications. Methods 187:44–53
22. Kent WJ, Zweig AS, Barber G et al (2010) BigWig and BigBed: enabling browsing of large distributed datasets. Bioinformatics 26: 2204–2207. https://doi.org/10.1093/bioinformatics/btq351
23. Quinlan A, Hall IM (2010) BEDTools: a flexible suite of utilities for comparing genomic features. Bioinformatics 26(6):841–842. https://doi.org/10.1093/bioinformatics/btq033
24. Liang K, Keleş S (2012) Normalization of ChIP-seq data with control. BMC Bioinformatics 13:199. https://doi.org/10.1186/1471-2105-13-199
25. Diaz A, Park K, Lim DA et al (2012) Normalization, bias correction, and peak calling for ChIP-seq. Stat Appl Genet Mol Biol 11(3). https://doi.org/10.1515/1544-6115.1750
26. Xiang G, Keller CA, Giardine B et al (2020) S3norm: simultaneous normalization of sequencing depth and signal-to-noise ratio in epigenomic data. Nucleic Acids Res 48(8):e43. https://doi.org/10.1093/nar/gkaa105
27. Egan B, Yuan CC, Ml C et al (2016) An alternative approach to ChIP-seq normalization enables detection of genome-wide changes in histone H3 lysine 27 trimethylation upon EZH2 inhibition. PLoS One 11:e0166438. https://doi.org/10.1371/journal.pone.0166438
28. Love MI, Huber W, Anders S (2014) Moderated estimation of fold change and dispersion for RNA-seq data with DESeq2. Genome Biol 15:550. https://doi.org/10.1186/s13059-014-0550-8
29. Robinson MD, McCarthy DJ, Smyth GK (2010) edgeR: a bioconductor package for differential expression analysis of digital gene expression data. Bioinformatics 26(1): 139–140. https://doi.org/10.1093/bioinformatics/btp616
30. Thomas R, Thomas S, Holloway AK (2017) Features that define the best ChIP-seq peak calling algorithms. Brief Bioinform 18(3): 441–450. https://doi.org/10.1093/bib/bbw035

31. Nakato R, Shirahige K (2017) Recent advances in ChIP-seq analysis: from quality management to whole-genome annotation. Brief Bioinform 18(2):279–290. https://doi.org/10.1093/bib/bbw023
32. Pepke S, Wold B, Mortazavi A (2009) Computation for ChIP-seq and RNA-seq studies. Nat Methods 6(11 Suppl):S22–S32. https://doi.org/10.1038/nmeth.1371
33. Zhang Y, Liu T, Meyer CA et al (2008) Model-based analysis of ChIPSeq (MACS). Genome Biol 9:R137. https://doi.org/10.1186/gb-2008-9-9-r137
34. Zang C, Schones DE, Zeng C et al (2009) A clustering approach for identification of enriched domains from histone modification ChIP-Seq data. Bioinformatics 25:1952–1958. https://doi.org/10.1093/bioinformatics/btp340
35. Li Q, Brown J, Huang H, Bickel P (2011) Measuring reproducibility of high-throughput experiments. Ann Appl Stat 5:1752–1779. https://doi.org/10.1214/11-AOAS466
36. Zhu L, Gazin C, Lawson N et al (2010) ChIPpeakAnno: a Bioconductor package to annotate ChIP-seq and ChIP-chip data. BMC Bioinformatics 11(1):237. https://doi.org/10.1186/1471-2105-11-237
37. Yu G, Wang LG, He QY (2015) ChIPseeker: a R/Bioconductor package for ChIP peak annotation, comparison and visualization. Bioinformatics 31(14):2382–2383. https://doi.org/10.1093/bioinformatics/btv145
38. Heinz S, Benner C, Spann N et al (2010) Simple combinations of lineage-determining transcription factors prime cis-regulatory elements required for macrophage and B cell identities. Mol Cell 38(4):576–589. https://doi.org/10.1016/j.molcel.2010.05.004
39. McLean C, Bristor D, Hiller M et al (2010) GREAT improves functional interpretation of cis-regulatory regions. Nat Biotechnol 28:495–501. https://doi.org/10.1038/nbt.1630
40. ENCODE Project Consortium, Moore JE, Purcaro MJ et al (2020) Expanded encyclopaedias of DNA elements in the human and mouse genomes. Nature 583(7818):699–710. https://doi.org/10.1038/s41586-020-2493-4
41. Ernst J, Kellis M (2020) ChromHMM: automating chromatin-state discovery and characterization. Nat Methods 9:215–216. https://doi.org/10.1038/nmeth.1906
42. Hoffman MM, Buske OJ, Wang J et al (2012) Unsupervised pattern discovery in human chromatin structure through genomic segmentation. Nat Methods 9:473–476. https://doi.org/10.1038/nmeth.1937
43. Hoffman MM, Ernst J, Wilder SP et al (2013) Integrative annotation of chromatin elements from ENCODE data. Nucleic Acids Res 41(2):827–841. https://doi.org/10.1093/nar/gks1284
44. Chen L, Wang C, Qin ZS et al (2015) A novel statistical method for quantitative comparison of multiple ChIP-seq datasets. Bioinformatics 31(12):1889–1896. https://doi.org/10.1093/bioinformatics/btv094
45. Yang Y, Fear J, Hu J et al (2014) Leveraging biological replicates to improve analysis in ChIP-seq experiments. Comput Struct Biotechnol J 9(13):e201401002. https://doi.org/10.5936/csbj.201401002
46. Steinhauser S, Kurzawa N, Eils R et al (2016) A comprehensive comparison of tools for differential ChIP-seq analysis. Brief Bioinform 17(6):953–966. https://doi.org/10.1093/bib/bbv110
47. Stark R, Brown G (2011) DiffBind: differential binding analysis of ChIP-Seq peak data. http://bioconductor.org/packages/release/bioc/vignettes/DiffBind/inst/doc/DiffBind.pdf
48. Ross-Innes CS, Stark R, Teschendorff AE et al (2012) Differential oestrogen receptor binding is associated with clinical outcome in breast cancer. Nature 481:389–393. https://doi.org/10.1038/nature10730
49. Lun AT, Smyth GK (2016) csaw: a Bioconductor package for differential binding analysis of ChIP-seq data using sliding windows. Nucleic Acids Res 44(5):e45. https://doi.org/10.1093/nar/gkv1191
50. Lun ATL, Smyth GK (2015) From reads to regions: a Bioconductor workflow to detect differential binding in ChIP-seq data. F1000 Res 4:1080. https://doi.org/10.12688/f1000research.7016.2
51. Benjamini Y, Hochberg Y (1995) Controlling the false discovery rate: a practical and powerful approach to multiple testing. J R Stat Soc Series B Stat Methodol 57(1):289–300
52. Skene PJ, Henikoff S (2017) An efficient targeted nuclease strategy for high-resolution mapping of DNA binding sites. eLife 6:e21856. https://doi.org/10.7554/eLife.21856
53. Kaya-Okur HS, Wu SJ, Codomo CA et al (2019) CUT&Tag for efficient epigenomic profiling of small samples and single cells. Nat Commun 10(1):1930. https://doi.org/10.1038/s41467-019-09982-5
54. Zhu Q, Liu N, Orkin SH et al (2019) CUT&RUNTools: a flexible pipeline for

CUT&RUN processing and footprint analysis. Genome Biol 20(1):192. https://doi.org/10.1186/s13059-019-1802-4

55. Meers MP, Tenenbaum D, Henikoff S (2019) Peak calling by Sparse enrichment analysis for CUT&RUN chromatin profiling. Epigenetics Chromatin 12:42. https://doi.org/10.1186/s13072-019-0287-4
56. Meers MP, Bryson TD, Henikoff JG et al (2019) Improved CUT&RUN chromatin profiling tools. eLife 8:e46314. https://doi.org/10.7554/eLife.46314
57. Adey A, Morrison HG, Asan et al (2010) Rapid, low-input, low-bias construction of shotgun fragment libraries by high-density in vitro transposition. Genome Biol 11, R119. https://doi.org/10.1186/gb-2010-11-12-r119
58. Ewels PA, Peltzer A, Fillinger S et al (2020) The nf-core framework for community-curated bioinformatics pipelines. Nat Biotechnol 38:276–278. https://doi.org/10.1038/s41587-020-0439-x
59. Patel H, Wang C, Ewels P et al (2019) nf-core/chipseq: nf-core/chipseq v1.2.1—Platinum mole (Version 1.2.1). Zenodo. https://doi.org/10.5281/zenodo.3966161
60. Di Tommaso P, Chatzou M, Floden E et al (2017) Nextflow enables reproducible computational workflows. Nat Biotechnol 35:316–319. https://doi.org/10.1038/nbt.3820
61. Afgan E, Baker D, Batut B et al (2018) The Galaxy platform for accessible reproducible and collaborative biomedical analyses: 2018 update. Nucleic Acids Res 46(W1):W537–W544. https://doi.org/10.1093/nar/gky379
62. Jiang S, Mortazavi A (2018) Integrating ChIP-seq with other functional genomics data. Brief Funct Genomics 17(2):104–115. https://doi.org/10.1093/bfgp/ely002

Part VI

Discovery of Histone Methyltransferase Substrates and Methylated Histone Interactors

Chapter 14

Characterization of SET-Domain Histone Lysine Methyltransferase Substrates Using a Cofactor S-Adenosyl-L-Methionine Surrogate

Alexandre Désert, Karine Guitot, Audrey Michaud, Daniel Holoch, Raphaël Margueron, Fabienne Burlina, and Dominique Guianvarc'h

Abstract

Identification of histone lysine methyltransferase (HKMT) substrates has recently benefited from chemical-biology-based strategies in which artificial *S*-adenosyl-L-methionine (SAM) cofactors are engineered to allow substrate labeling using either the wild-type target enzyme or designed mutants. Once labeled, substrates can be selectively functionalized with an affinity tag, using a bioorthogonal ligation reaction, to allow their recovery from cell extracts and subsequent identification. In this chapter, we describe steps on how to proceed to set up such an approach to characterize substrates of specific HKMTs of the SET domain superfamily, from the characterization of the HKMT able to accommodate a SAM surrogate containing a bioorthogonal moiety, to the proteomic analysis conducted on a cell extract. We focus in particular on the controls that are necessary to ensure reliable proteomic data analysis. The example of PR-Set7 on which we have implemented this approach is shown.

Key words Histone lysine methyltransferase, *S*-adenosyl-L-methionine surrogate, Bioorthogonal chemistry, Proteomics

Abbreviations

ACN	Acetonitrile
BSA	Bovine serum albumin
CuAAC	Cu(I)-catalyzed alkyne-azide cycloaddition
DTT	Dithiothreitol
EDTA	Ethylenediamine tetraacetic acid
HKMT	Histone lysine methyltransferase
LC	Liquid chromatography
MALDI TOF	Matrix assisted laser desorption ionization time-of-flight
MS	Mass spectrometry
PEG	Polyethylene glycol
PTM	Post-translational modification
SAM	*S*-Adenosyl-L-methionine

Raphaël Margueron and Daniel Holoch (eds.), *Histone Methyltransferases: Methods and Protocols*, Methods in Molecular Biology, vol. 2529, https://doi.org/10.1007/978-1-0716-2481-4_14,

SDS-PAGE	Sodium dodecyl sulfate-polyacrylamide gel electrophoresis
SET	Su(var)3–9, Enhancer-of-zeste and Trithorax
TFA	Trifluoroacetic acid
THPTA	Tris(3-hydroxypropyltriazolylmethyl)amine
XIC	Extracted ion chromatogram

1 Introduction

Post-translational modification (PTM) of histones by methylation of lysine residues is a crucial chromatin modification that is known to influence many biological processes including cell-cycle regulation, development, or differentiation. It is tightly regulated in humans with over 50 enzymes named lysine methyltransferases (KMTs) controlling the position and degree of methylation. Most KMTs belong to the SET-domain containing protein family [1]. This domain (roughly 130 amino acids) is responsible for the transfer of the methyl group from the cofactor SAM (*S*-adenosyl-L-methionine) to the ε-amino group of the targeted lysyl residue. While the role of histone lysine methylation in the regulation of chromatin structure is now well documented, our knowledge of non-histone lysine methylation remains very incomplete and many nuclear and cytoplasmic substrates of human lysine methyltransferases remain uncharacterized [2]. Several HKMTs, including EZH2 and G9a/GLP, have been found to be involved in pathologies including cancer, and furthermore, numerous small molecule inhibitors have been developed to target HKMTs as attractive drug targets for cancer therapy [3]. An exhaustive characterization of the substrates of these methyltransferases is therefore important to better understand their physiological roles and assess their potential as therapeutic targets by mapping the pathways they regulate. Indeed, there is now evidence that specific lysine methylation of non-histone proteins is involved in crucial signaling events [4]. Until recently, research on lysine methylation on non-histone proteins was limited because there was no strategy to identify lysine methylation throughout the entire proteome, and substrates have often been found through candidate approaches. More recently, high-throughput approaches have also been developed through the use of peptide and protein arrays (*see* Chapter 15) [5]. The main challenge to overcome in the discovery of lysine methyltransferase substrates lies in the small size and inert chemical nature of the transferred methyl group. These features make it difficult to use classical biochemical approaches (i.e., 2D gels) for separating methylated substrates from their unmethylated counterparts. Indeed, detecting changes in methylation patterns has generally required using radioactively labeled SAM allowing the detection of methylated peptides or proteins in in vitro assays. Pan-methyl antibodies were also used, but they suffer from poor selectivity. The increased sensitivity of mass spectrometry and recently developed methods using SILAC and quantitative proteomics now make it more straightforward to identify sites of lysine methylation [6]. However, because this protein modification tends to occur at low abundance, approaches based on mass spectrometry are still limited by the lack of proper methods to enrich methylated peptides. Several technologies to map the protein methylome have emerged in the last few years [7]. In the last decade, chemical biology-based strategies have been successfully developed for the investigation of several posttranslational modifications (PTMs) [8–10]. For lysine methylation, artificial SAM-like molecules can be used to deliver chemical building blocks for labeling protein substrates. A series of analogs functionalized with a reactive bioorthogonal moiety (mainly alkyne or azide) have been described, and their synthesis usually requires only a few steps [11, 12]. These bulkier molecules are generally not well tolerated by native methyltransferases, and a specially engineered mutant protein methyltransferase is needed to transfer a chemical tag from a given SAM derivative to a target residue of the protein substrates. Once labeled, substrates can be selectively functionalized with an affinity tag, using a bioorthogonal ligation reaction, to allow their recovery and identification. This approach has been implemented by the group of M. Luo on several HKMTs and notably on GLP, which belongs to the SET-domain protein superfamily [13]. The catalytic SET domain being conserved, the approach can be transposed to other enzymes of the SET superfamily, the design of their mutants being relatively simple. The strategy described for GLP has the advantage of differentiating methylation by the enzyme of interest from background methylation by the other native endogenous enzymes, because only the engineered enzyme of interest is able to use the SAM analog. This ideal situation, where an orthogonal pair of engineered enzyme-SAM analogs can be used, greatly simplifies the proteomic study. However, the identification of such an orthogonal pair is not always possible. It is not absolutely required to allow a rigorous characterization of the substrates of a new HKMT of interest, but its absence imposes additional controls to be performed in the proteomic experiment.

Here, we describe the protocol we used to implement this substrate profiling strategy for the HKMT PR-Set7 to profile non-histone substrates using a SAM surrogate, named ProSeAM (propargylic *Se*-adenosyl-*L*-selenomethionine), bearing a propargyl moiety instead of the methyl to be transferred, and a selenium instead of the sulfur [14, 15]. The general procedure is presented in Fig. 1 and examples of SAM analogs are shown in Fig. 2. The ProSeAM is processed by the wild-type PR-Set7 and even better by a gain-of-function mutant PR-Set7 Y334F. These enzymes are thus able to transfer the propargyl group from ProSeAM to lysine 20 of

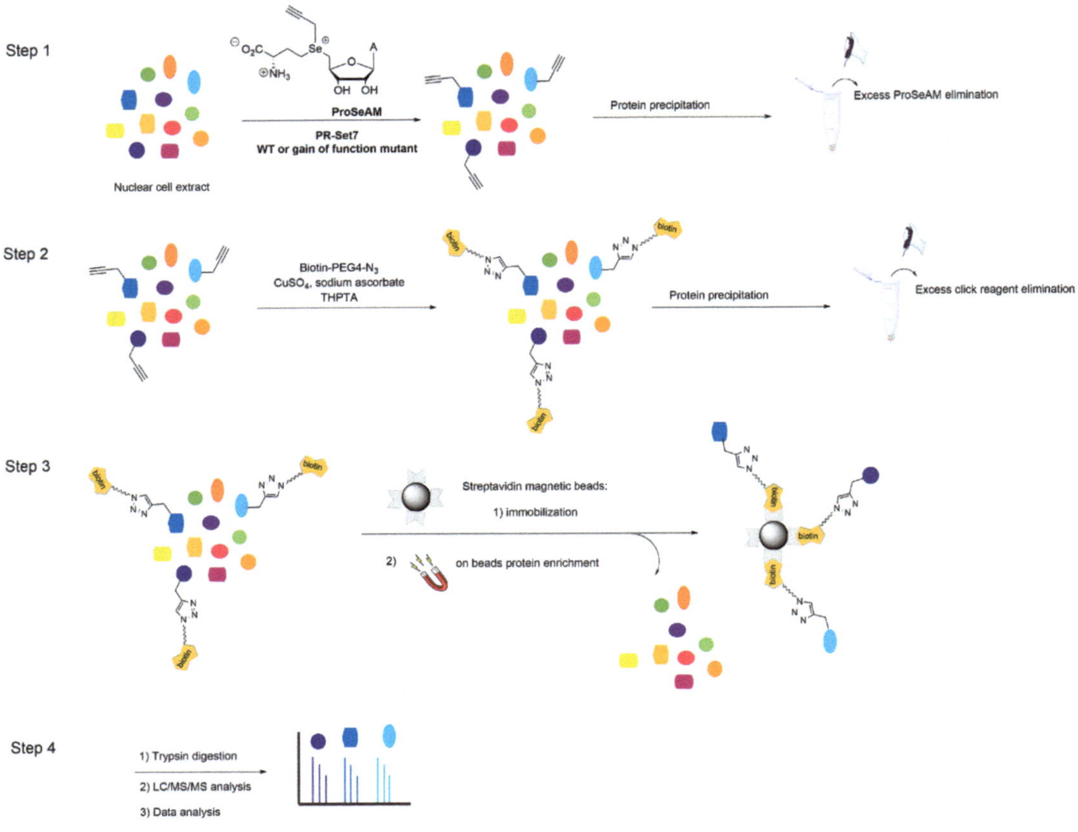

Fig. 1 Schematic description of the chemoenzymatic labeling strategy to identify new HKMT substrates

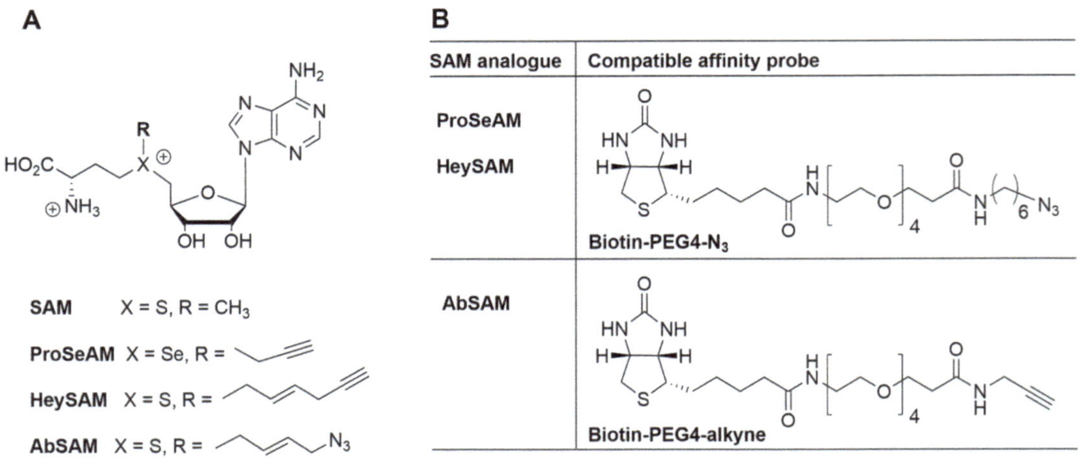

Fig. 2 (**a**) Native SAM and analogs cited in this study. (**b**) Complementary affinity probe used in the click reaction

histone H4 and potentially to their other substrates in cell extracts. Once installed on proteins, the alkyne moiety can then be selectively functionalized by an affinity probe such as a biotin with an appropriate linker bearing an azido group using click chemistry.

The tagged proteins can then be recaptured and enriched with streptavidin-coated magnetic beads and characterized by proteomic analysis. Depending on the studied enzyme, it is also possible to use mutants that are able to process larger cofactor analogs such as AbSAM or HeySAM (Fig. 2) [13, 16] which are not recognized by the wild-type enzyme, thus leading to higher enrichment rates. Such analogs however generally imply comprehensive studies (e.g., by molecular docking) for the design of the enzyme mutants, and the tolerance of these mutants remains difficult to predict, thus requiring in vitro assays before their application in proteomic studies.

To circumvent this potential pitfall, the use of the minimalist ProSeAM analog can be considered an alternative to mutant design since in several cases it does not require mutation of the HKMT of interest [15]. In this chapter, we will focus on the design and characterization of the most suitable enzyme/cofactor analog pair to perform the strategy. The technical precautions to be taken and the controls that are essential in a proteomic study using this methodology are discussed so that it can be implemented for other HKMTs.

2 Materials

1. PyMOL software.
2. *S*-adenosyl-L-methionine surrogate-ProSeAM:

   Chemically synthesized in house as previously reported [14] (*see* **Notes 1** and **2**). Into the resulting concentrated solution, a small amount of trifluoroacetic acid (TFA) was added to adjust pH to ~2. The concentration of ProSeAM was determined by its UV absorption ($\varepsilon_{260} = 15{,}400\ M^{-1}\ cm^{-1}$). The stock solution (2 mM) is divided into several small aliquots to prevent freezing/thawing processes that could cause degradation of the cofactor analog. The aliquots are stored at −20 °C.
3. Six-histidine-tagged PR-Set7 human histone-lysine *N*-methyltransferase and its mutant:

   Obtained from an overproducing strain (*E. coli* BL21 (DE3)/pLysS/pET28b(+)H6-setd7).
4. Methyltransferase substrate: Nucleosome core particles [17].
5. Glycine reaction buffer: 50 mM glycine, pH 9.8, 2 mM dithiothreitol, 25 μg/mL BSA, 10% glycerol.
6. Western blot reagents and buffers:
   (a) 5× SDS loading buffer: 10% SDS, 250 mM Tris-HCl pH 6.8, 50% glycerol, 0.25% bromophenol blue, 10% β-mercaptoethanol.
   (b) 4–15% polyacrylamide gel (Biorad).

(c) Running buffer 1×: 25 mM Tris base, 192 mM glycine, 0.1% SDS.

(d) PVDF membrane.

(e) Transfer buffer 1×: 25 mM Tris base, 192 mM glycine, 20% methanol.

(f) Ponceau staining solution: 0.5% (w/v) Ponceau S, 1% acetic acid.

(g) PBST: 1× PBS (137 mM NaCl, 2.7 mM KCl 10 mM Na_2HPO_4, 1.76 mM KH_2PO_4), 0.1% Tween.

(h) Blocking buffer: PBST supplemented with 3% BSA.

(i) 1:5000 Streptavidin–HRP conjugate solution: dilute streptavidin horseradish peroxidase stock solution (GE Healthcare) just before use in PBST 1×, 0.2% BSA.

(j) SuperSignal West Pico PLUS Chemiluminescent substrate was from Thermo Fisher Scientific.

(k) ChemiDoc Imaging System (Biorad).

7. Human embryonic kidney (HEK)-293T nuclear extracts:

   The cell line was obtained from Invitrogen and grown according to the manufacturer's instructions. Cells were incubated with 200 μL buffer A (10 mM HEPES pH 7.9, 2.5 mM $MgCl_2$, 0.25 M sucrose, 0.1% NP-40, 0.5 mM DTT, 1 mM PSMF) for 10 min on ice, centrifuged at 7000 × *g* for 10 min, resuspended in 200 μL buffer B (25 mM HEPES pH 7.9, 1.5 mM $MgCl_2$, 700 mM NaCl, 0.5 mM DTT, 0.1 mM EDTA, 20% glycerol), sonicated and centrifuged at 21,000 × *g* for 15 min.

8. Methanol.

9. Cu(I)-catalyzed alkyne-azide cycloaddition (CuAAC) reagents:

   (a) Tris(3-hydroxypropyltriazolylmethyl)amine (THPTA) was from Sigma Aldrich. A 200 mM stock solution in MilliQ water was prepared and stored at −20 °C.

   (b) $CuSO_4$ was from Sigma Aldrich. A 100 mM stock solution in MilliQ water was prepared just before use.

   (c) Sodium ascorbate was from Sigma Aldrich. A 100 mM stock solution in water was prepared just before use.

   (d) Biotin-PEG4-N_3 (100 μM) was from Thermo Fisher Scientific. A 5 mM solution in dimethylsulfoxide (DMSO) was prepared and stored at −20 °C.

10. Reagents and buffers for protein enrichment:

    (a) Resuspension buffer 1: 1× PBS (137 mM NaCl, 2.7 mM KCl 10 mM Na_2HPO_4, 1.76 mM KH_2PO_4), 0.5% SDS.

    (b) Dynabeads M-280 streptavidin magnetic beads with a streptavidin monolayer covalently coupled to the surface are from Thermo Fisher Scientific.

(c) PBST: 1× PBS (137 mM NaCl, 2.7 mM KCl, 10 mM Na_2HPO_4, 1.76 mM KH_2PO_4), 0.1% Tween.
(d) Resuspension buffer 2: PBST supplemented with 3% SDS, 10 mM DTT, 1 mM EDTA.
(e) PBS supplemented with 500 mM NaCl.
(f) PBS supplemented with 1% NP-40.
(g) PBS supplemented with 4 M urea.

11. Reagents and materials for sample preparation for mass spectrometry:
(a) 25 mM NH_4HCO_3 pH 8.
(b) Trypsin/LysC (Promega).
(c) Homemade C18 StageTip.
(d) Water/acetonitrile (ACN)/formic acid, 60:40:0.1 mixture.
(e) Vacuum concentrator.
(f) 0.3% trifluoroacetic acid (TFA) in water.

12. Liquid chromatography/tandem mass spectrometry setup:
RSLCnano system coupled online to an Orbitrap Exploris 480 mass spectrometer (Thermo Scientific) or an Ultimate3000 nano rapid separation LC system (Dionex) coupled to an LTQ 92 Velos mass spectrometer (Thermo Fisher Scientific) or similar instrument.

13. Software for mass spectrometry data processing:
(a) Proteome Discoverer (version 2.2).
(b) myProMS (version 3.9.2).
(c) MassChroQ (version 2.2.1).

3 Methods

The use of ProSeAM, due to the small size of the methyl surrogate (i.e., a propargyl group), has the advantage of being well tolerated by a subset of wild-type methyltransferases, and the labeling protocol can thus be carried out without the need to design HKMT mutants [15]. However, it should be kept in mind that strict controls will have to be carried out during the proteomic study because of the lack of specificity of this cofactor for the studied methyltransferase. If using ProSeAM, go directly to Subheading 3.2. If using bulky SAM analogs such as HeySAM or AbSAM, start at Subheading 3.1.

3.1 Design and Characterization of the HKMT Mutants

Mutants capable of handling large methyl surrogates which are not recognized by the wild-type target enzyme and other methyltransferases present in cell extracts can be designed, but it is difficult to predict the mutation that will allow proper handling of these SAM analogs depending on the HKMT being studied. However, given the relatively well-conserved SET domain of HKMTs, it is possible to build on previous results obtained by Luo's team with GLP that was already engineered to accept several SAM analogs [13, 18]. Thus, by superimposing with PyMOL software the ternary structure of the catalytic domain of the HKMT of interest (if known) with the structure of GLP (code PDB 2RFI [19]), amino acids of interest can be identified for site-directed mutagenesis. Different aromatic amino acids of the active site (especially amino acids which are superimposed with F1209 and Y1211 in GLP) can then be replaced with smaller amino acids (generally an alanine or a glycine). This should create HKMT variants able to accept non-natural SAM analogs by enlarging the SAM-binding cavity, without impairing binding of the histone tail and thereby potentially compromising substrate specificity.

3.2 Evaluation of the Processing of SAM Analogs by the Wild-Type or Engineered HKMT

In order to validate the transfer of the methyl surrogate from cofactor analog by the enzyme, regardless of whether it is mutated, a preliminary enzymatic assay must be carried out on a known substrate of the HKMT. Depending on the robustness of the enzyme in an in vitro assay, this substrate could be a peptide containing the target histone sequence of the studied HKMT. The functionalization can then be characterized by mass spectrometry, which is practical for a robust enzyme such as GLP [13]. For validation on peptides, MALDI-TOF MS analysis is particularly suitable [20]. The substrate can also be a tetramer or octamer of histones or a reconstituted nucleosome. The latter was used to validate the enzyme/SAM analog pair in the context of our study with PR-Set7. The characterization of the functionalization should then be done by Western blot analysis after a reaction of CuAAC with a biotin probe complementary to the chemical tag deposited on the protein (Fig. 2) and HRP-streptavidin readout.

1. Incubate 5 μg of PR-Set7 (WT or mutant) with 5 μg nucleosome core particles [17] and 100 μM ProSeAM (or other cofactor analogs) for 3 h at 30 °C in Glycine reaction buffer (*see* **Notes 2–4**).
2. Add the "click" reagents in the following order: 2 mM $CuSO_4$, 400 μM THPTA, 4 mM sodium ascorbate, and finally 100 μM biotine-PEG4-azide [or the appropriate probe functionalized with biotin depending on the cofactor analog used (*see* **Note 5**)] to the resuspended proteins.
3. Gently shake (300 rpm) for 2 h at 30 °C in the dark.
4. Stop the reaction by adding 0.25 volumes of 5× SDS loading buffer to the sample.
5. Denature the proteins by heating at 95 °C for 5 min.

6. Deposit the samples on a 4–15% polyacrylamide gel and perform the protein migration applying a 300 V potential for 15 min.
7. Transfer the protein onto a PVDF membrane and verify the transfer with Ponceau staining.
8. Rinse the membrane once with PBST and then incubate in Blocking buffer at room temperature for 1 h on an orbital shaker. Rinse once with PBST.
9. Incubate in 1:5000 Streptavidin–HRP conjugate solution for 1 h with gentle agitation.
10. Develop with SuperSignal West Pico PLUS Chemiluminescent substrate according to the manufacturer's instructions. The chemiluminescence was visualized with a ChemiDoc Imaging System.
11. Select the appropriate enzyme/cofactor analog pair leading to the best signal in comparison with the controls without enzyme. In our study, the pairs ProSeAM/PR-Set7 WT and, to a greater extent, ProSeAM/PR-Set7 Y334F, allowed the most robust labeling of H4 (*see* **Note 6**).

3.3 Enzymatic Functionalization with SAM Analog in the Presence of Cell Extracts

Once the SAM analog/enzyme derivative pairs have been validated on a known substrate of the HKMT of interest, they can be applied to cell extracts for substrate labeling. Carry out the different enzymatic assays and controls in five replicates (*see* **Note 7**).

1. Mix 20 μg of PR-Set7 with 200 μg of nuclear extract and 100 μM ProSeAM for 3 h at 30 °C in 200 μL Glycine reaction buffer (*see* **Note 4**).
2. Stop the reaction by adding four volumes of cold methanol.
3. Precipitate the proteins overnight at −80 °C.
4. Centrifuge at 20,000 × *g* for 30 min at 4 °C.
5. Remove the supernatant without disturbing the pellet.
6. Dissolve the pellet in 200 μL Resuspension buffer 1.

3.4 Click Chemistry Reaction

1. Add the "click" mixture (4 μL of a 5 mM biotine-PEG4-azide stock solution, 4 μL of a 100 mM $CuSO_4$ stock solution, 8 μL of a 100 mM sodium ascorbate stock solution and 0.5 μL of a 200 mM THPTA stock solution) to the resuspended proteins.
2. Gently shake (300 rpm) for 2 h at 30 °C in the dark.
3. Stop the reaction by adding four volumes of cold methanol.
4. Precipitate the proteins overnight at −80 °C.
5. Centrifuge at 20,000 × *g* for 30 min at 4 °C.
6. Remove the supernatant without disturbing the pellet.

3.5 On-Bead Protein Enrichment

1. Simultaneously with the protein precipitation step following the click chemistry reaction, place in a microtube (1.5-mL capacity) 100 μL of the commercial solution of streptavidin-coated magnetic beads (Dynabeads M-280) and immobilize the beads using a magnet. Remove the supernatant and wash the beads once with 500 μL 1× PBST.
2. Incubate the beads overnight with 500 μL Blocking solution with gentle agitation on a slowly rotating wheel.
3. Resuspend the previously precipitated proteins (*see* Subheading 3.4) in 100 μL Resuspension buffer 2.
4. Wash the magnetic beads twice with 200 μL 1× PBST, then resuspend the beads in 200 μL Resuspension buffer 2 (to yield a 5 mg/mL bead suspension) and mix with the solubilized proteins, incubate 2 h at 4 °C with up and down agitation.
5. Wash the magnetic beads three times with 500 μL of each of the following buffers: (1) PBST, (2) PBS 500 mM NaCl, (3) PBS 1% NP-40, (4) PBS 4 M urea.
6. Wash the magnetic beads once with 500 μL PBST, once with 500 μL PBS and twice with 500 μL 25 mM NH_4HCO_3 pH 8 (*see* **Note 8**).

3.6 On-Bead Protein Digestion and Mass Spectrometry

1. Resuspend the magnetic beads in 200 μL 25 mM NH_4HCO_3 pH 8.
2. Digest samples on the beads by adding 0.2 μg of trypsin/LysC for 1 h at 37 °C.
3. Load the resulting peptide mixtures onto a homemade C18 StageTip for desalting.
4. Elute the peptides from the beads using a 60:40:0.1 mixture of water/ACN/formic acid.
5. Concentrate under vacuum to dryness and resuspend in 10 μL 0.3% TFA in water prior to LC-MS/MS analysis.
6. Analyze eluted peptides by liquid chromatography–tandem mass spectrometry (LC-MS/MS).

3.7 Data Processing and Label-Free Quantification

For protein identification, the data were searched against the *Homo sapiens* one gene one protein (UP000005640_9606) UniProt database using Sequest HT through Proteome Discoverer (version 2.2). Enzyme specificity was set to trypsin and a maximum of two missed cleavages sites were allowed. Oxidized methionine, *N*-terminal acetylation, and loss of *N*-terminal methionine were set as variable modifications. Maximum allowed mass deviation was set to 10 ppm for monoisotopic precursor ions and 0.02 Da for MS/MS peaks.

The resulting files were further processed using myProMS (version 3.9.2). FDR calculation used Percolator and was set to 1% at the peptide level for the whole study. The label-free quantification was performed by peptide Extracted Ion Chromatograms (XICs) computed with MassChroQ (version 2.2.1) [21]. For protein quantification, XICs from proteotypic peptides shared between compared conditions (TopN matching) and missed cleavages were allowed. Median and scale normalization was applied on the total signal to correct the XICs for each biological replicate.

To estimate the significance of the change in protein abundance, a statistical test based on a linear model adjusted on peptides and biological replicates was performed, and *p*-values were adjusted using the Benjamini–Hochberg FDR procedure with a control threshold set to 0.05.

3.8 Proteomic Data Analysis

The data analysis consists of several comparisons between pairs of conditions, each performed in five replicates to establish lists of enriched proteins. A protein is significantly enriched in a condition if its fold change (FC) is greater than 2, and its adjusted *p*-value is less than 0.05.

The method described below is suitable in cases in which ProSeAM is used as a cofactor and presents the PR-Set7 study in which both PR-Set7 WT and gain-of-function mutant were able to process ProSeAM (*see* **Note 9**).

1. Compare the results obtained with the conditions "PR-Set7 gain-of-function mutant" and "no PR-Set7" (Fig. 3, green arrow) to obtain protein list A (*see* **Note 10**), the first list of enriched proteins with the gain-of-function mutant.
2. Compare the results obtained with the conditions "PR-Set7 gain-of-function mutant" and "PR-Set7 WT" (Fig. 3, blue arrow) to obtain protein list B (*see* **Note 11**), the second list of enriched proteins with the gain-of-function mutant.
3. Compare the results obtained with the conditions "PR-Set7 gain-of-function mutant" and "PR-Set7 loss of function mutant" (Fig. 3, red arrow) to obtain protein list C (*see* **Note 12**), the third list of enriched proteins with the gain-of-function mutant.
4. Combine the three lists of potential substrates and keep the proteins found in all three lists to obtain the maximum of PR-Set7 potential substrates (*see* **Notes 13–16**).

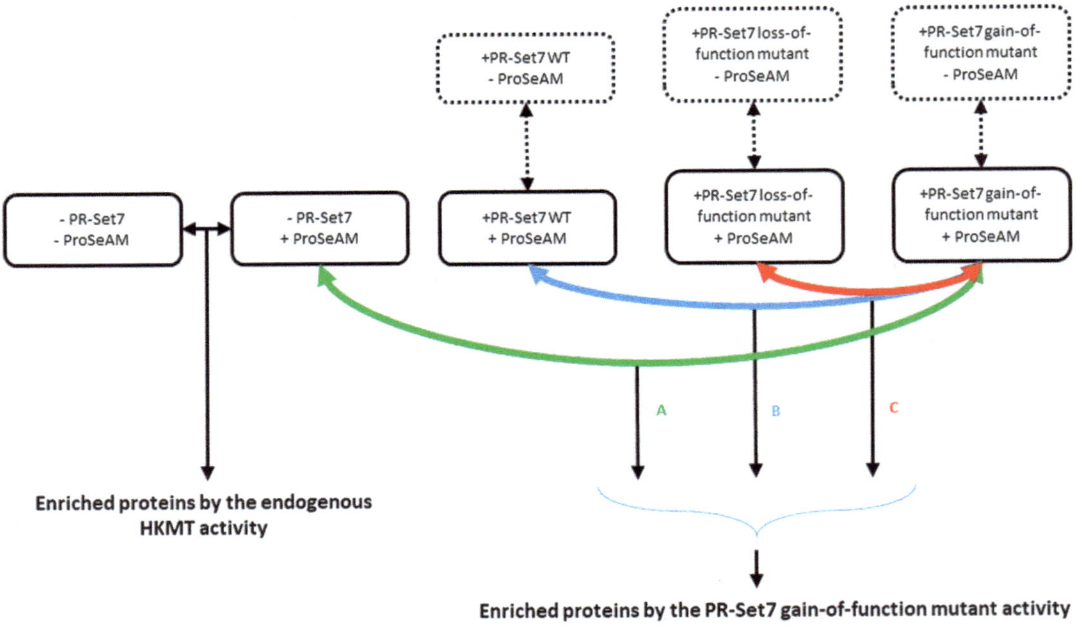

Fig. 3 Schematic workflow for the analysis of the proteomics data with different controls

4 Notes

1. Since ProSeAM is not commercially available, it requires several synthetic steps, but its synthesis is described in the literature [14]. Biologists with no expertise in organic chemistry should call upon a chemist collaborator or a custom synthesis company. Alternatively, it can be provided by Guianvarc'h's Lab under material transfer agreement.
2. Alternatively, bulky SAM analogs have previously been used to label substrates of specific protein methyltransferases engineered to process these cofactor analogs. It would be advantageous to use them if a mutant of the studied HKMT was designed that is able to use these cofactors. Among them, AbSAM [22] with an azidobutenyl group or HeySAM [18] with a hexenynyl group (Fig. 2) are efficient cofactors that are compatible for example with the GLP Y1211A mutant. These analogs can be synthesized according to protocols described in the literature in one step from commercially available *S*-adenosyl-L-homocysteine (SAH).
3. Alternatively, depending on the studied HKMT, 2.5 μg of histone tetramer or octamer should be used.
4. These conditions were optimized for PR-Set7 in this study, but the buffer should be considered on a case-by-case basis depending on the studied HKMT.

5. The CuAAC reaction involves the reaction between an azide and an alkyne forming a covalent chemical bond (a triazole). If the studied enzyme transfers an azide moiety from the SAM analog, then the affinity probe should be equipped with an alkyne function such as Biotin-PEG4-alkyne; if the enzyme transfers an alkyne moiety from the SAM analog, then the affinity probe should be equipped with an azide function such as Biotin-PEG4-N_3 (Fig. 2).
6. PR-Set7 and its mutants cannot process bulky SAM analogs, unlike GLP Y1211A.
7. If using ProSeAM and wild-type or gain-of-function enzyme as in the case of PR-Set7, several conditions should be included for a reliable characterization of enriched proteins: (1) without PR-Set7, (2) with PR-Set7 WT, (3) with PR-Set7 gain-of-function mutant, (4) with PR-Set7 loss-of-function mutant. Each of these four conditions should be done with or without ProSeAM, leading to a total of eight conditions.

   If using a HKMT mutant that is able to accommodate a bulky SAM analog that is incompatible with the native enzyme, which represents the ideal configuration of an orthogonal enzyme/cofactor pair, only four conditions are necessary: HKMT WT and mutant, with or without cofactor surrogate.
8. The beads should be trypsin-digested immediately after washing.
9. If an orthogonal pair of bulky cofactor/engineered enzyme was characterized (as in the case of GLP), a single comparison of two conditions (HKMT WT vs mutant) in the presence of cofactor will be sufficient to get a list of enriched proteins.
10. This comparison is expected to give a first list of proteins enriched by the action of PR-Set7 gain-of-function. Due to the large quantity of PR-Set7 added to the sample, the difference in the total amount of protein incubated with the magnetic beads is significant, and the identification of proteins by this comparison must therefore be supported by additional comparisons.
11. This comparison is expected to give a second list of proteins enriched by the action of PR-Set7 gain-of-function. Since ProSeAM is also accommodated to a lesser extent by PR-Set7 WT, this comparison leads to few enriched proteins.
12. This comparison is expected to give a third list of proteins enriched by the action of PR-Set7 gain-of-function and is generally the best control. It could be tempting to only use this condition to characterize PR-Set7 substrates but we found that the other conditions are useful to confirm the most relevant substrates.

13. We found H4 enriched in all three lists, which is a good indicator of confidence in the analysis performed. The known substrates of the studied HKMT should be retrieved in these lists.
14. The comparison between the two conditions (with or without cofactor) in the absence of the exogenous protein PR-Set7 (Fig. 3, black arrow) is expected to give a long list of enriched proteins since ProSeAM is a small cofactor that can be used by the many endogenous HKMTs in the cell extracts. A bulky cofactor analog with a better specificity for an appropriate mutant should instead lead to a short list of proteins for this comparison.
15. The comparison between enzymes with or without ProSeAM (Fig. 3, black dashed arrows) does not produce lists of relevant proteins with a non-exclusive cofactor/enzyme pair and can be omitted.
16. For the biological validation of the enriched proteins, several strategies could be envisioned depending on whether one wishes to adopt a biased (with some ideas for candidate substrates) or unbiased approach (the most enriched proteins are first selected).

Acknowledgments

We thank the Ministère de la Recherche for a grant to A.D. We also thank Damarys Loew and Berangere Lombard from the Institut Curie Mass Spectrometry and Proteomics facility for help with proteomic study. The present work was supported by l'Agence Nationale de la Recherche (AMetHist, ANR-17-CE12-0028).

References

1. Dillon SC, Zhang X, Trievel RC, Cheng X (2005) The SET-domain protein superfamily: protein lysine methyltransferases. Genome Biol 6:227. https://doi.org/10.1186/gb-2005-6-8-227
2. Zhang X, Huang Y, Shi X (2015) Emerging roles of lysine methylation on non-histone proteins. Cell Mol Life Sci 72:4257–4272. https://doi.org/10.1007/s00018-015-2001-4
3. Song Y, Wu F, Wu J (2016) Targeting histone methylation for cancer therapy: enzymes, inhibitors, biological activity and perspectives. J Hematol Oncol 9:49. https://doi.org/10.1186/s13045-016-0279-9
4. Carlson SM, Gozani O (2016) Nonhistone lysine methylation in the regulation of cancer pathways. Cold Spring Harb Perspect Med 6: a026435. https://doi.org/10.1101/cshperspect.a026435
5. Levy D (2019) Lysine methylation signaling of non-histone proteins in the nucleus. Cell Mol Life Sci 76:2873–2883. https://doi.org/10.1007/s00018-019-03142-0
6. Lund PJ, Lehman SM, Garcia BA (2019) Quantitative analysis of global protein lysine methylation by mass spectrometry. Methods Enzymol 626:475–498. https://doi.org/10.1016/bs.mie.2019.07.036
7. Carlson SM, Gozani O (2014) Emerging technologies to map the protein methylome. J Mol

Biol 426:3350–3362. https://doi.org/10.1016/j.jmb.2014.04.024

8. Luo M (2012) Current chemical biology approaches to interrogate protein methyltransferases. ACS Chem Biol 7:443–463. https://doi.org/10.1021/cb200519y
9. Chuh KN, Batt AR, Pratt MR (2016) Chemical methods for encoding and decoding of posttranslational modifications. Cell Chem Biol 23:86–107. https://doi.org/10.1016/j.chembiol.2015.11.006
10. Conibear AC (2020) Deciphering protein post-translational modifications using chemical biology tools. Nat Rev Chem 4:674–695. https://doi.org/10.1038/s41570-020-00223-8
11. Dalhoff C, Lukinavicius G, Klimasăuskas S, Weinhold E (2006) Direct transfer of extended groups from synthetic cofactors by DNA methyltransferases. Nat Chem Biol 2:31–32. https://doi.org/10.1038/nchembio754
12. Peters W, Willnow S, Duisken M, Kleine H, Macherey T, Duncan KE, Litchfield DW, Luescher B, Weinhold E (2010) Enzymatic site-specific functionalization of protein methyltransferase substrates with alkynes for click labeling. Angew Chem Int Ed Engl 49:5170–5173. https://doi.org/10.1002/anie.201001240
13. Islam K, Chen Y, Wu H, Bothwell IR, Blum GJ, Zeng H, Dong A, Zheng W, Min J, Deng H, Luo M (2013) Defining efficient enzyme-cofactor pairs for bioorthogonal profiling of protein methylation. Proc Natl Acad Sci U S A 110:16778–16783. https://doi.org/10.1073/pnas.1216365110
14. Willnow S, Martin M, Lüscher B, Weinhold E (2012) A selenium-based click AdoMet analogue for versatile substrate labeling with wild-type protein methyltransferases. Chembiochem 13:1167–1173. https://doi.org/10.1002/cbic.201100781
15. Bothwell IR, Islam K, Chen Y, Zheng W, Blum G, Deng H, Luo M (2012) Se-adenosyl-L-selenomethionine cofactor analogue as a reporter of protein methylation. J Am Chem Soc 134:14905–14912. https://doi.org/10.1021/ja304782r
16. Wang R, Islam K, Liu Y, Zheng W, Tang H, Lailler N, Blum G, Deng H, Luo M (2013) Profiling genome-wide chromatin methylation with engineered posttranslation apparatus within living cells. J Am Chem Soc 135:1048–1056. https://doi.org/10.1021/ja309412s
17. Dyer PN, Edayathumangalam RS, White CL, Bao Y, Chakravarthy S, Muthurajan UM, Luger K (2003) Reconstitution of nucleosome core particles from recombinant histones and DNA. Methods Enzymol 375:23–44. https://doi.org/10.1016/S0076-6879(03)75002-2
18. Islam K, Zheng W, Yu H, Deng H, Luo M (2011) Expanding cofactor repertoire of protein lysine methyltransferase for substrate labeling. ACS Chem Biol 6:679–684. https://doi.org/10.1021/cb2000567
19. Wu H, Min J, Lunin VV, Antoshenko T, Dombrovski L, Zeng H, Allali-Hassani A, Campagna-Slater V, Vedadi M, Arrowsmith CH, Plotnikov AN, Schapira M (2010) Structural biology of human H3K9 methyltransferases. PLoS One 5:e8570. https://doi.org/10.1371/journal.pone.0008570
20. Guitot K, Scarabelli S, Drujon T, Bolbach G, Amoura M, Burlina F, Jeltsch A, Sagan S, Guianvarc'h D (2014) Label-free measurement of histone lysine methyltransferases activity by matrix-assisted laser desorption/ionization time-of-flight mass spectrometry. Anal Biochem 456:25–31. https://doi.org/10.1016/j.ab.2014.04.006
21. Valot B, Langella O, Nano E, Zivy M (2011) MassChroQ: a versatile tool for mass spectrometry quantification. Proteomics 11:3572–3577. https://doi.org/10.1002/pmic.201100120
22. Islam K, Bothwell I, Chen Y, Sengelaub C, Wang R, Deng H, Luo M (2012) Bioorthogonal profiling of protein methylation using azido derivative of S-adenosyl-L-methionine. J Am Chem Soc 134:5909–5915. https://doi.org/10.1021/ja2118333

Chapter 15

Specificity Analysis of Protein Methyltransferases and Discovery of Novel Substrates Using SPOT Peptide Arrays

Sara Weirich and Albert Jeltsch

Abstract

Posttranslational methylation of amino acid side chains in proteins mainly occurs on lysine, arginine, glutamine, and histidine residues. It is introduced by different protein methyltransferases (PMTs) and regulates many aspects of protein function including stability, activity, localization, and protein/protein interactions. Although the biological effects of PMTs are mediated by their methylation substrates, the full substrate spectrum of most PMTs is not known. For many PMTs, their activity on a particular potential substrate depends, among other factors, on the peptide sequence containing the target residue for methylation. In this protocol, we describe the application of SPOT peptide arrays to investigate the substrate specificity of PMTs and identify novel substrates. Methylation of SPOT peptide arrays makes it possible to study the methylation of many different peptides in one experiment at reasonable costs and thereby provides detailed information about the specificity of the PMT under investigation. In these experiments, a known substrate sequence is used as template to design a SPOT peptide array containing peptides with single amino acid exchanges at all positions of the sequence. Methylation of the array with the PMT provides detailed preferences for each amino acid at each position in the substrate sequence, yielding a substrate sequence specificity profile. This information can then be used to identify novel potential PMT substrates by in silico data base searches. Methylation of novel substrate candidates can be validated in SPOT arrays at peptide level, followed by validation at protein level in vitro and in cells.

Key words Protein methyltransferase, Protein methylation, Substrate specificity, Peptide array, SPOT peptide synthesis

1 Introduction

In 1992, Roland Frank developed the SPOT array peptide synthesis method. With this method, multiple peptide chains with a maximum length of around 20 amino acids are synthesized on a cellulose membrane at defined positions [1]. The peptide chains fixed to the cellulose fibers membrane via a polyethylene glycol (PEG) linker can either be used as complete peptide array or be cleaved from the membrane and are then available as dissolved peptides

Raphaël Margueron and Daniel Holoch (eds.), *Histone Methyltransferases: Methods and Protocols*,
Methods in Molecular Biology, vol. 2529, https://doi.org/10.1007/978-1-0716-2481-4_15,

[2, 3]. In most cases, whole peptide arrays are preferred, since they are useful for a wide range of biochemical experiments, like antibody characterization, protein–protein interaction studies or investigation of peptide-modifying enzymes [4–9].

In the SPOT synthesis, a cellulose membrane is used with covalently bound PEG linkers with lengths of 8–10 ethylene glycol units. This linker ends with a free amino function which is used to couple the first amino acid, followed by the successive addition of the next amino acids. Peptide bond formation requires the activation of the carboxy group of the incoming amino acid, which is usually achieved by standard Fmoc chemistry [2, 3, 10, 11]. In contrast to the naturally produced peptide chains in ribosomes, the growth of the synthesized peptide chain starts at the C-terminus and proceeds toward the *N*-terminus. After the first amino acid is linked to the cellulose membrane, the synthesis follows a periodic process, where one amino acid after the other is attached to the C-terminal amino acid, upon building of a peptide bond. To prevent side chain reactions protection groups are necessary, which are divided in temporary and permanent protection groups. The α-amino group of the newly incorporated amino acids is protected by 9-fluorenyl-methoxycarbonyl (Fmoc). This ensures that in each step only a single building block is coupled to the growing peptide chain. After coupling to the peptide chain, the Fmoc protection group is removed to enable further chain growth. In comparison, the side chains of the amino acids contain permanent protection groups, which are only removed at the end of the synthesis. The SPOT peptide synthesis approach explained here uses an automated pipetting robot. Each single activated amino acid is delivered as a droplet of solvent to a defined spot on the cellulose membrane. The arrangement of multiple spots next to each other is called a peptide array. The size and the distance of the spots as well as the peptide length are variable. Using peptide arrays produced by the SPOT synthesis technology has the advantage that up to 400 peptides with different sequences are synthesized on a cellulose membrane comprising 10 × 15 cm in parallel and at comparably low cost. One important application is the identification of the substrate sequence specificity of protein methyltransferases (PMTs), the topic of this protocol.

Posttranslational methylation of amino acid side chains in proteins has been observed on lysine, arginine, glutamine, histidine, methionine, and cysteine residues [12, 13]. It regulates many aspects of protein function including stability, activity, localization, and protein/protein interactions, and it is introduced by different PMT families, which all employ S-Adenosyl-L-methionine (AdoMet) as donor of an activated methyl group. Among them, protein lysine methyltransferases (PKMTs) and protein arginine methyltransferases (PRMTs) are well studied, while protein glutamine methyltransferases (PQMTs) and protein histidine methyltransferases (PHMTs) are much less investigated. In the following, we will focus on the PKMTs, although the protocol described here is applicable to all types of PMTs.

PKMTs are enzymes that catalyze the methylation of lysine residues, widely found at the *N*-terminal tail of histones, but also in several non-histone proteins [14, 15]. Histone lysine methylation is involved in the regulation of gene expression and chromatin remodeling [14]. It also has important roles in human diseases like cancer [16], where the PKMTs are often found to contain somatic mutations [17]. Despite many examples documenting very important biological roles of protein lysine methylation, the analysis of the substrate spectrum of PKMTs is still in its infancy [13, 18–20]. Current proteomics approaches have detected several protein lysine methylation events, but in most instances, we are still not able to connect them directly with the specific PKMT responsible for their generation. The same is true, if we look in the opposite direction, because for most PKMTs the full spectrum of their cellular substrates is unknown. Hence, a huge demand exists for methods identifying PKMTs for known methylation events and more substrates for known PKMTs [20]. Therefore, finding ways to connect PKMTs with specific lysine methylation events in a systematic fashion is a technological challenge that many research groups are currently tackling. It is an essential research task, as the biological outcome of protein methylation is mediated by changes of critical properties of PKMT substrate proteins upon methylation, and therefore, the biological functions of PKMTs can only be fully understood in the context of their substrate profile.

The question if a particular protein can be lysine methylated by a given PKMT depends on several critical parameters: (a) both proteins must interact, (b) the region of the substrate protein containing the putative methylation target must be accessible and face the PKMT active site, and (c) the peptide segment containing the putative target lysine must fit into the active site pocket of the PKMT. Specificity analyses of PKMTs focus on the last condition. For this, peptide SPOT arrays were introduced as methylation substrates [21], because they make it possible to study the methylation of many different peptide substrates in one experiment at reasonable costs and thereby provide detailed information about the specificity of the PKMT under investigation (Fig. 1).

In this approach, a known substrate sequence is used as template to design a SPOT peptide array containing peptides with single amino acid exchanges at all positions of the sequence. Methylation of the array with the PKMT under investigation provides the detailed preferences for each amino acid at each position in the substrate sequence yielding a substrate specificity profile. This information is then used to identify potential PKMT substrates by in silico searches among the proteins shown to interact with the PKMT or among known lysine methylation sites in human proteins listed in Proteomics databases. Up to now, this method has been applied to 17 human and nonhuman PMTs (Table 1), and in most

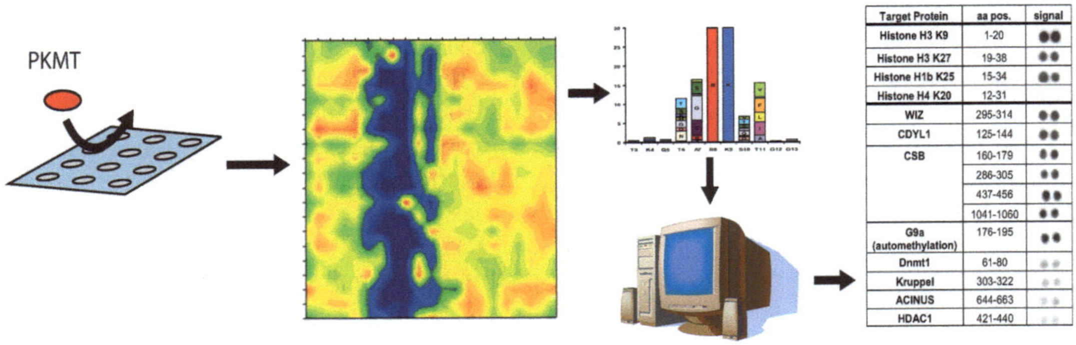

Fig. 1 General strategy for the identification of non-histone substrates of PKMTs, here illustrated with results obtained in the investigation of G9a [22]. Initially, a specificity analysis is performed using SPOT arrays. The specificity profile is combined with protein/protein interaction data to identify potential substrates. Methylation of these potential targets is then investigated at peptide and protein level first in vitro, then in vivo

Table 1
Summary of the application of specificity analysis of PMTs using peptide SPOT arrays. In the fourth column, the best known or initially described substrate of the corresponding PMT is listed

Enzyme	Type	Species	Substrate	Reference(s)
G9a	PKMT	Human	H3K9	[22]
Dim5	PKMT	*N. crassa*	H3K9	[21]
SET7/9	PKMT	Human	H3K4	[23]
SET8	PKMT	Human	H4K20	[24]
NSD1	PKMT	Human	H3K36	[8]
SUV39H2	PKMT	Human	H3K9	[25]
SMYD2	PKMT	Human	p53	[26, 27]
SUV420H1	PKMT	Human	H4K20	[28]
SUV420H2	PKMT	Human	H4K20	[28]
HEMK2	PQMT	Human	eRF1-Q235	[29]
SUV39H1	PKMT	Human	H3K9	[30]
Clr4	PKMT	*S. pombe*	H3K9	[31]
PRC2/EZH2	PKMT	Mouse	H3K27	[32]
RomA	PKMT	*L. pneumophila*	H3K14	[33]
METTL13	PKMT	Human	–	[34]
SETD2	PKMT	Human	H3K36	[35]
METTL9	PHMT	Human	–	[36]

instances, important specificity determinants were identified and novel PMT peptide and protein substrates were discovered. In addition, oriented peptide libraries have been applied to map the sequence preferences of PRMTs [37].

2 Materials

2.1 Synthesis of Peptide SPOT Arrays

1. MultiPep RS spotter (Intavis) operated by MultiPep Rsi Software (Intavis).
2. Amino-PEG cellulose membrane (Intavis).
3. *N,N*-Dimethylformamide.
4. 100% ethanol.
5. Fmoc-protected amino acids dissolved in *N*-Methyl-2-Pyrrolidone at final concentrations of 0.5 M (prepare freshly before setting up the synthesis).
6. Base: 1 M Oxyma Pure in *N*-Methyl-2-Pyrolidone.
7. Activator: 16% *N,N′*-Diisopropylcarboiimid in *N*-Methyl-2 Pyrolidone.
8. Capping mixture: 20% acetic acid in *N,N*-Dimethylformamide.
9. Piperidine: 20% piperidine in *N,N*-Dimethylformamide.
10. 1% bromophenol blue in *N,N*-Dimethylformamide.
11. Side-chain deprotection solution: 95% trifluoracetic acid, 3% triisopropylsilan, 2% ddH_2O in a chemically resistant box.
12. Dichloromethane.

2.2 Peptide Array Methylation

1. Active protein methyltransferases: can be commercially obtained, recombinantly expressed in bacteria and purified, or purified from eukaryotic cells.
2. Incubation buffer: depends on enzyme.
3. Methylation buffer: Incubation buffer supplemented with respective enzyme and labeled [methyl-^{3}H]-AdoMet (PerkinElmer).
4. Wash buffer: 1 mM NH_4HCO_3 supplemented with 1% SDS.
5. Amplify NAMP100V (GE Healthcare).
6. Hyperfilm™ high-performance autoradiography films (GE Healthcare).
7. Autoradiographic cassette.
8. −80 °C freezer.
9. Developing machine.

2.3 Data Analysis

1. Scanner.
2. Software for Densitometry (e.g., ImageJ [38]).
3. Microsoft Excel.

3 Methods

3.1 Design of the Substrate Sequence Specificity Array

1. To program the peptide sequences for the SPOT synthesis, use the MultiPep Rsi software.
2. Typically, 15 amino acid long peptide sequences with the target lysine in the middle are used as template sequence. Write the sequence in a single letter code, where each letter defines a specific amino acid.
3. Each amino acid of the template sequence can be exchanged by a selected amino acid of the natural available 20 amino acids with the specific command ".replace, A" (e.g., for an alanine walk screen).
4. Type the template sequence and finish the command with a semicolon.
5. Similar commands should be used for all naturally available amino acid residues (*see* **Note 1**).

3.2 Preparation of the MultiPep RS Spotter

1. Dissolve an appropriate amount of Fmoc-protected amino acids in *N*-Methyl-2 Pyrolidone to a final concentration of 0.5 M and transfer 5 ml of each into the corresponding amino acid vials in the derivative rack.
2. Fill up the 1-l diluter reservoir bottle for the syringe with synthesis grade *N,N*-Dimethylformamide.
3. Fill up the 2-l bottle with synthesis-grade *N,N*-Dimethylformamide and ethanol (*see* **Note 2**).
4. Prepare 30 ml base, 30 ml activator, 40 ml capping mixture, and 200 ml piperidine and place them at their defined positions in the MultiPep RS spotter machine (*see* **Note 3**).
5. Pre-wash the amino-PEG membrane in 20 ml *N,N*-Dimethylformamide for 15 min by shaking and transfer the membrane on the SPOT synthesizer frame (*see* **Note 4**).
6. Before spotting of the amino acids wash the membrane two times with ethanol.

3.3 Spotting of the First Amino Acid

1. Activate the carboxy group of the first amino acid by mixing 0.05 μl base, 0.05 μl activator, 0.001 μl *N*-Methyl-2-Pyrolidone with 0.1 μl amino acid for each coupling. The exact volumes will be calculated by the machine depending on the programmed peptide sequences and will include a small excess.

2. The programmable robot spots the desired volume (typically 0.2 μl) of the activated amino acid at the dedicated position onto the amino-PEG functionalized cellulose membrane and thereby generates an amide bond between the free amino group on the membrane and the activated C-terminus of the first amino acid.
3. Incubate the membrane for 10 min and repeat **step 2** (*see* **Note 5**).

3.4 Blocking of the Free Amino Acids

1. Flush the membrane with capping solution to block free amino groups between the spots and incubate for 5 min (*see* **Note 6**).
2. Wash the membrane eight times with *N,N*-Dimethylformamide and four times with ethanol.

3.5 Peptide Chain Elongation

1. Incubate the membrane with 20% piperidine for 5 min to remove the amino-protecting Fmoc groups of the coupled amino acid.
2. Repeat this step to improve the deprotection of the amino groups.
3. Wash the membrane eight times with *N,N*-Dimethylformamide, eight times with ethanol and let the membrane dry.
4. Activate the carboxy group of the next incoming amino acid (*see* Subheading 3.3 **step 1**) and perform coupling of this activated amino acid to the deprotected amino acid of the growing peptide chain coupled to the membrane.
5. Repeat Subheading 3.4, **steps 1** and **2**, and Subheading 3.5, **steps 1–4** until the desired peptide length is achieved (*see* **Note 7**).
6. After each five cycles, the MultiPep RS spotter machine makes a pause. This pause is used to refill the 1-l diluter reservoir bottle for the syringe with synthesis grade *N,N*-Dimethylformamide, the 2-l bottle with synthesis grade *N,N*-Dimethylformamide and ethanol, the piperidine and to prepare fresh capping solution. Afterwards restart the synthesis for the next five cycles.

3.6 Terminating the Peptide Synthesis and Quality Control

1. Optional: Terminate the last chain elongation by a capping step (*see* Subheading 3.4, **step 1**). This terminal acetylation of the free amine will generate a closer mimic of an internal peptide segment in a protein.
2. Wash the membrane first 10 times with *N,N*-Dimethylformamide and then 10 times with ethanol.
3. Let the membrane dry completely for 15 min.
4. Add 1% bromophenol blue solution manually on the membrane and incubate for 10 min. Afterwards use the

programmable robot again and start the program in the last cycle to repeat Subheading 3.5, **steps** 7 and **8**.

5. Take out the membrane from the synthesizer and mark the edges of the peptide array or individual peptide SPOT for further usage (*see* **Note 8**).

3.7 Side Chain Deprotection

1. Incubate the membrane in 30 ml side chain deprotection mixture for 1 h in a chemical-resistant box while shaking (*see* **Note 9**).
2. Wash the membrane five times for 2 min with 20 ml dichloromethane and two times for 2 min with ethanol.
3. Dry the membrane overnight at room temperature.
4. The membrane can be stored at room temperature for a few days and at −20 °C for long-term storage.

3.8 Peptide Array Methylation

1. Preincubate the peptide array in the respective incubation buffer, which has the same composition as the methylation buffer without enzyme and labeled [methyl-3H]-AdoMet, for 5 min on a shaker. The array should be completely covered with buffer (*see* **Note 10**).
2. Remove the incubation buffer and add methylation buffer, containing the respective enzyme and labeled [methyl-3H]-AdoMet (*see* **Note 11**).
3. Incubate the peptide array for at least 1 h at room temperature on a shaker.
4. Discard the methylation buffer in a radioactive waste container and wash the array 5 times for 5 min with 100 mM NH_4HCO_3 supplemented with 1% SDS to remove the bound protein. Dispose the wash solution into the radioactive waste.
5. After the last washing step, remove the washing buffer carefully and incubate the array in Amplify NAMP100V solution for 5 min.
6. Transfer the array into a sealed plastic bag and put it in an autoradiographic cassette.
7. Expose the peptide array to Hyperfilm™ high-performance autoradiography films in the dark at −80 °C for a defined time (*see* **Note 12**).
8. Afterwards develop the film and analyze the results.

3.9 Analysis of the Substrate Specificity Array

1. Scan the Hyperfilm™ high-performance autoradiography film with a conventional scanner and analyze the spot intensities by densiometry, for example, using ImageJ [38].

2. Subtract the background signal from each spot and normalize the intensities of each spot by setting the intensity of a reference substrate as 1.0 (*see* **Note 13**).
3. To quantify the signals and determine error margins, the peptide array methylation experiment needs to be repeated at least in triplicates. Data of different experiment can be normalized on the average activity after background correction. Average the normalized data for the individual spots and report them as mean values and standard deviations.
4. To indicate the activity, plot the data in 2D using a grayscale.
5. To analyze the quality of the data, prepare a plot of the distribution of the standard deviation of the repeated experiments.
6. Calculate the relative preference of the methyltransferase for each amino acid i at position x by a discrimination factor $D_{x;i} = \langle v_{j \neq i} \rangle / v_i - 1$.

   where v_i is the rate of methylation of the peptide variant carrying amino acid i at position x, and $\langle v_{j \neq i} \rangle$ is the average rate of methylation of all other peptides carrying a different amino acid $j \neq i$ at position x (including the wild-type sequence). This allows the accuracy of the recognition of each residue in the substrate peptide to be compared and displayed quantitatively.
7. Based on these data, a substrate sequence specificity motif for each methyltransferase can be created (*see* **Note 14**).

3.10 Search for Novel Histone Methyltransferase Substrates

1. After analyzing the substrate specificity profile and determining the substrate specificity motif, several proteomic databases are available to search for human proteins with known methylation sites like Phosphosite Plus [39] or proteins shown to interact with the PKMT (e.g., using String [40]). Proteins can be searched for the specificity profile using the Scansite database [41].
2. After identifying potential non-histone targets, design a new peptide array, which contains 15 amino acid long sequences of the potential novel targets with the predicted target lysine in the middle. To prove that the target lysine is methylated, it is important to include a matching lysine-to-alanine mutation next to each single non-histone substrate candidate.
3. Perform peptide array methylation as described previously (*see* Subheading 3.8) (*see* **Note 15**).
4. After identification of novel peptide substrates, their methylation must be confirmed at the protein level. For this, the corresponding proteins (or protein domains) and the corresponding K-to-A mutant proteins must be purified and incubated with PKMT in buffer containing radioactively

labeled AdoMet (*see* **Note 16**). The mixtures are then separated by SDS-PAGE. Afterwards, the gel is incubated with Amplify NAMP100V solution and treated as described above in (*see* Subheading 3.8, **steps 6–8**).

4 Notes

1. Note that Cysteine can be oxidized and Tryptophan suffers from lower synthesis yields. Array may be prepared by omitting these residues.
2. Make sure that the filter of the aspiration tubes reaches the bottom of the bottles.
3. The required amounts of base, activator, capping mixture, and piperidine always depend on the synthesis scale.
4. To check if the pump is working and to be sure that no air bubbles or wrinkles are produced, add 100% ethanol and start the pump.
5. Repetition of this step is important, since it ensures better coupling of the first activated amino acid.
6. This step is important to block the amino groups on the membrane between the spots and also the amino groups within the spots that did not form a bond with the amino acid. This will avoid the coupling of amino acids in further cycles with the membrane instead of the growing peptide chain.
7. Addition of each amino acid is called a cycle.
8. Bromophenol blue staining can be used to confirm the successful synthesis of peptides. The intensity of the stained spot can vary depending on the peptide sequence and the content of modified amino acids.
9. 95% trifluoracetic acid cleaves the side chain protection groups while 2% water and 3% triisopropylsilan protect the side chains of amino acids from modifications.
10. The volume of the buffer in Subheading 3.8, **steps 1** and **2**, depends on the size of the array. Perform the peptide array methylation in a box with suitable size or in a sealed plastic bag to avoid the usage of excess enzyme and radioactively-labeled AdoMet. The composition of the buffer depends on the respective PKMT. The concentration of the PKMT and exposure time (*see* **Note 12**) depend on the activity of the PKMT. Optimal conditions should be established beforehand using small arrays containing only a few substrate and negative control spots.
11. Radioactive reagents are potentially dangerous. Follow all local rules regarding work with radioactive compound, self-

protection and waste disposal. For environmental reasons, keep the amount of radioactive AdoMet used per experiment as low as possible.

12. Choose several exposure times to avoid saturation of strongly methylated peptide substrates. The exposure times must be optimized for each preparation of a PMT.
13. It can be necessary to subtract nonspecific activity due to protein or AdoMet binding to the peptide spots. Potential background signals not due to enzymatic methylation can be determined at peptide spots in which the target lysine has been replaced by alanine.
14. Lysine residues placed at other positions of the target sequence can create additional methylation sites. PKMTs can also methylate cysteine residues [25].
15. Always include the template sequence as control at the beginning and end of the array. This will help to evaluate the signal intensity of the novel targets and also if a consistent methylation all over the array took place. Potential substrates should have a strong methylation signal and show a loss of methylation in the corresponding K-to-A mutant.
16. Many PKMTs show automethylation. To differentiate the automethylation signal of the enzyme and the methylation signal of the targets in protein methylation analysis by SDS-PAGE, always include one control which contains the enzyme without the addition of any target.

References

1. Frank R (1992) Spot-synthesis: an easy technique for the positionally addressable, parallel chemical synthesis on a membrane support. Tetrahedron 48(42):9217–9232. https://doi.org/10.1016/S0040-4020(01)85612-X
2. Hilpert K, Winkler DF, Hancock RE (2007) Peptide arrays on cellulose support: SPOT synthesis, a time and cost efficient method for synthesis of large numbers of peptides in a parallel and addressable fashion. Nat Protoc 2(6):1333–1349. https://doi.org/10.1038/nprot.2007.160
3. Winkler DF, Hilpert K, Brandt O, Hancock RE (2009) Synthesis of peptide arrays using SPOT-technology and the CelluSpots-method. Methods Mol Biol 570:157–174. https://doi.org/10.1007/978-1-60327-394-7_5
4. Leung GC, Murphy JM, Briant D, Sicheri F (2009) Characterization of kinase target phosphorylation consensus motifs using peptide SPOT arrays. Methods Mol Biol 570:187–195. https://doi.org/10.1007/978-1-60327-394-7_7
5. Reineke U, Volkmer-Engert R, Schneider-Mergener J (2001) Applications of peptide arrays prepared by the SPOT-technology. Curr Opin Biotechnol 12(1):59–64. https://doi.org/10.1016/s0958-1669(00)00178-6
6. Winkler DF, Andresen H, Hilpert K (2011) SPOT synthesis as a tool to study protein-protein interactions. Methods Mol Biol 723:105–127. https://doi.org/10.1007/978-1-61779-043-0_8
7. Bock I, Kudithipudi S, Tamas R, Kungulovski G, Dhayalan A, Jeltsch A (2011) Application of Celluspots peptide arrays for the analysis of the binding specificity of epigenetic reading domains to modified histone tails. BMC Biochem 12:48. https://doi.org/10.1186/1471-2091-12-48
8. Kudithipudi S, Lungu C, Rathert P, Happel N, Jeltsch A (2014) Substrate specificity analysis and novel substrates of the protein lysine

methyltransferase NSD1. Chem Biol 21(2): 226–237. https://doi.org/10.1016/j.chembiol.2013.10.016

9. Kungulovski G, Kycia I, Mauser R, Jeltsch A (2015) Specificity analysis of histone modification-specific antibodies or reading domains on histone peptide arrays. Methods Mol Biol 1348:275–284. https://doi.org/10.1007/978-1-4939-2999-3_24
10. Frank R (2002) The SPOT-synthesis technique. Synthetic peptide arrays on membrane supports--principles and applications. J Immunol Methods 267(1):13–26. https://doi.org/10.1016/s0022-1759(02)00137-0
11. Li SS, Wu C (2009) Using peptide array to identify binding motifs and interaction networks for modular domains. Methods Mol Biol 570:67–76. https://doi.org/10.1007/978-1-60327-394-7_3
12. Walsh CT, Garneau-Tsodikova S, Gatto GJ Jr (2005) Protein posttranslational modifications: the chemistry of proteome diversifications. Angew Chem Int Ed 44(45):7342–7372. https://doi.org/10.1002/anie.200501023
13. Clarke SG (2013) Protein methylation at the surface and buried deep: thinking outside the histone box. Trends Biochem Sci 38(5): 243–252. https://doi.org/10.1016/j.tibs.2013.02.004
14. Bannister AJ, Kouzarides T (2011) Regulation of chromatin by histone modifications. Cell Res 21(3):381–395. https://doi.org/10.1038/cr.2011.22
15. Zhao Y, Garcia BA (2015) Comprehensive catalog of currently documented histone modifications. Cold Spring Harb Perspect Biol 7(9): a025064. https://doi.org/10.1101/cshperspect.a025064
16. Greer EL, Shi Y (2012) Histone methylation: a dynamic mark in health, disease and inheritance. Nat Rev Genet 13(5):343–357. https://doi.org/10.1038/nrg3173
17. Weirich S, Jeltsch A (2017) Mutations in histone lysine methyltransferases and demethylases. In: Encyclopedia of cancer. Elsevier 538–550. https://doi.org/10.1016/B978-0-12-801238-3.65056-0
18. Zhang X, Huang Y, Shi X (2015) Emerging roles of lysine methylation on non-histone proteins. Cell Mol Life Sci 72(22):4257–4272. https://doi.org/10.1007/s00018-015-2001-4
19. Biggar KK, Li SSC (2015) Non-histone protein methylation as a regulator of cellular signalling and function. Nat Rev Mol Cell Biol 16(1):5–17. https://doi.org/10.1038/nrm3915
20. Kudithipudi S, Jeltsch A (2016) Approaches and guidelines for the identification of novel substrates of protein lysine methyltransferases. Cell Chem Biol 23(9):1049–1055. https://doi.org/10.1016/j.chembiol.2016.07.013
21. Rathert P, Zhang X, Freund C, Cheng X, Jeltsch A (2008) Analysis of the substrate specificity of the Dim-5 histone lysine methyltransferase using peptide arrays. Chem Biol 15(1): 5–11. https://doi.org/10.1016/j.chembiol.2007.11.013
22. Rathert P, Dhayalan A, Murakami M, Zhang X, Tamas R, Jurkowska R, Komatsu Y, Shinkai Y, Cheng X, Jeltsch A (2008) Protein lysine methyltransferase G9a acts on non-histone targets. Nat Chem Biol 4(6):344–346. https://doi.org/10.1038/nchembio.88
23. Dhayalan A, Kudithipudi S, Rathert P, Jeltsch A (2011) Specificity analysis-based identification of new methylation targets of the SET7/9 protein lysine methyltransferase. Chem Biol 18(1):111–120. https://doi.org/10.1016/j.chembiol.2010.11.014
24. Kudithipudi S, Dhayalan A, Kebede AF, Jeltsch A (2012) The SET8 H4K20 protein lysine methyltransferase has a long recognition sequence covering seven amino acid residues. Biochimie 94(11):2212–2218. https://doi.org/10.1016/j.biochi.2012.04.024
25. Schuhmacher MK, Kudithipudi S, Kusevic D, Weirich S, Jeltsch A (2015) Activity and specificity of the human SUV39H2 protein lysine methyltransferase. Biochim Biophys Acta 1849(1):55–63. https://doi.org/10.1016/j.bbagrm.2014.11.005
26. Lanouette S, Davey JA, Elisma F, Ning Z, Figeys D, Chica RA, Couture JF (2015) Discovery of substrates for a SET domain lysine methyltransferase predicted by multistate computational protein design. Structure 23(1):206–215. https://doi.org/10.1016/j.str.2014.11.004
27. Weirich S, Schuhmacher MK, Kudithipudi S, Lungu C, Ferguson AD, Jeltsch A (2020) Analysis of the substrate specificity of the SMYD2 protein lysine methyltransferase and discovery of novel non-histone substrates. Chembiochem 21(1–2):256–264. https://doi.org/10.1002/cbic.201900582
28. Weirich S, Kudithipudi S, Jeltsch A (2016) Specificity of the SUV4-20H1 and SUV4-20H2 protein lysine methyltransferases and methylation of novel substrates. J Mol Biol 428(11):2344–2358. https://doi.org/10.1016/j.jmb.2016.04.015
29. Kusevic D, Kudithipudi S, Jeltsch A (2016) Substrate specificity of the HEMK2 protein glutamine methyltransferase and identification

of novel substrates*. J Biol Chem 291(12): 6124–6133. https://doi.org/10.1074/jbc.M115.711952

30. Kudithipudi S, Schuhmacher MK, Kebede AF, Jeltsch A (2017) The SUV39H1 protein lysine methyltransferase Methylates chromatin proteins involved in heterochromatin formation and VDJ recombination. ACS Chem Biol 12(4):958–968. https://doi.org/10.1021/acschembio.6b01076
31. Kusevic D, Kudithipudi S, Iglesias N, Moazed D, Jeltsch A (2017) Clr4 specificity and catalytic activity beyond H3K9 methylation. Biochimie 135:83–88. https://doi.org/10.1016/j.biochi.2017.01.013
32. Ardehali MB, Anselmo A, Cochrane JC, Kundu S, Sadreyev RI, Kingston RE (2017) Polycomb repressive complex 2 methylates elongin A to regulate transcription. Mol Cell 68(5):872–884.e6. https://doi.org/10.1016/j.molcel.2017.10.025
33. Schuhmacher MK, Rolando M, Bröhm A, Weirich S, Kudithipudi S, Buchrieser C, Jeltsch A (2018) The legionella pneumophila methyltransferase RomA methylates also non-histone proteins during infection. J Mol Biol 430(13): 1912–1925. https://doi.org/10.1016/j.jmb.2018.04.032
34. Jakobsson ME, Małecki JM, Halabelian L, Nilges BS, Pinto R, Kudithipudi S, Munk S, Davydova E, Zuhairi FR, Arrowsmith CH, Jeltsch A, Leidel SA, Olsen JV, Falnes PØ (2018) The dual methyltransferase METTL13 targets N terminus and Lys55 of eEF1A and modulates codon-specific translation rates. Nat Commun 9(1):3411. https://doi.org/10.1038/s41467-018-05646-y
35. Schuhmacher MK, Beldar S, Khella MS, Bröhm A, Ludwig J, Tempel W, Weirich S, Min J, Jeltsch A (2020) Sequence specificity analysis of the SETD2 protein lysine methyltransferase and discovery of a SETD2 super-substrate. Commun Biol 3(1):511. https://doi.org/10.1038/s42003-020-01223-6
36. Davydova E, Shimazu T, Schuhmacher MK, Jakobsson ME, Willemen HLDM, Liu T, Moen A, Ho AYY, Małecki J, Schroer L, Pinto R, Suzuki T, Grønsberg IA, Sohtome Y, Akakabe M, Weirich S, Kikuchi M, Olsen JV, Dohmae N, Umehara T, Sodeoka M, Siino V, McDonough MA, Eijkelkamp N, Schofield CJ, Jeltsch A, Shinkai Y, Falnes PØ (2021) The methyltransferase METTL9 mediates pervasive 1-methylhistidine modification in mammalian proteomes. Nat Commun 12(1):891. https://doi.org/10.1038/s41467-020-20670-7
37. Gayatri S, Cowles MW, Vemulapalli V, Cheng D, Sun Z-W, Bedford MT (2016) Using oriented peptide array libraries to evaluate methylarginine-specific antibodies and arginine methyltransferase substrate motifs. Sci Rep 6(1):28718. https://doi.org/10.1038/srep28718
38. Schneider CA, Rasband WS, Eliceiri KW (2012) NIH image to ImageJ: 25 years of image analysis. Nat Methods 9(7):671–675. https://doi.org/10.1038/nmeth.2089
39. Hornbeck PV, Zhang B, Murray B, Kornhauser JM, Latham V, Skrzypek E (2015) PhosphoSitePlus, 2014: mutations, PTMs and recalibrations. Nucleic Acids Res 43(Database issue): D512–D520. https://doi.org/10.1093/nar/gku1267
40. Szklarczyk D, Gable AL, Lyon D, Junge A, Wyder S, Huerta-Cepas J, Simonovic M, Doncheva NT, Morris JH, Bork P, Jensen LJ, Mering CV (2019) STRING v11: protein-protein association networks with increased coverage, supporting functional discovery in genome-wide experimental datasets. Nucleic Acids Res 47(D1):D607–d613. https://doi.org/10.1093/nar/gky1131
41. Obenauer JC, Cantley LC, Yaffe MB (2003) Scansite 2.0: proteome-wide prediction of cell signaling interactions using short sequence motifs. Nucleic Acids Res 31(13):3635–3641. https://doi.org/10.1093/nar/gkg584

Chapter 16

Identifying Specific Protein Interactors of Nucleosomes Carrying Methylated Histones Using Quantitative Mass Spectrometry

Andrey Tvardovskiy, Nhuong Nguyen, and Till Bartke

Abstract

Chemical modification of histone proteins by methylation plays a central role in chromatin regulation by recruiting epigenetic "readers" via specialized binding domains. Depending on the degree of methylation, the exact modified amino acid, and the associated reader proteins histone methylations are involved in the regulation of all DNA-based processes, such as transcription, DNA replication, and DNA repair. Here we present methods to identify histone methylation readers using a mass spectrometry–linked nucleosome affinity purification approach. We provide detailed protocols for the generation of semisynthetic methylated histones, their assembly into biotinylated nucleosomes, and the identification of methylation-specific nucleosome-interacting proteins from nuclear extracts via nucleosome pull-downs and label-free quantitative proteomics. Due to their versatility, these protocols allow the identification of readers of various histone methylations, and can also be adapted to different cell types and tissues, and other types of modifications.

Key words Histone, Nucleosome, Chromatin, Methylation, Histone modification, Native chemical ligation, Nuclear extract, Affinity purification, Mass spectrometry, Proteomics

1 Introduction

Histone proteins are modified by a multitude of posttranslational modifications (PTMs). These modifications are deposited by modifying enzymes and removed by demodifying enzymes. Depending on the chemical nature of a modification and the exact modified amino acid, different modifications can have very diverse functions [1, 2]. Histone modifications act by either directly affecting chromatin structure or by recruiting specific "reader" proteins that mediate the downstream functions of a particular modification. These readers are therefore inextricably intertwined with the functions of the modifying and demodifying enzymes themselves as

Andrey Tvardovskiy and Nhuong Nguyen contributed equally to this work.

Raphaël Margueron and Daniel Holoch (eds.), *Histone Methyltransferases: Methods and Protocols*,
Methods in Molecular Biology, vol. 2529, https://doi.org/10.1007/978-1-0716-2481-4_16,

they convert the information of a modification into biological signals. It is thus important to not only understand the function of modifying and demodifying enzymes but also the readers of the marks that they catalyze.

Here, we present a detailed workflow to identify and study readers of nucleosomes incorporating methylated histones using biochemical approaches and quantitative mass spectrometry (Fig. 1). We provide streamlined protocols for the purification of recombinant human core histones and for the semisynthesis of histones carrying methylation marks in their N-terminal tails. We further describe a method for reconstituting these histones into modified dinucleosomes together with biotinylated DNA. These nucleosomes are then used as baits in nucleosome affinity purifications to isolate methylation-specific binding proteins from nuclear extracts. Finally, we provide a fast and reliable protocol for processing nucleosome pull-down samples for mass spectrometry (MS) and guidance on how to conduct the mass spectrometric measurements and analysis of the proteomics data.

We have used these protocols in different formats for quantitative proteomics-linked nucleosome pull-downs using SILAC- and TMT-labeling [3–5]. We routinely identify and quantify around 1500 proteins in each pull-down, with the number of modification-responsive proteins ranging from a handful to several hundreds, depending on the modification(s) installed on the nucleosomes. Here we have compiled an optimized and streamlined procedure adapted for label-free quantitative proteomics, which we are currently using in our lab. This protocol provides excellent sensitivity and is versatile with regard to sample input. It is therefore easy to customize for different questions and settings and should be useful also in laboratories that do not have a dedicated proteomics background.

2 Materials

Prepare all solutions using ultrapure water (prepared by purifying deionized water, to attain a resistivity of 18 MΩ/cm at 25 °C) and analytical-grade reagents. Prepare and store all reagents at room temperature, unless indicated otherwise. A number of harmful reagents are used in these protocols; diligently follow appropriate safety precautions when handling these reagents and adhere to all waste disposal regulations when disposing of waste materials.

2.1 Expression of Core Histones

1. pET21(+) expression plasmids (Novagen) for human histones H2A, H2B, H3, and H4.
2. BL21-CodonPlus (DE3)-RIL competent cells (Agilent Technologies).

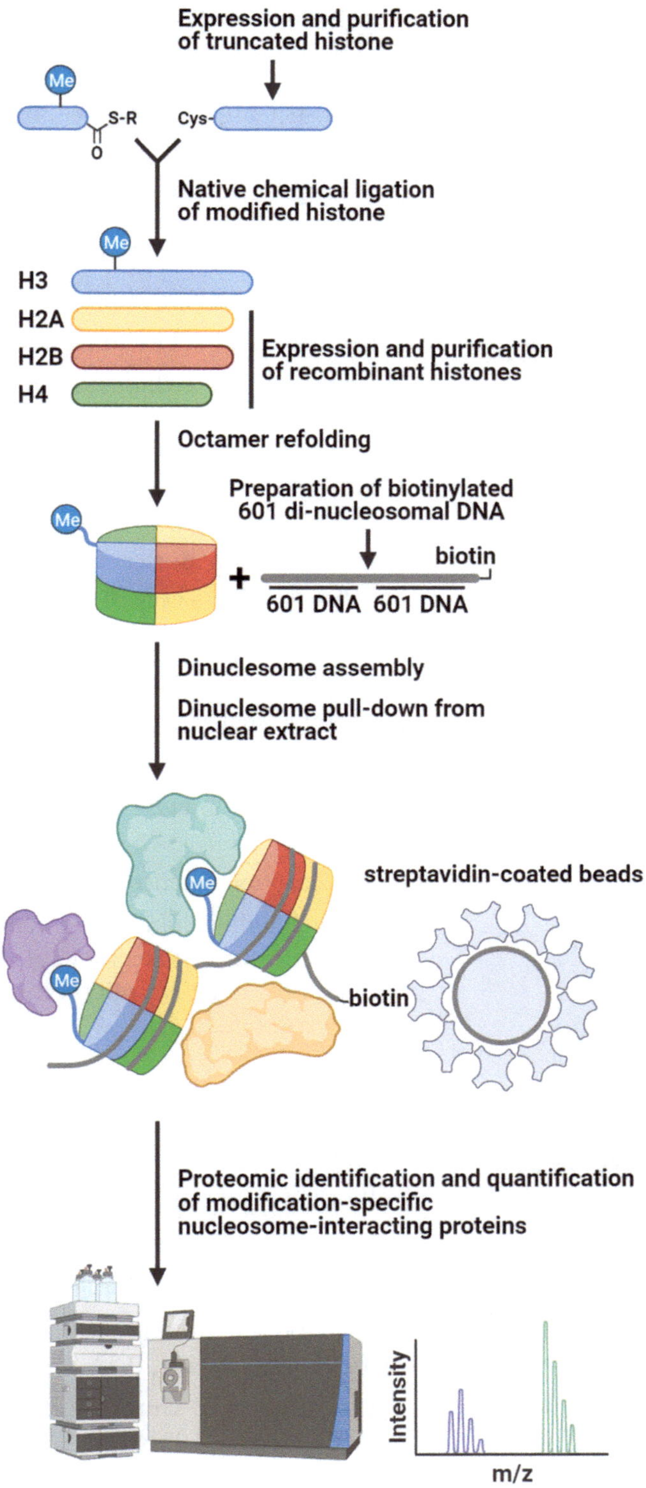

Fig. 1 Schematic illustration of the workflow for the identification of methylation-specific nucleosome-interacting proteins using mass spectrometry–linked dinucleosome pull-downs

3. Ampicillin (Amp): for a 1000× stock solution dissolve 100 mg/mL of ampicillin sodium salt powder in water, filter-sterilize through a 0.22 μm filter (prerinsed with water), and store in 1 mL aliquots at −20 °C.
4. Chloramphenicol (Cam): for a 1000× stock solution dissolve 34 mg/mL of chloramphenicol powder in ethanol, filter-sterilize through a 0.22 μm filter (prerinsed with ethanol), and store in 1 mL aliquots at −20 °C.
5. LB Ampicillin–Chloramphenicol ($LB_{Amp/Cam}$) agar plates.
6. LB Ampicillin–Chloramphenicol ($LB_{Amp/Cam}$) liquid medium.
7. Isopropyl-β-D-1-thiogalactopyranoside (IPTG): for a 1 M stock solution dissolve 2.38 g of Isopropyl-β-D-1-thiogalactopyranoside powder in 10 mL of water, filter-sterilize through a 0.22 μm filter (prerinsed with water), and store in 1 mL aliquots at −20 °C.
8. Histone Wash Buffer: 50 mM Tris–HCl (pH 7.5), 100 mM NaCl, 1 mM EDTA. Store at 4 °C.
9. Liquid nitrogen (LN_2) and dewar vessel (*see* **Note 1**).
10. Bacteriological incubator for agar plates.
11. Bacterial shaker incubator for liquid bacterial cultures.
12. Sterile conical flasks of various sizes (50 mL to 5 L).
13. Spectrophotometer and disposable 1.5 mL polystyrene cuvettes.
14. Tabletop microcentrifuge.
15. Floor-standing high-speed centrifuge with fixed angle rotors and centrifuge beakers suitable for large volumes up to 6× 1 L (e.g., Sorvall LYNX 6000 centrifuge with Fiberlite F9-6x1000 LEX rotor).

2.2 SDS Polyacrylamide Gel Electrophoresis (SDS-PAGE)

1. ProtoGel 40% (w/v) acrylamide–bisacrylamide stock solution (37.5:1). Once opened store at 4 °C protected from light (*see* **Note 2**).
2. Resolving gel buffer: 1 M Tris–HCl (pH 8.8).
3. Stacking gel buffer: 1 M Tris–HCl (pH 6.8).
4. 20% SDS stock solution in water (*see* **Note 3**).
5. SDS-PAGE running buffer: 25 mM Tris–HCl (pH 8.3), 0.192 mM glycine, 0.1% SDS (*see* **Note 4**).
6. 5× SDS-PAGE sample buffer: 312 mM Tris–HCl (pH 6.8), 50% glycerol, 10% SDS, 25% β-mercaptoethanol, and 0.01% bromophenol blue. Keep one aliquot at 4 °C for current use and store the remaining aliquots at −20 °C. Prepare 1× SDS-PAGE sample buffer by diluting the 5× buffer 1:5 with water immediately before use (*see* **Note 5**).

7. Ammonium persulfate (APS): 20% (w/v) solution in water. Keep one aliquot at 4 °C for current use and store the remaining aliquots at −20 °C.
8. *N,N,N′,N′*-tetramethylethylenediamine (TEMED): store at 4 °C.
9. Unstained protein molecular weight standard (e.g., Precision Plus protein standard, Bio-Rad).
10. Heat block or Thermomixer for 1.5 mL microtubes.
11. Mini-gel system with glass plates, combs, and gel pouring stand (e.g., Mini PROTEAN Tetra Cell, Bio-Rad).
12. Electrophoresis power supply.

2.3 Coomassie Stain

1. Coomassie staining solution: 25% (v/v) methanol, 10% (v/v) acetic acid, 0.1% Coomassie Brilliant Blue R-250 in water (*see* **Notes 6** and 7).
2. Coomassie destaining solution: 25% (v/v) methanol, 8% (v/v) acetic acid in water (*see* **Notes 6** and 7).
3. Plastic container with lid for staining/destaining gels.
4. Orbital shaker or rocking platform.

2.4 Histone Inclusion Body Preparation

1. Histone Wash Buffer: 50 mM Tris–HCl (pH 7.5), 100 mM NaCl, 1 mM EDTA. Store at 4 °C.
2. Triton X-100 stock solution: 20% (v/v) Triton X-100 in water.
3. Heated water bath.
4. Probe sonifier with a 3 mm (1/8 in.) tapered microtip probe.
5. 50 mL polycarbonate Oak Ridge centrifuge tubes.
6. High-speed centrifuge with fixed angle rotor for 50 mL Oak Ridge centrifuge tubes (e.g., Sorvall LYNX 6000 centrifuge with A27-8x50 rotor).

2.5 Purification of Histones

1. Unfolding Buffer: 20 mM Tris–HCl (pH 7.5), 7 M guanidine hydrochloride, 10 mM DTT. Prepare fresh using a 1 M dithiothreitol (DTT) frozen stock.
2. SAU-0 buffer: 20 mM NaAcetate (pH 5.2), 7 M urea, 1 mM EDTA, 5 mM β-mercaptoethanol. Prepare this buffer just before use with freshly prepared and deionized 8 M urea stock solution (*see* **Note 8**). Filter and degas the buffer before use.
3. SAU-200 buffer: 20 mM NaAcetate (pH 5.2), 7 M urea, 200 mM NaCl, 1 mM EDTA, 5 mM β-mercaptoethanol. Prepare this buffer just before use with freshly prepared and deionized 8 M urea stock solution (*see* **Note 8**). Filter and degas the buffer before use.

4. SAU-600 buffer: 20 mM NaAcetate (pH 5.2), 7 M urea, 600 mM NaCl, 1 mM EDTA, 5 mM β-mercaptoethanol. Prepare this buffer just before use with freshly prepared and deionized 8 M urea stock solution (*see* **Note 8**). Filter and degas the buffer before use.
5. SAU-1000 buffer: 20 mM NaAcetate (pH 5.2), 7 M urea, 1 M NaCl, 1 mM EDTA, 5 mM β-mercaptoethanol. Prepare this buffer just before use with freshly prepared and deionized 8 M urea stock solution (*see* **Note 8**). Filter and degas the buffer before use.
6. HU-1000 buffer: 20 mM HEPES–NaOH (pH 8.0), 7 M urea, 1 M NaCl, 1 mM EDTA, 5 mM β-mercaptoethanol. Prepare this buffer just before use with freshly prepared and deionized 8 M urea stock solution (*see* **Note 8**). Filter and degas the buffer before use.
7. Denaturing Lysis Buffer (for H2A): 100 mM sodium phosphate (pH 8.0), 8 M urea, 10 mM DTT. For this add 1.659 g of solid dibasic sodium phosphate dihydrate ($Na_2HPO_4 \cdot 2H_2O$), 0.106 g of solid monobasic sodium phosphate dihydrate ($NaH_2PO_4 \cdot 2H_2O$), and 1 mL of 1 M DTT to 100 mL of 8 M urea solution, and adjust the pH to 8.0 with HCl or NaOH if required. Prepare this buffer just before use from a freshly prepared and deionized 8 M urea stock solution (*see* **Note 8**).
8. Denaturing Wash Buffer (for H2A): 100 mM sodium phosphate (pH 8.0), 7 M urea, 10 mM DTT. Prepare this buffer just before use from a freshly prepared and deionized 8 M urea stock solution (*see* **Note 8**). For 100 mL buffer: add 1.659 g of solid dibasic sodium phosphate dihydrate ($Na_2HPO_4 \cdot 2H_2O$), 0.106 g of solid monobasic sodium phosphate dihydrate ($NaH_2PO_4 \cdot 2H_2O$), and 1 mL of 1 M DTT to 87.5 mL 8 M urea stock solution. Adjust the pH to 8.0 with HCl or NaOH if required, and then fill up to 100 mL with water.
9. Denaturing Elution Buffer (for H2A): 100 mM sodium phosphate (pH 8.0), 1 M NaCl, 7 M urea, 10 mM DTT. Prepare this buffer just before use by dissolving 2.92 g of solid NaCl in 50 mL of the Denaturing Wash Buffer.
10. Dithiothreitol (DTT): for a 1 M stock solution dissolve 1.54 g of DTT powder in 10 mL of water, filter-sterilize through a 0.22 μm filter (prerinsed with water), and store in 1 mL aliquots at −20 °C for up to a year.
11. Laboratory rotating mixer wheel.
12. 50 mL polycarbonate Oak Ridge centrifuge tubes.
13. Plastic syringes of various sizes (1–50 mL).
14. Disposable 0.45 μm syringe filter units (PES membranes).

15. Dialysis tubing with 12–14 kDa (e.g., Spectra/Por 4 Dialysis Membrane) or 6–8 kDa (e.g., Spectra/Por 1 Dialysis Membrane) molecular weight cutoffs (MWCO) and dialysis clamps.
16. Disposable gravity flow columns with frit (e.g., Qiagen 5 mL polypropylene columns, Cat. No. 34964). A laboratory stand with boss head and clamp will be required as column holder.
17. SP Sepharose Fast Flow resin (Cytiva).
18. 5 mL HiTrap SP Sepharose HP column (Cytiva).
19. HiPrep 26/60 Sephacryl S-200 HR column (Cytiva).
20. HiLoad SP Sepharose HP 16/10 column (Cytiva).
21. Probe sonifier with a 3 mm (1/8 in.) tapered microtip probe.
22. Vacuum pump with vacuum tubing and 0.22 μm filter assemblies for filtering and degassing chromatography buffers.
23. FPLC chromatography system (*see* **Note 9**) and accessories, including a 50 mL super-loop.
24. Freeze-dryer system.
25. High-speed centrifuge with fixed angle rotor for 50 mL Oak Ridge centrifuge tubes (e.g., Sorvall LYNX 6000 centrifuge with A27-8x50 rotor).
26. Tabletop centrifuge with swinging-bucket rotor and buckets suitable for 50 mL Falcon tubes (e.g., Eppendorf centrifuge 5810R).
27. UV spectrophotometer and 0.5 mL quartz cuvette.

2.6 Native Chemical Ligation of Modified Histones

1. pET21(+) expression plasmid (Novagen) for N-terminally truncated human histone. This needs to be cloned so that the expressed histone is tailored to the chosen ligation strategy (*see* Subheading 3.2).
2. Unfolding Buffer: 20 mM Tris–HCl (pH 7.5), 7 M guanidine hydrochloride, 100 mM DTT. Prepare this buffer fresh with solid DTT before use (*see* **Note 10**).
3. SAU-200 buffer: 20 mM NaAcetate (pH 5.2), 7 M urea, 200 mM NaCl, 1 mM EDTA. Prepare this buffer just before use with freshly prepared and deionized 8 M urea stock solution (*see* **Note 8**). Filter and degas the buffer before use. Do NOT add β-mercaptoethanol to this buffer (*see* **Note 11**)!
4. SAU-1000 buffer: 20 mM NaAcetate (pH 5.2), 7 M urea, 1 M NaCl, 1 mM EDTA. Prepare this buffer just before use with freshly prepared and deionized 8 M urea stock solution (*see* **Note 8**). Filter and degas the buffer before use. Do NOT add β-mercaptoethanol to this buffer (*see* **Note 11**)!
5. Native Chemical Ligation Buffer: 200 mM potassium phosphate (pH 7.9), 2 mM EDTA, 6 M guanidine hydrochloride.

Prepare this buffer from a 1 M potassium phosphate (pH 7.9) stock solution.

6. Acetonitrile, HPLC-grade (*see* **Note 12**).
7. Trifluoroacetic acid (TFA), HPLC-grade (*see* **Note 12**).
8. Dry ice.
9. Ethanol 100%.
10. Argon gas.
11. Tris(2-carboxyethyl)phosphine hydrochloride (TCEP) powder.
12. 5 M KOH solution (*see* **Note 13**).
13. 4-mercaptophenylacetic acid (MPAA) powder.
14. N-terminal histone thioester peptides. These need to be synthesized based on the chosen ligation strategy (*see* Subheading 3.2).
15. Dithiothreitol (DTT): for a 1 M stock solution dissolve 1.54 g of DTT powder in 10 mL of water, filter-sterilize through a 0.22 μm filter (prerinsed with water), and store in 1 mL aliquots at −20 °C for up to a year.
16. Acetic acid (*see* **Note 14**).
17. Methanol 100%, HPLC-grade (*see* **Note 15**).
18. Leak-proof container.
19. 5 mL round-bottom flask with ground glass joint.
20. Small stir bars.
21. Three-way vacuum/inert gas adapter (T-bore glass stopcock) with matching joint for the 5 mL round-bottom flask.
22. Party balloons.
23. Three-colour pH strips for neutral pH, resolution 0.5 pH units.
24. HiPrep 26/60 Sephacryl S-200 HR column (Cytiva).
25. SOURCE 15RPC column with 4 mL bed volume (Cytiva).
26. C8 reversed-phase chromatography HPLC column with 4 mL bed volume (e.g., PerkinElmer Aquapore RP-300 (C8) 250 mm × 4.6 mm i.d.).
27. HPLC or FPLC chromatography system (*see* **Notes 9** and **16**) and accessories, including a 2 mL sample loop and a 50 mL super-loop.
28. SpeedVac vacuum concentrator.
29. Thermomixer.
30. Teflon-coated freeze-dryer system with a condenser with low freezing point (−85 °C) suitable for solvents and a chemical resistant vacuum pump (*see* **Note 17**).

31. Vacuum pump with vacuum tubing.
32. Magnetic stirrer.
33. UV spectrophotometer and 0.5 mL quartz cuvette.

2.7 Histone Octamer Refolding

1. Lyophilized core histones H2A, H2B, H3, and H4 (*see* Subheadings 3.1 and 3.2).
2. Unfolding Buffer: 20 mM Tris–HCl (pH 7.5), 7 M guanidine hydrochloride, 10 mM DTT. Prepare fresh using a 1 M dithiothreitol (DTT) frozen stock.
3. Octamer Refolding Buffer: 10 mM Tris–HCl (pH 7.5), 2 M NaCl, 1 mM EDTA, 5 mM β-mercaptoethanol. Add β-mercaptoethanol just before use.
4. Dialysis tubing with 6–8 kDa MWCO (e.g., Spectra/Por 1 Dialysis Membrane) and dialysis clamps.
5. Centrifugal spin filtration unit with 10 kDa MWCO (e.g., Sartorius Vivaspin 4 with PES membrane).
6. Protein low-binding microtubes.
7. Superdex 200 16/60 Prep-grade gel-filtration column (Cytiva).
8. Tabletop microcentrifuge.
9. UV spectrophotometer and 0.5 mL and 0.1 mL quartz cuvettes.
10. Magnetic stirrer.
11. FPLC chromatography system and accessories, including a 2 mL sample-loop.
12. Refrigerated tabletop centrifuge with swinging-bucket rotor and buckets suitable for 15 mL Falcon tubes (e.g., Eppendorf centrifuge 5810R).
13. Refrigerated tabletop microcentrifuge.

2.8 Nucleosome Assembly

1. pUC19 plasmid containing eight repeats of the 601 dinucleosome sequence separated by *Eco*RV restriction sites (pUC19-8x601 di-Nuc).
2. pUC19 plasmid containing tandem repeats of the MMTV-A sequence separated by *Eco*RV restriction sites (pUC19-MMTV-A).
3. XL10 Gold competent cells (Agilent Technologies). Other equivalent high-quality *E. coli* cloning strains are also suitable.
4. Ampicillin (Amp): for a 1000× stock solution dissolve 100 mg/mL of ampicillin sodium salt powder in water, filter-sterilize through a 0.22 μm filter (prerinsed with water), and store in 1 mL aliquots at −20 °C.
5. LB Ampicillin (LB_{Amp}) agar plates with 50 μg/mL ampicillin.

6. LB Ampicillin (LB_{Amp}) liquid medium containing 50 µg/mL ampicillin.
7. 5× DNA loading dye: 0.5% SDS, 25% glycerol, 25 mM EDTA, and 0.25% Orange G. Keep one working aliquot at RT and store 1 mL aliquots frozen at −20 °C.
8. Tris-borate-EDTA electrophoresis buffer (TBE): for a 10× TBE buffer (900 mM Tris base, 900 mM boric acid, 20 mM EDTA) dissolve 109 g Tris base and 55.6 g boric acid in 900 mL of deionized water, add 40 mL of 0.5 M EDTA stock solution (pH 8.0), adjust volume to 1 L. To prepare 1× TBE gel running buffer dilute 1/10 in deionized water before use. Store at RT.
9. TE buffer: 10 mM Tris–HCl (pH 8.0), 1 mM EDTA.
10. HiTrap Q buffer A: 10 mM Tris–HCl (pH 8.0), 100 mM NaCl, 1 mM EDTA. Filter and degas the buffer before use.
11. HiTrap Q buffer B: 10 mM Tris–HCl (pH 8.0), 1 M NaCl, 1 mM EDTA. Filter and degas the buffer before use.
12. High-salt nucleosome reconstitution buffer (RB-high): 10 mM Tris–HCl (pH 7.5), 2 M KCl, 1 mM EDTA, 1 mM DTT. Make fresh and prechill at 4 °C. Add DTT fresh from a 1 M stock.
13. Low-salt nucleosome reconstitution buffer (RB-low): 10 mM Tris–HCl (pH 7.5), 250 mM KCl, 1 mM EDTA, 1 mM DTT. Make fresh and prechill at 4 °C. Add DTT fresh from a 1 M stock.
14. 2× native PAGE sample buffer: 10% sucrose in water, a small amount of bromophenol blue can be added for better visibility. Store at 4 °C.
15. *Eco*RV restriction enzyme (100 U/µL) and matching restriction buffer.
16. *Eco*RI restriction enzyme (100 U/µL) and matching restriction buffer.
17. Klenow Fragment (3′-> 5′ exo-) and matching reaction buffer.
18. Agarose, electrophoresis-grade or ultrapure.
19. Ethidium bromide or SYBR Safe DNA gel stain (*see* **Note 18**).
20. 5 M NaCl, autoclave before use.
21. 40% (w/v) PEG-6000 solution, autoclave before use.
22. 70% Ethanol.
23. 100% ethanol, ice-cold.
24. 3 M sodium acetate (pH 5.2).
25. dATP, PCR-grade, 100 mM stock solution. Store aliquoted at −20 °C.

26. Biotin-11-dUTP, PCR-grade, 1 mM or 10 mM stock solutions (*see* **Note 19**). Store aliquoted at −20 °C.
27. Phenol–chloroform–isoamyl alcohol (25:24:1, v/v) (*see* **Note 20**).
28. 20% ethanol, filtered and degassed for chromatography.
29. 30% (w/v) acrylamide–bisacrylamide stock solution (59:1): Dissolve 29.5 g of acrylamide powder and 0.5 g of bisacrylamide powder in a total volume of 100 mL of water. Filter through a 0.22 μm filter (prerinsed with water), and store at 4 °C protected from light (*see* **Note 2**). The 59:1 acrylamide to bisacrylamide solution is not commercially available and has to be prepared from solid acrylamide and bisacrylamide.
30. Ammonium persulfate (APS): 20% (w/v) solution in water. Keep one aliquot at 4 °C for current use and store the remaining aliquots at −20 °C.
31. *N,N,N′,N′*-tetramethylethylenediamine (TEMED): store at 4 °C.
32. Plasmid Gigaprep Kit (e.g., Qiagen Plasmid Giga Kit/tip 10,000 or equivalent).
33. 50 mL polypropylene Oak Ridge centrifuge tubes (keep separate—*see* Subheading 3.4.1).
34. Dialysis buttons with 3.5 kDa MWCO (e.g., Pierce Slide-A-Lyzer MINI dialysis units, 0.1 mL).
35. DNA low-binding microtubes.
36. Gel-loading tips (round).
37. HiTrap Q HP cartridge with 1 mL bed volume (Cytiva).
38. Bacteriological incubator for agar plates.
39. Bacterial shaker incubator for liquid bacterial cultures.
40. Sterile conical flasks of various sizes (50 mL to 5 L).
41. Floor-standing high-speed centrifuge with fixed angle rotors and centrifuge beakers suitable for large volumes up to 6× 1 L (e.g., Sorvall LYNX 6000 centrifuge with Fiberlite F9-6x1000 LEX rotor).
42. High-speed centrifuge with fixed angle rotor for 50 mL Oak Ridge centrifuge tubes (e.g., Sorvall LYNX 6000 centrifuge with A27-8x50 rotor).
43. Agarose gel electrophoresis chamber with gel trays and combs.
44. Electrophoresis power supply.
45. UV transilluminator with camera and printer.
46. UV spectrophotometer and 0.5 mL quartz cuvette.
47. Vortexer.

48. Tabletop microcentrifuge.
49. FPLC chromatography system and accessories, including 10 mL and 50 mL super loops.
50. Dual-channel peristaltic pump (alternatively two single-channel peristaltic pumps).
51. Mini-gel system with glass plates, combs, and gel pouring stand (e.g., Mini PROTEAN Tetra Cell, Bio-Rad). Glass plates with 1.5 mm spacers and 1.5 mm 10-well combs are required.

2.9 Nuclear Extract Preparation

1. HeLa S3 cells (ATCC No.: CCL-2.2/RRID: CVCL_0058).
2. HeLa S3 culture medium: RPMI 1640 medium supplemented with 10% fetal bovine serum, 1% (v/v) penicillin–streptomycin, and 2 mM L-glutamine.
3. Phosphate buffered saline (PBS): 137 mM NaCl, 2.7 mM KCl, 10 mM Na_2HPO_4, 2 mM KH_2PO_4. For 1 L dissolve 8 g of NaCl, 0.2 g of KCl, 1.44 g of Na_2HPO_4, and 0.24 g KH_2PO_4 in 800 mL of water, adjust the pH to 7.4 with HCl, and add water to 1 L. Aliquot and sterilize by autoclaving.
4. Buffer A: 10 mM HEPES–KOH (pH 7.9), 1.5 mM $MgCl_2$, 10 mM KCl.
5. Buffer C (no NaCl): 20 mM HEPES–KOH (pH 7.9), 25% glycerol, 1.5 mM $MgCl_2$, 0.2 mM EDTA. Add 0.1% NP-40, 1 mM DTT, 0.5 mM PMSF, and 1× Complete protease inhibitors w/o EDTA (Roche) just before use (*see* Subheading 3.5).
6. Buffer C (1000 mM NaCl): 20 mM HEPES–KOH (pH 7.9), 25% glycerol, 1.5 mM $MgCl_2$, 0.2 mM EDTA, 1 M NaCl. Add 0.1% NP-40, 1 mM DTT, 0.5 mM PMSF, and 1× Complete protease inhibitors w/o EDTA (Roche) just before use (*see* Subheading 3.5).
7. Trypan blue 0.4% solution.
8. NP-40 stock solution: 10% (v/v) NP-40 in water.
9. Dithiothreitol (DTT): for a 1 M stock solution dissolve 1.54 g of DTT powder in 10 mL of water, filter-sterilize through a 0.22 μm filter (prerinsed with water), and store in 1 mL aliquots at −20 °C for up to a year.
10. Phenylmethylsulfonyl fluoride (PMSF): for a 1 M stock solution dissolve 1.742 g PMSF in 10 mL of ethanol, and store at −20 °C for up to 6 months (*see* **Note 21**).
11. Complete, EDTA-free Protease Inhibitor Tablets (Roche): for a 50× stock solution dissolve one tablet in 1 mL of deionized water, and store at −20 °C for up to 2 months.
12. 5 M NaCl, autoclave before use.

13. RNase A, DNase, and protease-free.
14. RNasin Ribonuclease Inhibitor (Promega) or equivalent.
15. Bradford protein assay reagent (e.g., Bio-Rad Protein Assay, Cat. No. 5000006).
16. Bovine serum albumin (BSA) protein standard.
17. Liquid nitrogen (LN_2) and dewar vessel (*see* **Note 1**).
18. T175 tissue culture flasks.
19. 5 mL open-top thin wall polypropylene centrifuge tubes (Beckman Cat. No. 326819 or equivalent).
20. 0.5 mL protein low-binding microtubes
21. 1 L and 6 L glass tissue culture spinner flasks.
22. 15 mL Dounce tissue homogenizer set with tight and loose pestles (Wheaton).
23. Laboratory stand with boss head and clamp.
24. Laboratory tube roller or laboratory rotating mixer wheel.
25. Tissue culture spinner flask stirring platform and control unit.
26. Tissue culture CO_2 incubator.
27. Tissue culture setup including biological safety cabinet and vacuum aspirator.
28. Automated cell counter or inverted tissue culture microscope and Neubauer counting chamber.
29. Floor-standing high-speed centrifuge with fixed angle rotors and centrifuge beakers suitable for large volumes up to 6× 1 L (e.g., Sorvall LYNX 6000 centrifuge with Fiberlite F9-6x1000 LEX rotor).
30. Refrigerated tabletop centrifuge with swinging-bucket rotor and buckets suitable for 15 mL and 50 mL Falcon tubes (e.g., Eppendorf centrifuge 5810R).
31. Floor-standing ultracentrifuge with 5 mL swinging-bucket rotors (e.g., Beckman Optima XE-90 centrifuge with SW 55 Ti swinging-bucket rotor or equivalent).
32. Spectrophotometer and disposable 1.5 mL polystyrene cuvettes.

2.10 Nucleosome Affinity Purification

1. Low-salt nucleosome reconstitution buffer (RB-low): 10 mM Tris–HCl pH 7.5, 250 mM KCl, 1 mM EDTA. Make fresh and store at 4 °C.
2. Pull-down buffer: 20 mM HEPES–KOH pH 7.9, 150 mM NaCl, 0.2 mM EDTA, 10% glycerol. Make fresh and store at 4 °C.

3. Adjustment Buffer: 20 mM HEPES–KOH pH 7.9, 10 mM NaCl, 0.2 mM EDTA, 5% glycerol. Make fresh and store at 4 °C.
4. 5× SDS-PAGE sample buffer: 312 mM Tris–HCl pH 6.8, 50% glycerol, 10% SDS, 25% β-mercaptoethanol, and 0.01% bromophenol blue. Keep one aliquot at 4 °C for current use and store the remaining aliquots at −20 °C. Prepare 1× SDS-PAGE sample buffer by diluting the 5× buffer 1:5 with water immediately before use (*see* **Note 5**).
5. NP-40 stock solution: 10% (v/v) NP-40 in water.
6. Dithiothreitol (DTT): for a 1 M stock solution dissolve 1.54 g of DTT powder in 10 mL of water, filter-sterilize through a 0.22 μm filter (prerinsed with water), and store in 1 mL aliquots at −20 °C for up to a year.
7. Phenylmethylsulfonyl fluoride (PMSF): for a 1 M stock solution dissolve 1.742 g PMSF in 10 mL of ethanol, and store at −20 °C for up to 6 months (*see* **Note 21**).
8. Complete, EDTA-free Protease Inhibitor Tablets (Roche): for a 50× stock solution dissolve one tablet in 1 mL of deionized water, and store at −20 °C for up to 2 months.
9. Sodium butyrate: for a 1 M stock solution dissolve 1.1 g of sodium butyrate in 10 mL of water and store in 1 mL aliquots at −20 °C for up to 6 months.
10. Trichostatin A (TSA), 5 mM ready-made solution in DMSO (Sigma T1952) (*see* **Note 22**).
11. 1.5 mL DNA low-binding microtubes.
12. 1.5 mL protein low-binding microtubes.
13. Streptavidin Sepharose High Performance beads (Cytiva).
14. Refrigerated tabletop microcentrifuge.
15. Laboratory rotating mixer wheel.
16. Vortexer.
17. Heat block or Thermomixer for 1.5 mL microtubes.

2.11 Proteomic Analysis of Nucleosome Pull-Down Samples

1. Water, LC-MS-grade.
2. Triethylammonium bicarbonate buffer (TEAB), 1.0 M, pH 8.5 (Sigma-Aldrich).
3. UA buffer: 8 M urea, 100 mM TEAB (pH 8.5) in LC-MS-grade water. Prepare fresh with molecular biology-grade urea, ultrapure, >99.5%.
4. Wash Buffer: 50 mM TEAB in LC-MS-grade water.
5. Dithiothreitol (DTT): for a 1 M stock solution dissolve 1.54 g of DTT powder in 10 mL of LC-MS-grade water. Store in 1 mL aliquots at −20 °C for up to a year.

6. Iodoacetamide (IAA): for a 0.5 M stock solution dissolve 27.75 mg IAA in 0.3 mL LC-MS-grade water, cover with foil to protect from light, and store at −20 °C for up to 3 months.
7. Acetonitrile, LC-MS-grade (*see* **Note 12**).
8. Trifluoroacetic Acid (TFA), LC-MS-grade (*see* **Note 12**).
9. Lysyl Endopeptidase (Lys-C), mass spectrometry–grade.
10. Trypsin, mass spectrometry–grade.
11. 1.5 mL protein low-binding microtubes.
12. Centrifugal filter units with 30 kDa MWCO (Vivacon-500, Sartorius).
13. Parafilm.
14. pH test strips pH 0–14, resolution 1.0 pH unit.
15. Vortexer.
16. Thermomixer for 1.5 mL microtubes.
17. Tabletop microcentrifuge.

3 Methods

The procedures for preparing modified nucleosomes and for performing nucleosome affinity purifications are quite involved, and require a number of different skills. Good knowledge of purifying and handling proteins and DNA, and experience with different chromatography techniques and FPLC chromatography systems is required. Furthermore, high accuracy is essential at multiple steps; this is highlighted in the individual methods.

The final analysis of the nucleosome affinity purifications is carried out by high-performance Liquid Chromatography Tandem Mass Spectrometry (LC-MS/MS). In most settings this will be carried out by a dedicated mass spectrometry (MS) or proteomics facility or a specialized proteomics laboratory. Since each MS facility or laboratory has their own setup in terms of mass spectrometers, sample preparation procedures, and data analysis pipelines, we recommend working closely with the facility or laboratory that will carry out the MS measurements when establishing the proteomics procedures. We are therefore only providing a simple and robust protocol for the MS sample preparation and general advice on parameters for the MS measurements and tools and workflows for the analysis of the proteomics data. These can then be adapted to the local procedures.

3.1 Expression and Purification of Recombinant Human Core Histones

The first step in the assembly of modified nucleosomes is the purification of recombinant core histones from bacteria. Here we essentially follow the protocol developed by Karolin Luger's lab [6, 7] that uses an initial inclusion body preparation, followed by gel-filtration and ion-exchange chromatography steps. Background information on the expression and purification of recombinant histones can be found in these protocols. Since the human histones that we are using all display different purification behaviors, we have introduced several modifications to the original procedure to streamline the protocols for each individual human core histone. The purification deviates most substantially for histones H2A and H4; these procedures and where they connect to the general protocol are described in Subheadings 3.1.4 and 3.1.5.

3.1.1 Expression of Histones in Bacteria

The human core histones H2A, H2B, H3.1, and H4 are expressed and purified from bacteria. We use pET21(+) vectors, BL21-CodonPlus (DE3)-RIL *E. coli* cells, and standard LB medium for the expression. Other vectors, *E. coli* expression strains, and media can be used; however, the expression conditions need to be optimized (*see* **Note 23**). Here we describe the procedure for 1 L of expression culture (*see* Subheadings 3.1.2–3.1.4 for selection of appropriate expression culture volumes):

1. Transform the pET21(+) histone expression plasmid into BL21-CodonPlus (DE3)-RIL cells or streak cells from a transformed frozen stock on $LB_{Amp/Cam}$ agar plates and incubate the plate in a bacterial incubator at 37 °C overnight (O/N). Once colonies are visible store the plate at 4 °C.
2. The next day in the evening: inoculate 20 mL of $LB_{Amp/Cam}$ medium in a 100 mL conical flask with a handful of colonies and incubate O/N in a shaker incubator at approximately 250 rpm at 37 °C.
3. The next day in the morning: inoculate 1 L of $LB_{Amp/Cam}$ medium in a 5 L conical flask with 10 mL of the overnight culture and shake in a shaker incubator at approximately 200 rpm at 37 °C.
4. Regularly monitor the growth of the bacteria by measuring the optical density of the culture in a spectrophotometer at a wavelength of 600 nm (OD_{600}). When the OD_{600} reaches 0.6, transfer 1 mL of the culture to a 1.5 mL Eppendorf tube and pellet the cells in a tabletop microcentrifuge at full speed for 5 min. Remove and discard the supernatant, resuspend the cell pellet in 100 μL of 1× SDS-PAGE sample buffer, and lyse the cells by boiling in a heat block at 95 °C for 5 min. Let the sample cool and store on ice or at −20 °C until running the SDS-PAGE gel (*see* **step 10**). This is the "uninduced" sample.
5. While pelleting the cells, add IPTG to 0.2 mM final concentration to the remaining culture to induce the expression of the histone. Return the culture to the shaker incubator and shake at 37 °C for a further 2 h.

6. Measure the OD_{600} and transfer a volume corresponding to 0.6 OD_{600} (i.e., equivalent to the uninduced culture) to a 1.5 mL Eppendorf tube. Pellet and boil the cells in 100 μL of 1× SDS-PAGE sample buffer as described for the uninduced sample. This is the "induced" sample.
7. Harvest the cells by centrifugation in a large capacity centrifuge at 6000 × *g* for 10 min at 4 °C.
8. *For H2B, H3.1, and H4*: H2B, H3.1, and H4 are insoluble and the cells will be used to prepare histone inclusion bodies (*see* Subheading 3.1.2). Harvest the cells, remove the medium (*see* **Note 24**), and resuspend the pellet in 20 mL of Histone Wash Buffer per 1 L of culture. Transfer to a 50 mL Falcon tube, snap-freeze in LN_2 (*see* **Note 1**), and store at −20 °C until further processing (*see* **Note 25**).
9. *For H2A*: human H2A is mainly soluble and follows a different protocol. The harvested cell pellet can either be directly frozen and stored at −20 °C (without resuspension in any buffer) until further processing. Alternatively, the pelleted cells can be used immediately for the cation-exchange capturing step (*see* Subheading 3.1.4).
10. Check the expression level of the histone in the "uninduced" and "induced" samples by running a 17.5% SDS-PAGE gel and performing a Coomassie stain. Loading 5 μL of each sample should result in a good staining of the proteins and the expressed histone should be clearly visible in the "induced" sample as a major band migrating at a molecular weight of approximately 10–15 kDa, depending on the histone (*see* Table 1). If the histone is expressed well, proceed with the purification.

3.1.2 Histone Inclusion Body Preparation

When expressed in *E. coli* the human histones H3.1 and H4 are insoluble and are deposited in inclusion bodies. An inclusion body prep can therefore be used as a first purification step for these histones. Human histone H2B is partially soluble and in our experience it does not form inclusion bodies as efficiently as H3.1 and H4. While 1 L or 2 L of bacterial culture are generally sufficient to purify decent amounts of H3.1 and H4, a greater volume (6 L) is required as starting material to achieve good yields for H2B. The protocol below describes the histone inclusion body preparation procedure for a cell pellet resulting from 1 L of expression culture, for other volumes the procedure needs to be adjusted accordingly.

1. Thaw the frozen cells (resuspended in 20 mL of Histone Wash Buffer per 1 L of culture) from Subheading 3.1.1 (**step 8**) in a water bath at 30 °C. Once the cells are thawed place them on ice.

Table 1
Protein parameters of human histones and histone octamers. The table lists the molecular weight (MW), isoelectric point (pI), extinction coefficient at 280 nm (ε_{280nm}), and the absorbance of a 0.1% solution (Abs 0.1% [= 1 mg/mL]) at 280 nm with cysteines in the oxidized (ox.) or reduced (red.) form for the human core histones, the truncated forms of histones H3.1 and H4, and human histone octamers. Histones and octamers are purified and stored under reducing conditions; therefore, the absorbance in the reduced state should be used. The absorption values for ligated histones H3.1 and H4, and histone octamers reconstituted with ligated histones are very similar to the wild-type histones and octamers and therefore not listed. The values were obtained through the ProtParam (https:/web.expasy.org/protparam/) tool

Histone	MW [Da]	pI	ε_{280nm} [M^{-1} cm^{-1}]	Abs 0.1% [= 1 mg/mL]
Human histone H2A	13.976	10.90	4.470	0.320
Human histone H2B	13.775	10.32	7.540	0.541
Human histone H3.1	15.273	11.13	4.595 (ox.)/4.470 (red.)	0.301 (ox.)/0.293 (red.)
Human histone H4	11.236	11.36	5.960	0.530
H3.1 Δ1–31 T32C	12.056	10.17	4.595 (ox.)/4.470 (red.)	0.381 (ox.)/0.371 (red.)
H4 Δ1–28 I29C	8.356	10.53	5.960 (ox.)/5.960 (red.)	0.713 (ox.)/0.713 (red.)
WT histone octamer	108.394	10.98	44.950 (ox.)/44.700 (red.)	0.415 (ox.)/0.412 (red.)

2. Sonicate the cell suspension four times for 30 s (1 min at 1 s ON/1 s OFF cycles) each using a probe sonifier (*see* **Note 26**). Position the sonifier probe in the center to avoid frothing and place the cells on ice for 1 or 2 min between each sonication step to prevent heating of the sample.
3. Transfer the cell suspension to a 50 mL Oak Ridge centrifuge tube and spin for 20 min at 12,000 × *g* and 4 °C (e.g., Sorvall LYNX 6000 centrifuge with A27-8x50 rotor or equivalent). We use polycarbonate tubes as the pellets and suspensions are more visible through the clear plastic.
4. Discard the supernatant and resuspend the pellet in 20 mL of Histone Wash Buffer +1% Triton X-100. The pellet needs to be completely resuspended until particulates are no longer visible.
5. Transfer the suspension to a 50 mL Falcon tube and sonicate for 30 s using a probe sonifier as above.
6. Distribute the sample into two 50 mL Oak Ridge tubes and fill up each tube to approximately 45 mL with Histone Wash Buffer +1% Triton X-100. This is the first wash.
7. Spin for 20 min at 12,000 × *g* and 4 °C.
8. Discard the supernatant and resuspend the two pellets in a total of 20 mL Histone Wash Buffer +1% Triton X-100. The pellets need to be completely resuspended.

9. Combine the suspensions in a 50 mL Falcon tube and sonicate for 30 s using a probe sonifier as above.
10. Distribute the sample into two 50 mL Oak Ridge tubes and fill up each tube to approximately 45 mL with Histone Wash Buffer +1% Triton X-100. This is the second wash.
11. Spin for 20 min at 12,000 × *g* and 4 °C.
12. Discard the supernatant and resuspend the two pellets in a total of 20 mL Histone Wash Buffer (without Triton X-100). The pellets need to be completely resuspended.
13. Combine the suspensions in a 50 mL Falcon tube and sonicate for 30 s using a probe sonifier as above.
14. Distribute the sample into two 50 mL Oak Ridge tubes and fill up each tube to approximately 45 mL with Histone Wash Buffer (without Triton X-100). This is the third wash.
15. Spin for 20 min at 12,000 × *g* and 4 °C.
16. Discard the supernatant and resuspend the pellets in a total of 20 mL Histone Wash Buffer (without Triton X-100). The pellets need to be completely resuspended.
17. Combine the suspensions in a 50 mL Falcon tube and sonicate for 30 s using a probe sonifier as above. This is the fourth and final wash.
18. Transfer 10 μL of the suspension to a 1.5 mL microtube, mix with 70 μL of water and 20 μL of 5× SDS-PAGE sample buffer, and boil for 5 min at 95 °C. This is the "inclusion body" sample. Let it cool and store on ice or at −20 °C until running an SDS-PAGE gel (*see* step 21).
19. Transfer the inclusion body suspension to one 50 mL Oak Ridge tube, fill up to approximately 45 mL with Histone Wash Buffer (without Triton X-100), and spin for 20 min at 12,000 × *g* and 4 °C.
20. Completely remove and discard the supernatant. This pellet is the extracted inclusion body that will be used for the histone purification procedure (*see* Subheading 3.1.3). The pellet can either be frozen or used immediately, depending on the schedule (*see* **Note 27**). If freezing, tightly close the lid of the Oak Ridge tube, and store the tube at −20 °C until further processing.
21. Run 5 μL of the "inclusion body" sample on a 17.5% SDS-PAGE gel for quality control. The histone should be clearly visible as a major band at approximately 10–15 kDa (*see* Table 1).

3.1.3 Purification of Histones

All four core histones are expressed without a tag and purified via a denaturing gel-filtration chromatography followed by an ion-exchange chromatography step (*see* **Notes 9** and **28**). We have streamlined the protocols for each histone, so that this two-step purification procedure can be performed in two days. For good yields we use cation-exchange-captured H2A (*see* Subheading 3.1.4) from 4 L of culture or inclusion body pellets corresponding to 6 L (H2B), 1 L (H3.1), or 2 L (H4) of culture as starting material. We describe our "standard" protocol for histones extracted from inclusion bodies here. Modifications to the protocol introduced for H2A and H4 are highlighted in the procedure and described in Subheadings 3.1.4 and 3.1.5, respectively.

1. Starting in the morning, attach a HiPrep 26/60 Sephacryl S-200 HR column to an FPLC chromatography system and wash the column with >100 mL of filtered and degassed water.
2. While washing the column in water prepare 1.5 L of SAU-200 (for H2A, H2B, and H3.1) or SAU-1000 (for H4) gel-filtration buffer from a freshly prepared deionized 8 M urea stock solution (*see* **Note 8**).
3. When the buffers are ready, equilibrate the column with approximately 500 mL SAU-200 (H2A, H2B, and H3.1) or SAU-1000 (H4) buffer during the day.
4. In the afternoon while the column is equilibrating in the gel-filtration buffer: thaw the frozen inclusion body pellets (*see* Subheading 3.1.2 **step 20**) and thoroughly resuspend them in 25 mL of Unfolding Buffer, breaking up clumps with the pipette. Cysteines are reduced with 10 mM DTT. For H2B use inclusion body pellets corresponding to 6 L, for H3.1 to 1 L, and for H4 to 2 L of culture. H2A is extracted directly from the *E. coli* cells and captured on cation-exchange beads instead of an inclusion body extraction (*see* Subheading 3.1.4).
5. Rotate the inclusion body suspension in a 50 mL Oak Ridge tube for 1 h at RT on a rotating wheel to extract the histones and then centrifuge for 20 min at 20,000 × *g* and 25 °C (e.g., Sorvall LYNX 6000 centrifuge with A27-8x50 rotor or equivalent).
6. Transfer the supernatant (SN) containing the extracted histones to a new 50 mL Oak Ridge tube and centrifuge again for 20 min at 20,000 × *g* and 25 °C.
7. Transfer the SN to a 50 mL plastic syringe and filter through a 0.45 μm filter. This is the extracted inclusion body. The filter clogs easily due to residual particles, several filters might be necessary to filter all of the sample.
8. In the evening, once the column is equilibrated in the gel-filtration buffer: load the extracted inclusion body (H2B,

H3.1, and H4) sample or the cation-exchange-captured H2A (*see* Subheading 3.1.4) into a 50 mL super-loop and separate over the HiPrep 26/60 Sephacryl S-200 HR column. Run two consecutive gel-filtration runs in SAU-200 buffer (SAU-1000 for H4) at a flow rate of 1 mL/min overnight, injecting half of the sample (approximately 12 mL) in each run (*see* **Note 29**). Run both runs at RT (*see* **Notes 9** and **28**) and collect 5 mL fractions separately for each run.

9. The next day in the morning: check the fractions by running a 17.5% SDS-PAGE gel and performing a Coomassie stain. Combine the fractions containing the histone peak (usually 40–50 mL total volume from both runs). For histones H2A, H2B, and H3.1 continue with **step 10**. The purification of histone H4 differs in the ion-exchange chromatography step, continue with the procedure described in Subheading 3.1.5.
10. While running the SDS-PAGE gel prepare fresh SAU-200 and SAU-600 buffers using leftover deionized urea from preparing the gel-filtration buffer (*see* **Note 8**) and equilibrate a 20 mL HiLoad SP-Sepharose HP 16/10 column in SAU-200 buffer.
11. Load the combined histone fractions directly onto the equilibrated HiLoad SP Sepharose HP 16/10 column, wash the column with 5 column volumes (CV) of SAU-200 (i.e., 100 mL), and then elute with a gradient from 0% - 100% SAU-600 over 20 CV (Buffer A: SAU-200; Buffer B: SAU-600). Run the column at a flow rate of 2.5 mL/min and collect 5 mL fractions during the gradient.
12. Check the ion-exchange chromatography fractions by running a 17.5% SDS-PAGE gel and performing a Coomassie stain.
13. Combine the fractions containing the pure histone and dialyze extensively (at least three buffer changes) against water containing 1 mM DTT using dialysis tubing with a 12–14 kDa (H2A, H2B, H3.1) or 6–8 kDa (H4) molecular weight cutoff (MWCO). We usually do a first dialysis for >2 h in the afternoon/evening directly after the ion-exchange purification, a second dialysis overnight, and a third approximately 8 h dialysis the following day.
14. After the last dialysis, transfer the dialyzed histone into a 50 mL Falcon tube and determine the protein concentration by measuring the OD at 280 nm of the undiluted solution against water containing 1 mM DTT (we use an aliquot from the DTT-containing water put aside from the last dialysis step as the blank). For calculating the concentrations *see* Table 1.
15. Aliquot the histones into microtubes (*see* **Note 30**) and freeze-dry overnight. The freeze-dried histones are generally highly pure and can be stored at −80 °C for many years. Expected yields are approximately 20–25 mg for a 4 L of culture of H2A,

approximately 25–30 mg for a 6 L of culture H2B, >50 mg for a 1 L of culture of H3.1, and approximately 20–30 mg for a 2 L culture of H4.

3.1.4 Alternative Protocol for Capturing Histone H2A on Cation-Exchange Beads

In our experience the human histone H2A does not form inclusion bodies efficiently, and an inclusion body preparation is therefore impractical as a first enrichment step. Instead, we capture H2A from whole *E. coli* extracts by a quick denaturing ion-exchange chromatography step. This prefractionation step utilizes a cation-exchanger resin at a pH of 8.0, at which H2A will bind to the resin while most contaminating proteins will not. Captured H2A is then step-eluted and used for the histone purification instead of the extracted inclusion bodies. The protocol below is designed to work for 4 L of bacterial culture:

1. Harvest freshly grown or thaw frozen cells from 4 L of *E. coli* culture expressing H2A (*see* Subheading 3.1.1 **step 9**) and resuspend the pellet in 100 mL of Denaturing Lysis Buffer. This will result in a final concentration of approximately 7 M urea.
2. Distribute into six approximately 20 mL aliquots in 50 mL Falcon tubes and sonicate each aliquot four times for 30 s (1 min at 1 s ON/1 s OFF cycles) using a probe sonifier. Do not let the samples get hot to avoid decay of the urea (*see* **Note 28**).
3. Rotate the cell suspension for 1 h at RT on a rotating wheel to extract the H2A.
4. Transfer to 50 mL Oak Ridge tubes and centrifuge for 30 min at 20,000 × *g* and 25 °C (e.g., Sorvall LYNX 6000 centrifuge with A27-8x50 rotor or equivalent).
5. While the cell suspension is spinning equilibrate 8 mL (bed volume) of SP Sepharose Fast Flow beads in Denaturing Wash Buffer. Equilibrate the beads in batch in 50 mL Falcon tubes with at least three washes with 50 mL of Denaturing Wash Buffer. For each wash resuspend the beads in the buffer, spin down the beads at 1500 × *g* for 10 min at RT in a tabletop centrifuge, let the beads settle, and then discard the supernatant.
6. When the centrifugation step is finished add the cleared *E. coli* lysate to the 8 mL of SP Sepharose Fast Flow beads equilibrated in Denaturing Wash Buffer and rotate on a rotating wheel for 30 min at RT. If using 50 mL Falcon tubes this will have to be done in several tubes.
7. Spin down the beads at 1500 × *g* for 10 min at RT in a tabletop centrifuge and wash once in batch with 40 mL of Denaturing Wash Buffer.

8. Spin down the beads and load onto a gravity flow column preequilibrated in Denaturing Wash Buffer. Let excess buffer drain and elute 6× with 5 mL Denaturing Elution Buffer, collecting each fraction.
9. Combine all elution fractions (30 mL total) and filter through a 0.45 μm syringe filter. This is the cation-exchange-captured H2A that is used as input for the H2A purification instead of an extracted inclusion body (*see* Subheading 3.1.3 **step 8**).

3.1.5 Alternative Protocol for Purifying Histone H4

The purification of human histone H4 generally follows the "standard" purification protocol and utilizes a first enrichment step via an inclusion body preparation, followed by denaturing gel-filtration and ion-exchange steps. However, we found that H4 interacts very strongly with the chromatography media and the vast majority is retained on the columns under the standard purification conditions, drastically reducing yields (*see* **Note 31**). We therefore use modified conditions for the gel-filtration and ion-exchange chromatography to improve the yields. H4 is expressed and extracted from inclusion bodies as usual (*see* Subheadings 3.1.1 and 3.1.2); however, as described in the original protocol [6] the gel-filtration chromatography is performed in SAU-1000 buffer containing 1 M NaCl to decrease the interaction with the resin, and in the ion-exchange step we simultaneously increase the salt concentration and the pH which improves the elution efficiency from the SP Sepharose column.

1. Express and extract histone H4 from the inclusion bodies as described in Subheadings 3.1.1 and 3.1.2, and follow the gel-filtration purification protocol for H4 as described in Subheading 3.1.3. However, instead of equilibrating and running the HiPrep 26/60 Sephacryl S-200 HR column in SAU-200 buffer perform the gel filtration in SAU-1000 buffer (*see* **Notes 9** and **28**).
2. In the morning when the overnight gel-filtration runs are finished: check the fractions by running a 17.5% SDS-PAGE gel and performing a Coomassie stain. Combine the fractions containing the H4 peak (usually 40–50 mL total volume from both runs).
3. While running the SDS-PAGE gel prepare fresh SAU-0, SAU-200, and HU-1000 buffers using leftover deionized urea from preparing the gel-filtration buffer (*see* **Note 8**) and equilibrate a 5 mL HiTrap SP Sepharose HP column in SAU-200 buffer.
4. Combine the positive fractions and dilute 1:5 in SAU-0 to adjust the NaCl concentration to 200 mM.
5. Load the diluted fractions onto the equilibrated 5 mL HiTrap SP Sepharose HP column (*see* **Note 32**), wash the column with 5 CV (25 mL) of SAU-200, and then elute with a gradient

from 0% - 100% HU-1000 over 10 CV (Buffer A: SAU-200; Buffer B: HU-1000). Run the column at a flow rate of 1 mL/min and collect 2 mL fractions during the gradient (*see* **Note 33**).

6. Check the ion-exchange chromatography fractions by running a 17.5% SDS-PAGE gel and performing a Coomassie stain.
7. After combining the ion-exchange chromatography fractions containing pure histone H4 the protocol follows the standard histone purification protocol. Continue from Subheading 3.1.3 **step 13**. Due to the smaller size of H4 the dialysis is carried out with 6–8 kDa MWCO dialysis membranes (e.g., Spectra/Por 1) instead of 12–14 kD MWCO membranes. When freeze-drying, aliquot the H4 into 0.4 or 0.8 mg aliquots (*see* **Note 30**). H4 tends to aggregate. If the H4 precipitates during the final dialysis steps into water, reduce the amount of input material and/or perform a first dialysis into deionized water containing 350 mM NaCl and 0.1% acetic acid (*see* **Note 14**) overnight followed by three extensive dialysis steps into water containing 0.1% acetic acid.

3.2 Generation of Post-translationally Modified Histones by Native Chemical Ligation

The key component for the identification of methylation-specific nucleosome binding proteins are histones that carry defined methylation marks at specific positions in their sequence. There are multiple ways to produce histones that incorporate modified amino acids [8]. In our lab we employ the native chemical ligation (NCL) strategy [9] to install methylated lysines or arginines in the N-terminal tails of histones. Ligation of full-length modified histones is achieved by fusing a synthetic modified N-terminal histone tail peptide containing a C-terminal thioester to a truncated histone core lacking the N-terminal tail that is expressed and purified from bacteria and that contains a cysteine at its N-terminus (Fig. 2). Once a reliable ligation strategy has been identified, this procedure is very robust. In addition, since the N-terminal fragment is generated through solid-phase peptide synthesis any modification can be incorporated at any position within the chosen histone tail, and no artificial linkages or nonnatural amino acids are introduced. However, it is important to note that the native chemical ligation necessarily leaves behind a cysteine at the ligation site and therefore introduces a mutation in the histone sequence if an amino acid that is not naturally a cysteine is chosen as the ligation site (Fig. 2).

The main challenge in this procedure is the expression of the truncated histone core containing the N-terminal cysteine, and its expression and purification need to be optimized. The easiest way to position a cysteine at the N-terminus is to express the truncated histone recombinantly in *E. coli* and engineer the expression construct such that the methionine translated from the start codon is directly followed by the cysteine of the chosen ligation site. The *E. coli* methionyl-aminopeptidase then cleaves the N-terminal

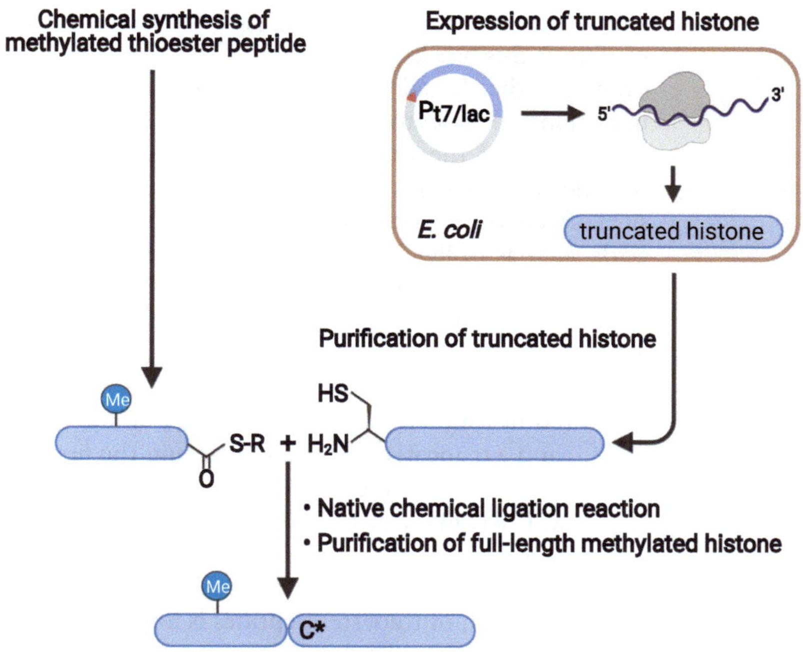

Fig. 2 Schematic illustration of the procedure for generating core histones with site-specific N-terminal methyl modifications. Modified histone N-terminal tail peptides containing a C-terminal thioester are generated through solid-phase peptide synthesis. Truncated histones lacking the N-terminal tail and containing an N-terminal cysteine (indicated by C*) at the desired ligation site are expressed in bacteria and purified by gel-filtration and reversed-phase chromatography (*see* Fig. 3). The full-length modified histones are then obtained by native chemical ligation and separated from unligated truncated histones by reversed-phase chromatography (*see* Fig. 4)

formyl-methionine exposing the cysteine at the N-terminus [10]. In our hands, the removal of the formyl-methionine works very efficiently and we employ this strategy to express truncated histone H4 [5]. Another way of positioning a cysteine at the N-terminus is to express a truncated histone with an N-terminal tag containing a protease cleavage site that leaves a cysteine in front of the histone core sequence, and to process the histone with the corresponding protease. We employ such a strategy to express truncated histone H3 [3, 4, 11] by processing it with coexpressed TEV protease which accepts a cysteine in its P1' site [12].

Another critical aspect is the choice of the ligation site and thus a ligation strategy that matches the experimental question. Long thioester peptides are challenging to synthesize, and they are therefore usually limited in length to around 30 to 40 amino acids. The native chemical ligation strategy is therefore most suitable for studying modifications in the N-terminal tails. If purchased from a commercial supplier, these peptides can also be quite costly, so the use of shorter peptides is advised wherever possible. There are a number of restrictions when it comes to the selection of the ligation site. A first parameter when expressing truncated histones in which the cysteine is directly positioned after the start codon is that the amino acid following the cysteine (the P2' site) must not be a proline as this blocks the cleavage of the N-terminal formyl-methionine by the *E. coli* methionyl-aminopeptidase [10]. A second important consideration is that the side chain of the C-terminal amino acid of the thioester peptide (i.e., the amino acid in the histone sequence preceding the cysteine placed at the ligation site) affects the speed of the ligation reaction, with certain amino acids such as isoleucine, valine, or proline reacting very slowly or not at all [13]. The ligation site must therefore be chosen with care based on the amino acids surrounding the cysteine. In order to be able to introduce methylations at all critical lysines in the histone H3 (H3K4, H3K9, H3K27) and H4 (H4K20) tails, we use a truncated histone H3 lacking the first 31 amino acids (H3 Δ1–31 T32C) and a truncated histone H4 lacking the first 28 amino acids (H4 Δ1–28 I29C), and corresponding N-terminal tail peptides (H3: a.a. 1–31, H4: a.a. 1–28) containing C-terminal benzyl thioesters (-S-Bzl).

There are additional aspects that should be mentioned. Firstly, since the length of thioester peptides is limited the native chemical ligation strategy is most suited for modifications in the N-terminal tails and not the method of choice for modifications in the histone core fold. Although sequential ligation of histones from multiple peptides is possible, e.g., to install lysine 36 methylation [14, 15], directly installing modified amino acids via genetic code expansion [16] might be a better strategy for targeting modifications in the histone core. Here, incorporation of phosphoserine and subsequent dehydroalanine conversion has emerged as an interesting and versatile method to install a variety of modifications [17, 18]. Placing modifications in the C-terminal part of a histone is possible via a native chemical ligation strategy using C-terminal intein fusions of histones and C-terminal histone peptides carrying an N-terminal cysteine [19]. Secondly, the modified N-terminal tail peptides can also be synthesized with chemistries other than C-terminal thioesters, e.g., a C-terminal hydrazide [15]. These can then be converted into a reactive thioester in the ligation reaction. Lastly, the cysteine that is introduced into the histone at the ligation site can be desulfurized and thereby converted into an alanine [20]. If the ligation site is chosen such that the cysteine replaces an alanine in the native histone sequence, the ligation can effectively be made traceless leaving no artificial linkages or mutations behind. We are omitting this step since human histone H3 contains two additional cysteines in its peptide sequence that we wish to preserve, and there is no suitable alanine ligation site in the histone H4 N-terminus that fulfils the requirements for ligation as detailed above and allows incorporation of H4K20 methylation. Instead we have opted for ligation sites in H3 and H4 where the cysteines constitute "conserved" mutations that are assumed not to interfere with the function of the proteins. In this section we will describe the procedures for preparing truncated histones (Fig. 3),

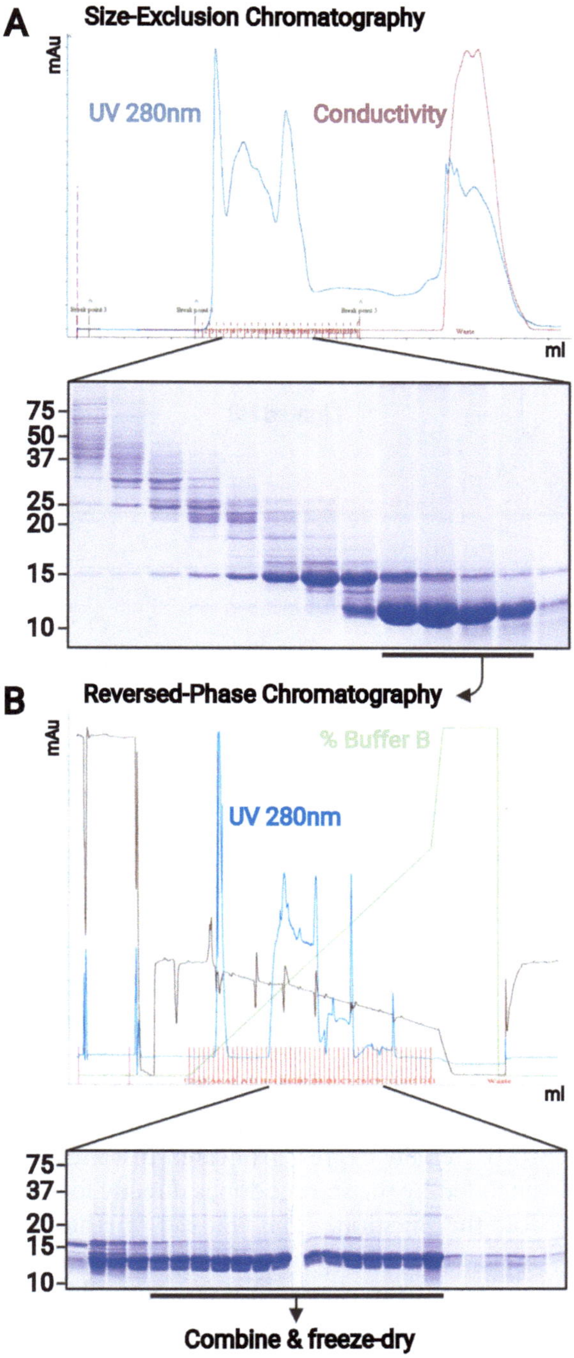

Fig. 3 Representative chromatograms and Coomassie-stained SDS-PAGE gels for the purification of truncated histone H3 for native chemical ligations. (**a**) A representative UV absorbance profile (UV 280 nm) of the gel-filtration (size-exclusion) chromatography purification of truncated histone H3 using a HiPrep 26/60 Sephacryl S-200 HR column. Selected fractions were resolved by SDS-PAGE and stained with Coomassie. Fractions containing the truncated H3 were combined and further purified by reversed-phase chromatography. (**b**) A representative UV absorbance profile (UV 280 nm) of the reversed-phase chromatography purification of truncated histone H3 using a 4 mL SOURCE 15RPC column. The solvent gradient is indicated in green (% Buffer B). Selected fractions were resolved by SDS-PAGE and stained with Coomassie. Fractions containing pure truncated H3 were combined and lyophilized as indicated. Molecular weight markers (kDa) are indicated on the left of the gels

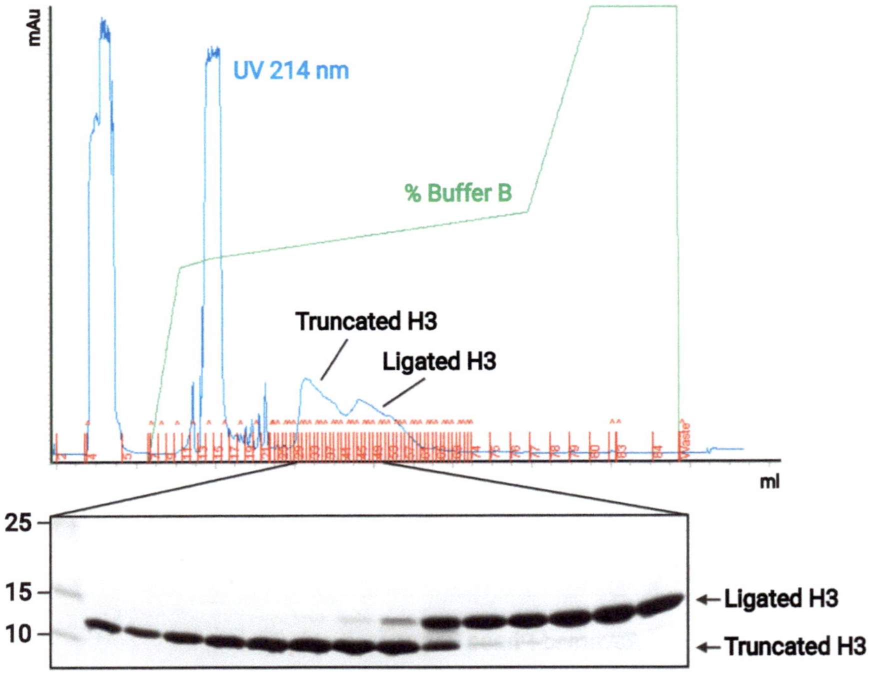

Fig. 4 Purification of ligated modified histone H3. A representative UV absorbance profile (UV 214 nm) of a reversed-phase chromatography purification of ligated lysine 9 tri-methylated histone H3 (H3K9me3) using a PerkinElmer Aquapore RP-300 (C8) 250 mm × 4.6 column. The solvent gradient is indicated in green (% Buffer B). Selected fractions were resolved by SDS-PAGE and stained with Coomassie. Fractions containing full-length ligated H3 were combined and lyophilized. Molecular weight markers (kDa) are indicated on the left of the gel

the native chemical ligation procedure (Fig. 2), and the final purification of ligated modified full-length histones (Fig. 4).

3.2.1 Expression and Purification of N-Terminally Truncated Histones

The best way to make truncated histones lacking the N-terminal tail and containing an N-terminal cysteine at the desired ligation site is to bacterially express truncated histones in which the methionine encoded by the start codon is directly followed by the cysteine used for the ligation. The *E. coli* methionyl-aminopeptidase then removes the N-terminal formyl-methionine thereby exposing the cysteine at the N-terminus (*see* **Note 34**). Truncated histones produced in this way can be expressed following the procedure described in Subheading 3.1.1 (for materials and reagents *see* Subheading 2.1). Our protocol for generating modified histones by NCL works similarly well for both histones H3 and H4. Both truncated histones are insoluble when expressed in *E. coli* and are deposited in inclusion bodies. The first step in the purification procedure therefore is an inclusion body preparation which follows the protocol described in Subheading 3.1.2 (for materials and reagents *see* Subheading 2.4). Using the inclusion bodies as starting material the protocol for purifying the truncated histones follows a two-step procedure (Fig. 3) similar to the standard histone purification protocol (*see* Subheading 3.1.3); however, the ion-exchange chromatography step following the denaturing gel filtration is replaced by a reversed-phase chromatography (*see* **Note 35**). The protocol below describes the procedure for purifying truncated histones from inclusion bodies obtained from 2 L of bacterial culture (*see* **Note 36**).

1. Starting in the morning, attach a HiPrep 26/60 Sephacryl S-200 HR column to an FPLC chromatography system and wash the column with >100 mL of filtered and degassed water.
2. While washing the column in water prepare 1.5 L of SAU-200 (for truncated H3) or SAU-1000 (for truncated H4) gel-filtration buffer from a freshly prepared deionized 8 M urea stock solution (*see* **Note 8**).
3. When the buffers are ready, equilibrate the column with approximately 500 mL SAU-200 (truncated H3) or SAU-1000 (truncated H4) buffer during the day. The gel-filtration buffers for the truncated histones are prepared without β-mercaptoethanol (*see* **Note 11**). As explained in Subheading 3.1.5 we use a dedicated column for H4 purifications (*see* **Note 31**).
4. In the afternoon while the column is equilibrating in the gel-filtration buffer: thaw the frozen inclusion body pellets (*see* Subheading 3.1.2) and thoroughly resuspend them in 25 mL of Unfolding Buffer, breaking up clumps with the pipette. The cysteines are reduced with 100 mM DTT (*see* **Notes 10** and **37**). For the purification of truncated histones for NCL we use inclusion body pellets corresponding to 2 L of culture.
5. Rotate the suspension in a 50 mL Oak Ridge tube for 1 h at RT on a rotating wheel to extract the histones and then centrifuge for 20 min at 20,000 × *g* and 25 °C (e.g., Sorvall LYNX 6000 centrifuge with A27-8x50 rotor or equivalent).
6. Transfer the supernatant (SN) containing the extracted truncated histones to a new 50 mL Oak Ridge tube and centrifuge again for 20 min at 20,000 × *g* and 25 °C.
7. Transfer the SN to a 50 mL plastic syringe and filter through a 0.45 μm filter. This is the extracted inclusion body. The filter clogs easily due to residual particles, several filters might be necessary to filter all of the sample.
8. In the evening, once the column is equilibrated in the gel-filtration buffer: load the extracted inclusion body sample into a 50 mL super-loop and separate over the HiPrep 26/60 Sephacryl S-200 HR column. Run two consecutive gel-filtration runs in SAU-200 buffer (for truncated H3) or

SAU-1000 (for truncated H4) at a flow rate of 1 mL/min overnight, injecting half of the sample (approximately 12 mL) in each run (*see* **Note 29**). Run both runs at RT (*see* **Notes 9, 11**, and **28**) and collect 5 mL fractions separately for each run.

9. The next day in the morning: check the gel-filtration fractions (Fig. 3A) by running a 17.5% SDS-PAGE gel and performing a Coomassie stain. Combine the fractions containing the truncated histone peak (usually 40–50 mL total volume from both runs).
10. While running the SDS-PAGE gel prepare fresh 0.1% trifluoroacetic acid in water (Buffer A) and 90% acetonitrile/0.1% trifluoroacetic acid in water (Buffer B) (*see* **Notes 12** and **16**) and equilibrate a 4 mL SOURCE 15RPC column (*see* **Note 36**) with at least two cycles of Buffer A and Buffer B (*see* **Note 38**).
11. Using a 50 mL super-loop load the combined fractions directly onto the equilibrated 4 mL SOURCE 15RPC column at a flow rate of 0.4 mL/min, wash the column with 5 CV (20 mL) of Buffer A, and then elute with a gradient from 0% - 65% Buffer B over 20 CV. Run the column at a flow rate of 0.5 mL/min and collect 2 mL fractions during the gradient. Clean the column after the purification run by two additional up-down gradient cycles between Buffer A and Buffer B, then wash the column in water and store in 20% Ethanol (*see* **Note 12**).
12. To check the reversed-phase chromatography fractions (Fig. 3b), dry 12.5 μL of each fraction in a SpeedVac concentrator or by letting the liquid evaporate in a fume hood. Add 15 μL of 1× SDS-PAGE sample buffer to the dried sample and shake at 850 rpm for 10 min in a Thermomixer at 65 °C, then boil at 95 °C for 5 min. Separate 5 μL of each sample on a 17.5% SDS-PAGE gel and perform a Coomassie stain. The truncated histones do not elute as a sharp peak but rather elute as a "smear" over many fractions. However, most of these fractions contain essentially pure histones (*see* **Note 35**).
13. Combine the fractions containing pure truncated histones and freeze-dry them overnight (*see* **Notes 17**, **39**, and **40**).
14. The next day: resuspend the dried histones in 20 mL of purified degassed water and determine the protein concentration by measuring the OD at 280 nm of the undiluted solution against water (*see* **Note 41**).
15. Aliquot 1 mg aliquots into microtubes and freeze-dry again overnight. The freeze-dried truncated histones are generally highly pure and can be stored at −80 °C for several years. The expected yields are approximately 25–30 mg for 2 L of culture.

3.2.2 Native Chemical Ligation and Purification of Modified Histones

For the preparation of modified histone H3 we use the native chemical ligation strategy [9] to ligate N-terminal H3 peptides (a.a. 1–31) to truncated H3 (H3 Δ1–31 T32C), and for the preparation of modified histone H4 we ligate N-terminal H4 peptides (a.a. 1–28) to truncated H4 Δ1–28 I29C. The ligated histones are then separated from the unligated truncated histones via a reversed-phase chromatography purification using a C8 HPLC column (Fig. 4). We routinely use peptides containing a C-terminal benzyl thioester (-S-Bzl), and our histone H4 peptides are N-terminally acetylated, mimicking cellular H4 which exists mainly in N-terminally blocked form [21]. The protocol describes the ligation procedure for a final yield of around 2 mg of ligated modified histones.

1. Add 1 mL of Native Chemical Ligation Buffer into a 5 mL round-bottom flask with ground glass joint.
2. Connect the 5 mL round-bottom flask to a vacuum pump via a three-way vacuum/inert gas adapter (T-bore glass stopcock). Switch on the vacuum and degas for about 5 min while stirring the buffer with a small stir bar (*see* **Note 42**). Switch off the vacuum pump and turn the valve to fill the flask with argon gas via the inert gas inlet (*see* **Note 43**).
3. Transfer the degassed buffer into a microtube containing 20 mg of TCEP powder and carefully resuspend. TCEP is the reducing agent for the native chemical ligation reaction.
4. Adjust pH to 7.5 with 5 M KOH using pH strips. For measuring the pH use a pipette to take out a small amount of buffer (10–20 μL) and drip onto the pH strip.
5. Transfer the buffer to the 5 mL round-bottom flask and degas again (repeat **step 2**).
6. Transfer the degassed buffer into a microtube containing 25 mg of MPAA powder and resuspend. MPAA is the thiol catalyst for the native chemical ligation reaction (*see* **Note 44**).
7. Adjust pH to 7.5 again with 5 M KOH using pH strips (*see* **step 4**).
8. Transfer the buffer to the 5 mL round-bottom flask again and degas (repeat **step 2**). This is the active Native Chemical Ligation Buffer.
9. Use 550 μL of the degassed Native Chemical Ligation Buffer to carefully dissolve 1 mg of N-terminal histone thioester peptide (e.g., H3: a.a. 1–31; or H4: a.a. 1–28; dependent on ligation strategy) containing the desired modifications without introducing any bubbles (*see* **Note 45**).
10. Use the resulting peptide solution to carefully dissolve 4 mg of truncated histone (e.g., H3 Δ1–31 T32C or H4 Δ1–28 I29C;

dependent on ligation strategy) prepared as described in Subheading 3.2.1 without introducing any bubbles. Make sure the tube is sealed.

11. Incubate the reaction in a Thermomixer at 40 °C overnight (without shaking).
12. The next day: quench the overnight ligation reaction by adding 60 μL of 1 M DTT and 700 μL of 0.5% acetic acid.
13. Centrifuge the reaction at maximum speed in a tabletop microcentrifuge for 10 min to remove any precipitates. Transfer the supernatant into a 2 mL microtube and top up to 2 mL with 0.5% acetic acid.
14. Prepare fresh 0.1% trifluoroacetic acid in water (Buffer A) and 90% acetonitrile/0.1% trifluoroacetic acid in water (Buffer B) (*see* **Notes 12** and **16**) and equilibrate a 4 mL C8 reversed-phase chromatography HPLC column (e.g., PerkinElmer Aquapore RP-300 (C8) 250 mm × 4.6 mm i.d. or similar) with at least two cycles of Buffer A and Buffer B (*see* **Note 38**).
15. Using a 2 mL sample loop load the cleared ligation reaction from **step 13** onto the equilibrated 4 mL reversed-phase chromatography HPLC column at a flow rate of 0.5 mL/min, wash the column with 5 CV (20 mL) of Buffer A, and then gradient-elute the ligated histones. For ligated histone H3 we use a gradient from 45–55% Buffer B over 10 CV and for ligated histone H4 we use a gradient from 35–45% Buffer B over 10 CV. Run the column at a flow rate of 0.5 mL/min and collect 1 mL fractions during the gradient.
16. Clean the column after the purification run by two additional up–down gradient cycles between Buffer A and Buffer B, then wash the column in water and store in 100% Methanol (*see* **Notes 12** and **15**).
17. To check the reversed-phase chromatography fractions (Fig. 4), dry 10 μL of each fraction in a SpeedVac concentrator or by letting the liquid evaporate in a fume hood. Add 10 μL of 1× SDS-PAGE sample buffer to the dried sample and shake at 850 rpm for 10 min in a Thermomixer at 65 °C, then boil at 95 °C for 5 min. Separate 5 μL of each sample on a 17.5% SDS-PAGE gel and perform a Coomassie stain. There are two major broad peaks corresponding to the ligated histones and the truncated histones.
18. Combine the fractions containing pure ligated histones and freeze-dry them overnight (*see* **Notes 17**, **39**, and **40**).
19. The next day: resuspend the ligated histones in 3 mL of purified water and determine the protein concentration by measuring the OD at 280 nm of the undiluted solution against water (For calculating the concentrations *see* Table 1).

20. Prepare 0.5 mg aliquots for ligated histone H3 or 0.4 mg aliquots for ligated histone H4 in microtubes and freeze-dry again overnight (*see* **Note 30**). The expected yield of this procedure is around 2 mg of pure modified histones. The freeze-dried modified histones can be stored at −80 °C for several years.

3.3 Refolding of Histone Octamers

Following the purification of the four core histones H2A, H2B, H3.1, and H4, they are refolded into histone octamers. For this we again essentially follow the protocol developed by Karolin Luger's lab [6, 7]. Here recombinant histones (*see* Subheading 3.1) or post-translationally modified histones generated by native chemical ligation (*see* Subheading 3.2) can be used to refold octamers in any desired combination. Histone mutants or modified histones produced by other techniques can also be used. The human histones H2A, H2B, H3.1, and H4 that are stored separately in lyophilized form at −80 °C are first unfolded, mixed in equimolar ratios, and then dialyzed into octamer refolding buffer. Refolded octamers are subsequently separated from H3/H4 tetramers and H2A/H2B dimers by gel-filtration chromatography (Fig. 5). The formation of octamers is favored over H3/H4 tetramers by adding a 10% excess of H2A and H2B. Here we describe the procedure for an input of around 4 mg of total histone protein. The method works equally well for larger or smaller amounts; however, the size of the gel-filtration column needs to be adjusted accordingly (*see* **Note 46**). In our experience there is a significant loss of histone proteins on the column, and only 25–30% of the input material is recovered as refolded octamers. For 4 mg of histone input a yield of around 1 mg of purified octamer is expected.

1. In the morning: collect 1 mg of each of the lyophilized histones H2A, H2B, and H3.1 and 0.8 mg of lyophilized H4 from the −80 °C freezer and let the tubes warm up to RT. In the meantime, prepare 10 mL of Unfolding Buffer.
2. Resuspend the histones separately in Unfolding Buffer to a concentration of approximately 2 mg/mL. The histones will dissolve immediately, but make sure that you have resuspended all of the protein in the tube by pipetting up and down several times and washing the walls of the tubes with the Unfolding Buffer. Do not vortex.
3. Let the histones unfold at RT for 1–2 h (longer unfolding achieves higher octamer yields).
4. In the meantime, prepare 2 L of Octamer Refolding Buffer and keep it at 4 °C (*see* **Note 47**).
5. Remove any undissolved matter by centrifugation, transfer the supernatants to new tubes, and determine the concentration of the individual unfolded histone proteins by measuring the

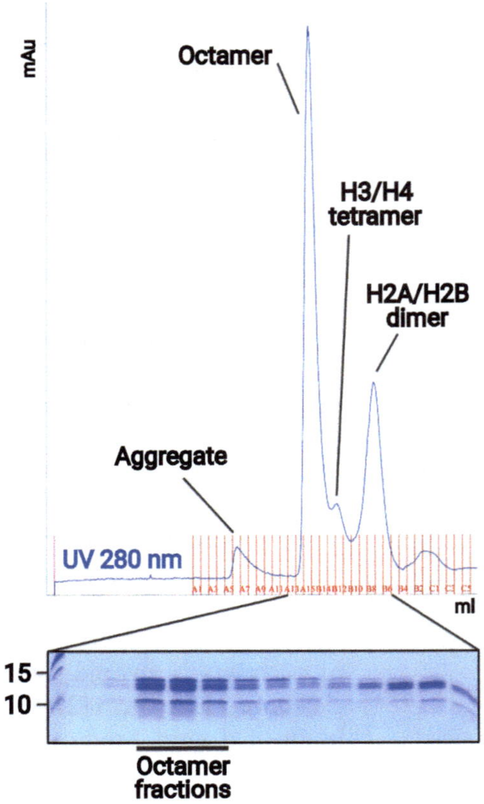

Fig. 5 Purification of histone octamers. A representative UV absorbance profile (UV 280 nm) of a gel-filtration chromatography purification of wild-type histone octamers using a Superdex-200 16/60 Prep-grade column. Selected fractions were resolved by SDS-PAGE and stained with Coomassie. Fractions containing pure histone octamers were combined, diluted with glycerol to a final concentration of 50% (v/v), and stored at −20 °C. Molecular weight markers (kDa) are indicated on the left of the gel

OD_{280nm} of the undiluted solutions against Unfolding Buffer. For calculating the concentrations *see* Table 1.

6. Combine 1 mg of each histone H2A, H2B, and H3.1 and 0.74 mg of H4 (*see* **Note 48**) and adjust the final protein concentration to 1 mg/mL with Unfolding Buffer.
7. Transfer into dialysis tubing (6–8 kD MWCO), and dialyse against at least three changes of 500 mL Octamer Refolding Buffer (*see* **Note 49**). The dialyses are done in the cold room at 4 °C.
8. In the evening: install a 120 mL Superdex 200 16/60 Prep-grade gel-filtration column on an FPLC system and equilibrate overnight with 120 mL (1 CV) of filtered and degassed water followed by 180 mL of filtered and degassed Octamer Refolding Buffer (*see* **Note 50**).

9. The next day: put aside an "input" sample of 4 μL from the contents of the dialysis bag and store it on ice; transfer the rest to a centrifugal spin filtration unit with 10 kDa MWCO (e.g., Sartorius Vivaspin 4 or similar) and concentrate to approx. 1 mL (it is essential to set aside an "input" prior to concentration as the sample is too concentrated for the gel after the concentration step).
10. Centrifuge the concentrated sample in a cooled tabletop microfuge at full speed at 4 °C for 10 min.
11. Transfer the supernatant to a new tube and spin again for 10 min at full speed in a tabletop microfuge at 4 °C. The two centrifugation steps will remove any remaining particles from the sample.
12. Load the sample into a 2 mL sample-loop and separate over the Superdex-200 16/60 Prep-grade gel-filtration column equilibrated in Octamer Refolding Buffer. Run the gel filtration at 0.8 mL/min and collect 2 mL fractions. The gel filtration should preferably be done at 4 °C but also works at RT (*see* **Note 50**).
13. Check the gel-filtration fractions on a 17.5% SDS-PAGE gel. For the gel it is important to dilute the NaCl concentration in the samples to below 1 M in order to avoid distortion of the bands due to the high salt content. For this transfer 4 μL of each fraction to a microtube and add 1 μL of 5× SDS-PAGE sample buffer followed by 5 μL of 1× SDS-PAGE sample buffer. Mix, boil for 5 min at 95 °C, and load 5 μL per lane. Treat the input sample in the same way and load as a marker.
14. Combine the octamer-containing fractions (Fig. 5) that are indicated by equimolar amounts of the four core histones (*see* **Note 51**).
15. The octamers can be stored in the refolding buffer at 4 °C for immediate use. For this the protein concentration is determined by measuring the OD_{280nm} of the undiluted octamer solution against the Octamer Refolding Buffer. The absorbance at 280 nm of a 0.1% solution (Abs 0.1% [1 mg/mL]) of wild-type human histone octamers is 0.412 (*see* also Table 1).
16. Alternatively the octamers can be diluted with glycerol to a final concentration of 50% (v/v) for long-term storage at −20 °C. For this concentrate the combined octamer fractions to approximately 2 mg/mL (can also be higher) using a centrifugal spin filtration unit (e.g., Sartorius Vivaspin 4 or equivalent). Measure the protein concentration as in step 15. and determine the precise volume of the concentrated histone octamer solution. Add the exact same volume of 100% glycerol and an adequate volume of a 1 M DTT stock solution for a final

concentration of 1 mM DTT, mix and store at −20 °C (*see* **Note 52**). This reduces the octamer concentration by half. Histone octamers can be stored like this for several years; however, they should be stored in protein low-binding microtubes to avoid any protein losses through adsorption to the plastic.

3.4 Assembly of Biotinylated Dinucleosomes

The modified histone octamers are reconstituted into nucleosomes together with nucleosomal DNA. For the nucleosome assembly we use a continuous salt gradient dialysis protocol to deposit the histone octamers onto the 601 nucleosome positioning sequence [22] that allows the reconstitution of homogenous nucleosome particles by directing the octamers to a single position on the DNA. We again follow the general protocol described for the reconstitution of mononucleosomes [6, 7]. However, we have adjusted the procedure in order to assemble nucleosomes with an end-biotinylated 601 DNA so that the nucleosomes can be immobilized on streptavidin-coupled beads for pull-down reactions (*see* Subheading 3.6). Furthermore, we found that affinity purifications with dinucleosomes, in which two octamers are assembled on a DNA sequence that contains two 601 sequences separated by a 40–50 bp spacer, are more robust than nucleosome pull-downs with mononucleosomes. We have therefore modified the protocol to include a 147 bp MMTV-A competitor DNA in the assembly reaction, similar to the assembly of chromatin arrays [23, 24]. While the octamer: DNA ratio needs to be precisely titrated for mononucleosomes, addition of the MMTV-A DNA creates an "octamer buffer" that allows slight overtitration with the octamers to saturate both 601 sites. This makes the assembly of dinucleosomes more consistent and the pull-downs are therefore more reproducible. In this section we describe the preparation of the biotinylated 601 dinucleosome DNA (Fig. 6) and the MMTV-A competitor DNA, the reconstitution of biotinylated 601 dinucleosomes, and the quality control of the nucleosomal assembly reaction by native PAGE (Fig. 7).

3.4.1 Preparation of Biotinylated 601 Dinucleosome DNA

Biotinylated nucleosomal DNA can be generated in different ways depending on the amounts and designs needed (*see* **Note 53**). We typically require milligram quantities of the final biotinylated 601 dinucleosome DNA. We therefore prepare the DNA from a pUC19 plasmid containing eight tandem repeats of a 601 dinucleosome sequence (*see* Fig. 6 and **Note 54**). The repeats are separated by *Eco*RV restriction sites and contain an *Eco*RI site at the opposite end. The dinucleosome DNA is excised from the vector backbone by a large-scale *Eco*RV digest and then separated from the vector DNA by a size-selective precipitation with polyethylene glycol (PEG). The isolated dinucleosome DNA is then further digested with *Eco*RI, and the *Eco*RI site is finally filled in with biotin-dUTP

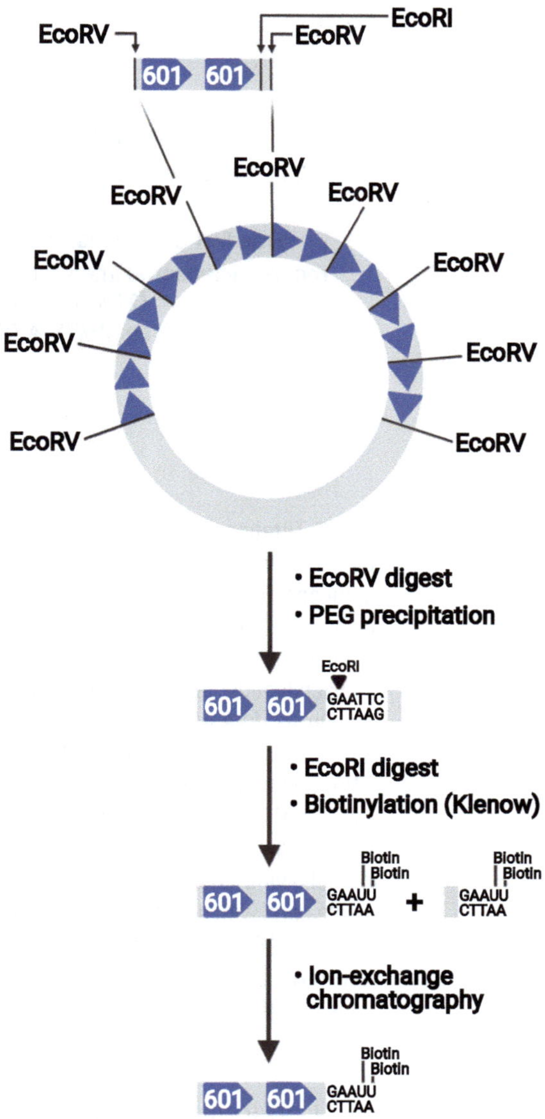

Fig. 6 Schematic illustration of the biotin-601 dinucleosome DNA purification strategy

using Klenow polymerase. This generates a dinucleosomal 601 DNA that is biotinylated at one end and blunt-ended at the other (Fig. 6). Here we describe the procedure for 10 mg of plasmid DNA as starting material which yields about 2–3 mg of final biotinylated 601 dinucleosome DNA. This protocol can also be used to prepare mono- and tetranucleosomal DNAs; however, the design of the plasmids, and the conditions for the digests, PEG precipitation, and the biotinylation reaction need to be adjusted.

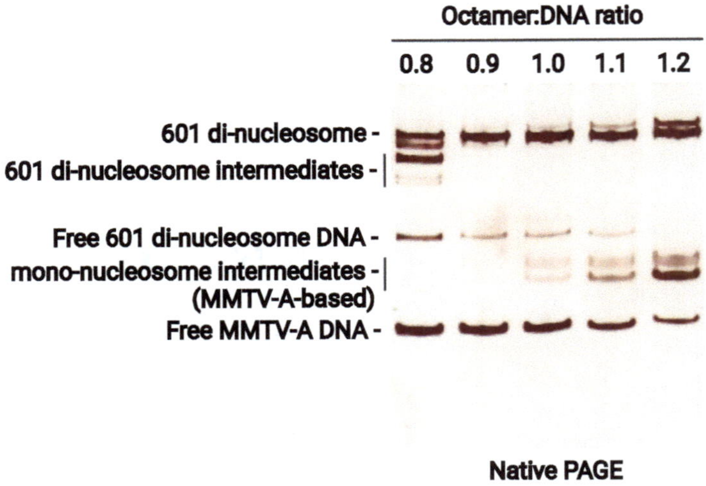

Fig. 7 Titration series of H3K9me3-modified histone octamers and biotin-601 dinucleosome DNA to determine the optimal conditions for dinucleosome reconstitutions. Nucleosomes were assembled with a constant amount of 601 dinucleosome DNA and MMTV-A competitor DNA, and increasing amounts of histone octamer as indicated. Reconstituted dinucleosomes were resolved by 5% native polyacrylamide gel electrophoresis and visualized using ethidium bromide. The amount of free 601 dinucleosome DNA and 601 dinucleosome intermediates decreases with an increasing ratio of histone octamer over the 601 dinucleosome DNA, and homogenous 601 dinucleosomes are formed. At higher octamer:DNA ratios MMTV-A-based mononucleosomes start to form indicating saturation of the 601 dinucleosome DNA. The optimal octamer:DNA ratio is selected based on the following criteria: (1) most of the 601 dinucleosomes should appear as a single band with no 601 dinucleosome intermediates present; (2) a small amount of the MMTV-A-based mononucleosomes should be visible; (3) the amount of free 601 dinucleosome DNA should be minimal. Most importantly, reconstitutions of different nucleosomes for the same experiment must be consistent

1. Transform the "pUC19-8x601 di-Nuc" plasmid into XL10 Gold cells and plate on LB_{Amp} agar plates containing 50 μg/mL ampicillin (*see* **Note 55**). Incubate the plate in a bacterial incubator at 37 °C overnight.
2. The next day around noon: inoculate 10 mL of LB_{Amp} medium (50 μg/mL ampicillin) in a 50 mL conical flask with a single small colony and incubate in a shaker incubator at approximately 250 rpm at 37 °C (*see* **Note 55**).
3. In the evening (around 6 or 7 pm); inoculate 6× 1 L LB_{Amp} medium (50 μg/mL ampicillin) in 6× 5 L conical flasks each with 1 mL of the preculture and shake in a shaker incubator at approximately 200 rpm at 37 °C overnight (*see* **Note 55**).
4. The next morning: harvest the cells by centrifugation in a large capacity centrifuge (e.g., Sorvall LYNX 6000 centrifuge with

Fiberlite F9-6x1000 LEX rotor) at 6000 × *g* for 10 min at 4 °C.

5. Purify the plasmid DNA using a Gigaprep Plasmid Kit according to the manufacturer's instructions (*see* **Notes 56** and **57**). Resuspend the final DNA pellet in water at a concentration of around 2 mg/mL (about 5 mL). The DNA can be stored at −20 °C until use.
6. Using all of the Gigaprep plasmid DNA (i.e., around 10 mg), set up an *Eco*RV digest at a DNA concentration of 1 mg/mL in a 15 mL Falcon tube. Use about 150 units of *Eco*RV per mg of DNA and incubate for at least 16 h (overnight) at 37 °C (*see* **Note 58**).
7. Check for completion of the digest by separating 2 μg of the DNA on a 1% agarose gel in 1× TBE buffer. If multimers of the dinucleosome fragment are still visible, add 50% more *Eco*RV and continue the digest for another 16 h at 37 °C. Check again on an agarose gel. Only proceed if the *Eco*RV digest is complete.
8. Once the digest is complete separate the excised *Eco*RV dinucleosome fragments from the linearised pUC19 vector backbone by PEG precipitation (*see* **Note 59**). Transfer the digest to a 50 mL polypropylene Oak Ridge centrifuge tube and adjust the solution with 5 M NaCl and 40% PEG-6000 stock solutions to a final concentration of 0.5 M NaCl and 5.5% PEG-6000 (*see* **Note 60**). Incubate on ice for 1 h and spin down the vector DNA for 20 min at 27,000 × *g* and 4 °C (e.g., Sorvall LYNX 6000 centrifuge with A27-8x50 rotor).
9. Transfer the supernatant containing the *Eco*RV-digested 601 dinucleosome fragments to a new polypropylene Oak Ridge centrifuge tube and set aside for step 11.
10. Wash the pellet with 70% ethanol and spin for 10 min at 20,000 × *g* and 4 °C. Remove the ethanol, air-dry the pellet, and resuspend the DNA in 5 mL of water. Determine the concentration and keep on ice—this is the "PEG pellet" sample.
11. Precipitate the *Eco*RV-digested 601 dinucleosome fragments from the supernatant by adding 2.5 volumes of ice-cold 100% ethanol and incubating the sample overnight at −20 °C. Depending on the volume the sample might have to be split into two centrifuge tubes.
12. Spin down the precipitated 601 dinucleosome fragments for 30 min at 20,000 × *g* and 4 °C. Discard the supernatant and wash the pellet with 5 mL of 70% ethanol. Spin again for 10 min at 20,000 × *g* and 4 °C. Remove the ethanol, air-dry the pellet, and resuspend the DNA in 2.5 mL of water.

Determine the concentration and keep on ice—this is the "PEG supernatant" sample.

13. Check 2 μg each of the precipitated "PEG supernatant" and the "PEG pellet" on a 1% agarose gel. There should be no contamination by the pUC19 vector in the *Eco*RV 601 dinucleosome fragment and only a small proportion of the fragment remaining in the PEG pellet. The recovery of the dinucleosome fragment should be around 80–90% of the expected maximum yield.
14. Keep 5–10 μg of the *Eco*RV-digested 601 dinucleosome fragment (i.e., the "PEG supernatant") on ice for a gel, and use the rest to set up an *Eco*RI digest at a DNA concentration of 1 mg/mL in a 15 mL Falcon tube. Use about 300 units of *Eco*RI per mg of DNA and incubate for at least 16 h (overnight) at 37 °C (*see* **Note 58**).
15. Check for completion of the digest by running a series of 0.5, 1, and 2 μg of the *Eco*RI-digested fragment next to the original (*Eco*RV-digested) dinucleosome fragment on a 2.5% agarose gel in 1× TBE buffer. The fragment should be completely shifted to a slightly smaller size, corresponding to the bases lost through the *Eco*RI digest. If some of the original dinucleosome fragment is still visible, add 50% more *Eco*RI and continue the digest for another 16 h at 37 °C. Check again on an agarose gel. Only proceed if the *Eco*RI digest is complete.
16. Transfer the *Eco*RI-digested 601 dinucleosome fragment to a polypropylene Oak Ridge centrifuge tube and ethanol-precipitate the DNA by adding 1/10 volume of 3 M sodium acetate (pH 5.2) and 2.5 volumes of ice-cold 100% ethanol and incubating the sample overnight at −20 °C.
17. Spin down the precipitated 601 dinucleosome fragment for 30 min at 20,000 × *g* and 4 °C. Discard the supernatant and wash the pellet with 5 mL of 70% ethanol. Spin again for 10 min at 20,000 × *g* and 4 °C. Remove the ethanol, air-dry the pellet, and resuspend the DNA in 2.5 mL of water. Determine the concentration and keep on ice.
18. Following the *Eco*RI digest, the 601 dinucleosome fragment is end-biotinylated by filling in the *Eco*RI overhangs with biotin-dUTP using Klenow Fragment (3′-> 5′ exo-) polymerase (*see* **Note 61**). Using all of the *Eco*RI-digested 601 dinucleosome fragment set up the biotinylation reaction at a final DNA concentration of 0.5 mg/mL in 1× Klenow reaction buffer containing 100 μM dATP, 40 μM biotin-11-dUTP, and 33.3 units/mL of Klenow Fragment (3′-> 5′ exo-). Let the biotinylation reaction proceed for 2 h at RT (25 °C) and then stop the reaction by adding 0.5 M EDTA (pH 8.0) to a final concentration of 20 mM.

19. To remove any traces of the small DNA fragment that is generated during the *Eco*RI digest (Fig. 6) we clean up the biotinylated 601 dinucleosome DNA via ion-exchange chromatography using an FPLC system. For this, first equilibrate a 1 mL HiTrap Q HP cartridge with HiTrap Q Buffer A. Using a 10 mL (or 50 mL) super loop load the biotinylation reaction onto the equilibrated 1 mL HiTrap Q HP column. Wash the column with 5 CV (5 mL) of Buffer A, and then gradient-elute the DNA with a gradient from 0%–100% HiTrap Q Buffer B over 20 CV (20 mL). Run the column at a flow rate of 0.5 mL/min and collect 1 mL fractions during the gradient. Measure the absorbance at 260 nm. In order to achieve a good separation of the biotinylated 601 dinucleosome DNA from the small DNA fragment the column must not be overloaded. If more than 3 mg of DNA needs to be separated it is advisable to split the purification into multiple runs, injecting no more than 2.5–3 mg of DNA per run. In this case the column should be cleaned with 5 mL of filtered 2 M NaCl and reequilibrated with 5 CV (5 mL) of Buffer A between runs. After the purification, clean the column with 5 mL of filtered 2 M NaCl, then wash the column in filtered and degassed water and store in 20% Ethanol.
20. Check 20 μL of the fractions on a 1.5% agarose gel, the biotinylated 601 dinucleosome DNA should elute as one major sharp peak just after the small fragment.
21. Combine the peak fractions containing the biotinylated 601 dinucleosome DNA (usually 3–4 mL) and then distribute into 2 mL microtubes. Phenol-extract the DNA with 50% of the volume of phenol–chloroform–isoamyl alcohol (25:24:1). For this, add an appropriate volume of phenol–chloroform–isoamyl alcohol (25:24:1) to each tube and vortex into an emulsion for 1 min. Spin at full speed for 1 min in a microfuge to separate the phases. Carefully transfer the upper aqueous phase containing the DNA to new 2 mL tubes, repeat three times in total (*see* **Note 20**).
22. Transfer the final aqueous phase to a 50 mL polypropylene Oak Ridge centrifuge tube and ethanol-precipitate the biotinylated dinucleosome DNA by adding 1/10 volume of 3 M sodium acetate (pH 5.2) and 2.5 volumes of ice-cold 100% ethanol and incubating the sample overnight at −20 °C.
23. Spin down the precipitated DNA for 30 min at 20.000 x g and 4 °C. Discard the supernatant and wash the pellet with 5 mL of 70% ethanol. Spin again for 10 min at 20,000 × *g* and 4 °C. Remove the ethanol, air-dry the pellet, and resuspend the DNA in 1 mL of water. This is the final biotinylated 601 dinucleosome DNA; determine the concentration and store in

aliquots at −20 °C (*see* **Note 62**). Aim for a concentration of around 2–3 mg/mL to facilitate the nucleosome reconstitutions. The expected yield is around 2–3 mg for a 10 mg input of the "pUC19-8x601 di-Nuc" plasmid.

3.4.2 Preparation of MMTV-A Competitor DNA

During the reconstitution of the biotinylated 601 dinucleosomes (*see* Subheading 3.4.3) the "nucleosome A" sequence of the 3′-LTR of the mouse mammary tumor virus (MMTV-A) is added as a competitor DNA to buffer excess octamers. This DNA is made from a pUC19 plasmid containing tandem repeats of the 147 bp MMTV-A sequence that are separated by *Eco*RV sites [23]. The procedure follows the protocol for making the 601 dinucleosome DNA; however, it omits the *Eco*RI digest and biotinylation reaction, and is therefore simpler and quicker. Here, we describe the procedure for about 10 mg of plasmid DNA (one Gigaprep) as starting material, which yields about 4 mg of final MMTV-A DNA.

1. Transform the "pUC19-MMTV-A" plasmid into XL10 Gold cells and plate on LB_{Amp} agar plates containing 50 μg/mL ampicillin (*see* **Note 55**). Incubate the plate in a bacterial incubator at 37 °C overnight.
2. The next day around noon: inoculate 10 mL of LB_{Amp} medium (50 μg/mL ampicillin) in a 50 mL conical flask with a single small colony and incubate in a shaker incubator at approximately 250 rpm at 37 °C (*see* **Note 55**).
3. In the evening (around 6 or 7 pm): inoculate 6× 1 L LB_{Amp} medium (50 μg/mL ampicillin) in 6× 5 L conical flasks each with 1 mL of the preculture and shake in a shaker incubator at approximately 200 rpm at 37 °C overnight (*see* **Note 55**).
4. The next morning: harvest the cells by centrifugation in a large capacity centrifuge (e.g., Sorvall LYNX 6000 centrifuge with Fiberlite F9-6x1000 LEX rotor) at 6000 × *g* for 10 min at 4 °C and purify the plasmid DNA using a Gigaprep Plasmid Kit according to the manufacturer's instructions (*see* **Notes 56** and **57**). Resuspend the DNA in water at a concentration of around 2 mg/mL (about 5 mL). The DNA can be stored at −20 °C until use.
5. Using all of the Gigaprep DNA (i.e., around 10 mg), set up an *Eco*RV digest at a DNA concentration of 1 mg/mL in a 15 mL Falcon tube. Use about 150 units of *Eco*RV per mg of DNA and incubate for at least 16 h (overnight) at 37 °C (*see* **Note 58**).
6. Check for completion of the digest by running 2 μg of the DNA on a 1% agarose gel in 1× TBE buffer. If multimers of the 147 bp MMTV-A fragment are still visible, add 50% more *Eco*RV and continue the digest for another 16 h at 37 °C.

Check again on an agarose gel. Only proceed if the *Eco*RV digest is complete.

7. Once the digest is complete separate the excised MMTV-A fragment from the linearised pUC19 vector backbone by PEG precipitation (*see* **Note 59**). Transfer the digest to a 50 mL polypropylene Oak Ridge centrifuge tube and adjust the solution with 5 M NaCl and 40% PEG-6000 stock solutions to a final concentration of 0.5 M NaCl and 7% PEG-6000 (*see* **Note 60**). Incubate on ice for 1 h and spin down the vector DNA for 20 min at 27,000 × *g* and 4 °C (e.g., Sorvall LYNX 6000 centrifuge with A27-8x50 rotor).
8. Transfer the supernatant containing the *Eco*RV-digested MMTV-A inserts to a new polypropylene Oak Ridge centrifuge tube and set aside for **step 10**.
9. Wash the pellet with 70% ethanol and spin for 10 min at 20.000 x g and 4 °C. Remove the ethanol, air-dry the pellet, and resuspend the DNA in 5 mL of water. Determine the concentration and keep on ice—this is the "PEG pellet" sample.
10. Precipitate the *Eco*RV-digested MMTV-A inserts from the supernatant by adding 2.5 volumes of ice-cold 100% ethanol and incubating the sample overnight at −20 °C. Depending on the volume the sample might have to be split into two centrifuge tubes.
11. Spin down the precipitated MMTV-A inserts for 30 min at 20,000 × *g* and 4 °C. Discard the supernatant and wash the pellet with 5 mL of 70% ethanol. Spin again for 10 min at 20,000 × *g* and 4 °C. Remove the ethanol, air-dry the pellet, and resuspend the DNA in 4 mL of water. Keep on ice—this is the "PEG supernatant" sample.
12. Check 2 μg of the precipitated "PEG supernatant" and the "PEG pellet" on a 1% agarose gel. There should be no cross-contamination between the pUC19 vector and the MMTV-A fragment.
13. Phenol-extract the MMTV-A DNA (i.e., the "PEG supernatant") with 50% of the volume of phenol–chloroform–isoamyl alcohol (25:24:1). For this, divide the reaction into 2 mL microtubes, add the appropriate volume of phenol–chloroform–isoamyl alcohol (25:24:1), and vortex into an emulsion for 1 min. Spin at full speed for 1 min in a microfuge to separate the phases. Carefully transfer the upper aqueous phase containing the DNA to new 2 mL tubes. Repeat three times in total (*see* **Note 20**).
14. Transfer the final aqueous phase to a 50 mL polypropylene Oak Ridge centrifuge tube and ethanol-precipitate the MMTV-A DNA by adding 1/10 volume of 3 M sodium acetate (pH 5.2)

and 2.5 volumes of ice-cold 100% ethanol and incubating the sample overnight at −20 °C.

15. Spin down the precipitated MMTV-A for 30 min at 20,000 × *g* and 4 °C. Discard the supernatant and wash the pellet with 5 mL of 70% ethanol. Spin again for 10 min at 20,000 × *g* and 4 °C. Remove the ethanol, air-dry the pellet, and resuspend the DNA in 2 mL of water. This is the final MMTV-A competitor DNA; determine the concentration and store in aliquots at −20 °C. Aim for a concentration of around 2–3 mg/mL to facilitate the nucleosome reconstitutions. The expected yield is around 4 mg for a 10 mg input of the pUC19-MMTV-A plasmid.

3.4.3 Reconstitution of Dinucleosomes for Affinity Purifications

Following the generation of the modified histone octamers (*see* Subheadings 3.1–3.3) and the biotinylated 601 dinucleosome DNA (*see* Subheading 3.4.1), these are reconstituted into dinucleosomes by salt gradient dialysis. Here, we essentially follow the protocol described for mononucleosomes [6] using the 601 sequence [22] to position two octamers on the dinucleosome DNA. We have found it difficult to reconstitute homogenous dinucleosomes from just octamers and dinucleosomal 601 DNA. To ensure full occupancy of both 601 nucleosome positions we therefore add a slight molar excess of octamers over the dinucleosome DNA, and buffer any surplus octamers with MMTV-A competitor DNA (*see* Subheading 3.4.2), similar to the protocols described for the reconstitution of nucleosomal arrays [23, 24]. The MMTV-A sequence has a weaker nucleosome-forming potential than the 601 sequence, therefore octamers will preferentially form nucleosomes on the 601 sequence during the salt gradient dialysis. Once all 601 sites are occupied nucleosomes will form on the MMTV-A DNA making sure that excess octamers do not lead to precipitation of the dinucleosomes. Since only the 601 dinucleosome DNA is biotinylated but not the MMTV-A DNA, only the 601 dinucleosomes will be immobilized on the streptavidin-coated beads for the nucleosome affinity purifications (*see* Subheading 3.5), and the MMTV-A DNA (and any nucleosomes formed from it) will be removed during the nucleosome immobilization and washing steps.

Similar to the reconstitution of mononucleosomes and nucleosomal arrays the optimal conditions for the assembly have to be determined experimentally, and the octamers and the 601 dinucleosome DNA need to be carefully titrated to find the best ratio for the reconstitution (Fig. 7). We usually reconstitute dinucleosomes at a scale of 25–50 μg in a volume of 50–100 μL in small dialysis buttons, and we usually use a 10–20% molar excess of octamers over the 601 nucleosome positioning sequences. The octamer: DNA ratio for optimal reconstitutions varies for different octamer and DNA preparations, mainly due to inaccuracies in the concentration measurements. It is therefore important to titrate each new batch of octamers and DNAs in order to get reliable reconstitutions (*see* **Note 63**). Once the optimal octamer:DNA ratio has been found the reconstitutions are usually very reproducible and consistent. As a general observation we find that the nucleosome assemblies work better at higher octamer and DNA concentrations. Here we describe the general reconstitution protocol; the exact amounts of octamers and DNAs and the volume of the reconstitution reactions need to be adjusted according to the results of the titrations and the experimental requirements.

1. Starting in the morning, hydrate 3.5 kDa MWCO dialysis buttons (or dialysis tubing for larger volumes) in water according to the manufacturer's instructions.
2. In the meantime prepare 400 mL of high-salt RB-high buffer in a 500 mL beaker (prechill at 4 °C) and equilibrate the dialysis buttons or dialysis tubing in RB-high buffer for a few minutes.
3. Assemble the nucleosome reconstitution reaction from the biotinylated 601 dinucleosome DNA, MMTV-A competitor DNA, histone octamers, and a 5 M NaCl stock solution. The optimal ratio of DNA vs. octamer concentration needs to be determined by titration (*see* Fig. 7 and **Note 63**). Use equal amounts of the biotinylated 601 dinucleosome DNA and MMTV-A competitor DNA in the reaction. First mix the DNAs and the 5 M NaCl solution, then fill up with an appropriate volume of TE buffer pH 7.5 to reach the calculated final volume and mix. Then add the desired amount of octamer (*see* **Note 64**). Make sure that the final NaCl concentration is 2 M. The final concentration of octamers should be around 2 to 4 μM. Mix thoroughly, and then carefully transfer the nucleosome assembly reaction to the dialysis buttons. Seal the buttons.
4. Place the buttons into the 400 mL RB-high buffer using a foam float and let them dialyse for around 30 min at 4 °C while stirring.
5. In the meantime prepare 2 L of RB-low (prechill at 4 °C). Transfer 1.6 L into a 2 L measuring cylinder for the nucleosome assembly. Save 400 mL for the final dialysis the next day, keep this at 4 °C.
6. Assemble a dual-channel pump setup for continuous exchange of buffer RB-high against RB-low as described in [6]. Set the pump to a flow rate of approximately 1.5 mL/min so that the RB-high buffer is completely exchanged against RB-low buffer over around 18 h.

7. Start the buffer exchange dialysis and let the nucleosome assembly proceed overnight at 4 °C (*see* **Note 65**). Check the pump flow rate every now and then and adjust if necessary.
8. The next day after the gradient dialysis has finished: place the dialysis buttons into 400 mL of RB-low buffer (1 mM DTT freshly added) and dialyse for another 2–4 h.
9. Carefully transfer the assembled dinucleosomes into DNA low-binding microtubes (*see* **Note 66**) and check the quality of the assembly reaction using a high-resolution native PAGE gel shift assay (*see* Subheading 3.4.4). The volume of the assembly reaction can change substantially during the dialysis—determine the volume and calculate DNA and octamer concentrations based on the new volume. The nucleosomes can either be stored at 4 °C for a few weeks or directly be immobilized on streptavidin beads (*see* Subheading 3.6.1). For other assays or for long-term storage the nucleosomes should be dialyzed into an appropriate low salt buffer overnight. The nucleosomes can also be concentrated in a centrifugal spin filtration unit without much loss.

3.4.4 Quality Control of Nucleosomal Assembly Reactions by Native PAGE

For analyzing the octamer vs. DNA titrations for finding the optimal conditions for the nucleosome assembly reactions (Fig. 7) and for determining the quality of the reconstituted dinucleosomes we use a high-resolution native PAGE gel shift assay [6, 7]. This assay is sensitive enough to detect any heterogeneity in the positioning of the histone octamers on the DNA. Homogenously assembled dinucleosomes are indicated by a single sharp 601 dinucleosome DNA band in the gel, while assembly reactions containing dinucleosomes with wrongly positioned or missing octamers are indicated by multiple or "fuzzy" bands. The following protocol describes the procedure for running the gel shift assay in a minigel setup.

1. Pour a nondenaturing 5% polyacrylamide gel (59:1 acrylamide–bisacrylamide) in 0.2× TBE buffer using a minigel assembly (10 × 8 cm) with 1.5 mm spacers and a 10-well comb. Let the gel polymerize overnight (*see* **Note 67**).
2. Prerun the gel before loading the samples for at least 1 h at 150 V and then mix the 0.2× TBE running buffer in the upper and lower gel chambers and redistribute (*see* **Note 68**).
3. Just before loading the samples, flush the wells thoroughly with 0.2× TBE (*see* **Note 68**).
4. Mix an appropriate volume of the nucleosome assembly reaction (from Subheading 3.4.3 **step 9**) with an equal volume of 2× native PAGE sample buffer. Load a volume corresponding to approximately 150–250 ng of 601 dinucleosome DNA. Load no more than 10 μL and layer the sample directly onto

the bottom of the well using a round gel-loading tip. Run the gel at 100 V in the cold room (*see* **Note 68**).

5. When the bromophenol blue has reached the bottom of the gel, disassemble the gel and place it into 50 mL of 0.2× TBE buffer. Stain with ethidium bromide (*see* **Note 18**) for around 10–15 min and visualize on a UV transilluminator. The majority of the 601 dinucleosome DNA should migrate as a single sharp band of a higher molecular weight than a faint band of the unincorporated DNA below. The MMTV-A buffer DNA will be visible as a major band of low molecular weight with several bands above indicating MMTV-A mononucleosomes (Fig. 7).

3.5 Preparation of HeLa S3 Nuclear Extracts

For the identification of methylation-specific nucleosome interactors nuclear proteins are isolated from nuclear extracts by affinity purifications with immobilized modified nucleosomes (*see* Subheading 3.6). Here we describe a simple protocol for preparing nuclear extracts for affinity pull-downs using sequential extraction of chromatin-bound proteins with increasingly higher salt concentrations (Fig. 8). Modifications to the standard procedure to probe RNA-dependent binding proteins or tightly bound chromatin proteins (*see* **Notes 69** and **70**) are indicated in steps 15 and 17 so that the protocol can be flexibly adjusted to different questions and conditions. This protocol uses HeLa S3 cells grown in regular tissue culture media and is therefore designed to be used for label-free quantification (*see* Subheading 3.7).

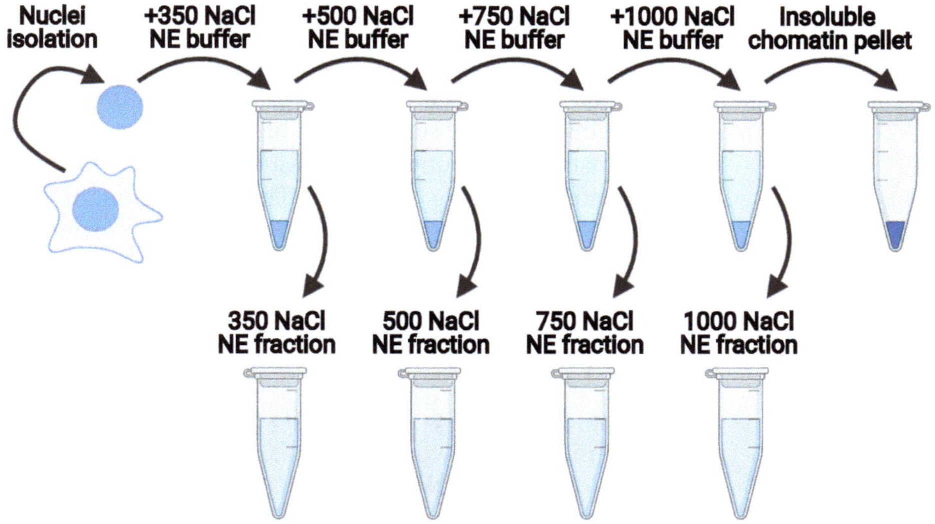

Fig. 8 Schematic illustration of the sequential nuclear extraction procedure. Nuclei are isolated using a gentle cell lysis method. Nuclear extract (NE) is then prepared by sequentially incubating nuclei/chromatin in buffers containing increasing concentrations of NaCl, ranging from 350 mM (350 NaCl NE buffer) to 1000 mM (1000 NaCl NE buffer)

Each preparation of a nuclear extract may differ in the exact protein composition; we therefore recommend to use aliquots from the same batch for each set of nucleosome pull-down experiments to achieve the best reproducibility and comparability between the samples. To this end, preparation of large quantities of nuclear extracts, i.e., 25–50 mg by the total protein amount, is recommended. This amount of nuclear extract is sufficient for 50–100 pulldown experiments and can be obtained by processing 4–5 L of HeLa S3 suspension culture at a cell density of 6–8 × 10^5 cells/mL. We also recommend preparing three independent extracts and mixing them prior to the final aliquoting to generate an "average" extract that evens out possible differences in the individual preparations. All steps during the extraction should be carried out with precooled centrifuges, pipettes, tubes, and so on, and the samples should be handled in the cold room and kept on ice throughout the procedure to preserve the integrity of the extracts.

1. Initiate the HeLa S3 culture by thawing a cryopreserved cell aliquot and adding the cells into a T175 cell culture flask containing 20 mL of prewarmed culture medium. Grow the cells at 37 °C and 5% CO_2 until they reach a density of 6–7 × 10^5 cells/mL.
2. Transfer the cells into a 1 L spinner flask containing 150 mL culture medium. Spin the cells at about 20–25 rpm to maintain good media aeration and prevent cell sedimentation. Do not spin the cells too fast to avoid damaging the cells.
3. Grow the cells until 6–8 × 10^5 cells/mL density is reached, then transfer them into a 6 L spinner flask containing 850 mL of prewarmed culture medium.
4. Expand the cell culture to the final volume of 5 L and grow the cells until they reach a density of 6–8 × 10^5 cells/mL. Check the cell viability by staining the cells with trypan blue using an automated cell counter or a tissue culture microscope and a Neubauer counting chamber. The proportion of dead cells should not exceed 10%. Do not let the cells grow to higher densities as this changes their metabolism and leads to inconsistencies in the nuclear extracts.
5. Harvest cells by centrifugation for 10 min at 1000 × *g* and 4 °C using 1 L centrifuge bottles.
6. Resuspend the cell pellets in 100 mL ice-cold PBS and split the suspension into two 50 mL Falcon tubes (use 20 mL of ice-cold PBS and two 15 mL Falcon tubes when processing cells obtained from <2.5 L of cell culture media at 6–8 × 10^5 cells/mL density).
7. Pellet the cells by centrifugation for 10 min at 1,000 × *g* and 4 °C and resuspend the cells in ice-cold PBS. Repeat two times.

8. Pellet the cells by centrifugation for 10 min at 1,000 × *g* and 4 °C. Discard the PBS, determine the cell pellet volume and resuspend the cells in 5× cell pellet volumes of ice-cold Buffer A. Let the cells swell for 10 min on ice.
9. Pellet the cells by centrifugation for 10 min at 1,000 × *g* and 4 °C. After incubating the cells in hypotonic Buffer A, the cell pellet volume should increase approximately two-fold.
10. Determine the cell pellet volume and resuspend the cells in 1× volume of ice-cold Buffer A supplemented with 0.2% NP-40, 1 mM DTT, 1 mM PMSF, and 1× complete protease inhibitors.
11. Lyse the cells with 10 strokes in a precooled Dounce homogenizer using a tight pestle. Move pestle up and down slowly; applying too much pressure during the lysis can rupture the nuclear envelope resulting in undesired chromatin release and partial nuclear protein loss. Check lysis efficiency by staining the cells with trypan blue (>90% cells should appear dead).
12. Pellet the nuclei by centrifugation in a conical Falcon tube for 15 min at 3,000 × *g* and 4 °C.
13. Discard the supernatant (cytoplasmic lysate) and gently resuspend the nuclear pellet in 10× pellet volumes of ice-cold Buffer A supplemented with 150 mM NaCl and 1 mM DTT.
14. Pellet the nuclei by centrifugation for 15 min at 3,000 × *g* and 4 °C.
15. Determine the nuclear pellet volume and gently resuspend in 2× pellet volumes of ice-cold Buffer C containing 420 mM NaCl prepared by mixing ice-cold Buffer C (no NaCl) with ice-cold Buffer C (1000 mM NaCl) in a 5.8: 4.2 ratio and supplemented with 0.1% NP-40, 1 mM DTT, 0.5 mM PMSF, and 1× complete protease inhibitors. The final NaCl concentration during the extraction is approximately 350 mM. Optionally: supplement Buffer C with RNase inhibitors or RNase A at the final contraction of 20 ng/μL (*see* **Note 69**).
16. Rotate the mixture for 1 h at 4 °C using a tube roller or a rotation wheel at low rpm settings (15–30 rpm) in the cold room.
17. Pellet the extracted nuclei by centrifugation for 15 min at 3,000 × *g* and 4 °C. Collect the supernatant and keep it at 4 °C until further processing. This is the 350 mM NaCl nuclear extract fraction that corresponds to nuclear extracts described by other procedures [25, 26]. If the nuclear extract is intended for pull-down experiments with nucleosomes decorated only with euchromatic modifications, such as H3K4me3 or H3 lysine acetylation, **steps 18–26** can be omitted, as most euchromatic proteins are extracted at this stage (*see* **Note**

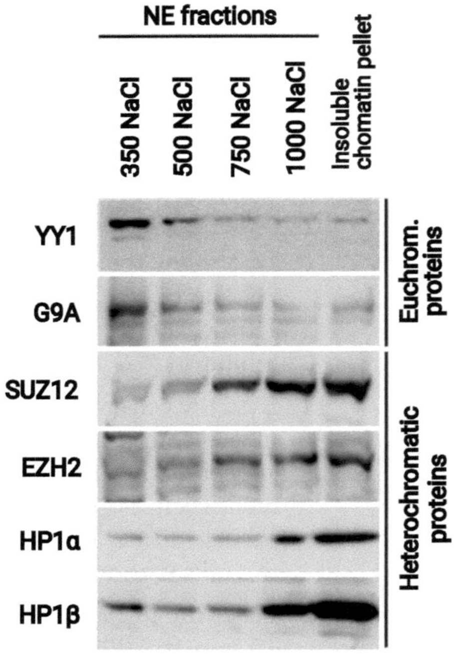

Fig. 9 The sequential nuclear extract preparation procedure allows extraction of both eu- and heterochromatic proteins. 20 μg of nuclear extract (NE) fractions obtained using extraction buffers with increasing NaCl concentrations ranging from 350 mM NaCl (NE fraction 350 NaCl) to 1000 mM NaCl (NE fraction 1000 NaCl) and the insoluble chromatin pellet (boiled in 1× SDS sample buffer) were resolved using SDS-PAGE, transferred to nitrocellulose membranes, and probed by western blot using antibodies against YY1, G9A, SUZ12, EZH2, HP1α, and HP1β. Heterochromatic proteins are preferentially released only at higher NaCl concentrations

70). If extraction of heterochromatic proteins is required (Fig. 9), continue with **step 18**.

18. Add 2× nuclear pellet volumes (determined in step 15) of ice-cold Buffer C containing 500 mM NaCl prepared by mixing ice-cold Buffer C (no NaCl) with ice-cold Buffer C (1000 mM NaCl) in 1:1 ratio and supplemented with 0.1% NP-40, 1 mM DTT, 0.5 mM PMSF, and 1× complete protease inhibitors. Optionally: supplement Buffer C with RNase inhibitors or RNase A at the final contraction of 20 ng/μL (*see* **Note 69**). The chromatin pellet looks like a gelatinous aggregate at this point, do not try to resuspend it in Buffer C as this will shear the chromatin, simply add the buffer to the pellet and proceed with **step 19**.
19. Rotate the mixture for 1 h at 4 °C using a tube roller or a rotation wheel at low rpm settings (15–30 rpm).

20. Pellet the chromatin by centrifugation for 15 min at 3,000 × *g* and 4 °C. Collect the supernatant and keep it at 4 °C until further processing. This is the 500 mM NaCl nuclear extract fraction.
21. Add 2× nuclear pellet volumes (determined in **step 15**) of ice-cold Buffer C containing 750 mM NaCl prepared by mixing ice-cold Buffer C (no NaCl) with ice-cold Buffer C (1000 mM NaCl) in 1:3 ratio and supplemented with 0.1% NP-40, 1 mM DTT, 0.5 mM PMSF, and 1× complete protease inhibitors. Optionally: supplement Buffer C with RNase inhibitors or RNase A at the final contraction of 20 ng/μL (*see* **Note 69**). As for **step 18**, do not try to resuspend the pellet in Buffer C, simply add the buffer and proceed with **step 22**.
22. Rotate the mixture for 2 h at 4 °C using a tube roller or a rotation wheel at low rpm settings (15–30 rpm).
23. Pellet the chromatin by centrifugation for 15 min at 3,000 × *g* and 4 °C. Collect the supernatant and keep it at 4 °C until further processing. This is the 750 mM NaCl nuclear extract fraction.
24. Add 2× nuclear pellet volumes (determined in **step 15**) of ice-cold Buffer C (1000 mM NaCl) supplemented with 0.1% NP-40, 1 mM DTT, 0.5 mM PMSF, and 1× complete protease inhibitors. Optionally: supplement Buffer C with RNase inhibitors or RNase A at the final contraction of 20 ng/μL (*see* **Note 69**). As for **step 18**, do not try to resuspend the pellet in Buffer C, simply add the buffer and proceed with **step 25**.
25. Rotate the mixture for 2 h at 4 °C using a tube roller or a rotation wheel at low rpm settings (15–30 rpm).
26. Pellet the chromatin by centrifugation for 15 min at 3,000 × *g* and 4 °C. Collect the supernatant and keep it at 4 °C until further processing. This is the 1000 mM NaCl nuclear extract fraction.
27. Transfer the nuclear extract fractions to fresh thin-walled ultracentrifuge tubes (5 mL, e.g., open-top thin wall polypropylene tube, Beckman Cat. No. 326819 or equivalent) and centrifuge for 1 h at 60,000 × *g* and 4 °C in an ultracentrifuge (e.g., Beckman Optima XE-90 centrifuge with SW 55 Ti rotor or equivalent—*see* **Note 71**). Keep the fractions separate at this point.
28. Transfer the supernatants to fresh 15 mL Falcon tubes and discard the pellets (*see* **Note 72**).
29. Determine the total protein concentration for each nuclear extract fraction (e.g., by Bradford assay). Chromatin-associated proteins elute differentially in the different fractions (Fig. 9). The fractions can either be kept and assayed separately or

combined if a "total" nuclear extract is desired. If combining the fractions, combine the supernatants from the 350, 500, 750, and 1000 mM NaCl nuclear extract fractions in a 2:1:1:1 ratio in a fresh 50 mL Falcon tube. Carefully mix by gently pipetting up and down a few times and then determine the total protein concentration of the mixture.

30. Aliquot the nuclear extracts (0.5 mg of total protein per aliquot) in 0.5 mL protein low-binding microtubes. Snap-freeze the aliquots in liquid nitrogen (*see* **Note 1**) and store them at −80 °C until further use.

3.6 Nucleosome Affinity Purification

The final experimental step in the identification of nucleosome-interacting proteins that recognize methylated histones is the nucleosome affinity purification experiment itself. In this procedure the modified dinucleosomes (prepared as described in Subheading 3.4) are immobilized on streptavidin-coated beads and incubated with HeLa S3 nuclear extracts (prepared as described in Subheading 3.5). Following the binding reaction and several washing steps to remove unbound proteins, the nucleosome-interacting proteins are eluted from the beads and subjected to mass spectrometric identification and quantification (*see* Subheading 3.7). In this protocol we describe a workflow for quantifying the binding of methylation-specific nucleosome-interacting proteins by label-free quantification of the mass spectrometry data. This allows nuclear extracts from unlabeled cells, grown in regular tissue culture media, to be used as extract input. However, to achieve sufficient statistical power for the quantification it is important to perform the pull-down experiments in at least three experimental replicates. Ideally, these should be carried out separately with three independent dinucleosome preparations to generate independent triplicates (*see* also Subheading 3.7.3). Due to the variability in the preparation of nuclear extracts each set of nucleosome pull-down experiments should be performed with aliquots from the same batch of nuclear extracts to achieve the best reproducibility and comparability between the samples (*see* also Subheading 3.5).

Furthermore, methylation-specific nucleosome-interacting proteins are identified by determining the extent of binding of a particular protein to a modified nucleosome in comparison to an unmodified nucleosome. It is therefore crucial to include the appropriate unmodified control nucleosome(s) in each set of pull-down experiments. For a nucleosome assembled with a modified histone generated by native chemical ligation it is critical to use a nucleosome incorporating the corresponding unmodified histone, generated in a similar way by ligating an unmodified histone tail peptide to the same histone core, as control. In our experience recombinant (unmodified) histones can behave differently from the ligated unmodified histones. Recombinant (unmodified) histones are therefore not the correct control for modified histones generated by native chemical ligation, and the corresponding ligated unmodified histone (prepared in the same way as the modified histone) should be used.

3.6.1 Immobilization of Nucleosomes on Streptavidin-Coated Beads

The first step of the nucleosome affinity purification procedure is the immobilization of the dinucleosomes on streptavidin-coated beads (*see* **Note 73**) via the biotinylated dinucleosomal DNA. This protocol describes the procedure for immobilizing dinucleosomes for one pull-down sample corresponding to 12 μg octamer and 10 μg of biotinylated 601 dinucleosome DNA (the exact values need to be determined by titration—*see* Subheading 3.4.3), which we find to work well for mass spectrometry–coupled nucleosome affinity purifications using 0.5 mg of nuclear extract. While we have found that using more nucleosomes reduces the number of identified proteins, downscaling the amounts of nucleosomes is possible but the results will strongly depend on the sensitivity of the mass spectrometer (*see* Subheading 3.7.2). All steps during the immobilization of the modified nucleosomes should be carried out with precooled centrifuges, buffers, tubes, and so on, and the samples should be kept on ice at all times. All incubations should be carried out in the cold room at 4 °C.

1. Continuing from Subheading 3.4.3 **step 9**, transfer an aliquot of the desired biotinylated dinucleosomes corresponding to 12 μg of histone octamer/10 μg of biotinylated 601 dinucleosome DNA into a precooled 1.5 mL DNA low-binding microtube. Top up the volume to 450 μL with ice-cold RB-low buffer and supplement with 0.1% NP-40 and 1 mM DTT (final concentrations). Keep the tube on ice.
2. Transfer an appropriate volume of Streptavidin Sepharose bead slurry (use 12 μL of slurry for immobilization of dinucleosomes corresponding to 12 μg of histone octamer/10 μg of biotinylated 601 dinucleosome DNA) into a 1.5 mL microtube. Top up to 1 mL with ice-cold water. Mix by inverting the tube 10–20 times.
3. Pellet the beads by centrifugation for 2 min at 300 × *g* and 4 °C. Remove the supernatant.
4. Equilibrate the beads with 1 mL of ice-cold RB-low buffer supplemented with 0.1% NP-40 by mixing on a rotating wheel for 5 min at 4 °C.
5. Pellet the beads by centrifugation for 2 min at 300 × *g* and 4 °C. Remove the supernatant.
6. Repeat **steps 4** and **5** two more times to completely equilibrate the beads in RB-low.

7. Add an appropriate volume of ice-cold RB-low buffer supplemented with 0.1% NP-40 and 1 mM DTT to the bead pellet (use 55 μL of the buffer for each dinucleosome sample corresponding to 12 μg of histone octamer/10 μg of biotinylated DNA). Thoroughly mix the beads by pipetting up and down several times.
8. Transfer 50 μL of the bead slurry to each dinucleosome sample.
9. Allow the dinucleosomes to bind to the beads for 1 h while mixing on a rotating wheel at 15–20 rpm in the cold room at 4 °C.
10. Spin down the beads, now containing the immobilized dinucleosomes, by centrifugation for 2 min at 300 × *g* and 4 °C. Remove the supernatant.
11. Add 1 mL of ice-cold RB-low buffer supplemented with 0.1% NP-40 and 1 mM DTT. Mix on a rotating wheel for 10 min at 15–20 rpm in the cold room at 4 °C.
12. Repeat **steps 10** and **11**.
13. Add 1 mL of ice-cold Pull-down Buffer supplemented with 0.1% NP-40 and 1 mM DTT. Let the beads equilibrate by rotating on a rotating wheel for 10–20 min at 15–20 rpm in the cold room at 4 °C. These are the immobilized dinucleosomes for the pull-down reactions.

3.6.2 Nucleosome Affinity Purification from Nuclear Extract

For the nucleosome affinity purifications, the immobilized dinucleosomes are incubated with nuclear extracts to capture proteins that bind to nucleosomes in a modification-specific manner. We usually use nuclear extracts from HeLa S3 cells since these can be generated in large quantities. However, due to the label-free mass spectrometric quantification protocol (*see* Subheading 3.7) nuclear extracts from any other cell type or tissue are also suitable and can be treated in the same manner. As described above, in order to identify proteins that specifically respond to methylation marks on histones it is important to include an unmodified control nucleosome in each experiment. Since at least three experimental replicates are needed for the modified and the unmodified nucleosomes (*see* Subheading 3.7.3), a typical nucleosome affinity purification experiment with label-free mass spectrometric quantification of the bound proteins will therefore consist of at least six samples. Below we describe the procedure for one nucleosome pull-down sample using 0.5 mg of nuclear extract. As for the immobilization (*see* Subheading 3.6.1), all steps during the nucleosome affinity purification procedure should be carried out with precooled centrifuges, buffers, pipette tips, and tubes, and the samples should be kept on ice or incubated in the cold room at 4 °C at all times. Precise pipetting and consistent incubation/rotation and centrifugation times and speeds between the replicate samples is essential for good reproducibility.

1. While the immobilized dinucleosomes are equilibrating in the Pull-down Buffer (*see* Subheading 3.6.1 **step 13**), thaw an appropriate number of aliquots of the nuclear extract (*see* Subheading 3.5). Use one 0.5 mg aliquot for each pulldown sample with dinucleosomes corresponding to 12 μg of histone octamer/10 μg of biotinylated 601 dinucleosome DNA, and thaw the extracts on ice or in an ice water bath to prevent any warming up of the proteins.
2. Transfer the nuclear extracts into a precooled 15 mL Falcon tube. While gently vortexing at low speed, add dropwise 3× nuclear extract volumes of Adjustment Buffer to reduce the NaCl concentration to 155 mM (*see* **Note 74**). Supplement the sample with 0.1% NP-40, 1 mM DTT, 0.5 mM PMSF, and 1× complete protease inhibitors (final concentrations). If the protein concentration exceeds 0.5 mg/mL dilute the nuclear extract further with an appropriate volume of ice-cold Pull-down Buffer supplemented with 0.1% NP-40, 1 mM DTT, 0.5 mM PMSF, and 1× complete protease inhibitors to the final protein concentration of 0.5 mg/mL. Optionally: If probing dinucleosomes decorated with histone lysine acetylation, add sodium butyrate to 5 mM and trichostatin A (TSA) to 250 nM final concentration (*see* **Note 22**). If probing RNA-dependent interactions, supplement with RNase inhibitors (*see* **Note 69**). Mix gently by pipetting up and down several times.
3. Transfer the nuclear extract into precooled 1.5 mL microtubes and spin at 16,000 × *g* for 10 min at 4 °C to remove any precipitates.
4. In the meantime pellet the beads with the immobilized dinucleosomes (*see* Subheading 3.6.1 **step 13**) by centrifugation for 2 min at 300 × *g* and 4 °C and remove the supernatant, leaving around 50 μL of buffer in the tube.
5. Add 950 μL of the cleared nuclear extracts from **step 3** to each tube containing the immobilized dinucleosomes on the beads.
6. Rotate the tubes on a rotating wheel for 4 h at 15–20 rpm in the cold room at 4 °C.
7. Spin down the beads by centrifugation for 4 min at 300 × *g* and 4 °C. Remove the supernatant.
8. Add 1 mL of ice-cold pull-down buffer and rotate the tubes on a rotating wheel for 10 min at 15–20 rpm in the cold room at 4 °C.
9. Repeat **steps** 7 and **8** two more times.

10. Spin down the beads by centrifugation for 4 min at 300 × *g* and 4 °C. Remove the supernatant.
11. Resuspend the beads in 40 μL of 1× SDS-PAGE sample buffer and elute the proteins by incubating the tubes at 95 °C in a Thermomixer while mixing at 250–300 rpm for 5 min (*see* **Note 75**).
12. Spin down the beads by centrifugation for 5 min at full speed and RT in a microcentrifuge. Transfer the supernatant containing the eluted modified histones and nucleosome-associated proteins to a fresh 1.5 mL protein low-binding tube. Store at −20 °C until further use.

3.7 Proteomic Analysis of Nucleosome Pull-Down Samples

The last step in the identification of nucleosome-interacting proteins that recognize methylated histones is the identification and quantification of the nucleosome-bound proteins by mass spectrometry to determine their enrichment on the modified nucleosome in comparison to the unmodified control nucleosome in the nucleosome affinity purification experiment (*see* Subheading 3.6). This procedure consists of: (1) processing of the proteins eluted from the nucleosomes for the mass spectrometric analysis (*see* Subheading 3.7.1), (2) the mass spectrometric measurements themselves (*see* Subheading 3.7.2), and (3) the processing and analysis of the mass spectrometry raw data (*see* Subheading 3.7.3). As this requires very expensive mass spectrometry equipment these steps will typically be carried out by a dedicated mass spectrometry (MS) or proteomics facility or a specialized proteomics laboratory, and we recommend working closely with the MS facility or laboratory when establishing the proteomics procedures. Since mass spectrometers, sample preparation procedures, and data analysis pipelines are different in each MS facility or laboratory we will keep the descriptions of the workflows for the MS measurements and the tools for the proteomics data analysis quite general. These are therefore meant to serve as guidelines to work out tailored procedures rather than specific protocols. We are, however, providing a fast and robust filter-aided sample preparation (FASP) protocol [27] for processing the eluted proteins from the dinucleosome pull-down samples into Lys-C and trypsin-digested peptides that can then be used for the MS measurements.

3.7.1 Sample Preparation for Liquid Chromatography Tandem Mass Spectrometry (LC-MS/MS) Analysis

LC-MS systems are not compatible with most laboratory detergents. Even trace amounts of some detergents, such as SDS, can greatly compromise the analysis. To this end, all buffers for proteomic sample preparation need to be prepared in fresh unused and clean laboratory glassware, rinsed with LC-MS-grade water. Avoid using any glassware that has been washed with detergents or autoclaved as this leads to contaminations.

Another common problem in MS analyses is sample contamination with keratins, which originate from skin, hair, dust, or clothing. If keratins are present in high concentrations, they can compromise MS analyses, resulting in low numbers of protein identifications and inaccurate quantifications. To this end, several precautions must be taken when preparing samples for LC-MS analysis: (1) perform all sample preparation steps in a laminar flow hood, if possible; (2) clean surfaces with water and ethanol before starting any work; (3) always use fresh gloves and wear a lab coat. If handling samples outside a laminar flow hood, wear a disposable head cover. To avoid protein or peptide losses during the procedure use only protein low-binding tubes.

1. Transfer 10 μL of nucleosome pull-down sample in 1× SDS-PAGE sample buffer (*see* Subheading 3.6.2 **step 12**) to a fresh 1.5 mL protein low-binding tube. Add 200 μL of Wash Buffer (50 mM TEAB) and 1 μL of 1 M DTT and mix the sample by vortexing. Place the tube in a Thermomixer and incubate for 30 min at 60 °C to reduce protein disulfide bonds.
2. Remove the tube from the Thermomixer and let it cool down to RT for about 15–20 min. Add 300 μL of UA buffer and 6 μL of 0.5 M iodoacetamide (IAA), and mix the sample by vortexing. Incubate for 30 min at RT in the dark to alkylate cysteine residues.
3. While the sample is incubating with IAA, preequilibrate a 30 kDa centrifugal spin filter unit (e.g., Sartorius Vivacon-500) by adding 100 μL UA buffer and centrifuging for 5 min at 14,000 × *g* and RT.
4. After 30 min of sample incubation with IAA, add 2 μL of 1 M DTT to quench the reaction and mix briefly by vortexing.
5. Transfer the sample to the preequilibrated 30 kDa centrifugal filter unit and centrifuge for 15 min at 14,000 × *g* and RT. The eluted proteins from the nucleosome pull-down are retained on the membrane.
6. Discard the flow-through and place the centrifugal filter unit back into the collection tube. Add 200 μL UA buffer and centrifuge for 15 min at 14,000 × *g* and RT. Discard the flow-through. Repeat this step two more times (for a total of three times).
7. Add 200 μL of Wash Buffer and centrifuge for 15 min at 14,000 × *g* and RT. Discard the flow-through. Repeat this step one more time (for a total of two times).
8. Place the centrifugal filter unit into a fresh collection tube. Add an appropriate volume of Wash Buffer into a fresh 1.5 mL protein low-binding tube; use 40 μL per sample. Add Lys-C to a final concentration of 12.5 ng/μL, mix by vortexing, and

spin briefly to return all liquid to the bottom of the tube. Transfer 40 μL into the centrifugal filter unit and incubate for 2 h at RT.

9. Add an appropriate volume of Wash Buffer into a fresh 1.5 mL protein low-binding tube; use 10 μL per sample. Add trypsin to a final concentration of 100 ng/μL, mix by vortexing, and spin briefly to return all liquid to the bottom of the tube. Transfer 10 μL into the centrifugal filter unit and wrap the top of the centrifugal filter unit tube with parafilm to minimize evaporation. Place the tube into a Thermomixer, mix at 450 rpm for 1 min, stop mixing, and incubate overnight at 37 °C.
10. The next morning: carefully remove the Parafilm and spin the centrifugal filter unit for 20 min at 16,000 × *g* and RT.
11. Add 20 μL of Wash Buffer containing 5% acetonitrile (*see* **Note 12**) onto the filter, centrifuge for 20 min at 16,000 × *g* and RT.
12. Transfer the flow through from the collection tube into a fresh 1.5 mL protein low-binding tube. Acidify by adding TFA (*see* **Note 12**) to a final concentration of 0.2% and mix briefly by vortexing. Check the pH by pipetting 1 μL onto a pH test strip. If the pH is >2, add another 0.25 μL of TFA to the sample and check the pH again. If necessary, repeat until the pH is 2. Store the samples at −20 °C until further use. These are the Lys-C and trypsin-digested peptides that will be used for the mass spectrometric measurements (*see* Subheading 3.7.2).

3.7.2 Mass Spectrometric Measurements

Identification of proteins whose binding to dinucleosomes is influenced by methylated histones (or other histone or DNA modifications of interest) relies on the accurate quantification of proteins bound to the modified nucleosome compared to an unmodified nucleosome using liquid chromatography-tandem mass spectrometry (LC-MS/MS) analysis. While various LC-MS/MS-based quantification strategies can be used for nucleosome affinity purification pulldown experiments [3, 4], we focus here solely on label-free protein quantification as a commonly used and widely available protein quantification technique, due to its flexibility and the lack of the necessity for protein labeling.

LC-MS/MS analysis: Due to the relatively high complexity and wide dynamic range of protein abundances in the dinucleosome pull-down samples, with many modification "readers" usually present in very low quantities, it is critical to use a robust and sensitive LC-MS/MS setup in order to achieve a comprehensive view of modification-specific dinucleosome binders. We therefore recommend using a MS instrument with high MS/MS acquisition rate coupled to a nano-flow high resolution C18 Reversed-Phase High Performance Liquid Chromatography (HPLC) system to achieve the best results. Examples of recommended MS instruments include but are not limited to: (1) Thermo Scientific QExactive™

Orbitrap mass spectrometers, or (2) Thermo Scientific Orbitrap Exploris™ mass spectrometers. The optimal LC-MS/MS method settings vary depending on the particular sample preparation method and the instrumental setup to be used, and therefore cannot be described here in detail. While it is difficult to accurately determine the absolute peptide amount in dinucleosome pull-down trypsin digests, we recommend starting the analysis by performing a LC-MS/MS test run using 3.5 μL out of the 70 μL (5% (v/v)) of a sample digest (*see* Subheading 3.7.1 **step 12**) and subsequently adjusting the optimal sample injection volumes based on the base peak chromatogram and total ion current (TIC) intensity. These optimizations should be carried out in close consultation with the MS facility or laboratory that carries out the LC-MS/MS analysis.

3.7.3 Processing and Analysis of Mass Spectrometry Data

The output from the LC-MS/MS measurements will be MS RAW data files. A variety of proteomic software packages can be used for RAW MS data post-processing, peptide search, and protein quantification, including but not limited to, (1) Proteome Discoverer™ Software, (2) MaxQuant Software, (3) Progenesis QI Software. The optimal RAW MS data post-processing and peptide search settings vary depending on the MS instrument resolution/mass accuracy used for the MS measurements and the data acquisition strategy (i.e., data-dependent acquisition [DDA] or data-independent acquisition [DIA]) and have to be chosen accordingly. For label-free protein quantification, we recommend using one of the well-established precursor-signal-intensity–based quantification algorithms, such as the "top three" method [28], the intensity-based absolute quantification (iBAQ) method [29], or a "Relative quantitation using nonconflicting peptides" method [30], in which the protein abundance in a run is calculated as the sum of all the unique peptide ion abundances corresponding to a particular protein.

For a threshold of ≥2 unique peptides identified per protein, the number of protein identifications for a single dinucleosome pull-down sample typically ranges between 1500 and 2000 proteins; however, this may vary depending on the sensitivity of the LC-MS/MS platform used for the analysis. To account for possible variation in total peptide loading between the samples, we recommend normalizing protein abundances in each sample by the sum of abundances of all quantified proteins in this sample. This works well for nucleosome pull-down samples since the background of nucleosome-binding proteins that do not respond to the modification(s) is very similar between different nucleosomes (*see* below). Importantly, the list of identified proteins usually contains some common contaminants, such as keratins, as well as streptavidin, which should be filtered out before the normalization procedure.

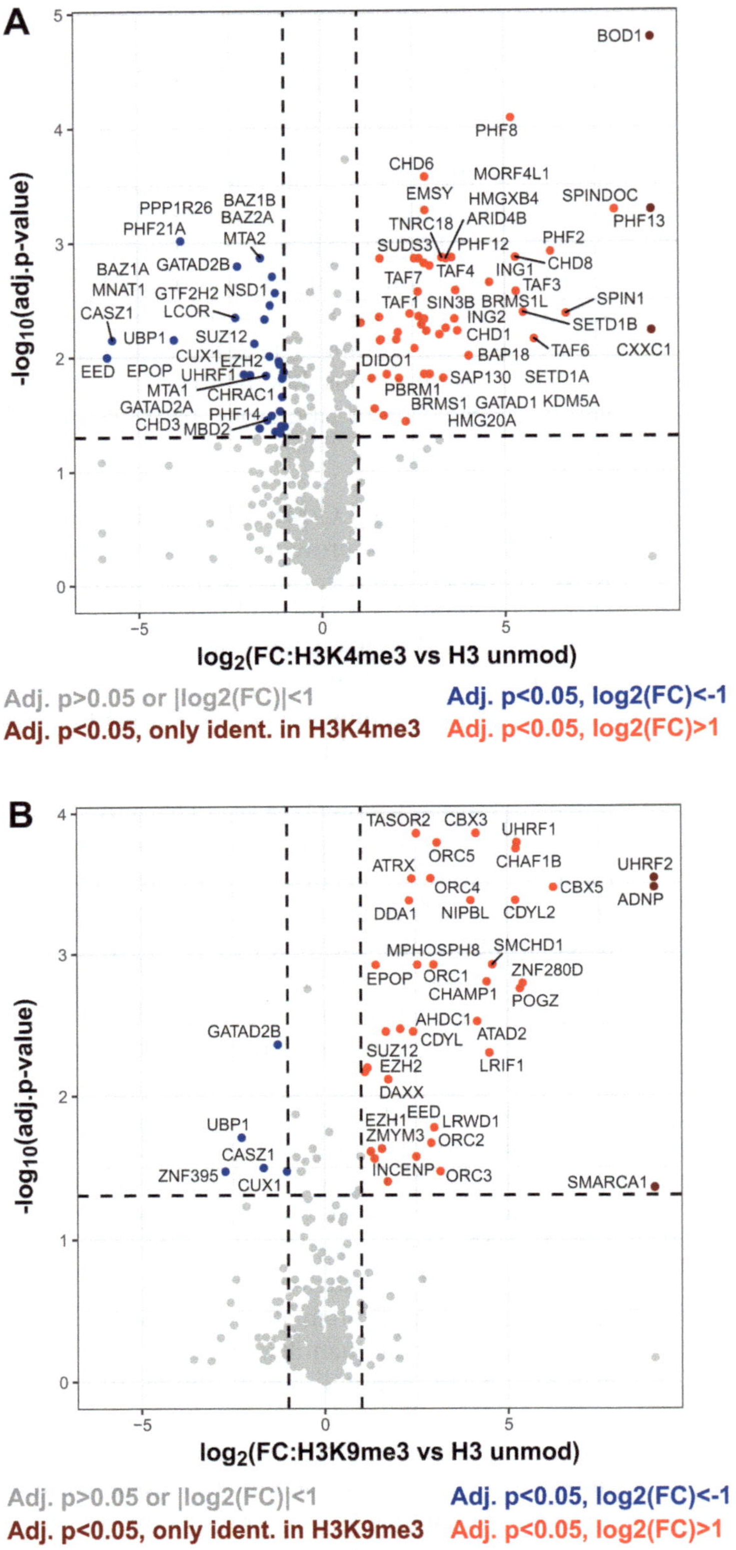

Fig. 10 Representative example of LC-MS/MS analysis results for H3K4me3 and H3K9me3 dinucleosome pull-down experiments. (**a**) Volcano plot depicting proteins whose binding to dinucleosomes is either increased

This filtering should again be done in consultation with the MS facility or laboratory that carries out the LC-MS/MS analysis, as this is a commonly encountered problem and the MS facility/laboratory will have an up-to-date list of common contaminants to be filtered out.

To allow for statistical evaluation of the protein binding responses to the nucleosomal modification(s) of interest, we recommend to include at least three technical replicates (performed starting from the dinucleosome reconstitution step) for each experimental condition (*see* also Subheading 3.6). Different statistical methods such as limma statistics [31] or Student's *t*-test can be used for statistical evaluation. It is, however, important to keep in mind that p-values obtained using *t*-test have to be adjusted for multiple testing, for example using the Benjamini-Hochberg method [32]. As mentioned above it is important to note that the binding of the vast majority of dinucleosome interacting proteins is usually not perturbed by the presence of histone or DNA methylation(s), which means that the normalized protein abundances for most of the quantified proteins should remain consistently similar between replicate experiments. The number of proteins exhibiting strong and reproducible binding responses to the dinucleosomal modification(s), i.e., at least a twofold difference in abundance compared to the unmodified control dinucleosome at a false discovery rate (FDR) of <0.05, can vary substantially for each particular modification, but usually ranges between a handful to several dozens (Fig. 10). Given that the dinucleosome pull-downs are performed under very mild physiological buffer conditions, many nuclear protein complexes remain intact during the purification, often resulting in similar binding responses for different subunits belonging to the same complex. It is therefore important to note that proteins demonstrating a response to a modification of interest might not necessarily directly interact with the modification, but might indirectly bind to the nucleosome via another protein or complex subunit that directly recognizes the modification. Mass spectrometry–linked nucleosome affinity purifications are therefore a powerful approach to identify modification-specific chromatin-binding proteins. Their mode of interaction, how a particular

Fig. 10 (continued) ($log_2FC > 1$, adjusted *p*-value <0.05) or decreased ($log_2FC < -1$, adjusted *p*-value <0.05) by the presence of H3K4me3. (**b**) Volcano plot depicting proteins whose binding to dinucleosomes is either increased ($log_2FC > 1$, adjusted *p*-value <0.05) or decreased ($log_2FC < -1$, adjusted *p*-value <0.05) by the presence of H3K9me3. Significantly regulated proteins that show a more than twofold increase or decrease in binding to the modified nucleosomes compared to the unmodified control nucleosomes are indicated in red or blue, respectively. The mass spectrometry data and the quantifications for the label-free H3-unmodified, H3K4me3, and H3K9me3 nucleosome pull-downs have been deposited to the ProteomeXchange Consortium via the PRIDE partner repository [33] under identifier PXD027238

protein recognizes the modified nucleosomes, and their biological functions, however, have to be verified in subsequent targeted validation experiments.

4 Notes

1. Care must be taken when handling liquid nitrogen (LN_2). Always wear waterproof cryogenic gloves, safety goggles or a face shield, and closed shoes when handling LN_2 to protect yourself against splashes or spills. If the optical and/or acoustic oxygen alarm goes off when decanting LN_2 switch off the LN_2 supply and leave the LN_2 storage room immediately as there is the danger of death through suffocation. Never travel in a lift with a vessel containing LN_2.
2. Unpolymerized acrylamide is a neurotoxin and carcinogen. Care must be taken to avoid skin contact. Always wear gloves when handling acrylamide and bisacrylamide. To avoid having to weigh in acrylamide and bisacrylamide we use a ready-made solution for preparing PAGE gels whenever possible. We pour PAGE gels in a fume hood. Once polymerized acrylamide–bisacrylamide is no longer harmful and can be disposed of via the regular waste. If weighing in acrylamide or bisacrylamide, wear gloves, safety goggles, and a face mask to avoid inhalation of the dust. Dissolve acrylamide–bisacrylamide in water inside a fume hood and mix on a stirrer placed inside the hood. To avoid exposing coworkers to acrylamide–bisacrylamide, cover the weigh boat containing the weighed acrylamide–bisacrylamide with another similar size weigh boat when transporting it to the fume hood.
3. When weighing SDS wear gloves, safety goggles, and a face mask to avoid inhalation of the dust. Dissolve SDS in water inside a fume hood and mix on a stirrer placed inside the hood. To avoid exposing coworkers to SDS, cover the weigh boat containing the weighed SDS with another similar size weigh boat when transporting it to the fume hood.
4. Simple protocol for preparing SDS-PAGE running buffer: First prepare 10× native buffer (0.25 M Tris–HCl, 1.92 M glycine): Weigh 30.3 g Tris–HCl and 144 g glycine, mix, and fill up to 1 L with water. To prepare 1× SDS-PAGE running buffer dilute 100 mL of 10× native buffer to 995 mL with water and add 5 mL of 20% SDS stock solution. Add SDS solution last, since it forms bubbles.
5. Protocol for preparing 5× SDS-PAGE sample buffer: For 10 mL dissolve 1 g of SDS powder (*see* **Note 3**) in 1.56 mL of 2 M Tris pH 6.8, 5 mL of glycerol, and 2.5 mL of

β-mercaptoethanol, and then adjust to 10 mL with water. Warming up a little helps to dissolve the SDS. Add a pinch of solid bromophenol blue last and adjust the amount, so that a 1× dilution has a sufficiently blue colour. SDS precipitates at 4 °C; therefore, the SDS-PAGE sample buffer needs to be warmed to RT and mixed prior to use to redissolve the SDS.

6. Coomassie staining and destaining solutions contain methanol and acetic acid. Methanol is highly flammable and toxic, and acetic acid is corrosive. Care must be taken to avoid skin contact and inhalation. Use gloves when preparing and handling Coomassie staining and destaining solutions. Prepare the solutions and perform the staining/destaining reactions in a fume hood. Keep storage bottles and gel containers tightly closed. Ethanol can be used as a less toxic alternative to methanol.
7. To save reagents the Coomassie staining solution can be collected and reused 2–3 times, and the destaining solution can be passed over an activated charcoal filter and recycled several times. However, efficiency of the stain/destain will decrease after multiple uses.
8. For deionizing the 8 M urea stock solution add 50 g Amberlite IRN-150 mixed-bed ion-exchange resin (Alfa Aesar Cat No. #40992) per 1 L of solution. Stir the solution for 1 h at RT and then remove the ion-exchanger beads by filtration through a muslin cloth filter. Leftover 8 M urea stock solution can be stored overnight at 4 °C and used the next day to prepare SAU-0, SAU-200, and SAU-600/HU-1000 for the ion-exchange purification.
9. The purification of the histones is carried out under denaturing conditions in 7 M urea to keep the histones unfolded. To prevent precipitation of the urea the gel-filtration and ion-exchange purifications for histones are performed at room temperature. We have a small ÄKTA prime plus FPLC system (Cytiva) that is dedicated to histone purifications and that is placed at RT. While this system is rather "primitive" it is very cost-effective and perfectly suitable for the histone purification procedures.
10. In the Unfolding Buffer for extracting truncated histones for native chemical ligations from inclusion bodies we use 100 mM dithiothreitol (DTT) instead of the 10 mM used in the standard unfolding buffer. For 30 mL dissolve 20.06 g of solid guanidine hydrochloride in water to a volume of approximately 28 mL, then add 0.6 mL of 1 M Tris–HCl (pH 7.5) and 0.463 g of solid DTT and rotate until the DTT is dissolved. Then adjust the pH and fill up with water to a final volume of 30 mL.

11. We do not add any reducing agents to the gel-filtration buffers for the purification of the truncated histones for native chemical ligations. DTT is not an effective reducing agent below a pH of 6.5, and β-mercaptoethanol blocks cysteines thereby reducing the reactivity of the N-terminal cysteine during the ligation reaction (*see* also **Note 35**). In our experience, if the buffers are degassed well to remove oxygen and if the purification procedure is carried out swiftly, the low pH of the buffers is sufficient to keep cysteines reduced and the truncated histones react well in the ligation reactions, even without the addition of reducing agents. TCEP could be used as it is an effective reducing agent over a broad pH range, but it is expensive for large volumes of buffer.
12. Always use HPLC-grade acetonitrile and trifluoroacetic acid for the reversed-phase chromatography buffers and LC-MS-grade acetonitrile and trifluoroacetic acid for mass spectrometry applications. Acetonitrile and trifluoroacetic acid are harmful chemicals. Acetonitrile is a toxic and flammable solvent and trifluoroacetic acid is highly corrosive. Gloves and safety goggles therefore must be worn when handling acetonitrile and trifluoroacetic acid. Buffers containing acetonitrile and trifluoroacetic acid need to be prepared and handled in a fume hood and bottles must always be tightly closed to avoid the release of harmful gases and aerosols. Waste containing acetonitrile and trifluoroacetic acid must be collected separately in tightly closed bottles, stored under a fume hood, and disposed of via the appropriate chemical waste route.
13. Be careful when handling 5 M KOH solution as it is highly corrosive. Always use gloves and safety goggles. Prepare the 5 M KOH solution in a fume hood and always add the KOH pellets to the water one after another to prevent splashing through rapid heat formation.
14. Acetic acid is corrosive. Care must be taken to avoid skin contact and inhalation. Use gloves and safety goggles and handle undiluted acetic acid in a fume hood. Bottles containing acetic acid must always be tightly closed to avoid the release of harmful gases and aerosols.
15. Methanol is highly flammable and toxic. Care must be taken to avoid skin contact and inhalation. Use gloves and safety goggles and handle methanol in a fume hood. Bottles containing methanol must always be tightly closed to avoid the release of harmful gases and aerosols.
16. For the reversed-phase chromatography step either a HPLC system or a FPLC system that can deliver high pressures is needed. FPLC systems used in biological applications are not normally designed for reversed-phase chromatography and

special care must be taken that the system is compatible with organic solvents and acids. The pumps and the tubing must not be exposed to the reversed-phase chromatography buffers for prolonged times and the solvent−/acid-containing buffers must be flushed from the system immediately after use.

17. For freeze-drying samples from reversed-phase chromatography purifications it is important to use a Teflon-coated freeze-dryer system and a chemical-resistant vacuum pump as harsh organic solvents and acids are used in the buffers. The solvents also have a low freezing point, and therefore a condenser capable of reaching temperatures below −80 °C must be used. The exhaust of the freeze-dryer should be connected to a chemical fume hood so that harmful gases and vapors are not expelled into the laboratory air.
18. Ethidium bromide is potentially cancerogenic. Always wear gloves when working with ethidium bromide, and use a dedicated ethidium bromide area and equipment. Adhere to all appropriate precautions and regulations when disposing of solid and liquid ethidium bromide waste.
19. Biotin-11-dUTP varies greatly in price and can be very expensive, depending on the supplier. It is worthwhile to shop around in order to find the supplier with the best price.
20. The phenol–chloroform–isoamyl alcohol mixture is a harmful chemical. Chloroform and isoamyl alcohol are toxic and flammable solvents and phenol is cancerogenic and causes severe burns. Gloves and safety goggles must be worn when handling phenol–chloroform–isoamyl alcohol solution. All procedures involving phenol–chloroform–isoamyl alcohol must be carried out in a fume hood. All phenol–chloroform–isoamyl alcohol waste must be collected separately in designated phenol waste containers stored under a fume hood, and disposed of via the appropriate chemical waste route.
21. PMSF is very toxic. Always wear gloves when handling PMSF-containing solutions. Care should be taken when weighing in PMSF, wear gloves and safety goggles, and preferably handle PMSF powder under a fume hood.
22. TSA is very toxic. Handle solutions containing TSA with gloves.
23. Depending on the exact cDNA sequence, *E. coli* expression strain, and medium (type and supplier) the expression level of the core histones can vary greatly and unpredictably. Expression conditions therefore need to be optimized. However, once optimal conditions have been found the expression is very robust and reproducible.

24. Before discarding used and contaminated LB medium, such as leftover bacterial cultures or supernatants, and bacterial agar plates, the media/plates need to be collected and the bacteria inactivated either by autoclaving or by chemical inactivation. Refer to your local rules and guidelines for the appropriate procedures.
25. Freezing the pellets helps to break open the cells and is recommended. The cells can be stored frozen in Histone Wash Buffer for several weeks (or even months).
26. We carry out the sonication in the 50 mL Falcon tube in which the cells were frozen. For cells from 1 L of culture (approximately 25 mL) we typically use a Branson 450 digital sonifier with a 3 mm (1/8 in.) tapered microtip probe set to approximately 25% amplitude and 1 s ON/1 s OFF cycles for 1 min (30 s total sonication time for each cycle). Place the sample in a soundproof cabinet or use ear protectors to protect yourself and others from ear damage through high energy ultrasound. The amplitude setting needs to be adjusted to achieve optimal cell lysis while keeping frothing to a minimum. Alternatively, the cells can be passed through a French press or a continuous cell disruptor if larger volumes need to be processed.
27. Inclusion body pellets can be stored at −20 °C for several months and aliquots used as needed. Freezing the inclusion bodies is recommended as it helps to resuspend them during the extraction procedure.
28. Urea in solution decays into ammonium cyanate. Isocyanic acid formed from the cyanate ions reacts with amino groups in proteins (N-terminus and side chains of Lys and Arg) leading to carbamylation of the histones over time if left in urea solutions. To prevent carbamylation we always use a freshly prepared and deionized urea stock solution (*see* **Note 8**), and we aim to complete the whole purification procedure (gel-filtration and ion-exchange steps) in less than 24 h in order to minimize the time that the histones are exposed to the urea buffers. Urea-containing solutions should also not be heated.
29. The maximum recommended volume that should be loaded onto the HiPrep 26/60 Sephacryl S-200 HR column is 15 mL. We find that the extraction of the inclusion bodies is more efficient if a larger volume of Unfolding Buffer (i.e., 25 mL) is used. In order to pass this volume over the column we split the gel filtration into two overnight runs. We therefore equilibrate the column during the day, extract the inclusion bodies in the afternoon, and run the two gel-filtration runs overnight. The runs will then be finished the next morning, so that the

ion-exchange (core histones) or RPC purification (truncated histones for NCL) can be performed the next day.

30. Aliquots of 0.5 mg or 1 mg for H2A, H2B, and H3.1, and 0.4 mg or 0.8 mg for H4 are most convenient as these amounts are needed for the octamer refolding reactions.
31. Due to the high degree of nonspecific interaction with the chromatography media we found that histone H4 negatively affects the separation of proteins on the Sephacryl S-200 HR gel-filtration column after several uses of the column. It also leaches out of the SP Sepharose ion-exchange column in subsequent purifications, thereby contaminating the other histones if the same columns are used for H2A, H2B, or H3.1 and the H4 purifications. We therefore have a separate set of chromatography columns exclusively reserved for H4 purifications.
32. We use a HiTrap SP Sepharose HP column with a 5 mL bed volume for the histone H4 ion-exchange chromatography step in order to reduce the available chromatography matrix that H4 can nonspecifically bind to. The capacity of the column is sufficient and the smaller size of the column also seems to help in eluting H4, probably due to electrostatic effects when H4 is bound to the resin at a high density.
33. During the ion-exchange purification for H4 we simultaneously increase the salt concentration from 200 mM to 1 M NaCl and we perform a buffer exchange from sodium acetate buffer at pH 5.2 to HEPES buffer at pH 8.0. The increased pH lowers the positive charge of H4 and thereby reduces the interaction with the negatively charged ion-exchange resin. Together with the steeper NaCl gradient and the smaller size of the 5 mL SP Sepharose column this improves the elution efficiency and thereby the overall yield of H4 significantly.
34. We have sequenced the purified truncated histones that we use for NCL by Edman degradation and found that the removal of the N-terminal formyl-methionine is very efficient.
35. For the NCL procedure to work efficiently it is important that cysteines are kept reduced. The reversed-phase chromatography has the advantage that it is carried out in a very low pH which keeps cysteines reduced, and that the truncated histones can be lyophilized immediately from the chromatography fractions. This avoids the lengthy dialysis step necessary after the ion exchange in the standard histone purification procedure during which the histones are exposed to oxygen and can potentially become oxidized. The reversed-phase chromatography therefore works well as a rapid final clean up step.
36. The purification procedure for truncated histones for NCL described here is designed so that commercial chromatography columns can be used. For purifying truncated histone H4

(H4 Δ1–28 I29C) we follow the protocol as described using a HiPrep 26/60 Sephacryl S-200 HR gel-filtration column and a homemade 4 mL bed volume Tricorn 5/200 column packed with SOURCE 15RPC media (Cytiva); however, a commercially available 3 mL RESOURCE RPC column (Cytiva) most likely also works. Truncated histone H3 (H3 Δ1–31 T32C) we usually purify in larger amounts, and we use inclusion bodies from 12 L of culture as starting material. Here we use a custom-made XK 50/100 Sephacryl S-200 HR gel-filtration column (Cytiva) with approximately 1.8 L bed volume and a homemade 10 mL bed volume Tricorn 10/100 SOURCE 15RPC column (Cytiva). The volumes of chromatography buffers and the chromatography protocols need to be adjusted accordingly. We have separate SOURCE 15RPC columns for truncated H3 and H4 to avoid cross-contamination. The SOURCE 15RPC Tricorn columns must be packed using solvent-resistant O-rings.

37. We use 100 mM DTT in the Unfolding Buffer for the truncated histones for NCL to ensure that the N-terminal cysteines are completely reduced during the inclusion body extraction step.

38. Reversed-phase chromatography columns must be cleaned and equilibrated before use. Run Buffer A (0.1% trifluoroacetic acid in water) for 2 CV at a flow rate of 0.5 mL/min. Then cycle at least 2× one gradient up (0–100% Buffer B over 20 min; Buffer B: 90% acetonitrile/0.1% trifluoroacetic acid in water) and one gradient down (100–0% B over 10 min) until the UV absorbance is stable and no more proteins are eluted. Equilibrate with another 5 CV of Buffer A.

39. Due to the high content of acetonitrile the combined reversed-phase chromatography fractions have a lower melting point than aqueous solutions. For effective freeze-drying the frozen sample must be very cold and have a large surface so that the sample can be cooled by the evaporation, otherwise it will melt and start to boil, and then "explode" in the freeze-dryer flask. This can be achieved by rotating the sample in a 50 mL Falcon tube or directly in a freeze-dryer flask in a cold bath made of dry ice in 100% ethanol. This should be done in a fume hood. Make sure to use a leak-proof container and be careful when handling the dry ice/ethanol bath as it is very cold. When finished let the dry ice evaporate and allow the ethanol to warm up before disposing of it in the sink. Never pour the cold ethanol into the sink as it will freeze and burst the pipes!

40. The collected ice from freeze-drying the reversed-phase chromatography fractions contains acetonitrile and trifluoroacetic acid. Once thawed collect the liquid and treat it like RPC

waste, dispose of it via the chemical waste route (*see* **Note 12**). Clean the freeze-dryer immediately after use.

41. The extinction coefficient of the truncated histone depends on the NCL strategy and the exact peptide sequence used. A good online tool for calculating extinction coefficients of proteins is ProtParam (https://web.expasy.org/protparam/). After entering the protein sequence of the truncated histone the tool will calculate various physical and chemical parameters, including the extinction coefficients and the corresponding absorbances at 280 nm. For truncated histone H3 (H3 Δ1–31 T32C) the absorbance of a 0.1% solution (Abs 0.1% [1 mg/mL]) is 0.371, and for truncated histone H4 (H4 Δ1–28 I29C) the Abs 0.1% [1 mg/mL] is 0.713 (*see* also Table 1).
42. To ensure full reactivity of the cysteines in the NCL reaction it is essential to remove any oxygen that is dissolved in the NCL buffer by thoroughly degassing the buffer. Residual oxygen oxidizes the cysteines leading to significantly reduced ligation efficiencies.
43. The inert argon gas prevents oxygen from dissolving again in the degassed NCL buffer. An easy way of flooding the system with argon is to inflate a balloon with argon from an argon gas cylinder and to then connect the filled balloon to the inert gas inlet of the T-bore stopcock.
44. There can be a small amount of precipitate, since MPAA is difficult to dissolve; however, it is important to use high enough amounts of TCEP and MPAA for the chemical ligation.
45. The N-terminal histone thioester peptides should be aliquoted into 1 mg aliquots before use. This can be done by dissolving the peptides in degassed water and freeze-drying them again in microtubes in 1 mg aliquots. The aliquots can then be stored at −20 °C or −80 °C.
46. We typically use a Superdex 200 10/300 GL column for octamer refolding reactions with up to 2 mg total protein and a Superdex 200 16/60 Prep-grade column for larger octamer refolding reactions. We have designated Superdex 200 columns that are exclusively reserved for purifying histone octamers.
47. Use 3× 500 mL of the Octamer Refolding Buffer for the dialyses and 1× 500 mL to equilibrate and run the gel-filtration column. Filter and degas the 500 mL gel-filtration buffer and add β-mercaptoethanol to all buffers just before use.
48. For successful refolding of octamers it is important to be precise at this step as the histones need to be mixed at exactly

equimolar ratios. If the ratios are incorrect the histones will form aggregates and the yield of refolded octamer will be low. To shift the equilibrium toward the formation of octamers a 10% excess of H2A and H2B is added to ensure that any H3/H4 tetramers are converted into octamers. If 1 mg of histone H3.1 is used as the reference point then 1 mg of each H2A and H2B and 0.74 mg of H4 need to be added to the refolding reaction, taking the molecular weights of the histones and the 10% excess of H2A and H2B into account. For larger or smaller refolding reactions the amounts need to be adjusted accordingly. In our experience, if the concentration measurement and aliquoting of the histones is precise and consistent during the purification procedure (*see* Subheadings 3.1 and 3.2) then it is not necessary to measure the histone concentrations again after the unfolding step, and the 1 mg (H2A, H2B, H3.1) or 0.8 mg (H4) aliquots can be used directly.

49. In our experience longer dialysis steps result in better octamer refolding, we usually do the first and second dialysis steps for at least 4 h, and the third dialysis step overnight.

50. If the gel filtration is done at RT make sure that the pH of the buffer is at 7.5.

51. The octamer peak should be symmetric. Only use the peak fractions and discard any trailing fractions. These might contain H3/H4 tetramers (visible as a small peak or "shoulder" in the chromatogram—*see* Fig. 5) which will interfere with nucleosome assembly later on.

52. The volume measurements and the pipetting of glycerol have to be precise at this step. It is therefore important to completely empty the pipette tips of the viscous glycerol solution and to mix well by thoroughly pipetting up and down multiple times. Inaccurate octamer concentrations or incomplete mixing will otherwise lead to inconsistencies in the nucleosome assemblies later on.

53. End-biotinylated nucleosomal DNAs can also be prepared by PCR using appropriate biotinylated primers and a plasmid containing a single repeat of the desired nucleosomal DNA as template. In this case, standard 100 μL PCR reactions using Taq polymerase are carried out in 96-well plates and the amplified DNA is subsequently purified by ion-exchange chromatography as described in Subheading 3.4.1 from **step 19** onward. In our hands the yield per plate (i.e., 10 mL of total PCR volume) is in the range of a few hundred micrograms. The PCR method is quicker, as it avoids the construction of the tandem repeat plasmids, and is suitable if small amounts or special designs of nucleosomal DNA are required. However, if larger amounts of specific nucleosomal DNAs are frequently

used it is worthwhile to invest into cloning the tandem repeat plasmids (*see* **Note 54**) as this will save time later on and reduce batch variations between different DNA preps.

54. The pUC19 plasmid containing eight tandem repeats of the 601 dinucleosome sequence (Fig. 6) is prepared as described in [7]. A single repeat of the 601 dinucleosome sequence containing *Eco*RV and *Eco*RI sites at the opposite ends is cloned into pUC19 flanked by KpnI/BglII and BamHI sites. The single repeat is then amplified to the desired number of copies using digests with KpnI and the compatible BglII and BamHI enzymes as described [7]. The design of such constructs is flexible and depends on the question to be addressed. For our biotinylated 601 dinucleosome DNA we use two 601 repeats separated by a 50 bp linker and flanked by 3′- and 5′-linkers of about 20 bp. Due to the high number of direct repeats these tandem 601 repeat plasmids tend to be unstable and lose repeats through recombination when propagated in *E. coli* (*see* also **Note 55**). We typically use constructs with no more than 16× 601 repeats in total, as these can be kept stable fairly easily and still result in decent yields of nucleosomal DNAs. Constructs with higher repeat numbers tend to be rather unstable. Other plasmid backbones can be used, although very-high-copy plasmids should be avoided as these tend to recombine at higher rates.
55. Due to the high number of identical repeats the tandem 601 and MMTV-A repeat constructs tend to be unstable and recombine out repeats during propagation in *E. coli* (*see* also **Note 54**). Therefore, special care needs to be taken when amplifying the "pUC19-8x601 di-Nuc" and "pUC19-MMTV-A" plasmids in order to prevent any loss of repeats. It is important to freshly transform the plasmids, as cells from frozen stocks or from older agar plates have often recombined the tandem repeats. It is further important to pick a rather small colony (i.e., average but not too small) from the plate, as large colonies indicate clones that have lost repeats and thereby gained a growth advantage. We have also found that reducing the selection pressure on the cells reduces the tendency of the plasmid to recombine. We achieve this by growing the cells with less antibiotics, in our case 50 μg/mL instead of 100 μg/mL of ampicillin that are normally used. Lastly, it is important to not let the culture overgrow, but to start with the DNA prep directly in the morning with fresh cells. Especially for the "pUC19-8x601 di-Nuc" plasmid we verify that the plasmid has not recombined before performing the large-scale DNA prep, and check the integrity of the 8× 601 dinucleosome insert in the Gigaprep DNA by a suitable restriction digest before proceeding with any further steps. If 601 repeats

have been lost then the 8× 601 dinucleosome insert will not have the same size as in the original plasmid. We discard cultures in which the 8× 601 dinucleosome insert has recombined. For the MMTV-A DNA this is less critical as it only serves as buffer DNA. Here we only ensure that the plasmid stocks are correct.

56. Communal centrifuge tubes, glass- and plasticware are sometimes contaminated by residual harsh detergents and/or other substances. These can interfere with the enzymatic reactions carried out for making the biotinylated nucleosomal DNA. We therefore thoroughly rinse out any glass- or plasticware used to make buffers before use, and we have a separate set of polypropylene Oak Ridge centrifuge tubes reserved for the DNA manipulation steps for nucleosomal DNA preps. We keep those apart from the common centrifuge tubes and wash them by hand with mild washing-up liquid, rinsing them well afterward with deionized water.

57. Purifying the "pUC19-8x601 di-Nuc" and "pUC19-MMTV-A" plasmid DNA using a Gigaprep kit is convenient and sufficient for our purposes. We normally use a standard Gigaprep kit with gravity flow columns for 10 mg of DNA—for 5–6 L of bacterial culture these typically yield between 10 and 15 mg of total DNA. A less pricey procedure to prepare very large amounts of plasmid DNA using "homemade" solutions is described in [7]. However, this procedure is quite involved and produces substantial amounts of phenol waste; we therefore prefer to use less toxic Gigaprep kits.

58. The number of units of the restriction enzyme required for a complete digest depends on the activity of the enzyme and the number of restriction sites that need to be cut. The amounts needed should be determined experimentally before performing the large-scale digests.

59. Separation of the small insert from the vector by PEG precipitation only works for DNA fragments with blunt ends, and will not work for DNA fragments with sticky ends.

60. The PEG precipitation is very sensitive to the PEG concentration and the size of the insert. The optimal PEG concentration that separates the excised insert from the pUC19 vector backbone therefore needs to be determined experimentally. For this we usually perform a series of small test precipitations with PEG concentrations ranging from 5–8%. We select the PEG concentration for the large-scale precipitation that completely precipitates the vector and leaves most of the *Eco*RV-digested fragment in the supernatant.

61. We use the Klenow Fragment (3′-> 5′ exo-) polymerase for the fill-in reaction as it is devoid of 5′-> 3′ and 3′-> 5′ exonuclease

activities. It will therefore only fill in the *Eco*RI overhang and not attack (and possibly biotinylate) the opposite *Eco*RV-digested blunt end.

62. The efficiency of the biotinylation should be tested in a streptavidin binding reaction. For this we usually incubate 25 μg of the biotinylated 601 dinucleosome DNA with 100 μL (bed volume) of Streptavidin Sepharose in 500 μL of TE buffer containing 500 mM NaCl and 0.05% NP40 by rotating on a wheel for 2 h at RT. We then spin down the streptavidin beads and separate 20 μL of the supernatant (equivalent to 1 μg of input DNA) on a 1% agarose gel. If the biotinylation was successful there is no DNA visible in the supernatant, compared to the input (20 μL taken out before addition of the streptavidin Sepharose).
63. Titrations are performed at a small scale (around 5–6 μg) by varying the molar ratio of histone octamers to DNA. We typically keep the amount of DNA constant and titrate the molar octamer:DNA ratio (i.e., octamer per 601 repeat) around 1:1 (e.g., 0.8:1, 0.9:1, 1:1, 1.1:1, 1.2:1 – *see* Fig. 7). The point at which uniform dinucleosomes are formed is indicated by the 601 dinucleosome DNA forming a single sharp band, and a small amount of mononucleosomes formed from the MMTV-A DNA being visible in the high-resolution gel shift assay (*see* Subheading 3.4.4).
64. Make sure that you add the octamers last, mixing octamers and DNA in the absence of NaCl results in precipitation of the octamers. If using octamer stock solutions that contain glycerol make sure to completely fill and empty the pipette tip as minor variations in octamer concentration in the reconstitution reaction will result in inconsistent nucleosome assemblies.
65. The continuous exchange of RB-high buffer against RB-low buffer results in a slow and gradual decrease of the KCl concentration over time. Reducing the KCl concentration too quickly leads to failed nucleosome assemblies.
66. Recombinantly assembled nucleosomes and chromatin have the tendency to adsorb to plastic. To avoid any losses of the dinucleosomes over time it is advised to use DNA low-binding microtubes for storage and experiments with nucleosomes and chromatin.
67. Pour the gel at RT (no stacking gel), let it polymerize, and then place it in the gel tank submerged in 0.2× TBE running buffer at 4 °C (cold room) until the next day. We find that complete polymerization overnight improves the resolution of the gel.
68. Mixing the running buffer from the upper and lower gel chambers, flushing the wells before loading the samples,

loading small sample volumes, and running the gel at 4 °C in the cold room significantly improves the resolution.

69. Binding of some proteins to the dinucleosomes in the affinity purification experiments can be greatly influenced by the presence of RNA. Freshly prepared nuclear extracts contain a substantial amount of nuclear and chromatin-associated RNAs, which can, however, be partially degraded during the storage of the nuclear extracts. This can result in inconsistencies between experiments performed with different extracts and when using freshly prepared extracts and extracts stored at −80 °C for several months. We therefore recommend either digesting all RNA using RNase A during the nuclear extract preparation or preventing RNA degradation during extract preparation and/or storage by the addition of RNase inhibitors.

70. Nuclear protein extraction using Buffer C containing 420 mM NaCl [25, 26] results in preferential extraction of euchromatic proteins, while most heterochromatic proteins are not fully released into the nuclear extract, and preferentially remain associated with the chromatin pellet. Heterochromatic protein extraction can be greatly improved by using stepwise extraction with Buffer C with increasing NaCl concentrations (described in Subheading 3.5, **steps 18–26**). For probing nucleosomes decorated with repressive modifications, such as H3K9me3 and H3K27me3, we therefore recommend following the stepwise nuclear extract preparation protocol to ensure more complete heterochromatic protein extraction (Figs. 8 and 9).

71. Ultracentrifuges are dangerous machines. Care must be taken when spinning the nuclear extracts in the ultracentrifuge. A swinging-bucket rotor should be used for this centrifugation step. Always make sure that the samples are precisely tared by weighing the tubes with the samples on a precision scale, and by loading them symmetrically into the rotor buckets. Rotor buckets in ultracentrifuge swinging-bucket rotors are matched and numbered; always place the correct bucket in the correct rotor position and make sure that the bucket is correctly mounted. Always observe the ultracentrifuge when ramping up speed, if there is the slightest sign of an imbalance immediately stop the run and rebalance/retare the samples and buckets.

72. When collecting the supernatants (i.e., the nuclear extract fractions) from the ultracentrifuge tubes, take care not to disturb and transfer any of the pellets or lipids (floating on top). This can best be achieved by carefully fixing the polypropylene tubes in a laboratory stand, punching through the wall of the tube with a needle a few millimeters above the pellet, and

carefully collecting the nuclear extract fractions with a fresh clean syringe.

73. Different types of streptavidin beads display different patterns of background protein binding. We have tested different streptavidin-coated beads and found Streptavidin Sepharose High Performance beads (Cytiva) to have the lowest background overall. However, the type of beads and the amount needed to immobilize the required/desired amounts of nucleosomes might have to be optimized for different questions and procedures.
74. This dilution ratio with the Adjustment Buffer applies if the 350, 500, 750, and 1000 mM NaCl nuclear extract fractions were mixed in a 2:1:1:1 ratio (*see* Subheading 3.5 **step 29**). If only the 350 mM fraction is used (*see* Subheading 3.5 **step 17** and **Note 70**) then the nuclear extract must be diluted with 1.25× nuclear extract volumes of Adjustment Buffer to reach a NaCl concentration of 155 mM.
75. The nucleosome-bound proteins are eluted from the beads in 1× SDS-PAGE sample buffer (*see* Subheading 3.6.2 **steps 11** and **12**). The eluted proteins can therefore directly be processed for the mass spectrometric analysis using a FASP protocol (*see* Subheading 3.7.1). Alternatively, the samples can be separated on an SDS-PAGE gel and the binding of a particular protein of interest to the modified nucleosome(s) can be probed by western blot, adding flexibility to this protocol.

Acknowledgements

We would like to thank Ann-Christine König from the Helmholtz Zentrum München Research Unit Protein Science for help with the quantification of the label-free mass spectrometry data for the H3K4me3 and H3K9me3 dinucleosome pull-down experiments. The development of these protocols was supported by funding from the Human Frontiers Science Program Organization, the Medical Research Council (Grant No. MC_UP_1102/2), the European Research Council (ERC StG No. 309952), the Deutsche Forschungsgemeinschaft (DFG Project No. 431163844 and 213249687/SFB 1064), and the Helmholtz Gesellschaft to T.B.

References

1. Kouzarides T (2007) Chromatin modifications and their function. Cell 128:693–705
2. Bannister AJ, Kouzarides T (2011) Regulation of chromatin by histone modifications. Cell Res 21:381–395
3. Bartke T, Vermeulen M, Xhemalce B, Robson SC, Mann M, Kouzarides T (2010) Nucleosome-interacting proteins regulated by DNA and histone methylation. Cell 143:470–484

4. Makowski MM, Gräwe C, Foster BM, Nguyen NV, Bartke T, Vermeulen M (2018) Global profiling of protein-DNA and protein-nucleosome binding affinities using quantitative mass spectrometry. Nat Commun 9:1653
5. Nakamura K, Saredi G, Becker JR, Foster BM, Nguyen NV, Beyer TE, Cesa LC, Faull PA, Lukauskas S, Frimurer T, Chapman JR, Bartke T, Groth A (2019) H4K20me0 recognition by BRCA1-BARD1 directs homologous recombination to sister chromatids. Nat Cell Biol 21:311–318
6. Luger K, Rechsteiner TJ, Richmond TJ (1999) Preparation of nucleosome core particle from recombinant histones. Methods Enzymol 304: 3–19
7. Dyer PN, Edayathumangalam RS, White CL, Bao Y, Chakravarthy S, Muthurajan UM, Luger K (2004) Reconstitution of nucleosome core particles from recombinant histones and DNA. Methods Enzymol 375:23–44
8. Musselman CA, Kutateladze TG (2019) Strategies for generating modified nucleosomes: applications within structural biology studies. ACS Chem Biol 14:579–586
9. Muir TW (2003) Semisynthesis of proteins by expressed protein ligation. Annu Rev Biochem 72:249–289
10. Wingfield PT (2017) N-terminal methionine processing. Curr Protoc Protein Sci 88: 6.14.1–6.14.3
11. Foster BM, Stolz P, Mulholland CB, Montoya A, Kramer H, Bultmann S, Bartke T (2018) Critical role of the UBL domain in stimulating the E3 ubiquitin ligase activity of UHRF1 toward chromatin. Mol Cell 72:739–752.e9
12. Tolbert TJ, Wong CH (2002) New methods for proteomic research: preparation of proteins with N-terminal cysteines for labeling and conjugation. Angew Chem Int Ed Engl 41:2171–2174
13. Hackeng TM, Griffin JH, Dawson PE (1999) Protein synthesis by native chemical ligation: expanded scope by using straightforward methodology. Proc Natl Acad Sci U S A 96: 10068–10073
14. Chen Z, Grzybowski AT, Ruthenburg AJ (2014) Traceless semisynthesis of a set of histone 3 species bearing specific lysine methylation marks. Chembiochem 15:2071–2075
15. Guidotti N, Lechner CC, Bachmann AL, Fierz B (2019) A modular ligation strategy for asymmetric bivalent nucleosomes trimethylated at K36 and K27. Chembiochem 20:1124–1128
16. Krauskopf K, Lang K (2020) Increasing the chemical space of proteins in living cells via genetic code expansion. Curr Opin Chem Biol 58:112–120
17. Yang A, Ha S, Ahn J, Kim R, Kim S, Lee Y, Kim J, Söll D, Lee HY, Park HS (2016) A chemical biology route to site-specific authentic protein modifications. Science 354:623–626
18. Wright TH, Bower BJ, Chalker JM, Bernardes GJ, Wiewiora R, Ng WL, Raj R, Faulkner S, Vallée MR, Phanumartwiwath A, Coleman OD, Thézénas ML, Khan M, Galan SR, Lercher L, Schombs MW, Gerstberger S, Palm-Espling ME, Baldwin AJ, Kessler BM, Claridge TD, Mohammed S, Davis BG (2016) Posttranslational mutagenesis: a chemical strategy for exploring protein side-chain diversity. Science 354:aag1465
19. McGinty RK, Kim J, Chatterjee C, Roeder RG, Muir TW (2008) Chemically ubiquitylated histone H2B stimulates hDot1L-mediated intranucleosomal methylation. Nature 453:812–816
20. Wan Q, Danishefsky SJ (2007) Free-radical-based, specific desulfurization of cysteine: a powerful advance in the synthesis of polypeptides and glycopolypeptides. Angew Chem Int Ed Engl 46:9248–9252
21. Song OK, Wang X, Waterborg JH, Sternglanz R (2003) An Nalpha-acetyltransferase responsible for acetylation of the N-terminal residues of histones H4 and H2A. J Biol Chem 278: 38109–38112
22. Lowary PT, Widom J (1998) New DNA sequence rules for high affinity binding to histone octamer and sequence-directed nucleosome positioning. J Mol Biol 276:19–42
23. Dorigo B, Schalch T, Bystricky K, Richmond TJ (2003) Chromatin fiber folding: requirement for the histone H4 N-terminal tail. J Mol Biol 327:85–96
24. Huynh VA, Robinson PJ, Rhodes D (2005) A method for the in vitro reconstitution of a defined "30 nm" chromatin fibre containing stoichiometric amounts of the linker histone. J Mol Biol 345:957–968
25. Dignam JD, Lebovitz RM, Roeder RG (1983) Accurate transcription initiation by RNA polymerase II in a soluble extract from isolated mammalian nuclei. Nucleic Acids Res 11: 1475–1489
26. Abmayr SM, Yao T, Parmely T, Workman JL (2006) Preparation of nuclear and cytoplasmic extracts from mammalian cells. Curr Protoc Mol Biol 75:12.1.1–12.1.10
27. Wiśniewski JR, Zougman A, Nagaraj N, Mann M (2009) Universal sample preparation

method for proteome analysis. Nat Methods 6: 359–362

28. Silva JC, Gorenstein MV, Li GZ, Vissers JP, Geromanos SJ (2006) Absolute quantification of proteins by LCMSE: a virtue of parallel MS acquisition. Mol Cell Proteomics 5:144–156
29. Schwanhäusser B, Busse D, Li N, Dittmar G, Schuchhardt J, Wolf J, Chen W, Selbach M (2011) Global quantification of mammalian gene expression control. Nature 473:337–342
30. Välikangas T, Suomi T, Elo LL (2018) A systematic evaluation of normalization methods in quantitative label-free proteomics. Brief Bioinform 19:1–11
31. Ritchie ME, Phipson B, Wu D, Hu Y, Law CW, Shi W, Smyth GK (2015) Limma powers differential expression analyses for RNA-sequencing and microarray studies. Nucleic Acids Res 43:e47
32. Benjamini Y, Hochberg Y (1995) Controlling the false discovery rate: a practical and powerful approach to multiple testing. J R Stat Soc B 57: 289–300
33. Perez-Riverol Y, Csordas A, Bai J, Bernal-Llinares M, Hewapathirana S, Kundu DJ, Inuganti A, Griss J, Mayer G, Eisenacher M, Pérez E, Uszkoreit J, Pfeuffer J, Sachsenberg T, Yilmaz S, Tiwary S, Cox J, Audain E, Walzer M, Jarnuczak AF, Ternent T, Brazma A, Vizcaíno JA (2019) The PRIDE database and related tools and resources in 2019: improving support for quantification data. Nucleic Acids Res 47:D442–D450

Part VII

Inheritance of Histone Methylation Patterns

Chapter 17

Investigating Mitotic Inheritance of Histone Posttranslational Modifications by Triple pSILAC Coupled to Nascent Chromatin Capture

Kyosuke Nakamura, Anja Groth, and Constance Alabert

Abstract

Pulse stable isotope labeling with amino acids in cell culture (pSILAC) coupled to mass spectrometric analysis is a powerful tool to study propagation of histone post-translational modifications (PTMs). We describe the combination of triple pSILAC with pulse-chase labeling of newly replicated DNA by nascent chromatin capture (NCC). This technology tracks newly synthesized and recycled old histones, from deposition to transmission to daughter cells, unveiling principles of histone-based inheritance.

Key words Histone, Posttranslational modifications, DNA replication, Mass spectrometry, SILAC, Nascent Chromatin Capture, Pulse-chase, Chromatin assembly

1 Introduction

The nucleosome is the basic unit of chromatin and consists of 147 base pairs of DNA wrapped around an octamer of histones. It comprises one histone H3–H4 tetramer flanked by two histone H2A–H2B dimers. Histones can carry a large variety of PTMs, such as phosphorylation, acetylation, ubiquitylation, and methylation, on the histone tails extruding from the nucleosome structure or within the histone core domain. Histone PTMs play fundamental roles in genome structure and function through various mechanisms [1–3]. To execute their function, several histone modifications, often acetylation, can directly affect the contact between the histone octamer and the DNA molecule, thereby altering intra- and inter-nucleosome contacts and chromatin compaction [4]. In addition, histone PTMs can serve as binding platforms for factors involved in a wide range of processes such as transcription, replication, DNA repair, and chromatin compaction. Although these factors often have additional means to access chromatin,

Raphaël Margueron and Daniel Holoch (eds.), *Histone Methyltransferases: Methods and Protocols*,
Methods in Molecular Biology, vol. 2529, https://doi.org/10.1007/978-1-0716-2481-4_17,

histone PTMs constitute an important level of chromatin regulation [1, 5, 6].

In S phase, the DNA replication machinery duplicates the entire genome nonlinearly, following a stable cell-type specific program [7]. At the molecular level, the progression of thousands of replisomes provokes the disassembly of nucleosomes ahead of replication forks followed by their reassembly on the two newly synthesized daughter DNA strands. To maintain the prereplication nucleosome density, new canonical histones are rapidly synthesized as cells enter S phase and assembled into nucleosomes onto newly replicated DNA. Remarkably, newly synthesized and recycled old histones are present in a 1:1 ratio on newly replicated DNA, distributed on both daughter strands and assembled into nucleosomes containing either new or old H3-H4 tetramers with very limited intranucleosome mixing [8–11]. Upon recycling, parental histones maintain their PTMs [12–14], while newly synthesized histones are largely unmodified except for the highly conserved diacetylation on histone H4 lysine 5 and 12 [11]. Once incorporated, histones are further modified so that the newly assembled composite of recycled and newly synthesized histones mirrors the prereplicative histone PTM state. This mechanism ensures the propagation of histone modifications through rounds of cell divisions [15].

Double and triple pSILAC have been used to distinguish recycled and newly synthesized histones [14, 16–18]. Both strategies rely on the substantial production of new canonical histones upon S-phase entry [19]. In double pSILAC, cells are grown in medium containing light (L)-arginine (R0) and arrested at the G1/S transition (Fig. 1). At this point, all histones present in the cells contain R0 (generation 1). Upon release into S phase, cells are switched into medium containing isotopically labeled arginine (R6). All newly synthesized histones produced and incorporated during S phase contain R6 (generation 2) and can thus be distinguished from preexisting histones containing R0 (generation 1). In triple pSILAC, after a couple of hours, cells are further switched into medium containing arginine labeled with a heavier isotope (R10) so that histones synthesized from this point are distinct from the two previous histone generations. Although triple pSILAC has proven to be a powerful tool to track in vivo histone fate through subsequent rounds of cell divisions [16, 17], it presents certain limitations for analyzing early stages of histone PTM transmission. First, throughout the first S phase, generation 1 histones are a mix of recycled and not-yet-recycled histones (Fig. 1). This limitation disappears when cells enter G2 phase, and 100% of generation 1 histones have now been recycled. The second limitation concerns the generation 2 histones. At the end of the R6 pSILAC labeling, new histones have been deposited on newly replicated DNA across the labeling period and thus represent an

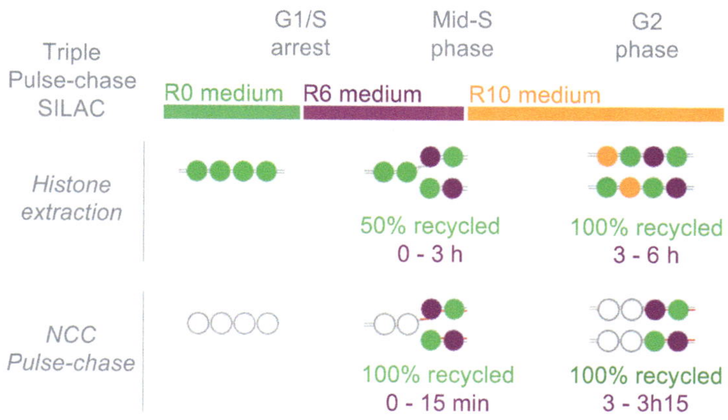

Fig. 1 Triple pSILAC (top) combined with histone extraction (middle) or Nascent Chromatin Capture (bottom). For parental histones (Generation 1, green), the percentage of recycled histones in shown. For newly synthesized histones (Generation 2, purple), the age of incorporated histones is shown. In the pSILAC-NCC setup, histones isolated and analyzed are shown in colours. The rest of the histone population is colourless

age-span ranging from a couple of minutes to hours. This wide age-range can mask early steps of PTM establishment on newly incorporated histones and remains a challenge throughout the experiment. To overcome these two limitations, we have combined triple pSILAC with pulse-chase labeling of newly replicated DNA by nascent chromatin capture (NCC) [12]. This design makes it possible, first, to refine the analysis of generation 1 histones to recycled histones only at any given time of the experiment, and second, to reduce the age-range of newly synthesized histones from a couple of hours to 15 min (Fig. 1). This jump in resolution for both recycled histones and newly synthesized histones has unveiled the true nascent state of chromatin, the T0 of chromatin assembly, and the distinct fate of newly synthesized and parental histones from the moment of their assembly onto newly replicated DNA.

2 Materials

All solutions should be freshly made.

1. HeLa S3 cells and cell culture equipment and supplies (sterile CO_2 incubator, spinner flasks, etc.)
2. *Complete SILAC medium* SILAC medium depleted for arginine and lysine is supplemented with 10% dialyzed FBS, 1% MEM nonessential amino acid mix, 1% GlutaMAX, 66 mg/L of arginine, heavy (Isotopically labeled arginine R6 or R10) or light (R0), and 100 mg/L of lysine (not isotopically labeled). Keep protected from light if heavy amino acids are added.

3. *20 mM Thymidine* in complete SILAC medium. Filter-sterilize prior to use.
4. PBS, pH 7.2.
5. *24 mM deoxycytidine* in prewarmed R6 complete SILAC medium.
6. *100 mM Biotin-16-dUTP* (IBA tagnology, on demand) in 10 mM Tris pH 8.0. Biotin-dUTP with shorter backbone length will affect NCC efficiency. 10 min prior to use, briefly spin the lyophilized biotin-dUTP and dissolve. Mix well and keep on ice until use. Use single-use aliquots of lyophilized biotin-dUTP as it is unstable upon freeze–defreeze cycles.
7. *KBH buffer* 50 mM KCl; 10 mM HEPES. To make 1 M HEPES stock dissolve HEPES powder to 1 M; do not adjust the pH.
8. *Cross-linking solution* 2% formaldehyde in PBS, diluted just before use from 37% solution methanol stabilizer (Sigma, cat. no. F8923).
9. *10% glycine* Prepared in autoclaved water at RT. Use at 1% final concentration (*see* Subheading 3).
10. *1 mM PMSF in PBS* Dilute PMSF in ice-cold PBS. Keep on ice before use.
11. *Sucrose buffer* 0.3 M sucrose; 10 mM HEPES–NaOH pH 7.9; 1% Triton X-100; 2 mM $Mg(OAc)_2$.
12. *15 mL Dounce homogenizer with tight pestle.*
13. *Glycerol buffer* 25% glycerol; 10 mM HEPES–NaOH pH 7.9; 0.1 mM EDTA; 0.1 mM EGTA; 5 mM $Mg(OAc)_2$.
14. *Sonication buffer* 10 mM HEPES–NaOH pH 7.9; 100 mM NaCl; 2 mM EDTA pH 8; 1 mM EGTA pH 8; 0.2% SDS; 0.1% sodium sarkosyl. Prior to use add PMSF to 1 mM final concentration.
15. Diagenode Bioruptor.
16. *Dilution buffer* 10 mM HEPES–NaOH pH 7.9; 100 mM NaCl; 2 mM EDTA pH 8; 1 mM EGTA pH 8. Prior to use add PMSF to 1 mM final concentration.
17. *Wash buffer* 10 mM HEPES–NaOH pH 7.9; 100 mM NaCl; 2 mM EDTA pH 8; 1 mM EGTA pH 8; 0.1% SDS. Prior to use add PMSF to 1 mM final concentration.
18. *4× LSB* 50 mM Tris–HCl pH 6.8; 100 mM DTT; 2% SDS; 8% glycerol; bromophenol blue. Aliquot and keep at −20 °C.
19. *Coomassie staining buffer* 0.25% Coomassie G250, 50% ethanol, 10% acetic acid.
20. *Coomassie destaining buffer* Dilute acetic acid to 10% final in water.

21. Liquid nitrogen.
22. MyC1 Streptavidin beads (Invitrogen, cat. no. 65001). Other types of beads will affect the NCC yield.
23. Magnetic stand.
24. Precast Bis-Tris gradient gel 4–12% 1.5 mm (Novex, cat. no. NP0335BOX)

    [Optional materials for testing biotin-dUTP labeling efficiency].
25. *CSK buffer* 10 mM PIPES pH 7, 100 mM NaCl, 300 mM sucrose, 3 mM $MgCl_2$.
26. *CSK-Triton* Add Triton X-100, 0.5% final, to CSK buffer.
27. Cytoslides (Thermo scientific, cat. no. 5991051).
28. Shandon filter cards (Thermo scientific, cat. no. 5991022).
29. Alexa Fluor 488—Streptavidin conjugated (Invitrogen, cat. no. S32354).

3 Methods

The method depicted here applies to HeLa S3 cells cultured in suspension. 2×10^8 cells is collected per time point and the experimental design is outlined in Fig. 2.

SILAC Labeling and Synchronization (48 h)

1. Label and expand cells in light SILAC medium (R0) in spinner flasks for 9 generations and up to 4×10^8 cells (*see* **Note 1**).
2. Count cells and adjust the concentration to 1.1×10^6 cells/mL in a total volume of 360 mL of R0 SILAC medium.
3. Add 40 mL of freshly made 20 mM thymidine dissolved in R0 SILAC medium and leave for 17 h in incubator.
4. Release cells from thymidine block by washing cells thoroughly twice with 200 mL of prewarmed PBS.

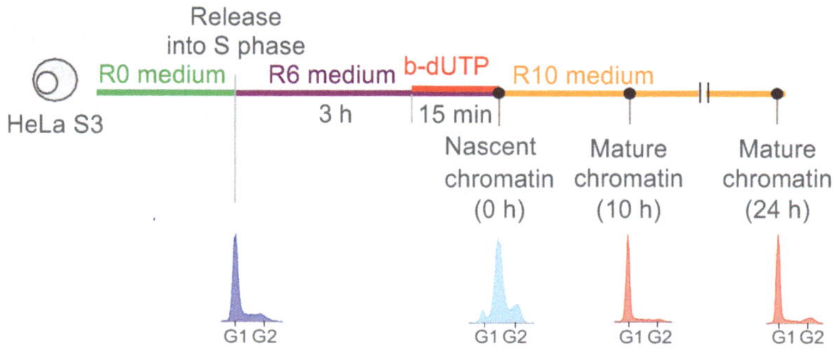

Fig. 2 Experimental design. (Figure adapted from [12])

5. Resuspend cells in 400 mL prewarmed R6 SILAC medium containing 24 mM deoxycytidine.
6. Transfer cells to the spinner flask and allow them to progress in S phase for 3 h in incubator (*see* **Note 2**).

Biotin-dUTP Labeling (40 min)

7. Spin 4 × 10^8 cells for 6 min at 300 × *g* at RT.
8. Remove the supernatant and resuspend in 40 mL of prewarmed KBH buffer. Pool the cells in one 50 mL tube (*see* **Note 3**).
9. Spin cells for 6 min at 300 × *g* at RT.
10. Remove the supernatant carefully (avoid disturbing the pellet) and resuspend in 8 mL of prewarmed KBH buffer containing 400 μL of 100 mM biotin-16-dUTP. Incubate for 5 min at 37 °C in a water bath with continuous and gentle shaking to prevent cells from sedimenting (*see* **Note 4**).
11. Resuspend cells in 200 mL of R6 SILAC medium and incubate at 37 °C for 15 min in a water bath. This step labels newly replicated DNA with biotin-dUTP (*see* **Note 5**). During this time, distribute cells into four 50 mL tubes.
12. Spin cells for 6 min at 300 × *g* at RT. Remove the supernatant and resuspend 2 cell pellets (2 × 10^8 cells) in 200 mL of R10 SILAC medium and transfer to a spinner flask for 2 h or 10 h incubation at 37 °C. This step chases the label to mark mature chromatin. Proceed to **step 14** with the two other cell pellets. This is the nascent sample.
13. After 2 h (mature, late S/G2) or 10 h (G1 phase), spin cells for 6 min at 300 × *g* at RT. Remove the supernatant and proceed to **step 14**.

Fixation and Nuclear Extraction (2 h)

14. Resuspend the two cell pellets from **step 12** (Nascent) or **step 13** (2 h, Mature; 10 h, G1 phase) in 5 mL Cross-linking solution each and pool in one 50 mL tube (10 mL total). Incubate immediately on a shaker at 36 rpm for 15 min at RT.
15. To stop cross-linking, add 2.8 mL of 10% glycine and incubate for 5 min at RT on a shaker at 36 rpm.
16. Spin cells for 10 min at 1000 × *g* at 4 °C. *From this step, work on ice.*
17. Wash the cell pellet twice in 50 mL of ice-cold PBS-PMSF.
18. Resuspend in 5 mL ice-cold Sucrose buffer and bring the volume to 50 mL.

19. Spin for 10 min at 1000 × *g* at 4 °C, resuspend each sample in 60 mL ice-cold Sucrose buffer and transfer to a 15 mL Dounce homogenizer on ice (*see* **Note 6**).
20. Dounce 20 times with a tight pestle. Repeat 3 times in order to Dounce the 60 mL samples.
21. Pool the dounced material in a new 50 mL tube and spin for 10 min at 2500 × *g* at 4 °C.
22. Discard the supernatant and resuspend the nuclear pellet in 50 mL ice-cold Glycerol buffer (*see* **Note 7**).
23. Spin for 10 min at 2500 × *g* at 4 °C and resuspend the nuclear pellet in 2 mL of ice-cold Glycerol buffer.
24. Snap-freeze in liquid nitrogen and store at −80 °C until use. This is the nuclear fraction. Nuclear fractions can be kept up to a year at −80 °C.

Preparation of Chromatin (2 h)

25. Thaw the nuclear fraction on ice and resuspend in 10 mL of ice-cold Sonication buffer.
26. Spin for 10 min at 2500 × *g* at 4 °C. Resuspend the nuclear pellet in Sonication buffer, 4 mL for 2 × 10^8 cells.
27. Aliquot into four 15 mL tubes and sonicate in the Diagenode Bioruptor at 4 °C. Bioruptor settings: High, 28 cycles of 30 s sonication and 90 s pause.
28. Spin for 15 min at 4500 × *g* at 4 °C and transfer the supernatant (solubilized chromatin) into a new tube (*see* **Note 8**). Solubilized chromatin can be kept at 4 °C up to 1 week.

Affinity Purification of Biotin-Tagged Chromatin (48 h)

29. Dilute solubilized chromatin 1:1 with Dilution buffer (*see* **Note 9**).
30. Save 100 μL of the input (add 100 μl of 4× LSB, boil 10 min and keep at −20 °C until **step 39**).
31. Add 200 μL of prewashed MyC1 Streptavidin beads, rotate end-over-end overnight at 4 °C. To prewash the beads, wash twice in PBS and once in Dilution buffer.
32. Place tubes on a magnetic stand and collect the unbound fractions. Keep at 4 °C until you reach **step 36**.
33. Wash the beads 5 times in 1 mL Wash buffer by inverting the tubes 10 times.
34. Add 80 μL of 2× LSB to the beads and boil for 40 min at 99 °C (*see* **Notes 10** and **11**).
35. Place the tube on a magnetic stand and collect the supernatant containing the purified chromatin. Store at −20 °C.

36. To improve the yield, process the unbound fraction from **step 32** for a second round of pull- down according to **steps 31** to **33**.
37. Add the supernatant from **step 35** and repeat boiling as in **step 34**.
38. Place the tube on a magnetic stand and collect the supernatant containing the final purified chromatin preparation. Proteins in LSB 2× can be stored at −20 °C until use.

Analysis of Histone Modifications Using Mass Spectrometry (48 h)

39. Run the samples on SDS-PAGE in a precast Bis-Tris gradient gel 4–12%, until histones are separated from bulk proteins (40 min, 120 V) and proceed to **step 40**. The intense band at 15 kDa corresponds to streptavidin released from the beads. For Western blotting, analyze the samples by standard electrophoresis conditions and proceed with immunoblotting (*see* **Notes 12** and **13**).
40. Stain the gel in colloidal blue, cut the histone bands and process for mass spectrometry analysis (*see* **Note 14**).

4 Notes

1. Several cell lines may stop growing after a couple of divisions when transferred from non-dialyzed serum–containing medium to dialyzed serum–containing SILAC medium, as dialyzed serum is depleted of small molecules. To overcome this problem, we recommend adapting cells to SILAC medium by maintaining cells in SILAC medium for 2 months. Once adapted, freeze separate vials of cells dedicated to SILAC experiments. If cells are kept in SILAC medium, the SILAC medium labeling for 9 generations step can be skipped.
2. The efficacy of biotin-dUTP labeling relies on efficient release of cells from the thymidine arrest into S phase. Monitor G1/S synchronization and release by collecting 0.5×10^6 cells prior to step 3 (asynchronous) and step 4 (G1/S arrest) and at each step of the time course. Perform a propidium iodide staining of DNA content for flow cytometry. If a substantial proportion of cells remains in G1 phase after the thymidine release, increase PBS washes at 37 °C and thoroughly resuspend the cell pellet.
3. Keep the KBH buffer at 37 °C in a water bath until use as this increases the biotin-dUTP incorporation efficiency.
4. Incorporation and labeling with biotin-dUTP is more efficient in a water bath compared to an incubator.
5. Biotin-dUTP labeling efficiency can be tested by immunofluorescence. Take 15,000 cells and cyto-spin on a coated glass slide

(500 rpm, 5 min). Wait 2 min to dry the slide and preextract cells with CSK-Triton for 5 min on ice to remove soluble proteins and excess biotin-dUTP. Fix with 2% formaldehyde for 15 min and analyze the biotin-dUTP incorporation by immunofluorescence using streptavidin conjugated to Alexa Fluor 488 (1:50). On average 80% of the cells should be biotin-dUTP-positive. If biotin-dUTP labeling is inefficient, cells may not replicate properly. Check S phase progression by propidium iodide–based FACS and DNA synthesis by EdU labeling. Alternatively, the KBH concentration may not be adapted to the inner cell concentration to allow optimal hypotonic shock and biotin-dUTP incorporation. KCl and HEPES concentration can be modified to improve this step [20].

6. Resuspend each sample carefully as cell aggregates will compromise homogenization.
7. Keep the supernatant if cytoplasmic material needs to be analyzed.
8. Fragment size can be estimated by taking 50 μL of soluble chromatin and adding 500 μL of 10 mM Tris–HCl pH 8. Shake overnight at 65 °C, 750 rpm to decrosslink. Then isolate DNA by classic phenol–chloroform extraction and analyze on an agarose gel. If DNA fragments are smeary, sonication may not have worked. Sonicate 5 cycles further or, alternatively, reduce cross-linking conditions to 1% formaldehyde.
9. This step reduces the SDS concentration, which affects the binding of biotin to streptavidin.
10. If beads stick to tube walls, tube may be *low DNA binding tubes* that have a special coating. Use uncoated tubes instead.
11. Include a brief vortex and short spin every 10 min to prevent beads from drying.
12. Smear bands detected for proteins higher than 110 kDa by Western blot can be due to inefficient decrosslinking. Digest samples with benzonase for 30 min at RT and reboil the samples for 10 min.
13. Unequal quantities of histones recovered between two conditions may be due to different amounts of biotin-dUTP incorporation between conditions. For pulse-chase samples, the efficiency of the chase (avoiding residual biotin-dUTP incorporation) can be improved by increasing the thymidine concentration. In unperturbed versus perturbed DNA replication conditions, extend the biotin-dUTP labeling time accordingly as in [21, 22].
14. If no band is visible in the pull-down, the release from streptavidin beads may not have worked. Increase the boiling time, up to 1 h.

Acknowledgments

C.A. is supported by CRUK-C57404/A21782 and ERC StG no. 715127. Work in the Groth lab was supported by the European Research Council (ERC StG no. 281765) and the Lundbeck Foundation (R198-2015- 269). Research at CPR is supported by the Novo Nordisk Foundation [NNF14CC0001].

References

1. Allshire RC, Madhani HD (2018) Ten principles of heterochromatin formation and function. Nat Rev Mol Cell Biol 19(4):229–244
2. Bannister AJ, Kouzarides T (2011) Regulation of chromatin by histone modifications. Cell Res 21(3):381–395
3. Stillman B (2018) Histone modifications: insights into their influence on gene expression. Cell 175(1):6–9
4. Hansen JC, Nyborg JK, Luger K, Stargell LA (2010) Histone chaperones, histone acetylation, and the fluidity of the chromogenome. J Cell Physiol 224(2):289–299
5. Talbert PB, Henikoff S (2021) The Yin and Yang of histone marks in transcription. Annu Rev Genomics Hum Genet 22:147–170
6. Zhou CY, Johnson SL, Gamarra NI, Narlikar GJ (2016) Mechanisms of ATP-dependent chromatin remodeling motors. Annu Rev Biophys 45:153–181
7. Marchal C, Sima J, Gilbert DM (2019) Control of DNA replication timing in the 3D genome. Nat Rev Mol Cell Biol 20(12):721–737
8. Gan H, Serra-Cardona A, Hua X, Zhou H, Labib K, Yu C et al (2018) The Mcm2-Ctf4-Polalpha axis facilitates parental histone H3-H4 transfer to lagging strands. Mol Cell 72(1):140–51 e3
9. Petryk N, Dalby M, Wenger A, Stromme CB, Strandsby A, Andersson R et al (2018) MCM2 promotes symmetric inheritance of modified histones during DNA replication. Science 361(6409):1389–1392
10. Yu C, Gan H, Serra-Cardona A, Zhang L, Gan S, Sharma S et al (2018) A mechanism for preventing asymmetric histone segregation onto replicating DNA strands. Science 361(6409):1386–1389
11. Annunziato AT (2015) The fork in the road: histone partitioning during DNA replication. Genes (Basel) 6(2):353–371
12. Alabert C, Barth TK, Reveron-Gomez N, Sidoli S, Schmidt A, Jensen ON et al (2015) Two distinct modes for propagation of histone PTMs across the cell cycle. Genes Dev 29(6):585–590
13. Scharf AN, Meier K, Seitz V, Kremmer E, Brehm A, Imhof A (2009) Monomethylation of lysine 20 on histone H4 facilitates chromatin maturation. Mol Cell Biol 29(1):57–67
14. Xu M, Wang W, Chen S, Zhu B (2011) A model for mitotic inheritance of histone lysine methylation. EMBO Rep 13(1):60–67
15. Stewart-Morgan KR, Petryk N, Groth A (2020) Chromatin replication and epigenetic cell memory. Nat Cell Biol 22(4):361–371
16. Alabert C, Loos C, Voelker-Albert M, Graziano S, Forne I, Reveron-Gomez N et al (2020) Domain model explains propagation dynamics and stability of histone H3K27 and H3K36 methylation landscapes. Cell Rep 30(4):1223–34 e8
17. Scharf AN, Barth TK, Imhof A (2009) Establishment of histone modifications after chromatin assembly. Nucleic Acids Res 37(15):5032–5040
18. Sweet SMM, Li M, Thomas PM, Durbin KR, Kelleher NL (2010) Kinetics of re-establishing H3K79 methylation marks in global human chromatin. J Biol Chem 285(43):32778–32786
19. Marzluff WF, Koreski KP (2017) Birth and death of histone mRNAs. Trends Genet 33(10):745–759
20. Koberna K, Stanek D, Malinsky J, Eltsov M, Pliss A, Ctrnacta V et al (1999) Nuclear organization studied with the help of a hypotonic shift: its use permits hydrophilic molecules to enter into living cells. Chromosoma 108(5):325–335
21. Nakamura K, Kustatscher G, Alabert C, Hodl M, Forne I, Volker-Albert M et al (2021) Proteome dynamics at broken

replication forks reveal a distinct ATM-directed repair response suppressing DNA double-strand break ubiquitination. Mol Cell 81(5): 1084–99 e6

22. Saredi G, Huang H, Hammond CM, Alabert C, Bekker-Jensen S, Forne I et al (2016) H4K20me0 marks post-replicative chromatin and recruits the TONSL-MMS22L DNA repair complex. Nature 534(7609): 714–718

Chapter 18

Investigating Mitotic Inheritance of Histone Modifications Using Tethering Strategies

Ajay Larkin, Amanda Ames, Melissa Seman, and Kaushik Ragunathan

Abstract

The covalent and reversible modification of histones enables cells to establish heritable gene expression patterns without altering their genetic blueprint. Epigenetic mechanisms regulate gene expression in two separate ways: (1) establishment, which depends on sequence-specific DNA- or RNA-binding proteins that recruit histone-modifying enzymes to unique genomic loci, and (2) maintenance, which is sequence-independent and depends on the autonomous propagation of preexisting chromatin states during DNA replication. Only a subset of the vast repertoire of histone modifications in the genome is heritable. Here, we describe a synthetic biology approach to tether histone-modifying enzymes to engineer chromatin states in living cells and evaluate their potential for mitotic inheritance. In *S. pombe*, fusing the H3K9 methyltransferase, Clr4, to the tetracycline-inducible TetR DNA-binding domain facilitates rapid and reversible control of heterochromatin assembly. We describe a framework to successfully implement an inducible heterochromatin establishment system and evaluate its molecular properties. We anticipate that our innovative genetic strategy will be broadly applicable to the discovery of protein complexes and separation-of-function alleles of heterochromatin-associated factors with unique roles in epigenetic inheritance.

Key words Lysine methyltransferase (KMT), Histone modifications, Epigenetics, *S. pombe*, Tethering

1 Introduction

Despite having identical genomes, our cells exhibit distinct phenotypic or epigenetic states that remain immutable throughout our lifetimes. Epigenetic mechanisms enable eukaryotic cells to forge heritable gene expression patterns without any alterations to their genetic blueprint [1]. Instead, cells rearrange histone-DNA complexes called chromatin to establish "open"'transcriptionally active euchromatin and "closed"'transcriptionally silent heterochromatin [2]. A unique repertoire of histone modifications distinguishes active from silent chromatin. Modifications such as histone H3

Ajay Larkin, Amanda Ames, and Melissa Seman contributed equally to this work.

Raphaël Margueron and Daniel Holoch (eds.), *Histone Methyltransferases: Methods and Protocols*,
Methods in Molecular Biology, vol. 2529, https://doi.org/10.1007/978-1-0716-2481-4_18,

lysine 4 methylation (H3K4me) or histone H4 lysine 16 acetylation (H4K16ac) mark actively transcribed genes while histone H3 lysine 9 or lysine 27 methylation (H3K9me or H3K27me) are associated with epigenetic silencing [1]. The sequence-specific recruitment of chromatin-modifying enzymes is essential to establish epigenetic states at unique genomic loci [3]. Once modification patterns are established, parental histones that are partitioned between daughter strands can potentially serve as templates to mark newly deposited histones following DNA replication [4]. The mitotic inheritance of chromatin states has far-reaching consequences for development, disease, and evolution [5]. However, studies in yeast and flies reveal that only a subset of histone modifications is heritable [6–8]. While repressive histone modifications such as histone H3 lysine 27 methylation (H3K27me) and histone H3 lysine 9 methylation (H3K9me) can be inherited during mitosis, histone modifications associated with active transcription are thought to lack epigenetic properties [9]. Deciphering the molecular mechanisms that promote the mitotic inheritance of chromatin states requires a genetic strategy that can uncouple their establishment from their subsequent maintenance.

The concerted activity of different chromatin-associated factors produces a complex landscape of histone modifications. For example, the silencing of *Hox* genes in development depends on the Polycomb group proteins consisting of two complexes—PRC1 and PRC2 [10]. PRC2 catalytically deposits H3K27me to enforce gene repression and reinforces this deposition through self-recruitment to H3K27me through the WD40 domain-containing EED [11–13]. Separately, PRC1 catalytic activity drives histone H2AK119 ubiquitination, which leads to chromatin compaction. In addition, histone modification-dependent positive feedback loops facilitate cross-talk between PRC1 and PRC2 complexes, thus producing a complex series of chromatin-dependent interactions that regulate gene expression [14]. The connection between different epigenetic modifications is not restricted to histones alone. The cross-talk between three different epigenetic components—DNA methylation, small RNA synthesis and H3K9 methylation—promotes epigenetic silencing in plants [15]. The increasing layers of complexity and cross talk become exceedingly difficult to parse with the level of precision that is possible in an in vitro biochemical assay.

Moreover, the current suite of in vitro assays is severely limited by the types of chromatin substrates that can be used. The majority of in vitro biochemistry studies focus on proteins binding to an individual nucleosome with appropriately modified histone subunits. The nucleosome substrates fail to capture the complexity of the chromatin environment in vivo consisting of nucleosomes marked with a broad spectrum of covalent modifications spanning several thousands of kilobases. Furthermore, many regulatory transcriptional changes are contextual, meaning they are strongly dependent on multiple factors such as genomic locus, protein effectors, histone PTMs, and DNA sequence. The requirement for cellular context can also complicate the in vitro biochemical characterization of chromatin modifiers in which chromatin has been extricated from its genomic neighborhood [16]. A crucial benefit to interrogating chromatin in cells is the ability to test the role of chromatin regulators, their effects on gene regulation, and their ability to confer gene expression memory within a native chromatin context. To address the difficulties we have highlighted, we have developed a system of tethering histone modifiers at a naïve genomic locus. By anchoring different alleles of chromatin modifiers, it is possible to identify more complex regulatory processes that control the mechanics of how a chromatin state can be established. Furthermore, tethering proteins in different mutant backgrounds enables us to decipher necessity and sufficiency criteria for diverse chromatin-mediated processes.

In *S. pombe*, Clr4 is the sole H3K9-specific methyltransferase that establishes H3K9 methylation at regions of constitutive heterochromatin [17]. This includes the pericentromeric repeats, the telomeres, the mating-type locus and the ribosomal DNA locus. A nuclear RNAi pathway is the primary mechanism that enables sequence-specific establishment of heterochromatin [18, 19]. The nuclear RNAi pathway in *S. pombe* leverages a complex series of positive feedback loops to produce programmable, complementary small RNAs which target specific genomic loci [20]. To bypass the complexity of endogenous recruitment mechanisms, we can simply tether Clr4 at an ectopic site, enabling us to measure all events that occur exclusively downstream of H3K9 methylation [21]. This includes the role of HP1 proteins Swi6 and Chp2, which are H3K9me readers that are essential to establish epigenetic silencing; histone deacetylases, which are thought to fortify the silent epigenetic state; and chromatin remodelers that reposition nucleosomes and alter chromatin structure in ways that ultimately exclude RNA polymerase II from heterochromatin [22, 23]. Reversible Clr4 tethering gives us the unique opportunity to measure the stability of the epigenetic state following establishment, quantify the rate of H3K9me decay, and determine the extent to which a 'memory' associated with gene silencing can be maintained across subsequent generations [7].

Based on our work developing synthetic approaches to interrogate mechanisms of epigenetic inheritance, we propose that all chromatin tethering systems must consider five key requirements: the chromatin enzyme (a) fused to a tethering protein whose recruitment to a target site is inducible (b) which targets a binding site (c) proximal to a reporter gene (d) inserted into a genomic locus (e) at a location agnostic to the chosen chromatin enzyme (Fig. 1).

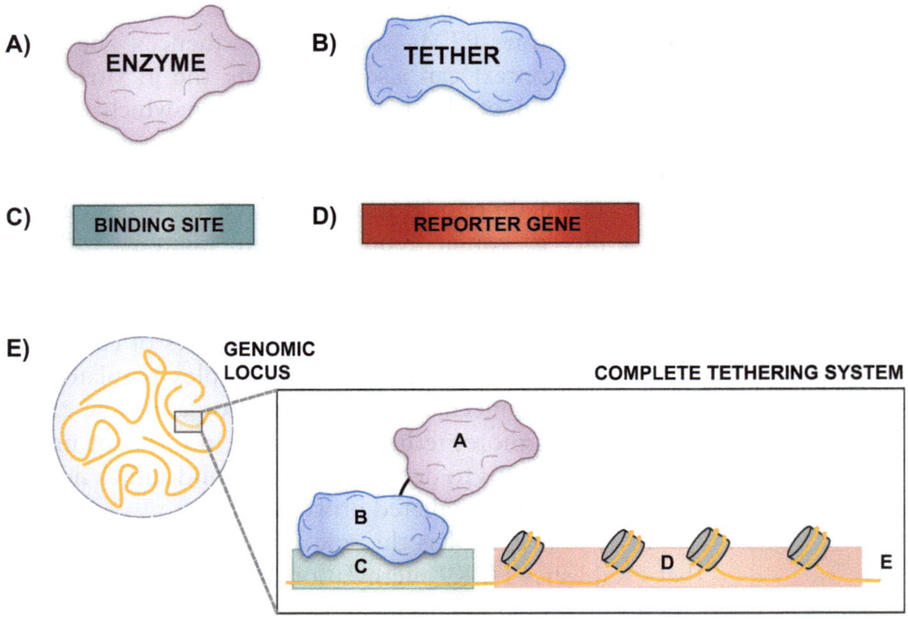

Fig. 1 Essential components for a synthetic chromatin tethering system. The requirements are (**a**) a chromatin modifier such as a histone methyltransferase, (**b**) a tethering protein that can be recruited to a specific sequence, (**c**) a tethering binding site that can be recognized by the tethering protein, (**d**) a reporter gene that provides a transcriptional readout, and (**e**) a genomic locus that is best suited to the experimental design

(a) *HISTONE-MODIFYING ENZYME:* The first component of the reporter system to select is a chromatin modifier. In our methods, we focus on Clr4, an H3K9 methyltransferase (*see* **Note 1**).

(b) *TETHERING/ANCHOR PROTEIN:* A tethering protein is required to recruit the enzyme to modify reporter gene expression. In our methods, we fused the bacterial tetracycline repressor protein (TetR) to the histone-modifying enzyme which is then recruited to the TetR binding sequence (*see* **Note 2**).

(c) *TETHER BINDING SEQUENCE:* To recruit the enzyme-tether to the reporter system, it must recognize a binding site. In our methods, the focus is on cloning the tetracycline operator sequence (TetO) which is bound by TetR. To allow for efficient recruitment of the enzyme-tether, we incorporate multiple copies (5–10) of the binding site (*see* **Note 3**).

(d) *REPORTER GENE:* This reporter gene should create an interpretable output in response to tethering a histone-modifying enzyme. For example, if evaluating whether a certain enzyme is capable of creating heterochromatin when tethered, the given reporter gene should be one that creates a phenotype when silenced. The focus of our methods is to utilize *ade6+* as a reporter of gene silencing, which causes yeast colonies to become red when silenced and remain white when actively expressed (*see* **Note 4**).

(e) *GENOMIC LOCUS:* A major consideration when designing a reporter system for heterochromatic gene silencing is selecting a genomic locus. The binding sequence can be inserted anywhere within the genome to observe the effects of recruiting the histone-modifying enzyme on a given locus. Our example system focuses on recruiting the tethering system at an ectopic locus to examine heterochromatin maintenance while excluding the effects of sequence-specific establishment pathways, such as RNAi, that establish heterochromatin at endogenous loci.

The modular nature of the chromatin tethering strategy we have outlined also allows the user to expand the scope of the tethering approach. This can be done by adding DNA-binding sites to recruit additional chromatin modifiers using an orthogonal DNA-binding domain. We have successfully used different iterations of the synthetic tethering strategy to decipher the molecular mechanisms of epigenetic inheritance [7, 24]. In this chapter, we highlight how a simple tethering module can be implemented in fission yeast to establish an H3K9me-dependent chromatin state, how the stability of the H3K9me-dependent state can be altered, and the conditions that are essential to study the mitotic inheritance of heterochromatin.

2 Materials

2.1 General Reagents and Equipment

1. Constant-temperature incubator.
2. Shaking incubator.
3. Spectrophotometer.
4. Petri dishes.
5. Thermal cycler.
6. Light microscope.
7. YEA (+Adenine) liquid media: 2.5 g yeast extract (Gibco catalog no. 211929, **Note 5**), 30 g glucose, and 225 mg of adenine hemisulfate salt per liter of media.
8. YEA (+Adenine) plates: 2.5 g yeast extract (Gibco catalog no. 211929, **Note 5**), 30 g glucose, 20 g agar, and 225 mg of adenine hemisulfate salt per liter of media.

2.2 Plasmid Cloning and Transformation

1. pFA6a vector [25],
2. pFA6a-*natMX6*-Pnmt1 plasmid (standard N-terminal tagging vector for subcloning),

3. BglII restriction enzyme (NEB R0144).
4. PacI restriction enzyme (NEB R0547).
5. AscI restriction enzyme (NEB R0558).
6. Custom plasmid, containing binding site: specific example is 10xTetO with 200 bp flanking of ura4 locus.
7. Custom plasmid, containing tethering enzyme: specific example is TetR.
8. Custom plasmid, containing reporter gene: specific example is ade6.
9. Custom plasmid, containing histone-modifying protein: specific example is *S. pombe* H3K9- methyltransferase Clr4ΔCD (deletion spanning amino acids 7–59).
10. Custom primers for amplification of cloning fragments, generating transformation products, screening transformed cells by colony PCR, sequence verification and qPCR analysis (available from many commercial vendors). Our primers for qPCR analysis can be found in Table 1.
11. Gibson Assembly Kit (NEB E5501S).
12. Custom-synthesized PCR-amplified transformation product, with ~200 bp overlap sequences to target genomic insertion site. These should be generated by PCR from the final Gibson products: specific examples are 10xTetO-reporter gene and TetR-Clr4-I gene insert.
13. Competent *E. coli* cells.
14. Materials for fission yeast transformations.
15. YEA (+Adenine) media.
16. YEA + NAT media: Add Nourseothricin to a final concentration of 100 μg/L, after autoclaving and cooling YEA media to 65–70 °C.

2.3 Silencing Assays

ade6+ colorimetric reporter plate silencing assay:

1. *YEA liquid media*
2. *YE plates: 5 g yeast extract and 30 g glucose, dissolved in 1 L water (see* **Note 5***)*
3. *YE+tet plates: Add anhydrotetracycline hydrochloride to a final concentration of 10 μM, after autoclaving and cooling media to 65–70° C.*
4. *Sterile 96-well plates*

Table 1
qPCR and primers for yeast transformation

QP1	GCCGTAAATGACGTTTTCGTCACC	SPCC330.06c fwd qPCR (ura4 proximal gene)
QP2	ACCTTGACAACCTTGCCATTCTCG	SPCC330.06c rev qPCR (ura4 proximal gene)
QP3	TACAGTGCCAGGCGAGGGTATTAT	ura4 fwd qPCR
QP4	TAGGAACCAGTAGCCAAAGAGCCT	ura4 rev qPCR
QP5	CAACCCTCAGCTTTGGGTCTTG	act1 fwd qPCR
QP6	TCCTTTTGCATACGATCGGCAATAC	act1 rev qPCR
QP7	AACGCTTGGCCATGGAATACACG	tub1 fwd qPCR
QP8	GAGAGGCGGTGATGGAAGAAACAAC	tub1 rev qPCR
QP9	CAAACCCTTGCCATGGATGC	ade6 fwd qPCR
QP10	ACCTGAATTGTGAGGCCGAG	ade6 rev qPCR
QP11	ACCCAGACCACATGAAGCAACAC	GFP fwd qPCR
QP12	CTTGTGACCCAAGATGTTACCGTC	GFP rev qPCR
QP13	GCAGTGGTGCTTCTGTCCTAACAGC	10XtetO fwd qPCR
QP14	GTCCGTGTAGCTCGAGCTCGACTT	10XtetO rev qPCR
QP15	GCGGTCATTTAAAGGCATAG	dg repeat fwd qPCR
QP16	CGACAACACGAATTTTACTTCTAGGTC	dg repeat rev qPCR
trpINS-F	AACGAAATCATAAAAATAATCTTTAA TTAGGCATAA AAAAATAATTAAAAAATTGGTGTTTCAA TTGTCGGT TCATTTAAGAATTCGAGCTCGTTTAAAC	Amplification of the TetR-Clr4 insert with the natMX6 marker for insertion at the trp1+ locus, forward primer
trpINS-R	CCAGAGTAAACAACAGCTATAGGATAC TTTTAAAAT CACTGTTTAGTCATTTGGTTTACAAC TTTATTATATT AACAGTTAGGTGTGGTCAATAAGAGCG	Amplification of the TetR-Clr4 insert with the natMX6 marker for insertion at the trp1+ locus, reverse primer

ura4+ live/dead reporter plate silencing assay:

1. *YEA liquid media*
2. YEA+FOA plates: Add 1 g of 5-fluoroorotic acid per liter of media, after autoclaving and cooling media to 65–70 °C.
3. YEA+FOA+tet plates: Add 1 g of 5-fluoroorotic acid per liter of media and anhydrotetracycline hydrochloride to a final concentration of 10 μM, after autoclaving and cooling media to 65–70 °C.

4. YEA plates
5. Sterile 96-well plates

GFP fluorescence reporter fluorescence assay:

1. *YEA liquid media*
2. *YEA+tet liquid media*
3. 70% ethanol
4. 1× Tris-buffered saline (TBS)
5. Flow cytometry instrument

2.4 Quantitative RT-PCR

1. YEA (+Adenine) media.
2. YEA + tet media.
3. Acidic phenol–chloroform–isoamyl alcohol (25:24:1).
4. 100% ethanol
5. 70% ethanol
6. Glycogen.
7. DNase I (NEB M0303).
8. RNeasy Mini Kit (Qiagen).
9. cDNA generation reagents: oligodT or gene specific primers, reverse transcriptase (such as SuperScript III Reverse Transcriptase, Invitrogen 18,080,044),
10. qPCR instrument,
11. qPCR Primers for gene of interest and a housekeeping gene such as tubulin,
12. Other qPCR reagents: polymerase, SYBR green mix, 25 mM dNTPs, qPCR plates.
13. Optional: BioAnalyzer (Agilent).
14. Optional: benchtop MiSeq (lllumina) or equivalent instrument.

2.5 Chromatin Immunoprecipitation

1. YEA (+Adenine) media.
2. YEA + tet media.
3. 37% Formaldehyde
4. 2.5 M Glycine
5. Tris-Buffered Saline (1×): 50 mM Tris-Cl, pH 7.5, 150 mM NaCl.
6. Lysis buffer: 100 mM HEPES–KOH pH 7.5, 280 mM NaCl, 2 mM EDTA, 1% Triton X-100, 0.1% Na-deoxycholate, 1 mM PMSF, protease inhibitors (leupeptin, pepstatin, aprotinin).
7. Dynabeads Protein A magnetic beads (Invitrogen 10001D) or Dynabeads M-280 Streptavidin (Invitrogen 11,205).

8. H3K9me2 antibody (Abcam ab1220), H3K9me3 antibody (Diagenode C15410193), or M2 FLAG antibody (Sigma: A8592).
9. Bead-beater for yeast cell lysis (Omni Bead Ruptor 24).
10. Glass beads (0.5 mm diameter).
11. Sonicator (Qsonica: Q800 Sonicator).
12. Bradford Reagent.
13. Plate reader.
14. Tube rotator.
15. Tris-EDTA buffer (1×): 10 mM Tris–HCl pH 8.0, 1 mM EDTA.
16. Hybridization oven.
17. Basic phenol–chloroform–isoamyl alcohol (25:24:1).
18. 100% ethanol
19. 70% ethanol
20. Glycogen.
21. qPCR instrument,
22. qPCR Primers for gene of interest,
23. Other qPCR reagents: polymerase, SYBR green mix, 25 mM dNTPs, qPCR plates.
24. Optional: BioAnalyzer (Agilent).
25. Optional: benchtop MiSeq (lllumina) or equivalent instrument.

3 Methods

3.1 Construction of a TetR-Clr4 Expression Plasmid

To piece together a chromatin-based tethering system, we have provided an algorithmic approach which can be applied to clone any histone-modifying enzyme fused to a tethering/anchoring protein. Our cloning method of choice is Gibson assembly, given that we can easily amplify, clone, and swap individual modules [26]. Here we will discuss how to clone the histone-modifying enzyme, Clr4, fused to the TetR tethering/anchoring protein while ensuring that the fusion protein is expressed at endogenous levels in *S. pombe* cells. The TetR-Clr4 plasmid consists of the endogenous *S. pombe clr4+ promoter*, followed by an nSV40 nuclear localization sequence and TetR DNA-binding protein (anchor/tether). The TetR DNA-binding protein is subsequently fused to an allele of Clr4 (histone-modifying enzyme) lacking the chromodomain (*clr4* Δ7–59) fused to a 2X-FLAG (DYKDDDDK) epitope tag (Fig. 2a). For modifying this protocol for other proteins, the promoter and *clr4*Δ*CD* allele are modular components that can be

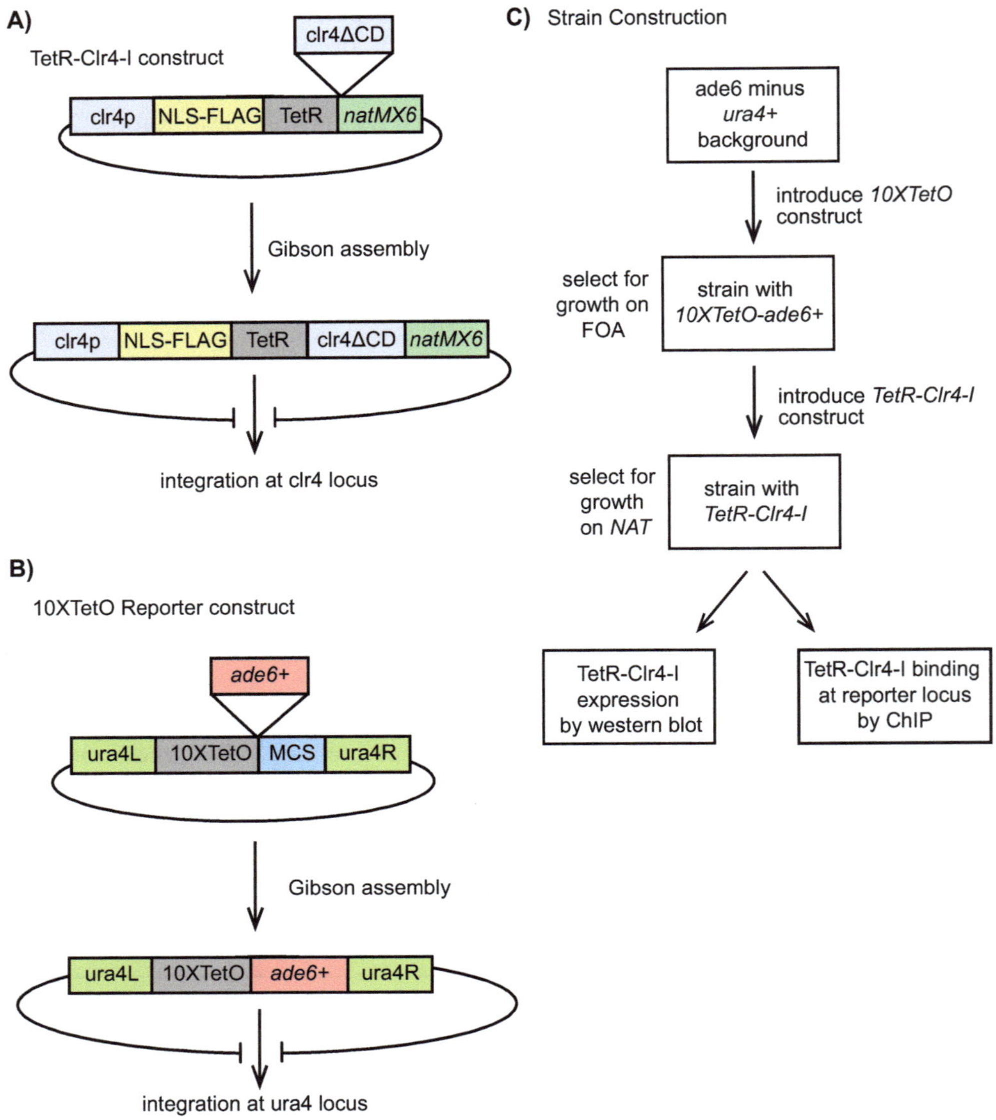

Fig. 2 Schematic of the cloning strategies utilized to implement a TetR-Clr4 tethering system in *S. pombe*. (**a**) The *TetR-Clr4-I* construct. The *TetR-Clr4-I* construct contains an nSV40 nuclear localization sequence for nuclear retention and a 2X-FLAG tag for downstream detection. (**b**) 1*0XTetO* reporter construct contains the tether binding sites and homology ends for insertion at the *ura4+* locus. (**c**) The constructs are transformed into *S. pombe* and colonies are selected using auxotrophic and antibiotic selection markers (FOA and nourseothricin in these experiments). Protein expression is validated using a Western blot and chromatin association is measured using chromatin immunoprecipitation (ChIP) assay

replaced with other histone-modifying enzymes as needed. The affinity tag is optional but helps with measuring expression levels of the fusion protein and for downstream chromatin immunoprecipitation (ChIP) applications.

1. To make the TetR-Clr4 plasmid, design primers that amplify each modular fragment with sequence overlaps of at least 30 bp

to adjacent PCR fragments. This should include primers to amplify (a) Clr4 without the H3K9 methylation binding domain chromodomain (CD), (b) the *clr4+* promoter to preserve endogenous levels of expression, (c) the TetR DNA-binding tether and the nSV40 nuclear localization signal, and, optionally, (d) include an epitope tag fused to the N-terminus of the *clr4ΔCD* allele.

2. Digest an N-terminal pFA6A plasmid vector with BglII and PacI. This digest leaves behind a resistance marker, nourseothricin (in the case of the TetR-Clr4 plasmid we use in this method), and an ADH terminator sequence.
3. Perform Gibson assembly with each of the PCR fragments (3 total fragments) and the restriction-digested plasmid backbone.
4. Sequence the plasmid to ensure proper Gibson assembly of the TetR-Clr4 sequence and make sure that the *clr4ΔCD* allele is in frame with the both the TetR tether and the optional FLAG epitope tag.

3.2 Construction of 10XtetO-Reporter Gene Plasmid

The next step is to design a plasmid that incorporates tether binding sites. The *tetO* operator sequence (40 bp) enables highly specific recruitment of TetR-fusion protein constructs. The operator sites are multiplexed to ensure that the locus is consistently being modified. Specifically, the *10xtetO*-reporter gene plasmid consists of 200 bp of homology to the 5′ end of the target locus (such as *ura4+*), 10 tandem *tetO* sequences, followed by the reporter gene with endogenous promoter and terminator sequences, followed by 200 bp of homology to the 3′ end of the target locus. We will discuss the use of *ade6+* and *ura4-GFP* as reporters at the endogenous *ura4+* locus although variations of this strategy may also be considered (Fig. 2b).

1. To make the tether binding site plasmid (*10XtetO-ade6+* or *10XtetO-ura4-GFP*), design primers that amplify each modular fragment with sequence overlaps of at least 30 bp to adjacent PCR fragments. This should include primers to amplify (a) *ura4+* right homology arms (200 bp), (b) 10XTetO sequences with unique spacer sequences to ensure that the repeats can be amplified, (c) reporter gene (*ade6+* or *ura4-GFP*), and (d) *ura4+* left homology arms (200 bp). The PCR fragments consist of regions that overlap with adjacent modules. The overlap must be at least 30 bp.
2. Digest a pFA6a plasmid vector with PacI and AscI.
3. Perform Gibson assembly with each of the Gibson fragments (4 fragments) and the cut plasmid backbone.
4. Sequence the plasmid to ensure proper *10xTetO-reporter gene* plasmid construction.

3.3 Strain Construction

Once the two plasmids are cloned, the next step is to introduce the two components into a wild-type yeast strain (Fig. 2c).

1. Amplify the 10xTetO-reporter gene plasmid with primers that include the cloned *ura4+* homology segments.
2. Transform *S. pombe* strains with the resulting PCR product using a standard fission yeast transformation protocol, such as a lithium acetate–based approach [27]. The strains used for this transformation must be *ura4+* since the homology arms that we have cloned in the *10XtetO-reporter gene* plasmid use an FOA-based selection assay (*see* **Note 6**).
3. After transformation, select for transformants that grow on media containing FOA. FOA enables selection of cells that lack a functional copy of the *ura4+* gene.
4. Check the positive clones for successful insertion using standard protocols for genotyping by colony PCR [28].
5. Next, amplify the TetR-Clr4 plasmid with homology arms that facilitate insertion of the fusion protein at either (a) the endogenous *clr4+* locus where the wild-type *clr4+* gene is replaced with TetR-Clr4 or (b) an ectopic site where the TetR-Clr4 gene is inserted as a second copy. The amplified PCR product must include the antibiotic resistance marker. As mentioned above, our TetR-Clr4-I plasmid has a nourseothricin resistance marker (*natMX6*). In our specific strain construction protocol, we used the *trp1+* locus as a favored site for ectopic insertion of the TetR-Clr4 fusion construct (*see* Table 1). The wild-type *clr4* + is essential for maintenance of heterochromatin. It is important to consider the downstream applications for choosing to replace the endogenous copy of *clr4+* (a) or to insert TetR-Clr4 (b) at an ectopic site.
6. Transform the yeast strain from **step 5** with the TetR-Clr4 PCR product from **step 6** using a standard fission yeast transformation protocol, such as lithium acetate [27].
7. Select for transformants that grow on media containing nourseothricin (YEA + NAT).
8. Check the positive clones for successful insertion using standard protocols for genotyping by colony PCR [20].
9. For quality control, measure 2XFLAG-TetR-Clr4 protein expression by preparing total protein cell lysates using a trichloroacetic acid (TCA) protein precipitation protocol followed by a western blot with a FLAG antibody [29].
10. Finally, to ensure that the tether protein (*TetR-Clr4-I)* occupies the tether binding sites (*10XTetO-reporter gene)*, use a standard chromatin immunoprecipitation (ChIP) protocol with a FLAG antibody, followed by quantitative real-time PCR (qPCR) [30].

3.4 Methods for Tracking Reporter Gene Silencing

As mentioned previously, our reporter system uncouples sequence-specific establishment processes from epigenetic inheritance (Fig. 3a, b). In our previous work we showed that deleting the H3K9me eraser, Epe1, in *S. pombe* promotes heterochromatin maintenance. However, replacing the wild-type Clr4 protein with an allele that lacks the H3K9me-binding chromodomain suppressed heterochromatin maintenance with no effect on establishment (Fig. 3c). Therefore, depending on the genetic context of the reporter assay (*epe1+* versus *epe1*Δ) we can identify enhancers or suppressors associated with the epigenetic inheritance of H3K9 methylation. Depending on the reporter gene that was used, the silencing assay can be qualitative (*ade6+*) or quantitative (GFP) (*see* **Note 4**). Establishment of the silent chromatin state can be measured by growing the yeast on the appropriate media (e.g., YE for *ade6+*, YEA + FOA for *ura4+*, adding anhydrotetracycline (AHT) to the media to release the TetR-Clr4 protein from the TetO binding site.

3.4.1 Colorimetric Output (ade6+ Reporter Silencing)

1. Inoculate yeast cultures in 3 mL of YEA. Grow the cultures overnight at 32 °C.
2. Measure OD600 of cultures.
3. Transfer exactly 1 OD600 of cells into a fresh 1.5 mL centrifuge tube. Remove the excess YEA medium after centrifugation. Wash cells in 1 mL of sterile water.
4. Transfer cells to a sterile 96-well plate. This simplifies subsequent steps of serial dilution.
5. Perform serial tenfold dilution of cells to generate five new dilutions, i.e., to obtain six cell densities in total.
6. Pipette 5 μl of each of the dilutions on YE or YE + tet plates. The low adenine content causes cells to accumulate a red pigment as a consequence of *ade6+* silencing. Cells remain white if *ade6+* is expressed (*see* **Note 7**). By plating the serial dilution of cells on YE and YE + tet, heterochromatin establishment and maintenance can be separately measured.
7. Incubate the plates for 2–3 days at 32 °C.
8. Cells turn red upon successful heterochromatin establishment. The pigment accumulation in the cells is accentuated when plates are kept at 4 °C for about 24 h before cataloging the resulting phenotypes.

3.4.2 GFP-Based Maintenance Assay

1. Inoculate yeast cultures in 2–3 mL of YEA. Grow cells overnight at 32 °C.
2. Ensure that strains are in logarithmic growth phase by diluting overnight cultures and growing cells for about 6–8 h.

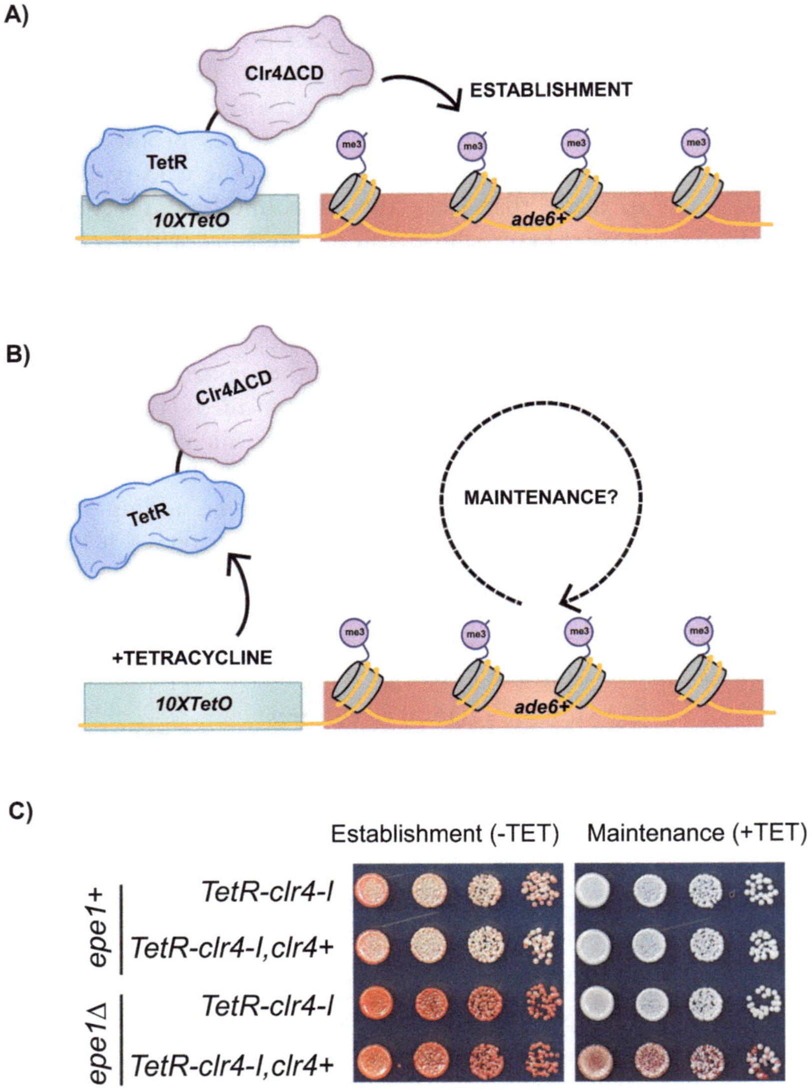

Fig. 3 Application of the TetR-Clr4 tethering system to observe establishment and epigenetic inheritance of H3K9 methylation. (**a**) Schematic of tethering TetR-Clr4-I upstream of the *ade6+* reporter gene. Under these conditions, Clr4 catalyzes H3K9me of nucleosomes associated with the *ade6+* reporter gene, leading to heterochromatin establishment (red colonies). (**b**) Schematic of TetR- Clr4-I dissociation after tetracycline addition. TetR-Clr4-I dissociation from the *10XTetO* binding sites enables us to measure initiator-independent epigenetic inheritance. If cells can maintain epigenetic silencing, colonies exhibit a red or sectored phenotype. In contrast, the loss of heterochromatin maintenance causes cells to turn white. (**c**) Tenfold serial dilution of cells to measure establishment (−tetracycline) and maintenance (+tetracycline). Maintenance is observed in cells that have an intact Clr4 chromodomain (*clr4+)* and lack Epe1 (*epe1Δ)*

3. For time point $t = 0$, centrifuge 1 OD600 of cells and wash with H2O.
4. Add 70% ethanol to the resulting cell pellet for 20 min to fix cells.

5. Wash cells twice with 1× Tris-buffered saline (TBS).
6. Finally, resuspend in 1 ml of 1× TBS and draw a desired volume to load cells on a flow cytometry instrument.
7. To measure maintenance, add 2.5 μg/mL of anhydrotetracycline (AHT) to media and repeat **steps 3–6** for a range of time points. A typical time-series to consider is intervals of 2–3 h, which corresponds to the cell-division time for *S. pombe.*
8. Use a typical FACS instrument-based data collection cycle to measure the fluorescence signal associated with approximately 20,000 cells.
9. Controls are essential in order to carefully set gates during data acquisition. Set forward and side scatter gates to exclude cell debris and dead cells.
10. Use GFP-positive and GFP-negative cells to gate cells by fluorescence emission to precisely assign the epigenetic status of the strain of interest.
11. Fit the resulting GFP fluorescence histograms to a Gaussian function, to measure the fraction of cells in the GFP-OFF and the GFP-ON state.

3.5 Quantitative RT-PCR Analysis to Measure RNA Levels

Once the ectopic tethering system has been characterized using a phenotypic assay, the next step is to correlate phenotypic changes with altered mRNA transcription. This provides a much more precise and quantitative view of the extent to which the tethering strategy affects gene expression. For our example system, it is recommended that such an analysis be performed before and after anhydrotetracycline addition to measure establishment and maintenance of the epigenetic state.

1. Grow 10 mL of fission yeast cultures to an OD600 between 0.8 and 1.2.
2. Pellet cells and extract RNA using a standard phenol–chloroform protocol [31]. Pelleted cells can be stored at −80 °C for RNA purification later.
3. Treat with DNase and purify RNA using columns from an RNA purification kit such as the Qiagen RNeasy kit. The use of columns simplifies the handling of RNA given its susceptibility to degradation.
4. Using gene-specific primers or oligodT, set up a reverse transcriptase reaction to produce cDNA from RNA.
5. Perform qPCR using the cDNA as a template to quantify target gene expression.
6. Optional: The RNA samples can also be used for RNA-seq to further investigate global differential expression changes. For example, RNA-seq could be performed to evaluate the extent

of spreading of gene silencing from the tethered locus. To perform RNA-seq, construct a library with total RNA from **step 3**, analyze the library for quality on a BioAnalyzer (Agilent), and sequence with a benchtop MiSeq (Illumina) or equivalent instrument.

3.6 Chromatin Immunoprecipitation Analysis to Measure Protein Occupancy

The primary advantage of using an ectopic tethering system is that histone modifications change in complex ways within a native chromatin context. To measure histone modification changes during establishment and maintenance, the next step is to perform a chromatin immunoprecipitation assay. Here, we outline a protocol with guidelines for antibody choice while performing these assays. Although we highlight measuring H3K9 methylation levels since we tether the histone modifier Clr4 in our methods, it is useful to also measure RNA polymerase and histone acetylation levels as well. Collectively, the ChIP measurements paint a molecular portrait of how H3K9 methylation leads to transcriptional gene silencing and its mitotic inheritance.

1. Grow 50 ml of yeast cultures to an OD600 between 1.2 and 2.0.
2. Add a final concentration of 1% formaldehyde to fix the cells and incubate for 15 min at room temperature.
3. Quench the formaldehyde using glycine at a final concentration of 125 mM. Wash twice with ice-cold 1xTBS.
4. Resuspend pellets in lysis buffer. Add glass beads. Lyse cells with a bead beater (4.84 m/s, pulse: 30 s on, 60 s off, 8 cycles). Check for full lysis of cells under a light microscope.
5. Remove glass beads with centrifugation.
6. Sonicate (amplitude: 70%, pulse: 15 s on, 30 s off, 30 min total) the cell extracts to fragment the DNA.
7. Centrifuge and move crude lysate (supernatant) to a new tube. Equalize the protein concentration using a Bradford assay.
8. Incubate the samples with the antibody for the histone modification of interest. For H3K9-di-methylation-(H3K9me2)-specific ChIP measurements, our antibody of choice is Abcam ab1220. Use 2 μl sample and rotate at 4 °C for 3 h. Other methylation states can be evaluated using H3K9me3- and H3K9me1-specific antibodies. However, the antibodies that are commercially available produce considerable background in our ChIP measurements.
9. Wash the samples with lysis buffer three times and once with 1xTE.
10. Reverse cross-links overnight by placing in a hybridization oven overnight at 65 °C.

11. Extract DNA using a standard phenol–chloroform protocol.
12. Finally, use ethanol precipitation to recover the DNA. Typically, ethanol precipitation at this stage is supplemented with a molecular crowding agent such as glycogen to ensure the highly efficient recovery of small amounts of nucleic acid material.
13. Perform qPCR with the DNA using primers to amplify the reporter gene or the genomic locus where the tethering system is located.
14. Optional: These ChIP samples can also be used for ChIP-seq to further investigate global transcription levels. For example, ChIP-seq could be performed to evaluate the extent of spreading of the modification mark from the tethered locus. To perform ChIP-seq, construct a library with pure DNA from **step 9**, analyze the library for quality on a BioAnalyzer (Agilent), and sequence with MiSeq (Illumina) or equivalent instrument.

3.7 Expanding the Scope of the Ectopic Tethering System

3.7.1 Methods for a Dual Chromatin-Modifying Enzyme Tethering Approach

To this point we have focused on making and investigating strains that have a single histone-modifying enzyme tethered. Specifically, we outlined how tethering the H3K9 methyltransferase, Clr4, using a tetracycline-inducible DNA-binding fusion enables us to investigate heterochromatin establishment and maintenance.

Because establishment of a gene expression state is a process that is complex and requires many changes within chromatin, it can be useful to tether two chromatin modifiers simultaneously to elucidate principles such as additivity and cooperativity in establishment and maintenance. For example, we have shown that when Clr4 is tethered, heterochromatin can be established but H3K9 methylation alone is insufficient for maintenance given the antagonistic functions of the H3K9me eraser, Epe1. However, by tethering a second histone-modifying enzyme, namely, the histone deacetylase Clr3, heterochromatin can be maintained despite the presence of the H3K9me eraser, Epe1 (Fig. 4) [24]. These observations underscore how the direct competition between Epe1 recruitment versus Clr3-mediated histone deacetylation ultimately determines the fate of heterochromatin maintenance.

We have successfully used the Gal4 DNA-binding domain as the tether and multiplexed UAS plasmids consisting of Gal4 DNA-binding sites as the tether binding sites. Cloning *10XTetO* and *10XGal4* binding sites in tandem creates a dual-tether binding site where proteins can be orthogonally recruited to an ectopic site. To make a system with a dual tether, follow Subheadings 3.1 and 3.2 with a second enzyme and a second orthogonal tether. The steps for transformation and characterization of strains are similar to those described in Subheadings 3.3–3.6, except that the strains now have two tethers and two tether binding sites.

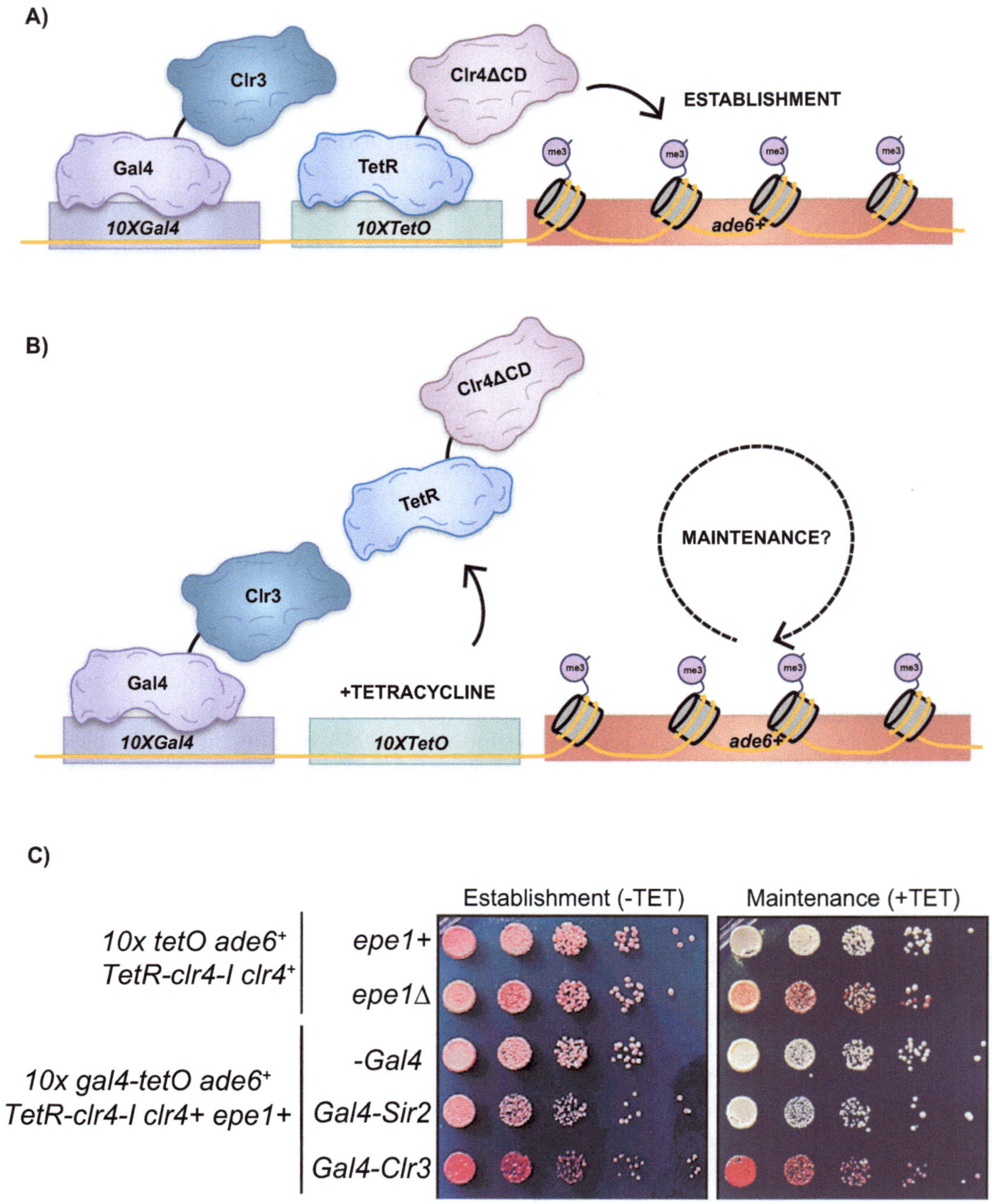

Fig. 4 Implementation of a dual-tethering approach to observe establishment and epigenetic inheritance of H3K9 methylation. (**a**) Schematic of tethering TetR-Clr4-I and Gal4-Clr3 upstream of the *ade6+* reporter gene. Under these conditions, Clr4 catalyzes H3K9me of nucleosomes associated with the *ade6+* reporter gene, leading to heterochromatin establishment (red colonies). (**b**) Schematic of TetR-Clr4-I dissociation after tetracycline addition while Gal4-Clr3 remains bound. TetR-Clr4-I dissociation from the *10XTetO* binding sites enables us to measure initiator-independent epigenetic inheritance. If cells are able to maintain epigenetic silencing, colonies exhibit a red or sectored phenotype. In contrast, the loss of heterochromatin maintenance causes cells to turn white. (**c**) Tenfold serial dilution of cells to measure establishment (−tetracycline) and maintenance (+tetracycline). Maintenance is observed in cells that have an intact Clr4 chromodomain (*clr4+)* in the presence of Gal4-Clr3. Continuous histone deacetylation circumvents the need to delete Epe1. Gal4-Sir2 tethering has no effect on heterochromatin maintenance

4 Notes

1. Since a broad spectrum of histone modifications influence transcription, we can study any histone-modifying enzyme in its native chromatin context using our tethering system.
2. Many options exist for the tether. Some options include DNA-binding constructs such as Gal4 or zinc finger binding proteins [32]. Another tethering strategy is to utilize a domain of a histone "reader" to recruit the histone-modifying enzyme of interest. For example, the chromodomain that binds H3K9me can be used as the tether to recruit the fusion protein of interest to all regions marked by H3K9me. Similarly, bromodomains or other reader domains can be utilized as the tether. More recently dCas9-based approaches have been used for tethering in fission yeast, providing a mechanism for the programmable tethering of histone-modifying enzymes [33].
3. The binding site is determined based on the choice of the tether. For DNA- or RNA-binding tethers, the binding site must be cloned at the locus of interest. For histone-modification-based binding tethers, no binding site needs to be cloned because the tether will be recruited directly to chromatin by epigenetic modifications. One example of a histone-modification-based recruitment approach is the fusion of two tandem Swi6 chromodomains to an RNAi adapter protein, Ers1, in *S. pombe*. The direct recruitment of Ers1 to pericentromeric repeats bypasses the need for Swi6 in recruiting the RNA- dependent RNA polymerase, Rdp1 [34].
4. Examples of genes that can be utilized to evaluate gene silencing or expression include *ade6+*, *ura4+*, and GFP. Using *ade6+* as a reporter is advantageous because when *ade6+* is silenced the yeast cells turn red. It is easy to visually distinguish red and white colony color qualitatively. *ura4+* is a useful reporter because when silenced, cells can grow on media containing FOA but not on media lacking uracil (−ura). Conversely, when *ura4+* is expressed, cells can grow on media lacking uracil and but fail to grow on FOA-containing media. This enables highly sensitive detection of silencing events that occur at a frequency as low as 1/10,000 [35]. One disadvantage of *ura4+* as a reporter gene is that FOA selection can be leaky and lead to the identification of false positives. Lastly, GFP as a reporter is advantageous because measurement of cells can be done in a high-throughput and quantitative manner. The disadvantage is that silencing cannot simply be viewed on a plate and requires the use of an expensive FACS instrument which may not be easily accessible in some laboratory settings.

5. For colorimetric selection on YE plates, we noticed that we get the most striking phenotypes using yeast extract from Gibco as specified in the Materials section.
6. Although we have outlined a strategy to clone homology arms that can facilitate the insertion of the tether binding sequence at the *ura4+* locus, a second and more convenient option is to use a new suite of *S. pombe* integration plasmids that enable insertions at *ura4+*, *ade6+*, *lys3+*, *his5+*, and *leu1+* loci [36]. These plasmids have built-in homology arms to enable high-throughput integration and selection with an auxotrophic or antibiotic marker.
7. Because epigenetic silencing is susceptible to stochastic change, cells often exhibit a variegated phenotype [37]. Due to the spontaneous loss of silencing, it is important to begin silencing assays with a single colony that has successfully established heterochromatin. For example, when using the *ade6+* reporter, starting the experiment with a colony that is red is important for examining epigenetic silencing.

Acknowledgments

A.L. and A.A. were supported by an NIH Predoctoral Training in Genetics T32GM007544. M.S. is supported by an NIH Cellular and Molecular Biology Predoctoral Training Grant T32GM007315. K.R. is supported by an NIH Award 5R35GM137832-02. The order of authorship was decided based on a game of rock, paper, scissors. No players were harmed during this process.

References

1. Strahl BD, Allis CD (2000) The language of covalent histone modifications. Nature 403(6765):41–45. https://doi.org/10.1038/47412
2. Kornberg RD, Lorch Y (1995) Interplay between chromatin structure and transcription. Curr Opin Cell Biol 7(3):371–375. https://doi.org/10.1016/0955-0674(95)80092-1
3. Moazed D (2011) Mechanisms for the inheritance of chromatin states. Cell 146(4): 510–518. https://doi.org/10.1016/j.cell.2011.07.013
4. MacAlpine DM, Almouzni G (2013) Chromatin and DNA replication. Cold Spring Harb Perspect Biol 5(8):a010207. https://doi.org/10.1101/cshperspect.a010207
5. Heard E, Martienssen RA (2014) Transgenerational epigenetic inheritance: myths and mechanisms. Cell 157(1):95–109. https://doi.org/10.1016/j.cell.2014.02.045
6. Audergon PN, Catania S, Kagansky A, Tong P, Shukla M, Pidoux AL, Allshire RC (2015) Epigenetics. Restricted epigenetic inheritance of H3K9 methylation. Science 348(6230): 132–135. https://doi.org/10.1126/science.1260638
7. Ragunathan K, Jih G, Moazed D (2015) Epigenetics. Epigenetic inheritance uncoupled from sequence-specific recruitment. Science 348(6230):1258699. https://doi.org/10.1126/science.1258699
8. Coleman RT, Struhl G (2017) Causal role for inheritance of H3K27me3 in maintaining the OFF state of a Drosophila HOX gene. Science

356(6333). https://doi.org/10.1126/science.aai8236
9. Reinberg D, Vales LD (2018) Chromatin domains rich in inheritance. Science 361(6397):33–34. https://doi.org/10.1126/science.aat7871
10. Francis NJ, Kingston RE (2001) Mechanisms of transcriptional memory. Nat Rev Mol Cell Biol 2(6):409–421. https://doi.org/10.1038/35073039
11. Cao R, Wang L, Wang H, Xia L, Erdjument-Bromage H, Tempst P, Jones RS, Zhang Y (2002) Role of histone H3 lysine 27 methylation in Polycomb-group silencing. Science 298(5595):1039–1043. https://doi.org/10.1126/science.1076997
12. Francis NJ, Saurin AJ, Shao Z, Kingston RE (2001) Reconstitution of a functional core polycomb repressive complex. Mol Cell 8(3): 545–556. https://doi.org/10.1016/s1097-2765(01)00316-1
13. Lavarone E, Barbieri CM, Pasini D (2019) Dissecting the role of H3K27 acetylation and methylation in PRC2 mediated control of cellular identity. Nat Commun 10(1):1679. https://doi.org/10.1038/s41467-019-09624-w
14. Kalb R, Latwiel S, Baymaz HI, Jansen PW, Muller CW, Vermeulen M, Muller J (2014) Histone H2A monoubiquitination promotes histone H3 methylation in Polycomb repression. Nat Struct Mol Biol 21(6):569–571. https://doi.org/10.1038/nsmb.2833
15. Matzke MA, Mosher RA (2014) RNA-directed DNA methylation: an epigenetic pathway of increasing complexity. Nat Rev Genet 15(6): 394–408. https://doi.org/10.1038/nrg3683
16. Park M, Keung AJ, Khalil AS (2016) The epigenome: the next substrate for engineering. Genome Biol 17(1):183. https://doi.org/10.1186/s13059-016-1046-5
17. Ivanova AV, Bonaduce MJ, Ivanov SV, Klar AJ (1998) The chromo and SET domains of the Clr4 protein are essential for silencing in fission yeast. Nat Genet 19(2):192–195. https://doi.org/10.1038/566
18. Hall IM, Shankaranarayana GD, Noma K, Ayoub N, Cohen A, Grewal SI (2002) Establishment and maintenance of a heterochromatin domain. Science 297(5590):2232–2237. https://doi.org/10.1126/science.1076466
19. Volpe TA, Kidner C, Hall IM, Teng G, Grewal SI, Martienssen RA (2002) Regulation of heterochromatic silencing and histone H3 lysine-9 methylation by RNAi. Science 297(5588): 1833–1837. https://doi.org/10.1126/science.1074973
20. Holoch D, Moazed D (2015) RNA-mediated epigenetic regulation of gene expression. Nat Rev Genet 16(2):71–84. https://doi.org/10.1038/nrg3863
21. Kagansky A, Folco HD, Almeida R, Pidoux AL, Boukaba A, Simmer F, Urano T, Hamilton GL, Allshire RC (2009) Synthetic heterochromatin bypasses RNAi and centromeric repeats to establish functional centromeres. Science 324(5935):1716–1719. https://doi.org/10.1126/science.1172026
22. Motamedi MR, Hong EJ, Li X, Gerber S, Denison C, Gygi S, Moazed D (2008) HP1 proteins form distinct complexes and mediate heterochromatic gene silencing by nonoverlapping mechanisms. Mol Cell 32(6):778–790. https://doi.org/10.1016/j.molcel.2008.10.026
23. Sugiyama T, Cam HP, Sugiyama R, Noma K, Zofall M, Kobayashi R, Grewal SI (2007) SHREC, an effector complex for heterochromatic transcriptional silencing. Cell 128(3): 491–504. https://doi.org/10.1016/j.cell.2006.12.035
24. Raiymbek G, An S, Khurana N, Gopinath S, Larkin A, Biswas S, Trievel RC, Cho US, Ragunathan K (2020) An H3K9 methylation-dependent protein interaction regulates the non-enzymatic functions of a putative histone demethylase. elife 9:e53155. https://doi.org/10.7554/eLife.53155
25. Bahler J, Wu JQ, Longtine MS, Shah NG, McKenzie A 3rd, Steever AB, Wach A, Philippsen P, Pringle JR (1998) Heterologous modules for efficient and versatile PCR-based gene targeting in Schizosaccharomyces pombe. Yeast 14(10):943–951. https://doi.org/10.1002/(SICI)1097-0061(199807)14:10<943::AID-YEA292>3.0.CO;2-Y
26. Gibson DG (2009) Synthesis of DNA fragments in yeast by one-step assembly of overlapping oligonucleotides. Nucleic Acids Res 37(20):6984–6990. https://doi.org/10.1093/nar/gkp687
27. Murray JM, Watson AT, Carr AM (2016) Transformation of Schizosaccharomyces pombe: lithium acetate/ dimethyl sulfoxide procedure. Cold Spring Harb Protoc 2016(4):pdb.prot090969. https://doi.org/10.1101/pdb.prot090969
28. Murray JM, Watson AT, Carr AM (2016) Colony polymerase chain reaction with Schizosaccharomyces pombe. Cold Spring Harb Protoc 2016(5). https://doi.org/10.1101/pdb.prot090993
29. Koontz L (2014) TCA precipitation. Methods Enzymol 541:3–10. https://doi.org/10.1016/B978-0-12-420119-4.00001-X

30. Egan ED, Braun CR, Gygi SP, Moazed D (2014) Post-transcriptional regulation of meiotic genes by a nuclear RNA silencing complex. RNA 20(6):867–881. https://doi.org/10.1261/rna.044479.114
31. Bahler J, Wise JA (2017) Preparation of total RNA from fission yeast. Cold Spring Harb Protoc 2017(4):pdb.prot091629. https://doi.org/10.1101/pdb.prot091629
32. Keung AJ, Bashor CJ, Kiriakov S, Collins JJ, Khalil AS (2014) Using targeted chromatin regulators to engineer combinatorial and spatial transcriptional regulation. Cell 158(1):110–120. https://doi.org/10.1016/j.cell.2014.04.047
33. Ishikawa K, Soejima S, Masuda F, Saitoh S (2021) Implementation of dCas9-mediated CRISPRi in the fission yeast Schizosaccharomyces pombe. G3 (Bethesda) 11(4):jkab051. https://doi.org/10.1093/g3journal/jkab051
34. Hayashi A, Ishida M, Kawaguchi R, Urano T, Murakami Y, Nakayama J (2012) Heterochromatin protein 1 homologue Swi6 acts in concert with Ers1 to regulate RNAi-directed heterochromatin assembly. Proc Natl Acad Sci U S A 109(16):6159–6164. https://doi.org/10.1073/pnas.1116972109
35. Iida T, Nakayama J, Moazed D (2008) siRNA-mediated heterochromatin establishment requires HP1 and is associated with antisense transcription. Mol Cell 31(2):178–189. https://doi.org/10.1016/j.molcel.2008.07.003
36. Vjestica A, Marek M, Nkosi PJ, Merlini L, Liu G, Berard M, Billault-Chaumartin I, Martin SG (2020) A toolbox of stable integration vectors in the fission yeast Schizosaccharomyces pombe. J Cell Sci 133(1):jcs240754. https://doi.org/10.1242/jcs.240754
37. Xu EY, Zawadzki KA, Broach JR (2006) Single-cell observations reveal intermediate transcriptional silencing states. Mol Cell 23(2):219–229. https://doi.org/10.1016/j.molcel.2006.05.035

Chapter 19

Investigating Histone Modification Dynamics by Mechanistic Computational Modeling

Govind Menon and Martin Howard

Abstract

The maintenance of transcriptional states regulated by histone modifications and controlled switching between these states are fundamental concepts in our understanding of nucleosome-mediated epigenetic memory. Any approach relying on genome-wide bioinformatic analyses alone offers limited scope for dissecting the molecular mechanisms involved in maintenance and switching. Mechanistic mathematical models—describing the dynamics of histone modifications at individual genomic loci—offer an alternative way to investigate these mechanisms. These models, in conjunction with quantitative experimental data—ChIP data, quantification of mRNA levels, and single-cell fluorescence tracking in clonal lineages—can generate predictions that drive more targeted experiments, allowing us to understand mechanisms that would be challenging to unravel by a purely experimental approach. In this chapter, we describe a generic stochastic modeling framework that can be used to capture histone modification dynamics and associated molecular processes—including transcription and read–write feedback by chromatin modifying complexes—at individual genomic loci. Using a specific example—transcriptional silencing by Polycomb-mediated H3K27 methylation—we demonstrate how to construct and simulate a stochastic histone modification model. We provide a step-by-step guide to programming simulations for such a model and discuss how to analyze the simulation output.

Key words Nucleosome-mediated epigenetic memory, Gene expression states, Maintenance and switching, Polycomb, Mechanistic modeling, Stochastic histone modification model, Bistability, Gillespie algorithm

1 Introduction

Regulation of chromatin states mediated by histone modifications has become central to our understanding of epigenetic switching and memory in a variety of developmental contexts. Bioinformatic analyses at the genome level have greatly contributed to our current picture of epigenetic regulation. However, the network of correlational relationships generated by such approaches is by itself insufficient to dissect the causal molecular mechanisms that drive switching and maintenance of transcriptional states. As an

Raphaël Margueron and Daniel Holoch (eds.), *Histone Methyltransferases: Methods and Protocols*,
Methods in Molecular Biology, vol. 2529, https://doi.org/10.1007/978-1-0716-2481-4_19,

alternative approach, mechanistic mathematical models describing the dynamics of histone modifications at individual genomic loci have been used to explain the causal basis of stable, mitotically heritable histone modification states [1, 2]. In conjunction with quantitative experimental data, such predictive models have also been used to study the establishment of heterochromatin domains [3] and to investigate switching and maintenance of epigenetic states during developmental transitions [4, 5], and whether such changes are gradual or "all or nothing." This methodology has greatly speeded our understanding of the complex underlying mechanisms in these cases. The predictions generated by these models have allowed for more targeted experiments, enabling us to understand mechanisms that would otherwise have been very challenging to systematically dissect by experiments alone.

Two broad classes of mechanistic mathematical models have been used to study the dynamics of histone modification based epigenetic systems—stochastic models that track modifications at the level of individual histones, and deterministic models that describe transitions between modification states over a whole locus at the level of a population average. The latter typically involve Ordinary Differential Equations (ODEs). Stochastic models in conjunction with quantitative experimental data have been used to study the cold-induced establishment and maintenance of H3K27me-mediated transcriptional silencing at the plant developmental regulator *FLOWERING LOCUS C* (*FLC*) [4, 6] (*see* Box 1), and H3K9me-mediated silencing of mating type loci in fission yeast [3] (*see* Box 2). A stochastic model combining H3K27me3 maintenance by Polycomb Repressive Complex 2 (PRC2)-mediated feedback, with transcriptional antagonism [1], has been shown to reproduce the experimentally observed dynamics of H3K27me3 recovery after replication described in [7]. Deterministic models have also been combined with quantitative experimental data to investigate developmental epigenetic switching in multiple contexts—activation of the T-cell fate commitment gene *Bcl11b* [8], cold-induced silencing dynamics in natural variants of *FLC* [5, 9], and the propagation of H3K27 and H3K36 methylation in mESCs [10].

Box 1 Modelling to Investigate Cold-Induced H3K27me3 Nucleation at FLC: Angel et al. [6]

Biological context

- Long-term cold exposure induces epigenetic silencing of floral repressor gene *FLC* in *Arabidopsis thaliana*
- Polycomb silencing through nucleation and spreading of H3K27me3
- Cell-autonomous, quantitative silencing - longer cold silences a greater fraction of loci [4]

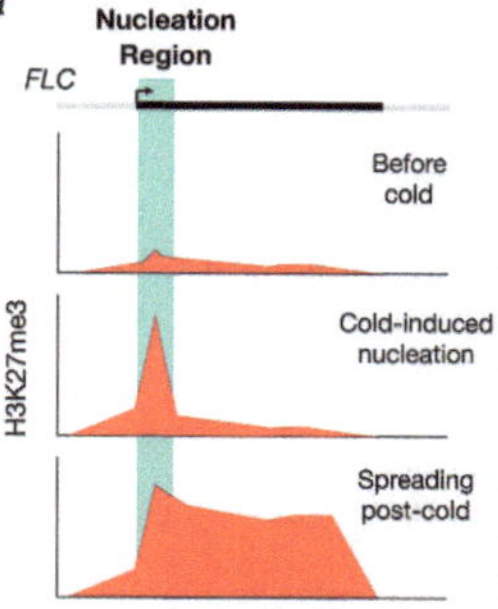

Model

- **M-U-A** type model for *FLC* locus: 35 nucleosome array, **M** marks represent H3K27me3
- Long-range interactions
- **Bistability** - transcriptionally active (high A) state and silenced (high M) state
- Different reaction rates in warm vs cold conditions
- Different reaction rates in Nucleation Region (3 nucleosomes) vs rest of locus

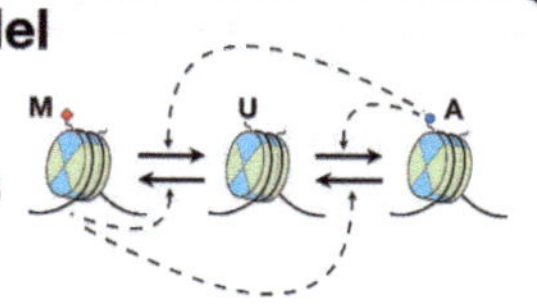

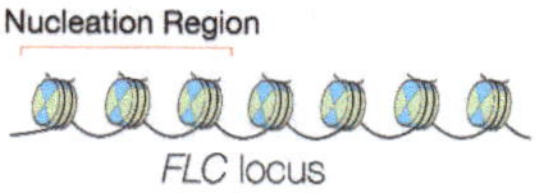

Mechanistic question: Is H3K27me3 accumulation in Nucleation Region all-or-nothing (digital) or gradual (analog)?

Model predictions/experimental validation

- Model used to examine three possible paradigms (a) of H3K27me3 build-up in Nucleation Region (NR) at individual loci
- In each case, model fitted to time course ChIP-qPCR profiles of H3K27me3 across *FLC* (b)
- Model predicts key differences between paradigms in terms of population averaged quantities:
  - H3K27me3 levels in NR required to induce switching to spread state after cold (c)
  - Fraction of cells silenced by exposure to interrupted cold
- Model predicted *FLC* expression compared to RT-qPCR quantified levels of spliced *FLC mRNA for* uninterrupted and interrupted cold (d):
  - Digital Class 1 (d1) model is the best fit
  - H3K27me3 nucleation at *FLC* is an all-or-nothing phenomenon

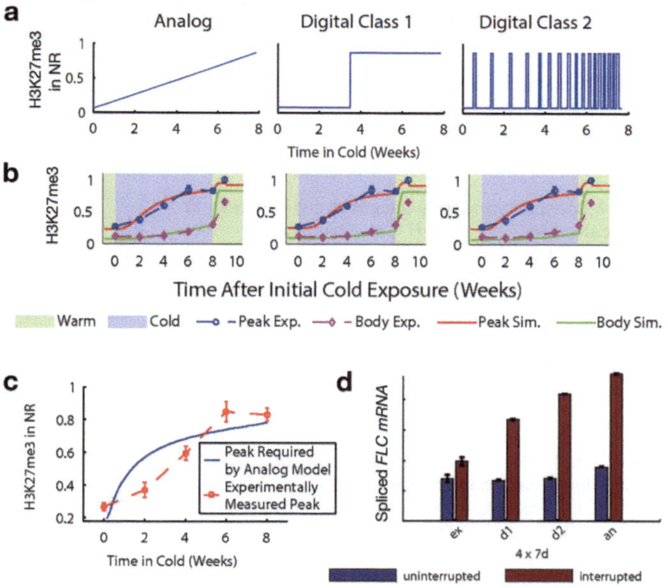

Figures a–d reproduced from [6].

Box 2 Modelling to Investigate Timing of Heterochromatin Formation in Yeast MatingType Region: Obersriebnig et al. [3]

Biological context

- A heterochromatin domain is naturally established at the mating-type region in *Schizosaccharomyces pombe*
- H3K9me silencing established by methyltransferase Clr4 (a) - nucleation and spreading of H3K9me
- Stochastic establishment and clonal inheritance studied using single-cell measurements of fluorescent reporters [3]

a

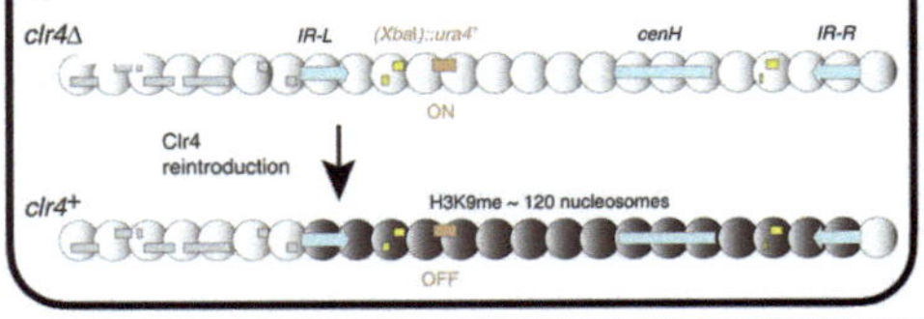

Model

- **M-U-A** type model for mating-type locus: 140 nucleosome array, **M** marks represent H3K9me
- Local and long-range interactions considered
- **Bistability** - transcriptionally active (high A) state and silenced (high M) state
- Spontaneous conversion of A to U in *cenH* region (29 nucleosomes)

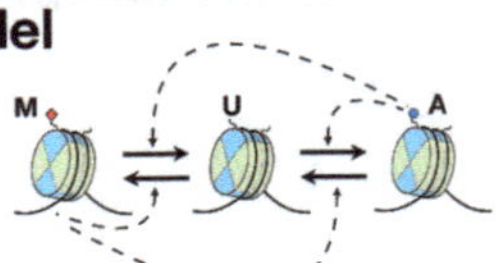

- Spreading to rest of locus through combination of local and global feedback interactions

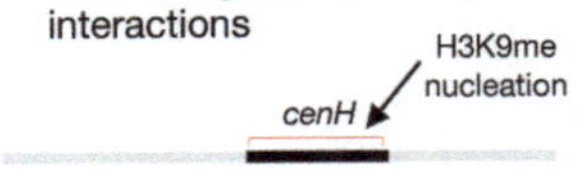

Mating-type locus

Mechanistic question: What types of mechanisms can generate both robust bistability and stochastic heterochromatin establishment that explains the large cell-to-cell variation in timing?

Model predictions/experimental validation

- Kinetics of nucleation and spreading measured at single-cell level using different fluorescent reporters within and outside *cenH* nucleation region (b, c)
- Fluorescence microscopy data from cell cultures indicates silencing at *cenH* first, which then facilitates silencing of the rest of the locus (c)
- Model used to examine different combinations of local and long-range recruitment mediated/read-write feedback interactions
- Model predicts that most of the interactions need to be local, with the M mediated A to U reactions being long-ranged (e), to achieve both robust bistability and match the observed kinetics of silencing: experimental data in (c) recapitulated in simulation results (d)

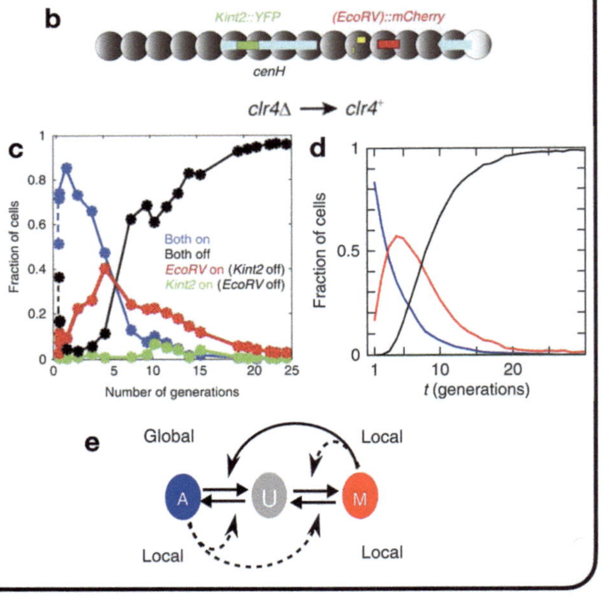

Figures a–e reproduced from [3].

Here, we demonstrate how to build and simulate a stochastic model of the type described in [1], capturing the dynamics of modifications at the level of individual histones, across a single genomic locus. The core components of the model, the assumptions used to capture these components, and the complete model structure are presented in the Materials section. Depending on the nature of the study, this modeling framework can be used to model either a single gene of interest or to model a whole class of genomic loci exhibiting similar modes of epigenetic regulation. The modeling framework is generic, and may be easily adapted to incorporate multiple histone modifications or molecular processes that may be of interest. We present a specific example here that focusses on transcriptional silencing at a locus by Polycomb-mediated H3K27 methylation, with the core components of the model being H3K27me3 maintenance by PRC2-mediated read–write feedback, H3K27me3-mediated transcriptional silencing, transcriptional antagonism of H3K27 methylation, and replicative dilution of H3K27 methylation. In the notes we discuss ways of incorporating other features of interest for adapting this framework to model other histone modification based epigenetic systems.

Simulating the model involves implementing an iterative Monte-Carlo approach or using the Gillespie algorithm [11] to compute the evolution of the histone modification state of a locus over time. In the Methods section, we present a detailed discussion on programming simulations for such a model and provide sample code. We describe model implementations using both the Gillespie algorithm and a Monte-Carlo approach. In addition, we present a brief description of a simpler "M-U-A" type stochastic model that has been used in several previous studies [2–4, 6]—this type of model describes transcriptionally repressive "M" modifications and activating "A" modifications, capturing their dynamics, including feedback interactions, at the level of individual histones. We discuss a Monte-Carlo implementation of this model. Simulations of a locus can be repeated multiple times, and the trajectories generated used to compute both averaged quantities, such as levels of histone modifications or transcription frequency, and to calculate probabilities and timescales for switching between states. To constrain and parametrize the model, these results can be compared to population level quantitative experimental data, such as ChIP data and quantification of mRNA levels, or switching data from single-cell fluorescence tracking in clonal lineages.

2 Materials

2.1 Building Blocks of the Model

We first describe the core components of a model that captures histone modification dynamics at the level of individual histones at a single genomic locus (Fig. 1). The model presented here focuses on

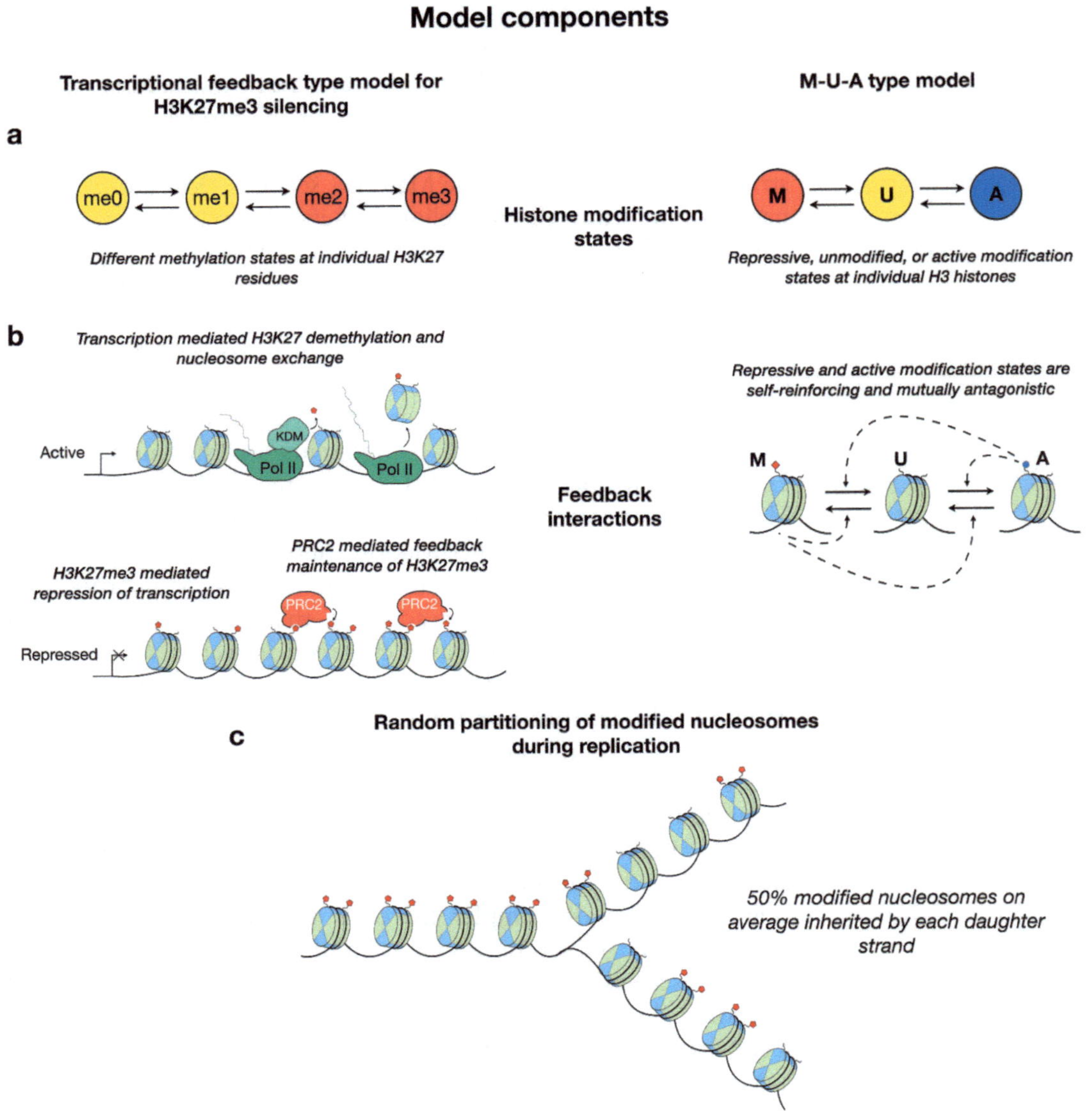

Fig. 1 Model components. (**a**) Histone modification states: the transcriptional feedback type model considered here incorporates different H3K27 methylation states of individual histones. The M-U-A type model, as used in multiple studies, incorporates repressive, unmodified, and active modification states of individual histones. (**b**) Feedback interactions: the transcriptional feedback type model incorporates mutual antagonism between transcriptional activity and PRC2-mediated repressive H3K27 methylation. The core of the M-U-A type model consists of self-reinforcing and mutually antagonistic feedback interactions between repressive and activating modification states. (**c**) Dilution of histone modifications at DNA replication—this is common to both types of models, forming a core feature of the modeling framework described here

transcriptional repression by Polycomb-mediated H3K27 methylation, together with transcriptional antagonism of H3K27 methylation. At the end of this section, we describe the full model structure, putting the different components together.

2.1.1 Histone Modifications

To model the histone modifications of interest, the following aspects need to be considered:

1. *Modification states represented*: In the case of lysine methylation at a particular residue, we can choose to represent only the unmethylated and fully methylated states (me0 and me3), or to represent all possible methylation states (me0, me1, me2, me3). In the current model, we represent all possible methylation states at H3K27.
2. *Transitions between modification states*: If partially modified methylation states are represented, we need to specify whether the methylation is processive, that is, direct transitions are possible between any two states (for instance from me0 directly to me2 or me3), or nonprocessive, that is, only transitions that add or remove a single methyl group are allowed. In the current model, we assume that both H3K27 methylation and demethylation are nonprocessive. The nonprocessive nature of methylation by PRC2 is supported by both in vitro [12] and in vivo data [13], while the H3K27 demethylase UTX has been shown to function nonprocessively in vitro [14].
3. *Combining mutually antagonistic modifications*: The modeling framework allows for mutually antagonistic modifications to be described in a single model, for instance activating H3K4 methylation and repressive H3K27 methylation (*see* [15] for a model that combines multiple activating and repressive modifications). In such cases, the molecular nature of the antagonism must be reflected in the set of allowed transitions between modification states. For example, if the antagonism of H3K27me is indirect, that is, through transcriptional activation by H3K4 methylation inducing H3K27 demethylation, the addition of methyl groups to H3K27 at a histone carrying H3K4me would in principle still be allowed. However, if the antagonism is direct [16], with the presence of H3K4me preventing PRC2 methylation of H3K27 on the same histone, for instance through steric hindrance, this transition would be disallowed.
4. *Experimental data*: As the model aims to describe the spatio-temporal evolution of modification states across a locus, quantification by ChIP-qPCR or ChIP-seq analysis for the modification of interest at the relevant locus/class of loci is a key experimental tool for constraining the model and validating model predictions. If the dynamics of the modification profile at a locus are slow enough, time course ChIP analysis can also provide valuable insights. It is important to note that these experimental methods generate averaged modification states at the population level while the simulation output consists of trajectories representing individual loci.

2.1.2 Read–Write Enzymes

A key feature in our current understanding of how histone modification states are established and maintained is the existence of feedback mechanisms for adding modifications. These feedback mechanisms are understood to be mediated by enzymatic complexes that combine "read" (recognition of a modified histone) and "write" (methyltransferase) functions from different subunits [17]. Such *cis*-acting positive feedback, where the presence of a modification promotes addition of the same modification to neighboring histones, is essential for a model to generate stable alternative epigenetic modification states of a locus.

To incorporate such read–write mechanisms into the model the following aspects need to be considered.

1. *Modifications that activate the enzymatic complex to "write" more modifications*: In the current model, based on in vitro data [18], we assume that only H3K27me3 and H3K27me2 (with tenfold reduced efficacy) can activate PRC2.
2. *The range of the feedback mechanism*: Both "local" read–write feedback (where existing modifications on a nucleosome can recruit enzymatic machinery to add modifications to neighboring nucleosomes) and "global" feedback mechanisms (where existing modifications on a nucleosome can recruit enzymatic machinery to add modifications to other nucleosomes irrespective of their relative locations) have been assumed in previous modeling studies [2]. In the current model, we limit the feedback-mediated H3K27 methylation to a nearest neighbor interaction between nucleosomes.
3. *The recruitment of the enzymatic complex to the locus*: In the current model, we describe the enzymatic complex PRC2 implicitly—its effect will be captured through the kinetic parameters that determine the rates of H3K27 methylation. PRC2 recruitment to the locus is not explicitly described, but the model provides a simple way to capture the strength of PRC2 activity at a locus.

2.1.3 Transcription

1. *The effect of transcription on modifications*: In the current model, we assume transcriptional antagonism of H3K27 methylation. This is assumed to occur in two ways—(1) through demethylation (motivated by experimentally observed association between transcription elongation factors and H3K27 demethylases at promoters of PRC2 target genes [19]), and (2) through nucleosome exchange during transcription (nucleosome exchange is found to correlate positively with transcriptional activity) [20].
2. *The effect of modifications on transcription*: Modifications are assumed to activate or repress transcriptional activity at the locus. The current model describes H3K27 methylation, which is understood to repress transcription. Our

understanding of transcriptional silencing by H3K27 methylation is based on multiple observations, including reduced levels of productive transcription [21], increased chromatin compaction [20], and antagonism of H3K27 acetylation [22] at H3K27me2 and H3K27me3 enriched loci.

2.1.4 Replication

Dilution of histone modification levels during DNA replication at a locus is an important feature of epigenetic dynamics in dividing cells. In our model we assume that replication occurs once per cell cycle and causes a random 50–50 partitioning of intact parental nucleosomes to daughter strands. For the single locus in our model, this means that during replication, each nucleosome can be lost with probability half. It is assumed that unmarked nucleosomes are inserted into the gaps.

2.1.5 Noise

Fluctuations at the molecular level—for instance in the number of modified histones at a locus, the number of transcription events occurring, or in the level of trans-acting factors—may be expected to play an important role in the epigenetic regulation of transcription. The stability of an epigenetically silenced state can depend on how robustly the PRC2-mediated feedback methylation of H3K27 can respond to fluctuations in H3K27me3 levels, for instance following DNA replication. We examine the effect of fluctuations in our model by adopting a stochastic modeling framework which naturally incorporates noise. The modeling framework allows noisy (background) methylation/demethylation reactions to be incorporated in addition to feedback driven methylation/demethylation reactions (generated, for instance, by PRC2 recruitment/transcription). The interplay between the two kinds of processes is important: for example, stronger background methylation/demethylation can dominate the feedback driven reactions, thus weakening feedback dependent behavior such as bistability.

2.2 Spatiotemporal Context

2.2.1 Spatial Extent and Boundaries

1. The size of the genomic locus being modeled depends on the specific context—for instance the model may be used to represent the epigenetic state of only a promoter element or only a gene body or even both of these elements together. The model describes the modification states of nucleosomes at a single genomic locus considered in isolation—that is, ignoring any higher-level interactions, for instance between chromatin domains. Therefore, the spatial extent of the locus is captured by the number of nucleosomes spanned by the locus.
2. Depending on the context, a rough estimate of this number may be enough—for instance, calculated using the genomic length of the locus and an average nucleosome spacing. Typical genomic loci of interest consist of tens of nucleosomes. For systems of this size, with the kinds of interactions modeled here, small differences in the number of nucleosomes are likely

to have only minor effects on the model predictions. However, when representing smaller genomic regions, for instance domains such as the H3K27me3 nucleation region at the *FLC* gene [6], which is estimated to involve only three nucleosomes, a difference of even a single nucleosome can significantly impact quantitative model predictions. *See* **Note 1** for further discussion on the size of the locus.

3. Domain boundaries, both for the whole locus and those potentially restricting the spread of modifications within the locus, are captured implicitly in the model: genomic regions outside the locus are not described and internal domain boundaries are represented by imposing spatial constraints on modification reactions (i.e., confining them to a specified domain).
4. We note that potential 3D spatial relationships between the nucleosomes within a locus, for instance through looping interactions, can be captured within our framework in a relatively simple way—by allowing long-ranged interactions, where the modifications in one region of the locus can influence modification/demodification rates in another region, not necessarily adjacent to the first. Long ranged interactions of this type can further buffer the system against fluctuations in modification levels and enable more robust epigenetic states.

2.2.2 Timescales and Kinetics

1. *Kinetics of reactions*: The modeling framework described here contains essentially four types of reactions: addition of modifications, removal of modifications, transcription and DNA replication. While the relationship between the modification state of a locus and the rates of these reactions can be specified mathematically based on the considerations outlined above (e.g., range of feedback interactions, transcription frequency determined by coverage of modifications), the absolute timescales of these reactions still need to be set. Values for the kinetic parameters (reaction rate constants) that determine these timescales can be chosen based on experimental quantification of histone modification kinetics (such as [23]), and for transcription, based on single cell quantification of transcription frequency (for instance, *see* [24]). Depending on the context, it may also be important to incorporate cell-cycle-dependent processes—where the kinetic parameters associated with certain processes depend on the current phase of the cell cycle (*see* **Note 2**).
2. *Cell cycle length*: Cell cycle length is a key model parameter that specifies the time interval between successive DNA replication events. As replication dilutes the level of modifications at the locus, this parameter can determine the stability of epigenetic states—longer cell cycles can favor stable maintenance by allowing more time for the restoration of modifications by feedback mechanisms. Also *see* **Note 3**.

3. *Duration of simulation*: The duration of simulation can be an important factor in assessing the stability of an epigenetic state—depending on the context, even certain unstable states may persist over multiple cell cycles. Typically, we would like to track the evolution of the modification state over ten cell cycles or more.

2.3 Choice of Stochastic Histone Level Model

Multiple studies have used deterministic models based on ODEs to quantitatively capture the experimentally observed dynamics of histone modification/transcription states at both single loci ([5, 8, 9]) and genome wide [10]. In [8], an ODE model—describing the time evolution of the fraction of T-cell *Bcl11b* loci at different states of cis and trans activation—was combined with single cell fluorescent imaging to examine the dynamics of epigenetic activation at this locus. Questa et al. [5] used a similar population level model describing the fraction of loci at different stages of H3K27me3 silencing, combined with time-course ChIP of histone modifications and quantification of transcript levels, to analyze the dynamics of cold induced silencing and reactivation at the plant *Arabidopsis thaliana* gene *FLC*. To study the same process at *FLC* in *Arabidopsis halleri*, Nishio et al. [9] used an ODE model that described the dynamics of the histone modification profile at a single locus—using a spatially compartmentalized description, where the locus is divided into domains, each domain having an associated (average) level of H3K27me3/H3K4me3. An ODE model describing average histone modification levels in distinct spatial domains at a genome wide level has also been used to analyze the propagation dynamics of H3K27 and H3K36 modifications through DNA replication [10].

While ODE based models are simpler to program and analyze, and may be suited to answering certain types of questions, there are other questions that arise in the same contexts which require "molecular level" stochastic models of the kind described here, which track histone modifications at the level of individual histones. For example, when examining the stochastic nature and timescale of switching between different epigenetic states at an individual locus (such as at *FLC* or *Bcl11b*), whether the switch involves a gradual build-up/removal of repressive modifications or is essentially an "all-or-nothing/digital" phenomenon, whether this happens within one cell cycle, or takes longer, with the locus having to go through multiple cycles of replication—these aspects cannot be captured by a model that describes only the population level fraction of loci in each modification state. In contrast, a molecular level model can capture the stochastic nature of the switching, reproduce these timescales and also capture how this timescale depends on the dynamics of other molecular processes at the locus. An important aspect of epigenetic modification states is their stability, and their response to molecular level noise. Fluctuations in the levels of modifications or transactivation factors are expected to have an important effect on stability and switching. Using a stochastic model provides a natural setting for examining the effect of noise [1, 2]. Other questions concerning the nature of the molecular processes, such as the range of read–write interactions required for stable silencing, or how H3K27me3 nucleated at a localized region can spread across a whole locus, can also be addressed using stochastic models [2–4].

2.4 Model Structure

2.4.1 Transcriptional Feedback Model

This model describes a generic PRC2 target locus represented by a fixed number of histones, and tracks H3K27 methylation status at each histone, as well as transcription events. The main features of this model are (Fig. 1):

1. Nonprocessive H3K27 methylation and demethylation: The three possible methyl groups at H3K27 residues are assumed to be added or removed one by one. This assumption is supported by experimental studies and is an essential feature for the model to exhibit robust epigenetic memory states.
2. Positive feedback in the PRC2-mediated methylation of H3K27: H3K27me2 and me3 modified histones are capable of activating PRC2 to modify a neighboring histone. The positive feedback is required to allow self-sustaining repressive states. The local mechanisms driving PRC2 recruitment (e.g., CpG islands) are described implicitly, captured by a parameter β. The rate of methylation is calculated as follows at the ith histone ($i \in \{1, 2, \ldots, N\}$):

$$r_i^{\mathrm{me}} = \beta\left(\delta_{S_i,\mathrm{me0}}(\gamma_{\mathrm{me0-1}} + k_{\mathrm{me0-1}}E_i) + \delta_{S_i,\mathrm{me1}}(\gamma_{\mathrm{me1-2}} + k_{\mathrm{me1-2}}E_i) + \delta_{S_i,\mathrm{me2}}(\gamma_{\mathrm{me2-3}} + k_{\mathrm{me2-3}}E_i)\right)$$

where the effect of modifications on neighboring nucleosomes is captured by

$$E_i = \sum_{j \in M_i} \left(\rho_{\mathrm{me2}}\delta_{S_j,\mathrm{me2}} + \delta_{S_j,\mathrm{me3}}\right)$$

Here, M_i is the set of indices corresponding to the nearest-neighbor histones to histone i. All histones except the ones at the boundary, that is, histones $2 < i < N - 1$, have five nearest neighbors—one on the same nucleosome, and two each on the two neighboring nucleosomes. Thus, for $2 < i < N - 1$,

$$M_i = \left\{ \begin{array}{l} \{i-3, i-2, i-1, i+1, i+2\} \text{ for } i \text{ even}, \\ \{i-2, i-1, i+1, i+2, i+3\} \text{ for } i \text{ odd} \end{array} \right\}.$$

As nucleosomes outside the locus are not represented in the model, the nucleosomes at the boundary are considered to have only one neighboring nucleosome. This means that for histones $i = \{1, 2, N - 1, N\}$,

$$M_i = \left\{ \begin{array}{l} \{i+1, i+2, i+3\} \text{ for } i = 1 \\ \{i-1, i+1, i+2\} \text{ for } i = 2 \\ \{i-2, i-1, i+1\} \text{ for } i = N-1 \\ \{i-3, i-2, i-1\} \text{ for } i = N \end{array} \right\}.$$

See Table 1 for a description of the parameters used above.

The above expression for methylation propensity combines contributions from both noisy (background) methylation (captured by rates $\gamma_{\text{me0}-1}$, $\gamma_{\text{me1}-2}$, $\gamma_{\text{me2}-3}$) and PRC2 recruitment driven feedback methylation (captured by rate parameters $k_{\text{me0}-1}$, $k_{\text{me1}-2}$, $k_{\text{me2}-3}$).

The subscripts denote the corresponding reaction—for instance $\gamma_{\text{me0}-1}$ is the rate of noisy conversion of H3K27me0 to H3K27me1.

$\delta_{m,n}$ denotes the Kronecker delta, defined as follows.

$$\delta_{m,n} = \left\{ \begin{array}{l} 1, m = n \\ 0, m \neq n \end{array} \right\}$$

where $m, n \in \{\text{me0}, \text{me1}, \text{me2}, \text{me3}\}$

3. Transcription-mediated antagonism of the repressed state: Transcription removes the K27 modifications in two ways—(1) through demethylation (motivated by the experimentally observed association between transcription elongation factors and H3K27 demethylases at promoters of PRC2 target genes), and (2) through nucleosome exchange during transcription. The model captures these effects as follows:
   (a) Whenever a transcription event occurs, each histone at the locus is demethylated with probability p_{dem}.
   (b) Whenever a transcription event occurs, both histones at each nucleosome are replaced by unmarked histones with probability p_{ex} per nucleosome (nucleosome exchange).
4. Noisy demethylation: Background demethylation at a rate γ_{dem} per histone, in addition to the transcription driven demethylation.
5. H3K27 methylation-based transcriptional repression: H3K27me2/me3 marks anywhere in the locus are allowed to have a repressive effect on transcription. This effect is captured by a function describing the transcription rate, as described below. The active transcriptional state is defined by low levels of H3K27me2/me3, and there are no "activating marks" in the model. The overall relationship between H3K27 methylation levels and transcription is quantified by the following mathematical relationship, describing the transcription frequency *f* at the locus:

$$f = \alpha\left(f_{\max} - \frac{P_{27\mathrm{me}2/3}}{P27_T}\left(f_{\max} - f_{\min}\right)\right) \text{ for } P_{27\mathrm{me}2/3} < P_{\mathrm{T}}$$

and

$$f = \alpha f_{\min} \text{ for } P_{27\mathrm{me}2/3} \geq P_{\mathrm{T}}$$

$P_{27\mathrm{me}2/3}$ is the level of K27me2/me3 coverage across the whole locus:

$$P_{27\mathrm{me}2/3} = \frac{\text{Total number of K27me2 or me3 modified histones at the locus}}{\text{Total number of histones at the locus}}$$

Note the threshold for full silencing, P_{T} can be chosen to be less than one, so that full H3K27me2/3 coverage is not needed for full silencing. A threshold less than one was chosen in [1], to fit experimentally measured recovery of H3K27me3 after replication. This assumption is also supported by recent experimental work [25] showing that partial H3K27me3 coverage may be enough to maintain full silencing.

The above relationship involves the parameters f_{max} and f_{min} representing the maximum (no H3K27me2/3) and minimum (fully silenced) rates of transcription.

The multiplicative factor α captures trans-factor mediated regulation of the transcription rate: $\alpha = 1$ is neutral, $\alpha > 1$ represents activation, and $\alpha < 1$ represents repression. Note that the above model for transcription rates represents transcription as essentially a Poisson process, with the probability per unit time of transcription being a function of the histone mark coverage. Transcriptional bursting behavior can also be incorporated within this framework—*see* **Note 4**.

6. DNA replication: Replication occurs once per cell cycle, with each nucleosome in the model replaced with a new unmodified nucleosome with probability half. Replication thus perturbs the repressed state by removing methylated H3K27 histones.

2.4.2 M-U-A Type Model

The M-U-A type model is an alternative, simpler model structure, that incorporates histone modification dynamics similar to the above model. The M-U-A model describes mutually antagonistic active and repressive marks (Fig. 1), rather than mutual antagonism between repressive marks and transcription [2]. As these models have been used in multiple studies in conjunction with quantitative experimental data [3, 4, 6], we also discuss this model structure here:

1. Similar to the transcriptional antagonism type model above, this model tracks the modification state of a system of histones spanning a single locus.
2. Each histone is assumed to be interconverted between three states—a repressive modification state (M), an unmodified state (U), and an active modification state (A).

3. Interconversion between modification states can be either feedback mediated (depending on the modification state of other histones at the locus) or feedback independent.
4. Previous studies have examined models with both short and long ranged feedback interactions—we will consider long ranged interactions here, where all histones at the locus can interact equally.
5. The effect of replication is captured in exactly the same way as in the above model—by the random removal of nucleosomes with 50% probability and replacement with unmodified nucleosomes, once per cell cycle.
6. Transcription and its potential antagonism of repressive marks are not directly captured by this model.

3 Methods

3.1 Programming the Simulation Algorithm

Here we describe how to implement the models presented above. We start with the transcriptional feedback type model (with nearest-neighbor read–write interactions), using the Gillespie algorithm. Next, we describe how to implement the same model using a simpler Monte-Carlo approach. While the Monte-Carlo implementation is simpler to program, the Gillespie algorithm implementation is more computationally efficient. Finally, we describe an implementation of the M-U-A type model (with long-ranged interactions), using a Monte-Carlo approach. Our description does not rely on a particular choice of programming language. In the discussion below, the model represents a genomic locus containing N histones (with N even), that is, $N/2$ nucleosomes.

Using a pseudo-random number generator

A core component of both the Gillespie algorithm and Monte-Carlo approaches is a subroutine that generates pseudo-random numbers from a uniform (0,1) probability distribution—sampling from such a distribution is required at multiple stages in both simulation approaches. The widely used and well-characterized Mersenne Twister algorithm [26] is a suitable pseudo-random number generator for this purpose. Implementations of this algorithm are available in several standard numerical libraries and scientific computing systems including Python, R, and MATLAB.

3.1.1 Transcriptional Feedback Type Model Using a Gillespie Exact Stochastic Simulation Algorithm

The Gillespie exact Stochastic Simulation Algorithm (SSA) [11] is a method to simulate the time evolution of a stochastic system of reactions. The input to the algorithm consists of three main components:

1. The starting state of the system: In our model, we want to explicitly track the methylation state of each H3 histone at the locus, so the starting state is an array S whose elements S_i represent the modification state of the "ith" histone.

2. A specified set of "reactions," each of which can change the state of the system in a specified way: In our model, there are *2 N + 1* possible reactions: *N* methylation reactions (one at each histone), N demethylation reactions (one at each histone), and transcription. Transcription is assumed to remove modifications through demethylation and nucleosome exchange.
3. A specified set of rules for calculating the "propensity" (essentially probability of occurrence per unit time) for each reaction. This involves using the mathematical relationships specified in the Materials section (Subheading 2.4.1), relating the modification state of the locus to the rates of reactions.

Capturing the effect of replication: The model also has to capture DNA replication as an event disrupting the modification state of the locus. A simple way to incorporate the effect of replication is to assume that replication occurs deterministically, that is, at fixed times determined only by a specified cell-cycle duration. Therefore, there is no propensity associated with replication—the times at which replication events occur are essentially specified at the beginning of the simulation. We also assume that replication affects the whole locus instantaneously—the simulations do not capture different stages of the passage of a replication fork through the locus, only the state of the whole locus following disruption by replication.

Gillespie Implementation

The Gillespie algorithm implementation of the transcriptional feedback type model above (Sample code can be accessed at https://github.com/gm1613/Mechanistic-Histone-Modification-Model.git) consists of the following steps (simulation output shown in Fig. 2):

1. Specifying the model:
   (a) Create an array *S* representing the *N* histone modification states and set the initial modification states. For ease of coding, we use {0, 1, 2, 3} to denote {me0, me1, me2, me3} states of H3K27 residues. Thus, each entry S_i can take the values {0, 1, 2, 3} . We use this {0, 1, 2, 3} notation throughout this section. The array *S* is used to store the current state of all *N* histones over the course of the simulation.
   (b) Specify the mathematical relationships for computing reaction propensities—as given under Subheading 2.4.1.
   (c) Specify values for all kinetic parameters and probabilities for the different molecular processes (listed in Table 1).
   (d) Specify the cell-cycle length.
2. Specify simulation parameters: Duration of simulation and number of trajectories to be simulated.

Table 1
Model parameters for transcriptional feedback type model of H3K27me silencing

Parameter	Description	Value used in example
N	Number of histones	60
k_{me}	PRC2-mediated methylation rate (histone^{-1} s^{-1})	8×10^{-6}
k_{me0-1}	PRC2-mediated methylation rate (me0 to me1) (histone^{-1} s^{-1})	$9k_{me}$
k_{me1-2}	PRC2-mediated methylation rate (me1 to me2) (histone^{-1} s^{-1})	$6k_{me}$
k_{me2-3}	PRC2-mediated methylation rate (me2 to me3) (histone^{-1} s^{-1})	k_{me}
γ_{me2-3}	Noisy methylation rate (me2 to me3) (histone^{-1} s^{-1})	$k_{me}/20$
γ_{me1-2}	Noisy methylation rate (me1 to me2) (histone^{-1} s^{-1})	$k_{me1-2}/20$
γ_{me0-1}	Noisy methylation rate (me0 to me1) (histone^{-1} s^{-1})	$k_{me0-1}/20$
β	Relative local PRC2-activity	1
ρ_{me2}	Relative activation of PRC2 by H3K27me2	0.1
γ_{dem}	Noisy demethylation rate (histone^{-1} s^{-1})	4×10^{-7}
α	Trans-acting gene activation	1
f_{min}	Minimum transcription initiation rate (s^{-1})	$f_{max}/40$
f_{max}	Maximum transcription initiation rate (s^{-1})	4×10^{-3}
P_T	H3K27me2/me3 coverage threshold for full repression of transcription	1/3
p_{dem}	Demethylation probability (histone^{-1} transcription^{-1})	4×10^{-3}
p_{ex}	Histone exchange probability (histone^{-1} transcription^{-1})	1×10^{-3}
	Cell-cycle length (hours)	22

3. To simulate a single trajectory, start by opening an output file and recording the initial time and state of the system:
   (a) Open a text file and record the current time and state of the system. The state of the system can be recorded by writing the whole array of modification values to the output file. This is essential if the aim is to analyze the time evolution of the full spatial profile of modifications. If the full profile is not needed for our analysis, for instance, if we are interested only in analyzing the time evolution of the fractional coverage of a modification, we can choose to record only some measures of the state of the locus, for instance the total fractional coverage of the different modification states. Note that this choice has no bearing on the model or the algorithm—in this step we merely record the current state of the system in an output file, while the algorithm still uses the full array of modification states for N histones.

4. Iteratively perform the following steps.

   (a) **Step 1:** For the current state of the system (as represented by the entries in the array of modification states S) compute propensities for all reactions (including transcription).

   (i) The methylation propensities at each histone i are computed as follows.

$$r_i^{\mathrm{me}} = \beta\left(\delta_{S_i,0}(\gamma_{\mathrm{me0-1}} + k_{\mathrm{me0-1}}E_i) + \delta_{S_i,1}(\gamma_{\mathrm{me1-2}} + k_{\mathrm{me1-2}}E_i) + \delta_{S_i,2}(\gamma_{\mathrm{me2-3}} + k_{\mathrm{me2-3}}E_i)\right)$$

where E_i represents the read–write feedback contribution from neighboring nucleosomes and is computed as follows.

$$E_i = \sum_{j \in M_i} \left(\rho_{\mathrm{me2}}\delta_{S_j,2} + \delta_{S_j,3}\right)$$

For interior histones, that is, for $2 < i < N - 1$,the set of nearest-neighbor indices, M_i is given by

$$M_i = \left\{ \begin{array}{l} \{i-3, i-2, i-1, i+1, i+2\} \text{ for } i \text{ even,} \\ \{i-2, i-1, i+1, i+2, i+3\} \text{ for } i \text{ odd} \end{array} \right\}.$$

For boundary histones, that is, for $i = \{1, 2, N - 1, N\}$, M_i is given by

$$M_i = \left\{ \begin{array}{l} \{i+1, i+2, i+3\} \text{ for } i = 1 \\ \{i-1, i+1, i+2\} \text{ for } i = 2 \\ \{i-2, i-1, i+1\} \text{ for } i = N-1 \\ \{i-3, i-2, i-1\} \text{ for } i = N \end{array} \right\}$$

Here $\delta_{m,\ n}$ denotes the Kronecker delta, defined for the $\{0, 1, 2, 3\}$ notation as

$$\delta_{m,n} = \left\{ \begin{array}{l} 1, m = n \\ 0, m \neq n \end{array} \right\}$$

where $m,\ n \in \{0, 1, 2, 3\}$.

   (ii) The demethylation propensities are all set to be equal to γ_{dem}.

   (iii) The transcription propensity is computed using the transcription firing function,

$$f = \alpha\left(f_{\max} - \frac{P_{27\mathrm{me2/3}}}{P27_T}(f_{\max} - f_{\min})\right) \text{ for } P_{27\mathrm{me2/3}} < P_{\mathrm{T}}$$

and

$$f = \alpha f_{\min} \text{ for } P_{27\mathrm{me2/3}} \geq P_{\mathrm{T}}$$

as described under Subheading 2.4.1.

(iv) These propensities are stored in an array r, with $(2\ N + 1)$ entries. The first N entries correspond to methylation reactions, corresponding to histones numbered $i \in \{1, 2, ..., N\}$, the next N entries correspond to demethylation reactions corresponding to histones numbered $i \in \{1, 2, ..., N\}$, and the last entry corresponds to a transcription event.

(v) Move to **step 2**.

(b) **Step 2:** Compute the total reaction propensity r_{total}, that is, the sum of all reaction propensities, and the array of cumulative propensities $r_{\text{cumulative}}$, with $(2\ N + 1)$ entries, where the kth entry is the sum of the first k entries of the array of propensities r. The array $r_{\text{cumulative}}$ will be used in **step 7** to probabilistically determine the next reaction to occur, based on the relative propensities of all the reactions. Move to **step 3**.

(c) **Step 3:** Compute the waiting time until the next reaction, τ:

(i) For the Gillespie exact SSA, this is done by sampling from an exponential distribution with the total reaction propensity as the decay constant.

(ii) To do this we first generate a pseudo-random number from a uniform distribution (0,1), using a reliable random number generator from the numerical library of choice. We call the number ξ_1.

(iii) The waiting time can then be computed as:

$$\tau = \frac{\ln(1/\xi_1)}{r_{\text{total}}}$$

(iv) Move to **step 4**.

(d) **Step 4:** Update the time (i.e., increment current time by τ) and check whether the simulation should terminate before this time or whether the next DNA replication event should happen before this time (the timing of the replication events is fixed by the cell-cycle lengths specified at the start of the simulation). If termination must occur, perform **Step 5**; if replication must occur, perform **Step 6**; otherwise perform **Step 7**.

(e) **Step 5:** Set the current time to the time limit of the simulation, record current time and system state in the output file, and terminate the algorithm.

(f) **Step 6:** Set the current time to the time of the next DNA replication event and change the state of the system to reflect the effect of replication on the histone marks.

(i) During and after replication, the algorithm tracks only one of the daughter strands.

(ii) To capture the effect of replication, we iterate over the entries of array S. For each nucleosome at the locus, its modification state is set to 0, that is, the pair of entries corresponding to the nucleosome is set to 0, with an inheritance probability of one half. We assume here inheritance of intact nucleosomes.

(iii) For each nucleosome, this is done by generating a pseudo-random number from a uniform distribution (0,1) and setting the corresponding histone modification state to 0 if the generated number is less than the inheritance probability 0.5.

(iv) Following this step, record the current time and system state in the output file and return to **Step 1**.

(g) **Step 7:** Choose the next reaction to occur based on relative propensities (reactions with higher propensity are more likely to occur) and update the system state accordingly.

(i) To do this we first generate a pseudo-random number from a uniform distribution (0,1). We call the number ξ_2.

(ii) Find the smallest $k \epsilon \{1, \ldots, 2N + 1\}$ such that $r^k_{\text{cumulative}}$

the kth entry of the cumulative propensity array is greater than the value $\xi_2 r_{\text{total}}$.

(iii) If $k < 2N + 1$ we update the system array S to reflect the occurrence of the reaction corresponding to index k: a methylation or a demethylation at the corresponding histone.

(iv) If $k = 2N + 1$ we update the system to reflect a transcription event. This involves carrying out two steps—nucleosome exchange and histone demethylation.

- Nucleosome exchange: Iterating over the entries of array S, for each histone, we generate a pseudo-random number from a uniform distribution (0,1) and set both entries corresponding to the current nucleosome to 0 if the generated number is less than p_{ex} (*see* Table 1).
- Demethylation: Iterating over the entries of array S, for each histone, we generate a pseudo-random number from a uniform distribution (0,1) and reduce the methylation state by one if the current state is {1, 2, or 3} AND the generated number is less than p_{dem} (*see* Table 1).

(v) Following the above steps, record the current time and system state in the output file and return to **Step 1**.

3.1.2 Transcriptional Feedback Type Model Using a Monte-Carlo Approach

A simpler Monte-Carlo approach to simulating stochastic histone modification models has also been used in previous studies ([2, 4])—here we discuss such an implementation for the transcriptional feedback type model. Similar to the Gillespie algorithm (*see* description under Subheading 3.1.1), the inputs required for the Monte-Carlo approach are the initial state, the specified reactions, and the rules for computing the propensity for each of these reactions. In addition to these inputs, we also need to specify a fixed time-step for updating the state of the system (see below).

The Monte-Carlo implementation consists of the following steps:

1. Specifying the model:
   (a) Create an array S representing the N histone modification states, setting the initial modification states. For ease of coding, we use {0, 1, 2, 3} to denote {me0, me1, me2, me3} states of H3K27 residues. Each entry S_i can take the values {0, 1, 2, 3} . We use this {0, 1, 2, 3} notation throughout this section.
   (b) Specify the mathematical relationship for computing the transcription rate based on the current system state—as given under Subheading 2.4.1.
   (c) Specify values for all kinetic parameters and probabilities for the different molecular processes (listed in Table 1).
   (d) Specify the size of time step Δt, and cell-cycle length. The size of the time step Δt is chosen based on two opposing considerations:
      (i) The probability for the fastest reaction in the system to occur in a time interval of size Δt should be low enough that it is sufficient for us to simulate at most a single occurrence of this reaction during this time interval. (In other words, Δt must be low enough that the possibility of this reaction occurring more than once in a time interval of size Δt can be ignored.)
      (ii) Δt should be large enough to allow the algorithm to terminate within a desired runtime. For simulations of a fixed duration, using a smaller Δt increases the number of time-steps taken by the algorithm and consequently increases the runtime.
2. Specify simulation parameters: Duration of simulation and number of trajectories to be simulated.
3. To simulate a single trajectory, start by opening an output file and recording the initial time and state of the system.
   (a) Open a text file and record the current time and state of the system. The state of the system can be recorded by writing the whole array of modification values to the

output file. This is essential if the aim is to analyze the time evolution of the full spatial profile of modifications. If the full profile is not needed for our analysis, for instance, if we are interested only in analyzing the time evolution of the fractional coverage of a modification, we can choose to record only some measures of the state of the locus, for instance the total fractional coverage of the different modification states. Note that this choice has no bearing on the model or the algorithm—in this step we merely record the current state of the system in an output file, while the algorithm still uses the full array of modification states for N histones.

4. Iteratively perform the following steps.
   (a) **Step 1:** Check if current time corresponds to DNA replication or end of simulation—if so, move directly to **step 6** or **step 7** respectively. Otherwise, for the current state of the system, compute the transcription propensity.

$$f = \alpha\left(f_{\max} - \frac{P_{27\mathrm{me}2/3}}{P27_T}(f_{\max} - f_{\min})\right) \text{for } P_{27\mathrm{me}2/3} < P27_{\mathrm{T}}$$

   and

$$f = \alpha f_{\min} \text{ for } P_{27\mathrm{me}2/3} \geq P27_T$$

   where the parameters are the same as in the Gillespie implementation (*see* Table 1). Move to **step 2**.

   (b) **Step 2:** Generate a pseudo-random number from a uniform distribution (0,1). We call the number ξ_1.

   (i) If $\xi_1 < f\Delta t$, we update the system to reflect a transcription event. This involves carrying out two steps—nucleosome exchange and histone demethylation.
      - Nucleosome exchange: Iterating over the entries of array S, for each histone, we generate a pseudo-random number from a uniform distribution (0,1) and set both entries corresponding to the current nucleosome to 0 if the generated number is less than p_{ex} (*see* Table 1).
      - Demethylation: Iterating over the entries of array S, for each histone, we generate a pseudo-random number from a uniform distribution (0,1), and reduce the methylation state by one if the current state is {1, 2, or 3} AND the generated number is less than p_{dem} (*see* Table 1).
      - Move to **step 3**.

   (ii) If $\xi_1 \geq f\Delta t$, we move directly to **step 3**.

(c) **Step 3:** Methylation reactions: Iterate over the array S of histone modification states. For histone i, the methylation state is updated in the following way.

(i) Choose a neighboring histone j at random—from the five nearest neighbors for histones on interior nucleosomes and from the three nearest neighbors for those on the boundary nucleosomes.

- To do this, generate a pseudo-random number from a uniform distribution (0,1). We call the number ξ_2.
- For interior histones $2 < i < N - 1$, find the smallest $k \epsilon \{1, \ldots, 5\}$ such that $\xi_2 < k/5$. The kth neighbor histone is chosen—the index j of this neighbor histone can be computed as follows.

$$j = \left\{ \begin{array}{ll} i - 4 + k & \text{for } i \text{ even}, k < 4 \\ i - 3 + k & \text{for } i \text{ even}, k \geq 4 \\ i - 3 + k & \text{for } i \text{ odd}, k < 3 \\ i - 2 + k & \text{for } i \text{ odd}, k \geq 3 \end{array} \right\}$$

- For boundary histones $i \in \{1, 2, N - 1, N\}$, find the smallest $k \epsilon \{1, \ldots, 3\}$ such that $\xi_2 < k/3$. The kth neighbor histone is chosen—the index j of this neighbor histone can be computed as follows.

$$j = \left\{ \begin{array}{ll} i + k & \text{for } i = 1 \\ i - k & \text{for } i = 2, k = 1 \\ i - 1 + k & \text{for } i = 2, k > 1 \\ i - 3 + k & \text{for } i = N - 1, k < 3 \\ i - 2 + k & \text{for } i = N - 1, k = 3 \\ i - 4 + k & \text{for } i = N \end{array} \right\}$$

(ii) After choosing a nearest-neighbor histone j, a methylation reaction is attempted for histone i as follows.

- Generate a pseudo-random number from a uniform distribution (0,1), ξ_3
- If histone i carries me0 and neighbor histone j carries me2 then S_i is converted to me1 if $\xi_3 < (\gamma_{\mathrm{me0}\,-\,1} + \mathrm{k}_{\mathrm{me0}\,-\,1}\,\rho_{\mathrm{me2}})\Delta t$
- If histone i carries me0 and neighbor histone j carries me3 then S_i is converted to me1 if $\xi_3 < (\gamma_{\mathrm{me0}\,-\,1} + \mathrm{k}_{\mathrm{me0}\,-\,1})\Delta t$
- If histone i carries me1 and neighbor histone j carries me2 then S_i is converted to me2 if $\xi_3 < (\gamma_{\mathrm{me1}\,-\,2} + \mathrm{k}_{\mathrm{me1}\,-\,2}\,\rho_{\mathrm{me2}})\Delta t$

- If histone i carries me1 and neighbor histone j carries me3 then S_i is converted to me2 if $\xi_3 < (\gamma_{me1-2} + k_{me1-2})\Delta t$
- If histone i carries me2 and neighbor histone j carries me2 then S_i is converted to me3 if $\xi_3 < (\gamma_{me2-3} + k_{me2-3}\,\rho_{me2})\Delta t$
- If histone i carries me2 and neighbor histone j carries me3 then S_i is converted to me3 if $\xi_3 < (\gamma_{me2-3} + k_{me2-3})\Delta t$
- Noisy methylation: If the neighbor histone j carries me0 or me1 then S_i is incremented by one if:

$$\xi_3 < \gamma_{me0-1}\Delta t \quad \text{AND} \quad S_i = 0$$

$$\text{OR}$$

$$\xi_3 < \gamma_{me1-2}\Delta t \quad \text{AND} \quad S_i = 1$$

$$\text{OR}$$

$$\xi_3 < \gamma_{me2-3}\Delta t \ \text{AND} \quad S_i = 2$$

- After repeating the above steps for each element of array S, move to **step 4**.

(d) **Step 4:** Noisy demethylation: Iterate over the array S. For histone i, the state is updated as follows.

(i) Generate a pseudo-random number from a uniform distribution (0,1). We call the number ξ_4.

(ii) If the current methylation state is me1, me2, or me3, that is, $S_i \in \{1, 2, 3\}$, AND $\xi_4 < \gamma_{dem}\Delta t$, then S_i is reduced by 1 (representing nonprocessive demethylation)

(iii) After repeating the above steps for each element of array S, move to **step 5**.

(e) **Step 5:** Increment time by Δt. Record current system state and time in output file. Return to **Step 1**.

(f) **Step 6:** Update system state (array S) to reflect the effect of DNA replication. This is done exactly as described under Subheading 3.1.1, **Step 6**. Increment time by Δt. Record current system state and time in output file. Return to **Step 1.**

(g) **Step 7:** Terminate algorithm (the current time and state would already have been recorded by **step 5** or **step 6**).

3.1.3 M-U-A Type Model Using a Monte-Carlo Approach

A Monte-Carlo implementation of the M-U-A type model consists of the following steps:

1. Specifying the model:
   (a) Create an array S representing the N histone modification states, setting the initial modification states. For ease of coding, we use $\{-1, 0, 1\}$ to denote {A, U, M} states of the histones. Each entry S_i can take the values $\{-1, 0, 1\}$. We use this $\{-1, 0, 1\}$ notation throughout this section.
   (b) Specify values for all kinetic parameters and probabilities for the different molecular processes (listed in Table 2).
   (c) Specify the size of time step Δt, and cell-cycle length. Δt is chosen as discussed under Subheading 3.1.2.
2. Specify simulation parameters: Duration of simulation and number of trajectories to be simulated.
3. To simulate a single trajectory, start by opening an output file and recording the initial time and state of the system:
   (a) Open a text file and record the current time and state of the system. The state of the system can be recorded by writing the whole array of modification values to the output file. This is essential if the aim is to analyze the time evolution of the full spatial profile of modifications. If the full profile is not needed for our analysis, for instance, if we are interested only in analyzing the time evolution of the fractional coverage of a modification, we can choose to record only some measures of the state of the locus, for instance the total fractional coverage of the different modification states. Note that this choice has no bearing on the model or the algorithm—in this step we merely record the current state of the system in an output file, while the algorithm still uses the full array of modification states for N histones.
4. Iteratively perform the following steps:

Table 2
Model parameters for M-U-A type model

Parameter	Description
N	Number of histones
k_{M}	Rate of recruited conversion toward M state (histone^{-1} s^{-1})
k_{A}	Rate of recruited conversion toward A state (histone^{-1} s^{-1})
γ_+	Noisy addition rate (U to M or A) (histone^{-1} s^{-1})
γ_-	Noisy removal rate (M or A to U) (histone^{-1} s^{-1})

(a) **Step 1:** Compute the current fractional coverages of the two modifications across the locus: $P_{\mathrm{M}} = \frac{N_{\mathrm{M}}}{N}$ for the M modifications, and $P_{\mathrm{A}} = \frac{N_{\mathrm{A}}}{N}$ for the A modifications, where N_{M} and N_{A} are the numbers of histones carrying the M and A modifications respectively. Check if current time corresponds to DNA replication or end of simulation—if so, move directly to **step** 7 or **step 8** respectively. Otherwise, move to **step 2**.

(b) **Step 2:** Stimulated (feedback) reactions toward the M state, mediated by long-ranged interactions (whole locus). Iterate over the array S of histone modification states. For histone i, the modification state S_i is updated in the following way:

  (i) Generate a pseudo-random number from a uniform distribution (0,1). We call the number ξ_{1i}.

  (ii) If histone i is in the A or U states, that is, $S_i \in \{-1, 0\}$, it is moved one step toward the M state if $\xi_{1i} < k_M P_M \Delta t$, where P_M is the fractional coverage of the M modification across the locus computed in **Step 1**. Note that P_{M} is updated only in **Step 1** and does not reflect the changes carried out in **Step 2**.

  (iii) After repeating the above steps for each element of array S, move to **step 3**.

(c) **Step 3:** Stimulated (feedback) reactions toward the A state, mediated by long-ranged interactions (whole locus). Iterate over the array S of histone modification states. For histone i, the modification state S_i is updated in the following way.

  (i) Generate a pseudo-random number from a uniform distribution (0,1). We call the number ξ_{2i}.

  (ii) If histone i is in the M or U states, that is, $S_i \in \{0, 1\}$, it is moved one step toward the A state if $\xi_{2i} < k_{\mathrm{A}} P_{\mathrm{A}} \Delta t$, where P_{A} is the fractional coverage of the A modification across the locus computed in **Step 1**. Note that P_{A} is updated only in **Step 1** and does not reflect the changes carried out in **Step 3**.

  (iii) After repeating the above steps for each element of array S, move to **step 4**.

(d) **Step 4:** Noisy addition. Iterate over the array S of histone modification states. For histone i, the modification state S_i is updated in the following way:

  (i) Generate a pseudo-random number from a uniform distribution (0,1). We call the number ξ_{3i}.

(ii) If histone i is in the U state, that is, $S_i = 0$, it is moved to the M state or the A state each with probability half, if $\xi_{3i} < \gamma_+ \Delta t$.

(iii) After repeating the above steps for each element of array S, move to **step 5**.

(e) **Step 5:** Noisy removal. Iterate over the array S of histone modification states. For histone i, the modification state S_i is updated in the following way:

(i) Generate a pseudo-random number from a uniform distribution (0,1). We call the number ξ_{4i}.

(ii) If histone i is in the M or A states, that is, $S_i \in \{-1, 1\}$, it is moved to the U state if $\xi_{4i} < \gamma_- \Delta t$.

(iii) After repeating the above steps for each element of array S, move to **step 6**.

(f) **Step 6:** Increment time by Δt. Record current system state and time in output file. Return to **Step 1**.

(g) **Step 7:** Update system state (array S) to reflect the effect of DNA replication. This is done exactly as described under Subheading 3.1.1, **Step 6**. Increment time by Δt. Record current system state and time in output file. Return to **Step 1.**

(h) **Step 8:** Terminate algorithm (the current time and state would already have been recorded by **step 6** or **step 7**).

3.2 Initial Conditions and Simulation Output

Initial conditions: The initial modification state of the locus is specified as an input to the Gillespie or Monte-Carlo algorithm. To do this, each element of the array S is set to one of {0,1,2,3} for the transcriptional feedback H3K27me silencing model, and to one of {−1,0,1} for the M-U-A type model. For a bistable system (see below), the choice of initial state can determine the steady state that the system eventually reaches (on average), so it is important to explore different initial conditions. For the basic analysis we show here (for the transcriptional feedback H3K27me silencing model), two initial states are considered (Fig. 2a)—the completely unmodified state (all S_i set to 0) and the fully methylated state (all S_i set to 3).

Simulation output: The simulation output generated by either the Gillespie algorithm or a Monte-Carlo approach (programmed as described here) can be interpreted as a table of entries. The first column contains the times at which reactions have occurred (including replication), and the remaining columns contain either the full state of the system (N columns corresponding to N histones) or measures of the state of the system at these time points (state of the system *after* the reaction has occurred). For instance, one column may contain the fraction of K27me2/me3

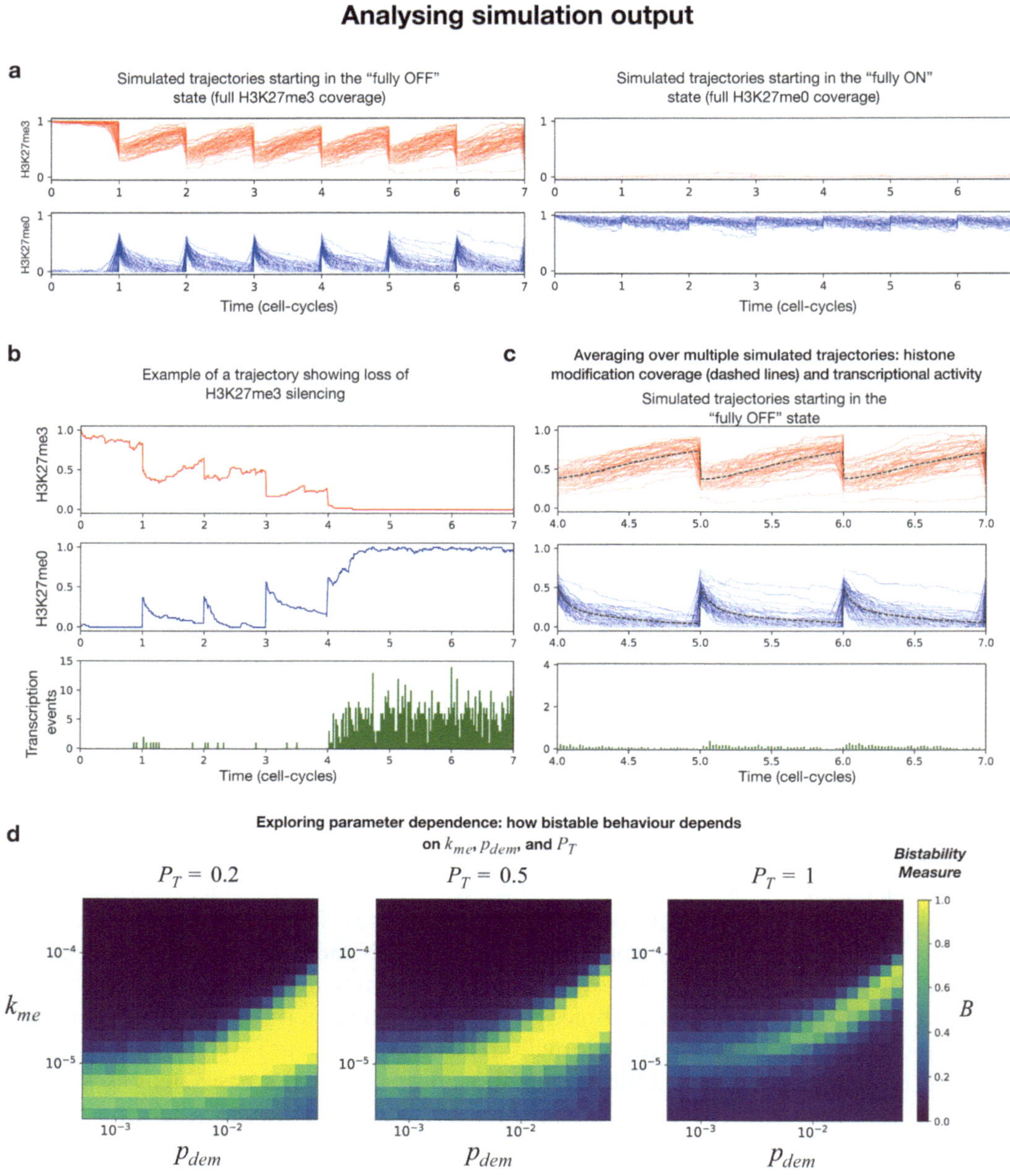

Fig. 2 Analyzing simulation output for the transcriptional feedback type model generated using the Gillespie algorithm. (**a**) Comparing time evolution of the modification state of the locus (fractional coverage of me3 and me0) between trajectories starting at different initial states (50 simulations starting at each initial condition—full me3 and full me0). (**b**) Example trajectory demonstrating loss of H3K27me3 silencing at a single locus (transcription-mediated demethylation rate tenfold higher than in (**a**)). The bottom row shows transcriptional activity over time at the locus (transcription events per 30 min). In this example, silencing is lost during the fifth cell cycle. (**c**) Averaging across multiple simulations: plot shows time-evolution over three cell cycles of H3K27me3, H3K27me0 (individual simulations: full lines; average: dashed lines) and transcription counts, averaged over 50 simulated trajectories starting in the full me3 state. (**d**) Exploring parameter dependence of bistable behavior: each heatmap shows how the bistability measure B (see heatmap scale on the right) depends on the PRC2-mediated H3K27 methylation rate k_{me} and the transcription-mediated H3K27 demethylation rate p_{dem}, for a particular level of the H3K27me2/me3 silencing threshold P_T. A lower silencing threshold allows more robust bistability—the bistability measure is high over a larger range of k_{me} and p_{dem} for $P_T = 0.2$ relative to when $P_T = 1$

coverage across the whole locus for the transcriptional feedback type model, while for the M-U-A type model, this could be the fractional coverage of M modifications.

3.3 Model Analysis: Time Averaging, Quantifying Bistability, First Passage Times

3.3.1 Time Averaging

Time averages of state-dependent quantities can be computed as follows.

$$\langle X \rangle_{\text{time}} = \sum_{i=0}^{K-1} X_i \frac{t_{i+1} - t_i}{t_K - t_0}$$

for a quantity X. For example, X may be the total fraction of H3K27me3 histones or the fractional coverage of M modifications.

3.3.2 Quantifying Bistability

The system is considered to be bistable if, for the same values of system parameters, the system can stably maintain two overall distinct modification states depending only on the initial state. As both states (OFF, high H3K27me2/3 or high M and ON, low H3K27me2/3 or high A) are subject to fluctuations, whether or not the system is in a particular state at a given time can only be defined using thresholds for coverage of the modifications.

Defining ON and OFF states: The two states can be distinguished using a state indicator at each time point: I_{OFF} which is 1 if the system is considered to be in the OFF state and 0 in the ON state, and I_{ON} which is 1 if the system is considered to be in the ON state and 0 in the OFF state. For the transcriptional feedback model, we can choose to compute I_{OFF} and I_{ON} as follows:

$$I_{\text{OFF}} = 1 \text{ if } \left(P_{27\text{me}2/3} > \frac{3(P27_{\text{T}})}{4} \right) \text{ and 0 otherwise}$$

$$I_{\text{ON}} = 1 \text{ if } \left(P_{27\text{me}2/3} < \frac{(P27_{\text{T}})}{4} \right) \text{ and 0 otherwise}$$

A similar approach can be used to compute I_{OFF} and I_{ON} for the M-U-A type model, for instance (following [27]):

$$I_{\text{OFF}} = 1 \text{ if } \left(N_{\text{M}} - N_{\text{A}} > \frac{N}{2} \right) \text{ and 0 otherwise}$$

$$I_{\text{ON}} = 1 \text{ if } \left(N_{\text{A}} - N_{\text{M}} > \frac{N}{2} \right) \text{ and 0 otherwise}$$

where N_{M} and N_{A} are the number of histones in the M and A states, respectively.

Using a bistability measure: One way of quantifying bistability is to use the bistability measure B, as in [1, 27], which is computed as:

$$B = 4 \langle I_{\text{OFF}} \rangle_{\text{time,trajectories}} \langle I_{\text{ON}} \rangle_{\text{time,trajectories}}$$

where I_{OFF} and I_{ON} (defined above) are averaged over time and over multiple trajectories and evaluated by simulating equal numbers of trajectories starting from both the uniform me0 (or A) and uniform me3 (or M) states of the locus. Values of B close to

1 indicate bistable behavior. The dependence of the bistable behavior on different model parameters can be assessed using the bistability measure (Fig. 2d). We note that there are other ways to assess the bistability of the system and assign a bistability score, such as by measuring the frequency of switching between OFF and ON states, or the time spent in intermediate states while switching between OFF and ON states [27].

3.3.3 Computing First Passage Times

Switching between transcriptionally active and repressed chromatin states may be an aspect of interest in a study using the above models (see simulated trajectory shown in Fig. 2b, where a locus switches from a repressed to an active state). Mean first passage times can be useful in characterizing the dynamics of switching between states. For the transcriptional feedback model, tFP_{me0} and tFP_{me3}, defined as the average time taken for the system to change to the opposite chromatin state, when initialized in the uniform me0 or me3 state, respectively, can be computed as

$$t\text{FP}_{\text{me0}} = \left\langle \min\left(t | P_{27\text{me}2/3} > \frac{3(P27_{\text{T}})}{4}\right)\right\rangle_{\text{trajectories}}$$

where the average is over trajectories starting in the uniform me0 state, and

$$t\text{FP}_{\text{me3}} = \left\langle \min\left(t | P_{27\text{me}2/3} < \frac{(P27_{\text{T}})}{4}\right)\right\rangle_{\text{trajectories}}$$

where the average is over trajectories starting in the uniform me3 state. A similar approach can be used to compute first passage times for the M-U-A type model.

3.4 Parameterizing the Model

Choosing parameter values for the kinetic parameters and timescales is an iterative process, with parameter values needing to be changed to enable the model to capture new experimental data—both quantitative and qualitative. Bulk population level quantification of histone modification levels at a single locus by ChIP-qPCR can give us spatially resolved modification levels across the gene. Furthermore, in the case of a system like *FLC*, where the switching occurs slowly at the population level [4], time-course ChIP experiments allow time-resolved measurements sufficient to address the dynamics of the switching process. These measurements can be used to constrain/parameterize our single locus chromatin-level models. Note that it is important to account for cell cycle synchronization, when comparing time-resolved population level measurements to averages over multiple simulated trajectories—*see* **Note 5**. However, direct experimental validation of the modification profile at a single locus in an individual cell, and its effect on transcription (including the all-or-nothing nature of the switch, the dynamics of the switching process, and the quantitative relationship between histone mark coverage and transcription kinetics) requires single cell assays capable of quantifying histone modifications across a specific locus and allowing live time-resolved measurements [28].

4 Notes

1. Locus-size dependence: Multiple aspects of the system behavior can depend on the number of nucleosomes at the locus. This includes the stability of an epigenetic state or even the possibility of bistable behavior—with the silenced state being less stable/more vulnerable to fluctuations at smaller sized loci. The dynamics of switching between states can also depend on the size of the locus and the range of read–write feedback interactions. Examining a range of locus-sizes using the above models can therefore yield potentially important insights—for instance size constraints for a particular class of loci subject to a common mode of epigenetic regulation.
2. Cell cycle–dependent processes: If the system of interest involves cell-cycle dependent effects independent of replication—for instance S-phase dependent incorporation of histone variants—such processes are easy to incorporate in the simulations. The propensities for the corresponding reactions can be made conditional on whether the current time lies within a certain interval, starting say at a fixed time before the next replication event. Similarly, developmental stage dependent processes can also be incorporated, for instance, by stopping replication after a fixed number of cell-cycle lengths or having the propensities of certain reactions dependent on how many cell-cycles have been completed.
3. Cell-cycle duration: Even a simple mechanism of cell cycle dependence, such as DNA replication–mediated disruption of the modification state, makes cell-cycle duration a key parameter that determines the behavior of the model. Longer cell cycles can favor stable maintenance of epigenetic states, while shorter cell cycles can be more disruptive. Cell-cycle duration can thus become an important factor in regulating switching of epigenetic states, with developmental stage dependent differences in cell-cycle length potentially having important consequences [29]. Such effects can be examined by changing a single parameter in the simulations described above. Variation in cell-cycle duration can also be incorporated by probabilistically choosing the length of each cell cycle over the course of a simulation, by sampling from a target distribution.
4. Transcriptional bursting: In the above model, we have assumed that transcription can be represented by essentially a Poisson process, with the probability per unit time of transcription

being a function of the histone mark coverage. However, transcriptional bursting, where bursts of high transcriptional activity are separated by intervals of low transcriptional activity, may be of interest in certain contexts. Such behavior can be captured within our modeling framework—having the transcription function depend on a promoter state which can stochastically switch, with the probability of switching potentially determined by trans-factor activity, is one way of achieving this [1].

5. When comparing averages over multiple simulated trajectories to the time evolution of population level experimental data, it may be important to ensure that the simulated trajectories reflect the presence or lack of cell-cycle synchronization in the experimentally observed population. The simulation output generated by following the above protocols correspond to all the simulated loci having synchronized cell cycles/replication. Lack of synchronization can be achieved by introducing a random shift in the start times corresponding to each simulation.

References

1. Berry S, Dean C, Howard M (2017) Slow chromatin dynamics allow polycomb target genes to filter fluctuations in transcription factor activity. Cell Syst 4(4):445–457.e8. https://doi.org/10.1016/j.cels.2017.02.013
2. Dodd IB, Micheelsen MA, Sneppen K, Thon G (2007) Theoretical analysis of epigenetic cell memory by nucleosome modification. Cell 129(4):813–822. https://doi.org/10.1016/j.cell.2007.02.053
3. Obersriebnig MJ, Pallesen EM, Sneppen K, Trusina A, Thon G (2016) Nucleation and spreading of a heterochromatic domain in fission yeast. Nat Commun 7:11518. https://doi.org/10.1038/ncomms11518
4. Angel A, Song J, Dean C, Howard M (2011) A polycomb-based switch underlying quantitative epigenetic memory. Nature 476(7358): 105–108. https://doi.org/10.1038/nature10241
5. Questa JI, Antoniou-Kourounioti RL, Rosa S, Li P, Duncan S, Whittaker C, Howard M, Dean C (2020) Noncoding SNPs influence a distinct phase of Polycomb silencing to destabilize long-term epigenetic memory at Arabidopsis *FLC*. Genes Dev 34(5–6):446–461. https://doi.org/10.1101/gad.333245.119
6. Angel A, Song J, Yang H, Questa JI, Dean C, Howard M (2015) Vernalizing cold is registered digitally at *FLC*. Proc Natl Acad Sci U S A 112(13):4146–4151. https://doi.org/10.1073/pnas.1503100112
7. Alabert C, Barth TK, Reveron-Gomez N, Sidoli S, Schmidt A, Jensen ON, Imhof A, Groth A (2015) Two distinct modes for propagation of histone PTMs across the cell cycle. Genes Dev 29(6):585–590. https://doi.org/10.1101/gad.256354.114
8. Ng KK, Yui MA, Mehta A, Siu S, Irwin B, Pease S, Hirose S, Elowitz MB, Rothenberg EV, Kueh HY (2018) A stochastic epigenetic switch controls the dynamics of T-cell lineage commitment. elife 7:e37851. https://doi.org/10.7554/eLife.37851
9. Nishio H, Buzas DM, Nagano AJ, Iwayama K, Ushio M, Kudoh H (2020) Repressive chromatin modification underpins the long-term expression trend of a perennial flowering gene in nature. Nat Commun 11(1):2065. https://doi.org/10.1038/s41467-020-15896-4
10. Alabert C, Loos C, Voelker-Albert M, Graziano S, Forne I, Reveron-Gomez N, Schuh L, Hasenauer J, Marr C, Imhof A, Groth A (2020) Domain model explains propagation dynamics and stability of histone H3K27 and H3K36 methylation landscapes. Cell Rep 30(4):1223–1234.e8. https://doi.org/10.1016/j.celrep.2019.12.060
11. Gillespie DT (1977) Exact stochastic simulation of coupled chemical reactions. J Phys Chem 81(25):2340–2361
12. McCabe MT, Graves AP, Ganji G, Diaz E, Halsey WS, Jiang Y, Smitheman KN, Ott HM, Pappalardi MB, Allen KE, Chen SB, Della

Pietra A 3rd, Dul E, Hughes AM, Gilbert SA, Thrall SH, Tummino PJ, Kruger RG, Brandt M, Schwartz B, Creasy CL (2012) Mutation of A677 in histone methyltransferase EZH2 in human B-cell lymphoma promotes hypertrimethylation of histone H3 on lysine 27 (H3K27). Proc Natl Acad Sci U S A 109(8):2989–2994. https://doi.org/10.1073/pnas.1116418109

13. Ferrari KJ, Scelfo A, Jammula S, Cuomo A, Barozzi I, Stutzer A, Fischle W, Bonaldi T, Pasini D (2014) Polycomb-dependent H3K27me1 and H3K27me2 regulate active transcription and enhancer fidelity. Mol Cell 53(1):49–62. https://doi.org/10.1016/j.molcel.2013.10.030
14. Agger K, Cloos PA, Christensen J, Pasini D, Rose S, Rappsilber J, Issaeva I, Canaani E, Salcini AE, Helin K (2007) UTX and JMJD3 are histone H3K27 demethylases involved in HOX gene regulation and development. Nature 449(7163):731–734. https://doi.org/10.1038/nature06145
15. Sneppen K, Ringrose L (2019) Theoretical analysis of polycomb-trithorax systems predicts that poised chromatin is bistable and not bivalent. Nat Commun 10(1):2133. https://doi.org/10.1038/s41467-019-10130-2
16. Voigt P, Tee WW, Reinberg D (2013) A double take on bivalent promoters. Genes Dev 27(12):1318–1338. https://doi.org/10.1101/gad.219626.113
17. Hyun K, Jeon J, Park K, Kim J (2017) Writing, erasing and reading histone lysine methylations. Exp Mol Med 49(4):e324. https://doi.org/10.1038/emm.2017.11
18. Margueron R, Justin N, Ohno K, Sharpe ML, Son J, Drury WJ 3rd, Voigt P, Martin SR, Taylor WR, De Marco V, Pirrotta V, Reinberg D, Gamblin SJ (2009) Role of the polycomb protein EED in the propagation of repressive histone marks. Nature 461(7265):762–767. https://doi.org/10.1038/nature08398
19. Chen S, Ma J, Wu F, Xiong LJ, Ma H, Xu W, Lv R, Li X, Villen J, Gygi SP, Liu XS, Shi Y (2012) The histone H3 Lys 27 demethylase JMJD3 regulates gene expression by impacting transcriptional elongation. Genes Dev 26(12):1364–1375. https://doi.org/10.1101/gad.186056.111
20. Deaton AM, Gomez-Rodriguez M, Mieczkowski J, Tolstorukov MY, Kundu S, Sadreyev RI, Jansen LE, Kingston RE (2016) Enhancer regions show high histone H3.3 turnover that changes during differentiation. elife 5:e15316. https://doi.org/10.7554/eLife.15316
21. Brookes E, de Santiago I, Hebenstreit D, Morris KJ, Carroll T, Xie SQ, Stock JK, Heidemann M, Eick D, Nozaki N, Kimura H, Ragoussis J, Teichmann SA, Pombo A (2012) Polycomb associates genome-wide with a specific RNA polymerase II variant, and regulates metabolic genes in ESCs. Cell Stem Cell 10(2):157–170. https://doi.org/10.1016/j.stem.2011.12.017
22. Pasini D, Malatesta M, Jung HR, Walfridsson J, Willer A, Olsson L, Skotte J, Wutz A, Porse B, Jensen ON, Helin K (2010) Characterization of an antagonistic switch between histone H3 lysine 27 methylation and acetylation in the transcriptional regulation of Polycomb group target genes. Nucleic Acids Res 38(15):4958–4969. https://doi.org/10.1093/nar/gkq244
23. Zee BM, Levin RS, Xu B, LeRoy G, Wingreen NS, Garcia BA (2010) In vivo residue-specific histone methylation dynamics. J Biol Chem 285(5):3341–3350. https://doi.org/10.1074/jbc.M109.063784
24. Ietswaart R, Rosa S, Wu Z, Dean C, Howard M (2017) Cell-size-dependent transcription of FLC and its antisense long non-coding RNA COOLAIR explain cell-to-cell expression variation. Cell Syst 4(6):622–635.e9. https://doi.org/10.1016/j.cels.2017.05.010
25. Jadhav U, Manieri E, Nalapareddy K, Madha S, Chakrabarti S, Wucherpfennig K, Barefoot M, Shivdasani RA (2020) Replicational dilution of H3K27me3 in mammalian cells and the role of poised promoters. Mol Cell 78(1):141–151.e5. https://doi.org/10.1016/j.molcel.2020.01.017
26. Matsumoto M, Nishimura T (1998) Mersenne twister: a 623-dimensionally equidistributed uniform pseudo-random number generator. ACM Trans Model Comput Simul 8(1):3–30. https://doi.org/10.1145/272991.272995
27. Sneppen K, Dodd IB (2012) A simple histone code opens many paths to epigenetics. PLoS Comput Biol 8(8):e1002643. https://doi.org/10.1371/journal.pcbi.1002643
28. Ludwig CH, Bintu L (2019) Mapping chromatin modifications at the single cell level. Development 146(12):dev170217. https://doi.org/10.1242/dev.170217
29. Reinig J, Ruge F, Howard M, Ringrose L (2020) A theoretical model of polycomb/trithorax action unites stable epigenetic memory and dynamic regulation. Nat Commun 11(1):4782. https://doi.org/10.1038/s41467-020-18507-4

Part VIII

Finding Inhibitors of Histone Methyltransferases

Chapter 20

Screening for Small-Molecule Inhibitors of Histone Methyltransferases

Nico Cantone, Richard T. Cummings, and Patrick Trojer

Abstract

Potent and highly selective small-molecule inhibitors are needed to unravel the biological complexities of histone methyltransferases and to reveal their therapeutic potential. A prerequisite to developing these inhibitors is the identification of validated chemical matter for initiating a medicinal chemistry campaign. For the most part, finding these initial starting points occurs through screening of large, unbiased compound libraries. The size and nature of these libraries, coupled with the complexities of the bisubstrate utilizing histone methyltransferases, necessitates that the primary screen and subsequent hit triage be carefully considered.

In this chapter, using EZH2 as a representative example, we describe a screening and hit triage campaign that identified validated chemical matter allowing initiation of medicinal chemistry studies. Moreover, we discuss a cell-based assay to support lead identification and optimization. The approach described here entailing a mixture of biochemical, biophysical and cell-based assays should be applicable to identifying validated starting points for other histone methyltransferases.

Key words Small molecule inhibitor, Histone methyltransferase, HMT, Epigenetic drug, Hit triage

1 Introduction

Histone methyltransferases (HMTs) catalyze the transfer of one, two, or three methyl groups to specific lysine residues, or one or two methyl groups to specific arginine residues on histone proteins, and universally use S-adenosyl-L-methionine (SAM) as methyl donor. Enzymes that catalyze histone lysine and arginine methylation are usually selective with respect to histone substrate, target amino acid residue, and the number of methyl groups that can be added to the target residue. Human HMTs comprise two enzyme families, lysine methyltransferases with 52 members [1, 2] and arginine methyltransferases with 9 members [3, 4]. In addition, the enzymes FBL and PCMT1 were identified to catalyze glutamine methylation of histone H2A [5] and aspartate methylation of histone H4 [6], respectively. There are additional methyltransferase

Raphaël Margueron and Daniel Holoch (eds.), *Histone Methyltransferases: Methods and Protocols*,
Methods in Molecular Biology, vol. 2529, https://doi.org/10.1007/978-1-0716-2481-4_20,

enzymes that catalyze nonhistone protein and nucleic acid methylation and many of the aforementioned HMTs also methylate nonhistone protein substrates, giving rise to a vastly complex and intriguing regulatory potential to control biological function.

HMTs have been of interest to the drug discovery community for decades given that many HMTs are dysregulated in cancer and promote tumor progression [2, 7, 8]. Despite substantial efforts in academia and industry, more than half of the human HMT enzymes remain untargeted with small molecules and the identification of HMT small-molecule inhibitors (SMIs) for this class of enzymes has been challenging. However, several HMTs have been successfully inhibited, and drug-like molecules targeting DOT1L, EZH2, PRMT5 and PRMT1 have advanced into clinical development [2, 9, 10]. The regulatory approval of the EZH2 inhibitor Tazemetostat (Tazverik™; [11]) as a therapeutic option for patients with epithelioid sarcoma and relapsed or refractory follicular lymphoma in 2020 represented a milestone for the HMT drug discovery community. The interest in targeting HMTs with SMIs will likely increase in coming years and here, using EZH2 as an example, we discuss hit finding and hit triage methodologies that can result in successful identification of HMT inhibitors.

Cell-based screening campaigns have proved difficult for the identification of EZH2 inhibitors and HMT inhibitors in general. Despite technical feasibility and impressive miniaturization of cellular readouts for compound-induced changes in global histone H3 lysine 27 trimethylation (H3K27me3) levels [12, 13], slow turnover of this histone modification and plasticity of the histone modification landscape in response to a vast variety of environmental input signal complicates hit identification and validation. Thus, medicinal chemistry campaigns have been advanced mostly through assays that monitor EZH2 inhibition in cells in a proximal manner. The most proximal readout is the detection of SMI binding to cellular EZH2 protein by cellular thermal shift assays (CETSA) as described previously [14]. While this technology was adapted to high-throughput dose–response formats that allowed for rank-ordering of compounds and correlation with other cellular readouts for inhibitors of the HMT SMYD3 [15], CETSA has not become a go-to methodology for high-throughput screening approaches. Rather, the product of EZH2's enzymatic reaction, H3K27me3, has become the standard of EZH2 inhibitor cell-based readouts, and several methodologies evolved that track reduction in global H3K27me3 levels as a means of assessing the degree to which EZH2 is inhibited. These methodologies include H3K27me3-specific antibody-based technologies (including ELISA, MSD-ELISA, AlphaLISA, IHC, IF and FACS) or LC-MS/MS based technologies. In Subheading 3.3, we describe an AlphaLISA based methodology to track EZH2 inhibition by changes in global H3K27me3 levels upon treatment of cells with

SMIs. While all these methodologies were used to identify EZH2 inhibitors they are also relevant for the discovery of other HMT inhibitors [16].

All clinical stage HMT inhibitors to date have been identified using biochemical assays and structure-based drug design or screening of large compound libraries. Biochemical screening for HMT inhibitors is subject to the same framework and strictures of screening other target classes [17]. Optimal screening of any enzyme requires an understanding of the enzymatic properties of the target. The in-depth concepts and subtleties of enzymology are well beyond the scope of this work but have been covered by multiple other sources [18–20]. From a practical viewpoint, the most critical parameters for screening are an understanding of k_{cat} and K_m. The turnover number, k_{cat} is the maximal rate of substrate conversion by the enzyme, typically expressed as the number of substrate molecules converted per enzyme molecule per unit time. This turnover number is determined when the enzyme is fully saturated with substrate and therefore at maximal velocity, V_{max}. Related to this, the Michaelis constant, K_m, is defined as the substrate concentration at ½ V_{max}. Histone methyltransferases are challenging in that, while they have comparatively low catalytic activities (k_{cat}s), they bind their substrates tightly with nM to low μM K_ms for the SAM cofactor and peptide/histone/nucleosome substrate [21, 22]. EZH2 is no exception to this with reported k_{cat}s of 0.10/min or 0.62/min for either HeLa nucleosomes ($K_m = 0.09$ μM) or unmethylated H3K27 peptide ($K_m = 4.2$ μM) respectively while binding SAM with K_ms of 0.10 μM (nucleosomes) and 0.34 μM (peptide) [23]. We have found similar catalytic properties with k_{cat}s of 25–40/h, substrate K_ms from mid-nM to low μM (oligonucleosomes and peptide respectively) and a SAM K_ms of ~400 nM for either substrate (unpublished observations and [24]). Given this, for EZH2 and other HMTs highly sensitive assays must be used, primarily those incorporating ^{3}H-SAM. Another challenge for identifying chemical matter is that, for the most part, validated screening hits and inhibitors for one HMT are unlikely to inhibit another, leaving little opportunity to take the quicker route of building selectivity into preexisting, validated chemotypes. This is particularly true for lysine methyltransferases including EZH2, [25–27].

Against this backdrop, we have found successful screening of HMTs requires (1) an unbiased screening collection, (2) a comparatively high compound concentration, and (3) screening under "balanced" conditions (i.e., at or near the cofactor/substrate K_ms so as to have a high likelihood of detecting all modes of inhibition) [20]. By employing such an approach, a relatively high number of compounds will be detected as positive, that is, modulating the apparent enzyme activity. However, in reality, most are artifactual, not inhibiting the target by directly and specifically engaging in a

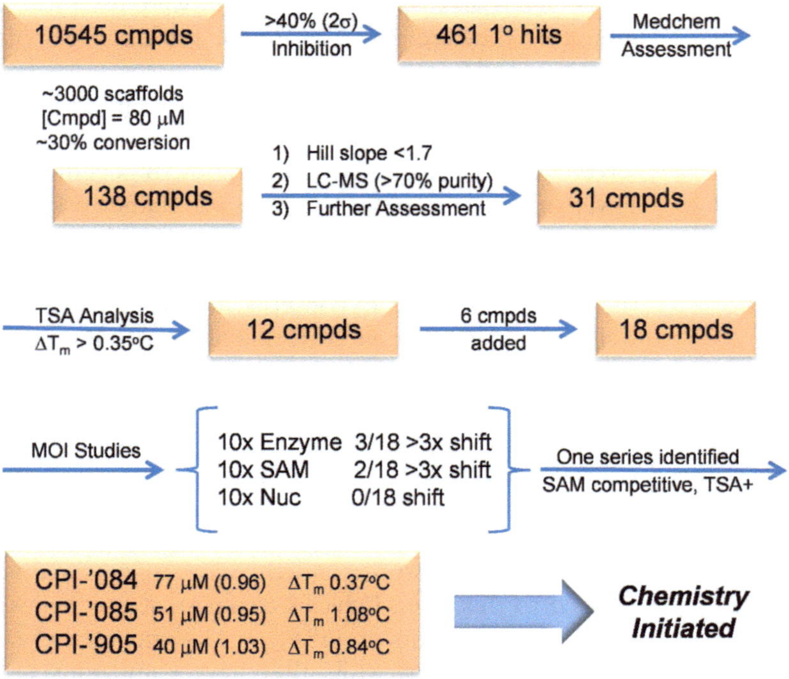

Fig. 1 Schematic representation of an EZH2 screening campaign. Number of compounds at each stage and criteria between stages are shown in boxes and arrows respectively. For the identified, validated starting points (last box) the parenthetical value after the biochemical IC_{50} is the Hill slope of the dose–response curve

stoichiometric manner but rather through some other means such as compound aggregation, spectral interference, or a host of other things [17]. Identification and elimination of these "bad actors" is a multifactorial/multistep process, typically taking significantly longer than the primary screen itself.

As a representative example, Fig. 1 shows a small-scale EZH2 screening campaign carried out in our laboratories. Obtaining catalytically active EZH2 required reconstitution of the multisubunit protein complex of which EZH2 is the catalytic component (Polycomb Repressive Complex 2 (PRC2); [28]). An unbiased set of 10,545 compounds, sourced from multiple vendors, with an emphasis on drug-like properties and containing ~3000 divergent scaffolds, was screened against pentameric PRC2 using PerkinElmer's streptavidin coated FlashPlates™ to capture the biotinylated oligonucleosomes after EZH2-mediated transfer of the tritiated methyl group from SAM to H3K27. The cofactor and substrate were both at 200 nM, or ~twofold below their measured K_ms (390 and 440 nM, respectively) and compound concentration was relatively high at 80 μM. Under these conditions the reaction proceeded to ~20–30% substrate consumption—sufficient to generate a robust signal over background (critical for identifying hits) but not so high as to be subject to a nonlinear response to

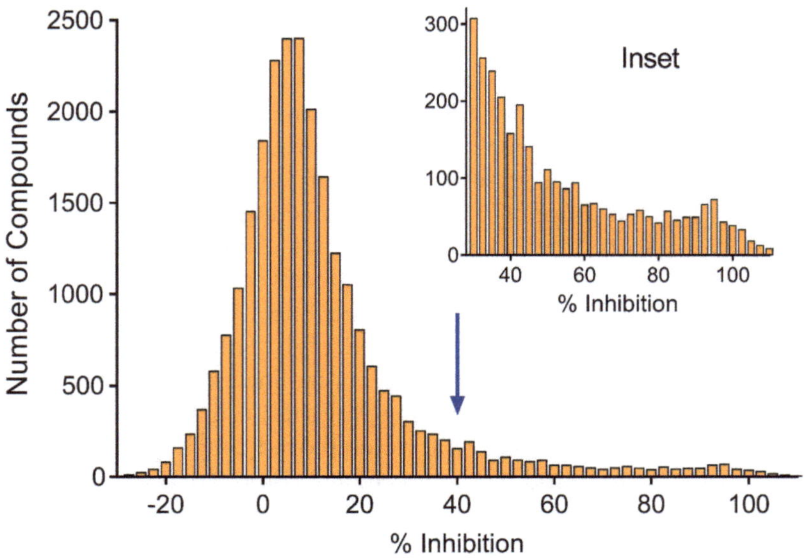

Fig. 2 Primary screen hit distribution. As expected, most compounds were inactive with the maximum centered near zero (mean + 6% inhibition). At greater percent inhibition the number of compounds diminished but remained relatively constant between 40% and 100% (see inset). A cutoff of 40% inhibition was deemed optimal and is indicated by the blue arrow

inhibition, which would lead to an underestimation of compound potency [29]. The resulting hit distribution is shown in Fig. 2. As expected for a screening campaign, the majority of compounds showed little to no inhibition and the mean inhibition (μ) across all compounds was 6.1% with a standard deviation (σ) of 17.3%. However, a closer examination of the distribution at higher % inhibition (Fig. 2, inset) did not show a continual gradual decrease as % inhibition increased. Rather, the compound frequency approached a constant or even slightly increased at the highest percent inhibition (~85% or higher). We attribute this result to the size and complexity of the oligonucleosome substrate, to the high concentration of compound, and the low salt buffers used in the biochemical assay (*see* Subheading 3). Nonetheless, while increasing the likelihood of artifactual hits, such an approach should not result in missing any weak, but tractable, hits and we set the cutoff at 40% inhibition or ~2σ, yielding 461 primary hits.

We have found medicinal chemistry analysis is critical for identifying desired chemotypes and for flagging compounds as likely being artifactual owing to undesired functional groups (e.g., covalent modifiers), or properties such as being redox active, or known frequent hitters. Using this filter prior to any triage narrowed the set down to 138 compounds. Ten-point dose–response titrations were then carried out starting at a top concentration of 160 μM. Compounds were assessed by potency and, more importantly, by the Hill coefficient for the dose–response curves. For a small molecule engaging the target in a 1:1 manner the Hill coefficient should ideally be 1 and deviations from this are likely inhibiting in an undesired manner [30]. In a practical sense, weak molecules may have issues with solubility or give incomplete dose–response curves, and a Hill coefficient cutoff of <1.7 was therefore employed. Mass spectral analyses were also carried out to assess compound integrity and those >70% pure and meeting the Hill criteria were subjected to another round of medicinal chemistry assessment and 31 compounds selected for additional work.

Despite the positive attributes of these hits, the likelihood they were artifactual remained high. Solubility issues, small molecule aggregation and trace contaminants from synthesis (e.g., metal ions, residual coupling reagents) can all result in apparent inhibition. At this point we employed biophysical analysis to ascertain compound binding to EZH2. Biophysical analysis does not rely on enzymatic activity but, rather, measures direct engagement of the small molecule and target, and potentially informing on affinity, stoichiometry, and binding kinetics [31, 32]. The large size of the PRC2 complex (MW >600 kDa) and the challenges in generating large amounts of it limit or preclude using many biophysical techniques (e.g., SPR, ITC, and NMR). However, we were able to employ Thermal Shift Analysis (TSA; also referred to as differential scanning fluorimetry or DSF) to EZH2. TSA employs fluorescent dyes (e.g., Sypro Orange) which bind to hydrophobic surfaces exposed as a protein denatures when subjected to a temperature gradient. The ability of a small molecule to stabilize the protein when bound results in a shift in the protein's melting temperature (T_m) and the magnitude of this shift (ΔT_m) often correlates with potency [33, 34]. Compounds were assessed at 160 μM and any that shifted $>2\sigma$ of the PRC2 control (0.34 °C) were deemed positive. These 12 compounds and several that either (1) could not be analyzed by TSA owing to high background fluorescence or (2) while synthetically attractive may not have sufficient potency were then subjected to mechanism of inhibition (MOI) studies.

MOI studies measure changes in affinity in dose–response titrations while varying a substrate's concentration relative to its K_m and, in accord with the Cheng–Prusoff relationship, provide insight into whether a compound is competitive, noncompetitive, or uncompetitive with substrate [20, 35]. While an observed MOI is not proof positive of a validated hit, it is strong evidence of a compound inhibiting the target in a well-understood manner and, if maintained as more potent analogs are made, is also an important consideration in cell-based assays (and beyond) where substrate concentrations may exceed K_ms significantly. The eighteen hits chosen for follow-up were subjected to three additional biochemical assays- two with either a tenfold increase in SAM or oligonucleosome or a third one with a tenfold increase in enzyme. The last assay is not an MOI assessment but serves as a test for aggregating compounds since an IC_{50} should not shift with higher enzyme given that [E] < <<[Cmpd] [30]. As shown in Fig. 1, two compounds gave evidence of a SAM-competitive MOI (a structurally similar analog had solubility issues precluding assignment), no compounds were substrate-competitive, and three compounds were likely promiscuous aggregators owing to their IC_{50} shift with higher EZH2.

From these studies three compounds were found to share a common chemotype, all were TSA-positive, two had an apparent SAM-competitive MOI and all had Hill coefficients very close to one. Although comparatively weak with affinities of 40–77 μM they were synthetically tractable and became the starting point for a medicinal chemistry campaign [14, 36].

In summary, through a multistep process including assessment of EZH2's enzymatic properties, the use of a diverse, unbiased screening set and a hit triage employing both biochemical and biophysical assessment and working closely with medicinal chemistry colleagues, we identified and validated advanceable chemical matter for a challenging and clinically important target. Screens employing a similar approach, potentially incorporating additional triage steps or biophysical techniques, should continue to provide medicinal chemists with suitable starting points for other HMTs and aid in developing additional molecules to assess and potentially treat unmet medical needs.

2 Materials

Prepare all solutions using ultrapure water and analytical grade reagents. Prepare and store all reagents at room temperature (unless indicated otherwise). Diligently follow all waste disposal regulations when disposing of waste materials.

2.1 EZH2 Biochemical Assay

1. Library of compounds.
2. Reaction Buffer: 50 mM Tris–HCl, pH 8.5, 1 mM dithiothreitol, 69 μM Brij-35, 5 mM $MgCl_2$, 0.1 mg/mL bovine serum albumin.
3. 2× enzyme solution: 5 nM PRC2, 400 nM nucleosomes (*see* **Note 1**), and 7.2 μM activator peptide in Reaction Buffer (*see* **Note 2**).
4. 2× cofactor solution: 400 nM ^{3}H-SAM
5. Stop Buffer: 50 mM Tris–HCl, pH 8.5, 200 mM ethylenediaminetetraacetic acid, 2 mM S-adenosyl-homocysteine.
6. Wash Buffer: 15 mM Tris–HCl, pH 8.5, 30 μM Tween 20.
7. PRC2 complex (*see* **Note 2**).
8. Oligonucleosomes containing biotinylated histone H3.

9. Activator peptide: H_2N-RKQLATKAAR(Kme3)SAPATGG VKKP-amide.
10. Tritiated S-adenosyl-L-methionine (^{3}H-SAM).
11. 384-well assay plates
12. 384-well white FlashPlate (PerkinElmer #SMP410J001PK)
13. Microplate sealing film.

2.2 TSA Assay

1. Library of compounds.
2. Buffer: 50 mM Tris–HCl, pH 8.5, 1 mM dithiothreitol.
3. Protein Thermal Shift Dye (Applied Biosystems).
4. Enzyme solution: 4× Protein Thermal Shift Dye, 100 μM activator peptide, and 0.6 μM PRC2 complex (*see* **Note 3**).
5. PCR plates.
6. Optical plate sealing film.
7. PRC2 complex.
8. Activator peptide: H_2N-RKQLATKAAR(Kme3)SAPATGG VKKP-amide.

2.3 H3K27me3 Cell Assay

1. Library of compounds.
2. HeLa cells.
3. Dulbecco's Modified Eagle Medium (DMEM).
4. White 384-well flat bottom plates.
5. AeraSeal Film (Sigma-Aldrich #A9224).
6. Aluminum plate sealing film.
7. Anti-H3K27me3 antibody (Cell Signaling Technologies #9733) custom-conjugated to AlphaLISA Acceptor Beads (PerkinElmer #677202).
8. Biotinylated anti-Histone H3 (C-terminal) antibody (PerkinElmer #AL118).
9. Streptavidin-conjugated AlphaLisa Donor Beads (PerkinElmer #6760002).
10. Cell-Histone Lysis Buffer (PerkinElmer #AL009F1).
11. Cell-Histone Extraction Buffer (PerkinElmer #AL009F2).
12. Cell-Histone Detection Buffer (PerkinElmer #AL009F3).
13. CellTiter-Glo® 2.0 Cell Viability Assay kit (Promega #G7571).

2.4 Equipment

1. Echo Acoustic Liquid Handler Model # 550 (Beckman Coulter Life Sciences).
2. Countess Automated Cell Counter (Life Technologies #C10281).
3. MicroFlo™ Select reagent dispenser (BioTek Instruments).

4. Plate shaker (ThermoScientific #4625-Q).
5. MultiDrop Combi reagent dispenser (ThermoScientific).
6. Envision Plate Reader (PerkinElmer #2104).
7. ELx 405 plate washer (BioTek Instruments).
8. Bravo liquid handler with 384-well head (Agilent Technologies).
9. ViiA 7 thermal cycler (Applied Biosystems).
10. TopCount scintillation counter (PerkinElmer).

3 Methods

3.1 Identification of EZH2 SMIs Using Biochemical Assays and Compound High Throughput Screening

1. Using the Echo, dispense compounds (as 10 mM DMSO stocks) to columns 3–22 of 384-well assay plates. Dispense an equal volume of DMSO to columns 1–2 and 23–24.
2. Using the MultiDrop Combi, add 12.5 μL of 2× enzyme solution to all wells of the assay plates.
3. Add 20 μL of Stop Buffer to columns 1 and 24. These pre-quenched wells will serve as the low signal controls; columns 2 and 23 will serve as the high signal controls.
4. Add 12.5 μL of 2× cofactor solution to all wells of the assay plates.
5. Allow the reaction to proceed for 60 min at room temperature.
6. Add 20 μL of Stop Buffer to columns 2–23.
7. Using the Bravo, transfer 35 μL of the quenched reaction mixtures from the assay plate to a FlashPlate (*see* **Note 4**).
8. Incubate FlashPlate for at least 1 h at room temperature to ensure complete capture of the biotinylated peptide.
9. Wash the FlashPlate twice by aspirating its contents using the plate washer, dispensing 95 μL Wash Buffer, aspirating again, dispensing another 95 μL Wash Buffer, and aspirating a final time (*see* **Note 5**).
10. Seal the FlashPlate and read on a scintillation counter for 1 min per well.
11. Normalize the signal from compound-containing wells to the signal from the low signal (100% inhibition) and high signal (0% inhibition) control wells.
12. Compounds with % inhibition above a defined cutoff may be considered hits.
13. Determine the IC_{50} of the hits by testing compounds at a range of concentrations and fitting the normalized data to a Hill model.

14. Determine modality of inhibition for the hits by repeating **step 15** with 10× higher concentration of enzyme, substrate, or cofactor (*see* **Note 6**).

3.2 Thermal Shift Analysis

1. Using the Echo, dispense compounds (as DMSO stocks) to columns 2–23 of 384-well PCR plates. Dispense an equal volume of DMSO to columns 1 and 24; these will serve as neutral controls.
2. Dispense 15 μL enzyme solution to each well of the compound plate.
3. Seal plate with optical plate sealing film.
4. Incubate plate for 60–90 min in the dark at room temperature.
5. Load plate into the thermal cycler and run a protocol that ramps the temperature from 25–85 °C at a rate of 0.1 °C per second.
6. Use Protein Thermal Shift software (Applied Biosystems) to determine the melting temperature of the protein in each well, either by the first derivative method or the Boltzmann fit method (*see* **Note 7**).
7. Compounds which increase the melting temperature of the protein to a degree at least two standard deviations above the neutral controls may be considered valid.

3.3 H3K27me3 Cell Assay

3.3.1 Cell Assay, Day 0: Plating and Compound Treatment

1. Use the Echo to dispense several concentrations (in a series of three-fold dilutions) of test compounds in duplicate 384-well tissue culture treated plates. Duplicate plates will be used for AlphaLISA and CellTiter-Glo cell viability assays, respectively (*see* **Note 8**). Dispense an equal volume of DMSO to 12 wells in the middle of the plate to serve as neutral controls. Leave the outer wells empty; these will serve as an edge-effect buffer for the rest of the plate.
2. Trypsinize HeLa cells grown according to standard procedures [37] and count cell suspension using a Countess® cell counter. Dilute cells to 67,000 per mL in 10% DMEM.
3. Use the BioTek MicroFlo™ to dispense 1000 cells (15 μL) per well in 10% DMEM into the compound-containing 384-well plate.
4. Centrifuge the plates at 250 x g for 1 min.
5. Seal the plate with AeraSeal film and incubate at 37 °C and 5% CO_2 for 72 h.

3.3.2 Cell Assay, Day 3: AlphaLISA H3K27me3 Readout

1. After incubation for 72 h, add 5 μL per well Cell-Histone Lysis Buffer to the plate destined for AlphaLISA assay.
2. Incubate for 30 min at RT on the microplate shaker at low speed.

3. Add 10 μL per well Cell-Histone Extraction Buffer.
4. Incubate for 30 min at RT for on the titer plate shaker at low speed.
5. Prepare 1× Cell-Histone Detection Buffer from the 10× stock provided in the kit.
6. Add 10 μL per well of 100 μg/mL anti-H3K27me3 Acceptor Beads (20 μg/mL final) and 15 nM biotinylated anti-Histone H3 (C-terminal) antibody (3 nM final) in 1× Cell-Histone detection buffer (*see* **Note 9**).
7. Seal the plate with aluminum plate sealing foil and incubate at 23 °C for 60 min.
8. Add 10 μL per well of 100 μg/mL Streptavidin Donor Beads (20 μg/mL final) in 1× Cell-Histone detection buffer.
9. Seal the plate with aluminum foil and incubate at 23 °C for 30 min.
10. Read plate using the EnVision plate reader.
11. Normalize the signal from compound-containing wells to the signal from neutral control wells and fit normalized data to a Hill model to determine IC_{50}.

3.3.3 Cell Assay, Day 3: Cell Viability

1. After incubation for 72 h, add 15 μL CellTiter-Glo to the plate destined for cell viability and shake on plate shaker.
2. Incubate at RT for 15–20 min on the microplate shaker at low speed.
3. Read plate using the EnVision plate reader.
4. Normalize the signal from compound-containing wells to the signal from neutral control wells and fit normalized data to a Hill model to determine IC_{50}.

4 Notes

1. Like many HMTs, PRC2 has nearly undetectable activity at physiological pH and NaCl concentration, therefore a screening assay requires a buffer with basic pH and no NaCl.
2. A detailed description of the production and purification of the PRC2 complex and oligonucleosome substrate may be found in Garapaty et al. [26].
3. The dye should be diluted in buffer prior to adding the protein. Exposure to concentrated dye could lead to protein denaturation.
4. PerkinElmer FlashPlates are no longer commercially available. Any method for measuring radioactive product with minimal

interference from unreacted ^{3}H-SAM should yield similar results.

5. Washing away the unreacted ^{3}H-SAM greatly reduces background signal, thereby widening the assay window.
6. If the stock concentration of ^{3}H-SAM is too low to achieve a concentration 10× above its K_m without the storage buffer interfering with enzyme activity, it can be supplemented with unlabeled SAM.
7. Large, multidomain proteins or protein complexes may show more than one transition in the melt curve, complicating the use of the first derivative to define the melting temperature. In this case, the Boltzmann fit method should be used.
8. The quality and health of the cells in response to EZH2 inhibition is important for interpreting the results of the H3K27me3 AlphaLISA assay. For this reason, a cell viability assay must be run in parallel. HeLa cells were chosen as an exemplary cell model that is inert to EZH2 loss or inhibition. Characterization of a given cell model is needed to ensure that cell target engagement can be disentangled from potential phenotypic consequences of target inhibition. As for EZH2 inhibitors and HeLa cells, only compounds that do not induce a reduction in cell viability are considered for assessment in the AlphaLISA readout.
9. The acceptor bead antibody's selectivity for H3K27me3 is essential. Before the antibody used in this assay, it should be thoroughly characterized to rule out cross-reactivity with other methylation states and other histone methylation sites.

Acknowledgments

We would like to take this opportunity to acknowledge and thank all our current and former Constellation colleagues who have contributed to our EZH2 program for 12+ years from inception to two clinical stage compounds.

References

1. McGrath J, Trojer P (2015) Targeting histone lysine methylation in cancer. Pharmacol Ther 150:1–22. https://doi.org/10.1016/j.pharmthera.2015.01.002
2. Rugo HS, Jacobs I, Sharma S, Scappaticci F, Paul TA, Jensen-Pergakes K, Malouf GG (2020) The promise for histone methyltransferase inhibitors for epigenetic therapy in clinical oncology: a narrative review. Adv Ther 37(7):3059–3082. https://doi.org/10.1007/s12325-020-01379-x
3. Jarrold J, Davies CC (2019) PRMTs and arginine methylation: cancer's best-kept secret? Trends Mol Med 25(11):993–1009. https://doi.org/10.1016/j.molmed.2019.05.007
4. Yang Y, Bedford MT (2013) Protein arginine methyltransferases and cancer. Nat Rev Cancer 13(1):37–50. https://doi.org/10.1038/nrc3409

5. Tessarz P, Santos-Rosa H, Robson SC, Sylvestersen KB, Nelson CJ, Nielsen ML, Kouzarides T (2014) Glutamine methylation in histone H2A is an RNA-polymerase-I-dedicated modification. Nature 505(7484):564–568. https://doi.org/10.1038/nature12819
6. Biterge B, Richter F, Mittler G, Schneider R (2014) Methylation of histone H4 at aspartate 24 by protein L-isoaspartate O-methyltransferase (PCMT1) links histone modifications with protein homeostasis. Sci Rep 4:6674. https://doi.org/10.1038/srep06674
7. Husmann D, Gozani O (2019) Histone lysine methyltransferases in biology and disease. Nat Struct Mol Biol 26(10):880–889. https://doi.org/10.1038/s41594-019-0298-7
8. Trojer P (2021) Histone methylation modifiers in medical therapeutics, chapter 37. In: Medical epigenetics, 2nd edn. Elsevier
9. Levell JR (2020) Recent progress in the discovery and clinical application of small-molecule inhibitors EZH2 and EED. Med Chem Rev 55(Chapter 13)
10. Stein EM, Garcia-Manero G, Rizzieri DA, Tibes R, Berdeja JG, Savona MR, Jongen-Lavrenic M, Altman JK, Thomson B, Blakemore SJ, Daigle SR, Waters NJ, Suttle AB, Clawson A, Pollock R, Krivtsov A, Armstrong SA, DiMartino J, Hedrick E, Lowenberg B, Tallman MS (2018) The DOT1L inhibitor pinometostat reduces H3K79 methylation and has modest clinical activity in adult acute leukemia. Blood 131(24):2661–2669. https://doi.org/10.1182/blood-2017-12-818948
11. Hoy SM (2020) Tazemetostat: first approval. Drugs 80(5):513–521. https://doi.org/10.1007/s40265-020-01288-x
12. Luense S, Denner P, Fernandez-Montalvan A, Hartung I, Husemann M, Stresemann C, Prechtl S (2015) Quantification of histone H3 Lys27 trimethylation (H3K27me3) by high-throughput microscopy enables cellular large-scale screening for small-molecule EZH2 inhibitors. J Biomol Screen 20(2):190–201. https://doi.org/10.1177/1087057114559668
13. Liu Y, Platchek M, Kement B, Bee WT, Truong M, Zeng X, Hung S, Lin H, Morrow D, Kallal LA, Xie Q, Agarwal P, Pope AJ, Wu Z (2014) A novel approach applying a chemical biology strategy in phenotypic screening reveals pathway-selective regulators of histone 3 K27 tri-methylation. Mol BioSyst 10(2):251–257. https://doi.org/10.1039/c3mb70413k
14. Bradley WD, Arora S, Busby J, Balasubramanian S, Gehling VS, Nasveschuk CG, Vaswani RG, Yuan CC, Hatton C, Zhao F, Williamson KE, Iyer P, Mendez J, Campbell R, Cantone N, Garapaty-Rao S, Audia JE, Cook AS, Dakin LA, Albrecht BK, Harmange JC, Daniels DL, Cummings RT, Bryant BM, Normant E, Trojer P (2014) EZH2 inhibitor efficacy in non-Hodgkin's lymphoma does not require suppression of H3K27 monomethylation. Chem Biol 21(11):1463–1475. https://doi.org/10.1016/j.chembiol.2014.09.017
15. McNulty DE, Bonnette WG, Qi H, Wang L, Ho TF, Waszkiewicz A, Kallal LA, Nagarajan RP, Stern M, Quinn AM, Creasy CL, Su DS, Graves AP, Annan RS, Sweitzer SM, Holbert MA (2018) A high-throughput dose-response cellular thermal shift assay for rapid screening of drug target engagement in living cells, exemplified using SMYD3 and IDO1. SLAS Discov 23(1):34–46. https://doi.org/10.1177/2472555217732014
16. Prabhu L, Chen L, Wei H, Demir O, Safa A, Zeng L, Amaro RE, O'Neil BH, Zhang ZY, Lu T (2017) Development of an AlphaLISA high throughput technique to screen for small molecule inhibitors targeting protein arginine methyltransferases. Mol BioSyst 13(12):2509–2520. https://doi.org/10.1039/c7mb00391a
17. Markossian S, Sittampalam GS, Grossman A et al (eds) (2004) Assay guidance manual. Eli Lilly & Company and the National Center for Advancing Translational Sciences, Bethesda, MD
18. Segel IH (1975) Enzyme kinetics: behavior and analysis of rapid equilibrium and steady state enzyme systems. Wiley, New York
19. Stein RL (2011) Kinetics of enzyme action: essential principles for drug hunters. Wiley, Hoboken, NJ
20. Copeland RA (2013) Evaluation of enzyme inhibitors in drug discovery: a guide for medicinal chemists and pharmacologists, 2nd edn. Wiley, Hoboken, NJ
21. Frankel A, Brown JI (2019) Evaluation of kinetic data: what the numbers tell us about PRMTs. Biochim Biophys Acta Proteins Proteom 1867(3):306–316. https://doi.org/10.1016/j.bbapap.2018.10.010
22. Kumar M, Zielinski T, Lowery RG (2015) Biochemical assay development for histone methyltransferases using a Transcreener-based assay for S-adenosylhomocysteine. Assay Drug Dev Technol 13(4):200–209. https://doi.org/10.1089/adt.2014.609
23. Diaz E, Machutta CA, Chen S, Jiang Y, Nixon C, Hofmann G, Key D, Sweitzer S, Patel M, Wu Z, Creasy CL, Kruger RG,

LaFrance L, Verma SK, Pappalardi MB, Le B, Van Aller GS, McCabe MT, Tummino PJ, Pope AJ, Thrall SH, Schwartz B, Brandt M (2012) Development and validation of reagents and assays for EZH2 peptide and nucleosome high-throughput screens. J Biomol Screen 17(10):1279–1292. https://doi.org/10.1177/1087057112453765

24. Stuckey JI, Cantone NR, Cote A, Arora S, Vivat V, Ramakrishnan A, Mertz JA, Khanna A, Brenneman J, Gehling VS, Moine L, Sims RJ 3rd, Audia JE, Trojer P, Levell JR, Cummings RT (2021) Identification and characterization of 2(nd) generation EZH2 inhibitors with extended residence times and improved biological activity. J Biol Chem 296:100349. https://doi.org/10.1016/j.jbc.2021.100349
25. Kaniskan HU, Martini ML, Jin J (2018) Inhibitors of protein methyltransferases and demethylases. Chem Rev 118(3):989–1068. https://doi.org/10.1021/acs.chemrev.6b00801
26. Garapaty-Rao S, Nasveschuk C, Gagnon A, Chan EY, Sandy P, Busby J, Balasubramanian S, Campbell R, Zhao F, Bergeron L, Audia JE, Albrecht BK, Harmange JC, Cummings R, Trojer P (2013) Identification of EZH2 and EZH1 small molecule inhibitors with selective impact on diffuse large B cell lymphoma cell growth. Chem Biol 20(11):1329–1339. https://doi.org/10.1016/j.chembiol.2013.09.013
27. Singh PK (2019) Histone methyl transferases: a class of epigenetic opportunities to counter uncontrolled cell proliferation. Eur J Med Chem 166:351–368. https://doi.org/10.1016/j.ejmech.2019.01.069
28. Cao R, Zhang Y (2004) SUZ12 is required for both the histone methyltransferase activity and the silencing function of the EED-EZH2 complex. Mol Cell 15(1):57–67. https://doi.org/10.1016/j.molcel.2004.06.020
29. Wu G, Yuan Y, Hodge CN (2003) Determining appropriate substrate conversion for enzymatic assays in high-throughput screening. J Biomol Screen 8(6):694–700. https://doi.org/10.1177/1087057103260050
30. Shoichet BK (2006) Interpreting steep dose-response curves in early inhibitor discovery. J Med Chem 49(25):7274–7277. https://doi.org/10.1021/jm061103g
31. Renaud JP, Chung CW, Danielson UH, Egner U, Hennig M, Hubbard RE, Nar H (2016) Biophysics in drug discovery: impact, challenges and opportunities. Nat Rev Drug Discov 15(10):679–698. https://doi.org/10.1038/nrd.2016.123
32. Folmer RH (2016) Integrating biophysics with HTS-driven drug discovery projects. Drug Discov Today 21(3):491–498. https://doi.org/10.1016/j.drudis.2016.01.011
33. Zhang R, Monsma F (2010) Fluorescence-based thermal shift assays. Curr Opin Drug Discov Devel 13(4):389–402
34. Pantoliano MW, Petrella EC, Kwasnoski JD, Lobanov VS, Myslik J, Graf E, Carver T, Asel E, Springer BA, Lane P, Salemme FR (2001) High-density miniaturized thermal shift assays as a general strategy for drug discovery. J Biomol Screen 6(6):429–440. https://doi.org/10.1177/108705710100600609
35. Cheng Y, Prusoff WH (1973) Relationship between the inhibition constant (K1) and the concentration of inhibitor which causes 50 per cent inhibition (I50) of an enzymatic reaction. Biochem Pharmacol 22(23):3099–3108. https://doi.org/10.1016/0006-2952(73)90196-2
36. Gehling VS, Vaswani RG, Nasveschuk CG, Duplessis M, Iyer P, Balasubramanian S, Zhao F, Good AC, Campbell R, Lee C, Dakin LA, Cook AS, Gagnon A, Harmange JC, Audia JE, Cummings RT, Normant E, Trojer P, Albrecht BK (2015) Discovery, design, and synthesis of indole-based EZH2 inhibitors. Bioorg Med Chem Lett 25(17):3644–3649. https://doi.org/10.1016/j.bmcl.2015.06.056
37. ATCC. https://www.atcc.org/products/all/CCL-2.aspx

INDEX

Raphaël Margueron and Daniel Holoch (eds.), *Histone Methyltransferases: Methods and Protocols*, Methods in Molecular Biology, vol. 2529, https://doi.org/10.1007/978-1-0716-2481-4,

R

S

T

U

W

X

Y

MIX
Papier aus verantwortungsvollen Quellen
Paper from responsible sources
FSC® C105338

If you have any concerns about our products,
you can contact us on
ProductSafety@springernature.com

In case Publisher is established outside the EU,
the EU authorized representative is:
Springer Nature Customer Service Center GmbH
Europaplatz 3, 69115 Heidelberg, Germany

Printed by Libri Plureos GmbH
in Hamburg, Germany